AF769024

ANALYSIS AND DESIGN OF VERTICAL CAVITY SURFACE EMITTING LASERS

WILEY SERIES IN LASERS AND APPLICATIONS

D. R. VIJ, Editor
Kurukshetra University

OPTICS OF NANOSTRUCTURED MATERIALS • Vadim Markel

LASER REMOTE SENSING OF THE OCEAN: METHODS AND APPLICATIONS • Alexey B. Bunkin and Konstantin Voliak

COHERENCE AND STATISTICS OF PHOTONS AND ATOMS • Jan Peřina, *Editor*

METHODS FOR COMPUTER DESIGN OF DIFFRACTIVE OPTICAL ELEMENTS • Victor A. Soifer

THREE-DIMENSIONAL HOLOGRAPHIC IMAGING • Chung J. Kuo and Meng Hua Tsai

ANALYSIS AND DESIGN OF VERTICAL CAVITY SURFACE EMITTING LASERS

S. F. Yu
School of Electrical & Electronic Engineering
Nanyang Technological University
Singapore

WILEY-
INTERSCIENCE

A JOHN WILEY & SONS, INC., PUBLICATION

CONTENTS

PREFACE

Design and fabrication of vertical cavity surface emitting lasers (VCSELs) requires an iterative process, which is extremely expensive and time-consuming. The use of computer-aided design (CAD) tools can help shorten the design cycle and speed up the development process. Laser models, which are found in the literature, can be used to implement CAD tools for the analysis and design of VCSELs. However, some comprehensive models, which perform sophisticated functions, are difficult to implement and show low computational efficiency. Other simplified models exhibit high computing speed but deliver inadequate descriptions of the observed effects. As a result, inconsistent conclusions may be obtained because different assumptions are applied. This book attempts to provide a guideline for the derivation of models based on appropriate assumptions for a particular problem so that the phenomena observed by the experiment can be easily explained. In fact, the objective throughout this book is to search for the simplest and most direct treatment for modeling VCSELs. The author believes that the laser models covered in this book can help the readers customize their CAD tools to fit into their applications. In addition, the readers should have no difficulty in implementing their own laser models.

The book is divided into nine chapters. Except for the first two introductory chapters, each chapter is independent and restricted to a particular subject to be studied. To the best of the author's knowledge, the most appropriate theories have been chosen to model the specific topic of VCSELs. In Chapters 3 and 4, theoretical models have been developed to analyze the modal profile and polarization, respectively, of VCSELs. The most popular structure of VCSELs is a cylindrical symmetric cavity, which is assumed in the derivation of the models. In addition, this configuration of VCSELs allows investigation of the modal profile and polarization separately such that the complexity of theoretical models can be reduced. In Chapter 3, different methods of solving the wave equation for the modal profile of VCSELs are discussed in detail. The reader can choose the most appropriate model with the required speed and accuracy to analyze the problems. In Chapter 4, two- and four-level models are described to study the polarization properties of the fundamental transverse mode. These simplified models can evaluate the dominant factors that control the polarization properties of VCSELs.

It must be noted that the investigation of VCSELs using cold cavity approximations is not realistic. This is so because most of the measurable data, such as threshold current, lasing wavelength, slope efficiency, and output power, all depend on the operating temperature of lasers. Furthermore, the optical behavior of VCSELs is affected by thermal lensing (i.e., self-focusing of transverse modes into the core region of the active layer). Therefore, the thermal properties of VCSELs are investigated in Chapter 5. The method of effective temperature using a simple rate equation model is presented. Effective thermal conductivity and heat generation rate are also derived. The objective in defining effective temperature is to simplify the study by using a rate equation model so that the computational efficiency can be improved. However, this approach will not provide detailed information on heat distribution. Detailed heat distribution inside the laser cavity is studied by solving the heat equation numerically. In this case, the influence of thermal lensing on the optical field profile can be evaluated.

Spatial hole burning of carrier concentration also has significant influence on the modal profile of VCSELs. Therefore, Chapter 6 describes the use of a simple rate equation to evaluate the distribution of carrier concentration inside the active region. In this case, self-consistent calculation of optical gain and carrier concentration (i.e., self-consistent calculation of the Poisson and Schrödinger equations) is ignored to simplify the calculation. Different methods for approximating the nonuniform distribution of carrier concentration are also discussed. On the other hand, nonuniform distributions of electric potential and current are required as the input parameters to calculate the heat distribution inside the laser cavity. They have to be solved numerically using the Poisson and continuity equations simultaneously with appropriate boundary conditions. The electric potential across the active layer and the corresponding carrier concentration can be linked together by a simple diode equation. This is so because the simplified relation between optical gain and carrier concentration has been assumed. The self-consistent calculation of optical field, heat, and electrical characteristics of VCSELs is also described in Chapter 6.

The dynamic response of VCSELs is analyzed in Chapter 7. Preliminary investigation of the dynamic response of VCSELs using a simple rate equation model is described. Hence, the time variation of carrier concentration and photon density inside the active layer can be calculated. Furthermore, detailed analysis of optical fields can be considered using the beam propagation method such that the influence of optical confinement on the dynamic response of VCSELs can be evaluated. However, detailed investigation of the transient response of heat and electrical properties is avoided in the self-consistent calculation. This is because the time variation of heat and voltage, which are related to heat and the Poisson equations, is much slower than that of photon density and carrier concentration. This assumption significantly reduces the computation time of the model without sacrificing the accuracy of the calculation. The influence of various transportation mechanisms inside the quantum well (QW) active region on the dynamic response of VCSELs is also discussed in this chapter.

The methods used to evaluate the spontaneous emission and linewidth of VCSELs are described in Chapter 8. Simple models have been developed to study these parameters quantitatively through the investigation of the spontaneous emission factor and linewidth enhancement factor. On the other hand, the magnitude of the spontaneous emission factor and linewidth enhancement factor is evaluated using rate equation model by empirically fitting the measurable data. Hence, design criteria to optimize the spontaneous emission of VCSELs are obtained. Other nonlinear features of VCSELs such as self-sustained pulsation, bistability, dual-wavelength operation, and wavelength tunability are studied in Chapter 9 using rate equation models. The advantage of using simple rate equation models is that the parameters that describe the nonlinear behavior of VCSELs can be easily extracted through some measurable data such as injection current and lasing power.

In conclusion, this book presents the most effective way to implement laser models of VCSELs, which the reader can easily understand. However, the readers are assumed to have the usual undergraduate background knowledge of electromagnetic theory and solid-state physics as well as basic computational skills. Materials of this research monograph concentrate on the evaluation of modeling techniques to analyze VCSELs under various operating conditions. As each chapter of this book is mostly independent of the other chapters, readers can selectively study any chapter for their own interest. Although this book is of most interest to the design engineer of VCSELs, it also provides valuable information to CAD tool designers in other fields of semiconductor lasers.

Siu Fung Yu

Singapore
April 2003

ACKNOWLEDGMENTS

The author is grateful to Abdul Al-Muhanna, Sebastiano Bocciardo, Liang-Chin Chen, Samuel Choi, Ciaran O Cochlain, Gerrit Hille, and Chia-Fu Hsu for their helpful advice, assistance, and encouragement in preparing part of this book during his time in Agere System (formerly Optoelectronic Center, Lucent Technologies Inc., USA). A special thanks is also given to our secretary Sally Haring for her care and love.

The author also thanks Joanne Zhong of Hong Kong University for her careful reading of the manuscript and P. C. Chui of Hong Kong University, Department of Electrical and Electronics Engineering, for his encouragement in writing this book.

Above all, the author would like to thank R. G. S. Plumb, his Ph.D. supervisor, and J. E. Carroll, the former head of the division at the Cambridge University Engineering Department for their guidance and supervision during his time in Cambridge.

CHAPTER 1

Vertical Cavity Surface Emitting Lasers — an Overview

This chapter reviews the most recent developments in vertical cavity surface emitting lasers (VCSELs) operating from infrared to ultraviolet. In addition, the upcoming business opportunities for VCSELs and their related products are described. The most recent progress in the numerical techniques and computer-aided design tools to analyze and design novel VCSELs is also reviewed. The possibility of realizing long-wavelength VCSELs operating at high power in stable single transverse mode is also studied.

1.1 INTRODUCTION

Vertical cavity surface emitting lasers (VCSELs) are made by sandwiching a light emitting layer (i.e., a thin semiconductor of high optical gain such as quantum wells) between two highly reflective mirrors. The mirrors can be dielectric multilayered or epitaxial growth mirrors of distributed Bragg reflectors (DBRs) with reflectivity greater than 99.9%. Light is emitted normally from the surface of the mirrors. The simple testing procedure is one of the merits of VCSELs even though the epitaxial growth of DBRs is required. This is because VCSELs allow manufacturers to carry out on-wafer testing prior to dicing and packaging so that the production cost is much lower than that of facet emitting lasers. In addition, the compact size of VCSELs (typically $400 \times 400\ \mu m^2$) yields more devices per wafer than do facet emitting lasers [1]. Hence, these unique characteristics of VCSELs allow manufacturing of low-cost semiconductor lasers in large quantities. Narrow beam divergence, low power consumption, high modulation bandwidth, and easy polarization control are the other advantages of VCSELs over facet emitting lasers. Therefore, VCSELs are ideal candidates for the light sources of optical fiber communication networks such as local area networks (LANs) and metropolitan networks. VCSELs also find application in optical storage, print heads, optical sensors, barcode scanners, digital displays, spatial light modulators, backplanes and smart pixels, and microscopes [2–5].

Analysis and Design of Vertical Cavity Surface Emitting Lasers By Siu Fung Yu
ISBN 0-471-39124-7

In this chapter, the most recent progress of VCSELs operating from infrared to ultraviolet is briefly described and the potential applications of VCSELs as light emitting sources in different commercial products are discussed. The reasons for the rapid commercialization of VCSELs and related products are also explained. The most recent progress of computational techniques and computer-aided design (CAD) tools used to analyze VCSELs are briefly described. Finally, the anticipated challenges in achieving high-speed and high-power VCSELs using novel structure and fabrication techniques are also explained.

1.2 RECENT DEVELOPMENT OF VCSELs

VCSEL was first proposed and fabricated by K. Iga and his colleagues at the Tokyo Institute of Technology, Japan in the late 1970s. They indicated that in order to realize low threshold current, VCSELs should have (1) extremely small cavity volume, (2) high optical gain, and (3) mirrors with extremely high reflectivity (>95%) [6]. At that time, it was difficult to obtain high optical gain in bulk materials, and it was a challenge to obtain mirrors of such high reflectivity. Despite these difficulties, they successfully demonstrated the first electrically pumped InGaAsP/InP VCSEL under pulse operation at 77 K in 1979 [7]. Several years later, they reported the achievement of an electrically pumped GaAs/AlGaAs VCSEL pulsing at room temperature [8,9]. Their early development of VCSELs usually involved a gold-based p-side reflector and n-side dielectric multilayered mirror (i.e., SiO_2/Si) in which light is emitted. However, further reduction of threshold current in GaInAsP/InP-based VCSELs was obstructed by the large Auger recombination and other nonradiative recombination inside such a small laser volume. Therefore, they concentrated on the development of low-threshold GaAs-based VCSELs. In 1987, they reported a GaAs VCSEL with threshold current of 6 mA under continuous-wave (CW) operation at 20.5°C [10]. Two years later, they also demonstrated the first room-temperature CW GaAs VCSEL emitting at 850 nm [11]. The achievement of K. Iga and his colleagues has initiated a worldwide competition in the realization of low-threshold, high-power VCSELs over a wide of wavelength operation range.

The rapid progress in the development of GaAs-based VCSELs is due to the promising properties of optical materials and the possibility of all-monolithic structures, including the epitaxial growth of DBRs. Therefore, metal-based mirrors were soon replaced by DBRs, which have been demonstrated successfully in GaAs/AlGaAs Fabry–Perot microcavities. The advantages of GaAs-based DBRs are (1) wide variation in refractive index between adjacent layers so that fewer layers are required to achieve high reflectivities [12] and (2) low electrical resistance, which can be easily obtained in DBRs by using the appropriate doping profile [13]. The application of DBRs in optically pumped VCSELs has shown significant reduction in threshold [14]. On the other hand, the use of optically transparent contacts, such as indium–tin–oxide (ITO), has improved the confinement of injection carrier concentration inside the laser cavity [15]. The optical transmittance of ITO is high (~92%), and ITO also forms good ohmic contacts to both p- and n-type contacts with specific contact resistances of 10^{-4} and $10^{-5}\ \Omega \cdot cm^2$, respectively. This optically transparent contact is commonly used for the early development of GaAs-based short-wavelength VCSELs. As mentioned before, relatively high gain is required to obtain CW operation at room temperature. The successful development of GaAs/AlGaAs quantum-well (QW) materials for the facet emitting lasers in the late 1980s provided the solution for VCSELs. Other researchers use GaAs/AlGaAs QWs as the active regions of GaAs-based VCSELs [16,17] while others have demonstrated that GaAs/AlGaAs QWs VCSELs, incorporated with graded heterojunction interfaces into the monolithic epitaxial DBRs [18], give a submilliampere threshold current as well as CW operation at room temperature [19,20].

Metal-based dielectric mirrors (i.e., such as MgF/ZnSe/MgF/Au dielectric mirror) are still used in GaAs-based 980-nm VCSELs to reduce the threshold current as well as enhance controllability of the optical mode [21]. The front and rear mirrors of this laser type are a post-growth-deposited MgF/ZnSe (e.g., 3.5 pairs) multilayered dielectric mirror and a n-type GaAs/AlAs (e.g., 25 pairs) DBR, respectively. A $\lambda/2$ cavity, which consists of an InGaAs QW active layer, GaAs spacers, and an oxide aperture, is sandwiched between the two mirrors. This type of device has demonstrated a threshold current of 91 μA with an aperture diameter of 3 μm. The use of oxide aperture is the result of the discovery of the Al containing III–V semiconductor [22]. The oxide aperture is obtained by oxidizing an $Al_xGa_{1-x}As$ layer so that a layer of native oxide is formed. The native oxide layer has a relatively low refractive index but high electrical resistivity [23]. Hence, a high-efficiency performance due to the precise confinement of optical mode and injection current inside the active region can be achieved. It has also been shown that VCSELs with front and rear AlGaAs/GaAs DBRs can also form oxide apertures inside the AlGaAs spacer layers to improve conversion efficiency [24].

The development of GaAs-based 980-nm VCSELs (i.e., using InGaAs/GaAs QWs) is based on the successful realization of 850-nm VCSELs (i.e., using GaAs/AlGaAs QWs) as both technologies for the fabrication of DBRs are identical (i.e., InGaAs is lattice-matched with GaAs at 980 nm). A typical 980-nm oxide aperture VCSEL has 30 pairs of p-type $Al_{0.9}Ga_{0.1}As$/GaAs layers and 20 layer pairs of n-type DBRs of the same composition as the p-type mirrors for light emission purposes. An active layer of 3 $In_{0.2}Ga_{0.8}As$/GaAs QWs is sandwiched between two spacer layers of AlAs from the DBRs to obtain an oxide aperture [25]. 980-nm VCSEL with GaAs/AlGaAs DBRs and selective oxide configuration has demonstrated a very low threshold current of 8.5 μA [26]. In addition, the maximum output power of similarly configured 980-nm VCSELs can be as large as 40 mW with an aperture diameter of only 16 μm under CW operation at 17°C maintaining a single-mode and single-lobe far-field profile [27]. Furthermore, a 980-nm VCSEL array (with 19 single lasers, where each VCSEL has a 50-μm-diameter aperture arranged in a honeycomblike layout to achieve maximum output power) can generate more than 1.4 W of output power under CW operation at 10°C heatsink temperature [28]. It is believed that the output power of a single-cavity 980-nm VCSEL will soon reach a level (i.e., several hundred milliwatts), which is compatible with that of a 980-nm high-power facet emitting laser. However, multi-transverse-mode operation in large-aperture VCSELs is a serious problem and has to be solved in order to replace facet emitting lasers. Two possible methods can be used to maintain single-transverse-mode operation in oxide aperture VCSELs with large cavity size: surface relief [29] and antiguiding effects in epitaxially regrown materials [30]. The basic idea of these methods is to introduce a spatial filter to eliminate the excitation of higher-order transverse modes.

The rapid development of GaAs-based lasers led to the commercialization of short-wavelength (i.e., 850–980-nm) VCSELs in 1997. Laser manufacturers sell short-wavelength VCSELs with ion implantation and oxide apertures worldwide. It must be noted that reliability is one of the most important factors of commercialized VCSELs. For example, transoceanic optical fiber communication systems based on semiconductor lasers are traditionally expected to require no more than three ship repairs in 25 years of operation. For the application of short-wavelength VCSELs in optical fiber computer networks such as LANs and metropolitan area networks (MANs), the required lifecycle should be much longer than the emergence of the next-generation products. It is noted that VCSELs with oxide apertures have a long working lifetime with median time to failure of $>10^7$ h. On the other hand, ion implanted VCSELs have an attractive selling price and acceptable reliability [31]. Therefore, oxide aperture and ion-implanted short-wavelength VCSELs have already been installed in most of the advanced LANs for application in gigabit Ethernet.

The realization of long-wavelength VCSELs using InP-based materials suffers from a significant drawback due to the (1) difficulty in obtaining highly reflective DBRs, (2) noticeable Auger recombination and inter-valence-band absorption, and (3) small conduction band offset. In addition, VCSELs with thick InGaAsP/InP DBRs (i.e., to produce high reflectivity) have high internal absorption loss and low thermal conductivity. In spite of these disadvantages, the 1300-nm InGaAsP VCSEL with n-type InGaAsP/InP mirror and SiO_2/Si mirror on the p side demonstrates a threshold current of 500 mA under pulsed operation at room temperature [32]. In fact, with appropriate n doping on InGaAsP/InP DBR, the corresponding electrical conductivity can be improved and the optical reflectivity can be maintained at 1500 nm [33]. Fabrication of 1550-nm InP QW VCSEL using n-type InGaAsP/InP DBRs on n-InP substrate has demonstrated 1-mW CW output power at room temperature [34]. Semiconductor/dielectric reflectors can be used as the high-reflectivity mirror on the p side of long-wavelength VCSELs. In early-stage 1300- and 1550-nm InGaAsP/InP VCSELs, MgO/Si and Al_2O_3/Si amorphous dielectric mirrors were used to provide sufficient optical reflection [35]. Low-loss p-GaAs/Al_xO_y dielectric mirrors also used in the development of the 1500-nm optically pumped VCSEL have shown a low pumping threshold of 1.4 mW [36]. Although these dielectric mirrors have good thermal conductivity and high reflectivity, they have a very high electrical resistance, making electrical excitation difficult and requiring a different injection method [37].

Fusion bonding of GaAs-based DBRs onto the AlGaInAs/InP or InGaAsP/InP QW active layer is an alternative method to realizing long-wavelength VCSELs [38–40]. Submilliampere and CW room-temperature operation of 1550-nm VCSELs can be easily obtained by wafer fusion of InGaAsP/InP QWs to two AlGaAs/GaAs DBRs [38]. This is because GaAs/AlGaAs DBR has high electrical and thermal conductivity. Double fusion of GaAs/AlAs DBRs to the AlGaInAs/InP QW active layer is also possible to fabricate for 1300-nm VCSELs [39]. The major drawback of using double fusion to produce long wavelength is the complicated fabrication procedures, which obstruct the commercialization of VCSELs because of the high production cost. The fusion technique can also been used to bond an AlGaAs/GaAs front DBR to an InGaAsP/InP half-cavity structure (i.e., n-doped InGaAsP/InP bottom DBR). VCSELs fabricated by this technique have demonstrated 1550-nm CW emission at 17°C [40]. This method can reduce the time required for the fabrication procedure to half, but fusion bonding is still required in the process. The growth of high-quality GaAs/AlAs DBRs or GaAlAs/GaAs metamorphic DBRs on InP-based materials may be an alternative way to realize long-wavelength VCSELs other than the wafer fusion technique [41,42]. However, this method may have the reliability problems associated with the heavily dislocated metamorphics mirrors due to the lattice mismatch (see Fig. 1.1).

The commercialization of GaAs-based short-wavelength VCSELs has been successfully realized using all-monolithic growth technology. This is because

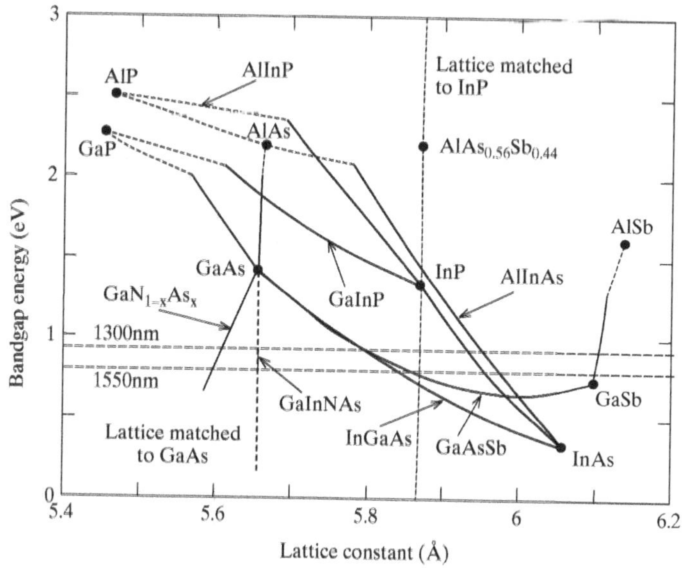

Figure 1.1 Energy band diagram versus lattice constant.

fabrication of lattice-matched systems can circumvent any reliability problems related to fused junction or metamorphic mirror techniques. In addition, high reproducibility and one-time calibration significantly reduce the production cost of GaAs-based VCSELs. Therefore, the technology of all-monolithic growth should be applied to fabricate long-wavelength VCSELs. In fact, it has been shown that using the InP lattice-matched AlGaInAs/InAlAs DBR system, all-monolithic growth VCSELs can be obtained [43,44]. This device consists of lattice-matched InGaAlAs/InAlAs n- and p-type DBRs grown on n-InP substrate. The n-type (typically ~43.5 periods) mirror is usually doped with silicon, and the p-type (typically ~35 periods) mirror is doped with carbon in order to increase the electrical conductivity. AlGaInAs bulk [43] and AlInGaAs strain-compensated QWs [44] have been used as the active layer of this long-wavelength VCSEL. The optimal threshold current and output power of the 1550-nm VCSEL at room temperature with aperture diameter of 25 μm are recorded to be ~6 mA and ~8 mW, respectively [44], and the corresponding differential efficiency is about 42%. This achievement of low threshold current is due to the high reflectivity of InGaAlAs/InAlAs DBRs at 1550 nm as the corresponding refractive index difference and characteristic temperature are about 0.31 ± 0.01 and 122 K, respectively. It is also shown in Figure 1.1 that the lattice of AlGaAsSb/AlAsSb combination matches that of InP. This may represent an alternative choice of DBRs to realize monolithic growth VCSELs [45,46]. In addition, the AlGaAsSb/AlAsSb mirror has a contrast of refractive indices higher than that of the InGaAlAs/InAlAs mirror (i.e., the use of alternated pairs can be reduced by 20%) so that the corresponding thermal and electrical conductivities can be improved. This VCSEL can be grown by molecular beam epitaxy (MBE) on an n-doped InP substrate. The front mirror consists of 30 periods of $AlAs_{0.56}Sb_{0.44}/Al_{0.2}Ga_{0.8}As_{0.58}Sb_{0.42}\lambda/4$ layers lattice-matched to InP plus a phase matching layer for the front gold contact (reflectivity >99.9%). The rear output DBR consists of the same material combination (reflectivity ~ 99.6%). The λ cavity active layer consists of five strain-compensated AlInGaAs QWs and a heavily doped tunnel junction sandwiched between the two mirrors [45]. This device, with an emission area of 25 μm diameter, demonstrates a threshold current of 7 mA and maximum output power of 2 mW at 17°C under electrically pulsed operation. The performance of this device can be further improved by forming an aperture inside the active region. Because of the different etching rates of the materials for active layer and DBRs, the active layer can be undercut using a mixture of 1 M citric acid and hydrogen peroxide as the AlAsSb-based DBR has not been etched significantly. Hence, the undercut active layer forms a thin airposted structure, which improves the confinement of optical field and injection current so that the corresponding threshold current can be reduced by more than 50% [46].

Figure 1.1 shows that the GaInNAs system is lattice-matched with GaAs. This implies that GaAs-based DBRs can be used to realize long-wavelength VCSELs using the GaInNAs system as the active layer. One of the possible active layers consists of $(Ga_{0.7}In_{0.3})(N_{0.004}As_{0.996})$ wells and GaAs barriers. In order to obtain long-wavelength VCSELs, the active layer is sandwiched between p- and n-type AlAs/GaAs DBRs consisting of 21 and 25.5 layer pairs, respectively. Doping can be applied to increase the electrical conductivity of DBRs. This type of VCSEL has been demonstrated at room-temperature operation under electrically pulsed modulation emitting at 1180 nm [47]. The merit of using GaInNAs/GaAs QWs as the active region is due to the technology of fabricating AlAs/GaAs DBRs that can be adopted directly from GaAs-based VCSELs. However, the electrical and optical characteristics of GaInNAs/GaAs QWs are still not fully understood, and further exploration is required to fabricate more reliable GaInNAs QWs with controllable electrical and optical performance [48]. Room-temperature CW operation of $Ga_{0.64}In_{0.36}N_{0.003}As_{0.997}$/GaAs QW VCSELs (with $Al_{0.7}Ga_{0.3}As$/GaAs DBRs and n-GaAs as substrate) has demonstrated maximum output power and threshold current of 1 mW and 2 mA, respectively, emitting at 1200 nm [49]. Another similar approach is to use $GaAs_{0.665}Sb_{0.335}$/GaAs QWs as the active layer, which is lattice-matched with AlAs/GaAs DBRs and n-GaAs substrate. This has also demonstrated the room-temperature CW operation of such an oxide aperture device emitting at 1.23 μm that has a threshold current of 0.7 mA [50]. Longer lasing wavelength can also be obtained by increasing the Sb content. GaAsSb VCSELs could be a viable low-cost light source for optical fiber data link systems. InGaAs/GaAs and InAs/InGaAs quantum dot (QD) active layers, which are lattice-matched with the GaAs-based materials, can be tuned to have maximum optical gain peak near 1300 nm [25,51]. This implies that long-wavelength VCSELs can also be realized using GaAs-based QD materials. In fact, InAs (pyramidal island)/$In_{Ga0.15}As_{0.85}$ (regrowth layer with GaAs barriers) QD VCSELs using AlAs/GaAs DBR on the p side and GaAs/AlO on the n side with n-GaAs substrate demonstrated 1300-nm emission under pulsed operation at 20°C [51]. Therefore, it can be concluded that using GaInNAs/GaAs QWs, GaAsSb/GaAs QWs, or GaAs-based QDs with GaAs/AlAs DBRs may be promising methods of realizing long-wavelength VCSELs. This is because all-monolithic growth, predicable reliability, and low production cost can be maintained during the fabrication of long-wavelength VCSELs.

Short-wavelength VCSELs in the visible range is another attractive topic to be studied. This is because visible VCSELs can be found in wide-ranging applications such as laser printing and scanning, plastic-fiber-based communications, optical data storage (CD, DVD, etc.), and display applications. Plastic optical fiber has minimum attenuation at 650 nm, which indicates that the development of high-speed VCSELs at such a wavelength is necessary to realize low-cost gigabit per second optical data links. It has shown that the conversion efficiency and maximum output power of a red VCSEL is 11% and 8 mW, respectively [52]. High modulation bandwidth can also be obtained in red VCSELs using oxide aperture structure. Usually, red (~670-nm) oxide aperture VCSELs consist of an $Al_{0.5}Ga_{0.5}As/Al_{0.9}Ga_{0.1}As$ DBR (i.e., DBRs typically have 72 and 44 pairs in the n and p sides, respectively) and InGaP/InGaAlP QW active layer. Standard oxide aperture can be fabricated by oxidizing a higher Al content layer inside the n-side mirror to an appropriate aperture size [53]. This device has been characterized with respect to its analog modulation performance and a free dynamic range of 92–100 dB $\cdot$ $Hz^{2/3}$ in the frequency range of 0.1–3 GHz has been measured from a VCSEL with 18-μm aperture diameter. However, it is difficult to manufacture visible wavelength lasers because the active region has a large energy bandgap and p-type doping is technically hard to perform. In addition, reliability is difficult to maintain if aluminum is included in the system, as the degradation due to Al oxidation is appreciable. Therefore, the commercialization of red VCSELs is still in progress. Green semiconductor lasers are very important for realizing optical storage, printing, and display applications. ZnSe is a possible material system to provide CW operation of green/blue and reliable semiconductor lasers. Blue-green optically pumped VCSELs have been demonstrated at room temperature [54]. The active region of this laser consists of three $Zn_{1-x}Cd_xSe$ QWs and Zn(S,Se) barrier layers. The QW region is sandwiched between ZnSe confinement layers. Low-loss multilayer SiO_2/HfO_2 stacks are deposited onto the QW active layer to form the front and rear mirrors, and the device is pumped at room temperature with output emission of 496.2 nm. It is believed that electrically pumped blue-green VCSELs can be obtained in a similar manner [55]. Alternatively, using GaN and its material systems as the active layer can realize green to UV emitting lasers. Reports on the reliability performance of GaN-based LEDs and facet emitting lasers indicate that blue or shorter wavelength emission can also be realized in VCSELs [56]. However, the challenges to achieve blue/violet GaN-based VCSELs require high-reflectivity mirror materials and the fabrication of microcavity resonator structure. It is possible to fabricate monolithic growth VCSELs using AlGaN/GaN DBRs, but their small difference in refractive index requires a large number of layer pairs to achieve the required reflectivity. Hence, it is difficult to apply in situ epitaxial growth technique to form DBRs. Nonetheless, violet VCSELs with AlGaN/GaN DBRs emitting at 401 nm under photopumped room temperature has been reported [57]. On the other hand, it may be possible to replace AlGaN/GaN DBRs by dielectric multilayered mirrors in blue/violet VCSELs. In fact, it has been shown that 10 pairs of SiO_2/HfO_2 dielectric layers can give more than 99% reflectivity between 400 and 450 nm [58,59]. The remaining challenge to realizing blue/violet VCSELs is the highly transparent carrier concentration (i.e., high threshold current density). This is because electron and hole masses of wide-bandgap materials are heavier than those of narrow-bandgap materials. In fact, transparent carrier concentration of wide-bandgap materials is reduced if the QW structure is used to realize the active region such as $GaN/Al_{0.1}Ga_{0.9}N$ QWs. Therefore, it is believed that low threshold current blue VCSELs can be realized using GaN QWs [60].

1.3 VCSELs APPLICATIONS

There are various commercial products in our everyday lives using VCSELs as the light emitting source such as optical fiber computer networks, high-density optical storage, and printing, optical sensing, and display systems. In addition,

VCSELs can be used in parallel optical interconnect and optical information processing, which are still under extensive research and development. In the following paragraphs, the potential applications of VCSELs as the light emitting sources in different commercial products are discussed.

Optical fiber communications have been studied extensively for several decades. In short-distance optical fiber computer networks, red/850/980-nm light emitting diodes (LEDs) are usually used as the light emitting sources in the first generation of LANs. Plastic-fiber-based fiber links (e.g., using graded-index 1-mm core fiber) using red LEDs as the light emitting sources because the emission wavelength of red LEDs matches the low-loss band of the fiber. On the other hand, 850/980-nm LEDs are found in multimode silica-based fiber computer networks because of their low production cost and reliable performance. However, the data transmission speed of LEDs is limited to several hundred megabits per second (Mbit/s). This problem can be resolved provided VCSELs are used to replace the LEDs in upgrading the existing optical fiber computer networks to the speed of gigabits per second (Gbit/s). This is possible because VCSELs have inherent high modulation speed but low production cost, which is compatible with LEDs. In fact, the merits of VCSELs over LEDs have been demonstrated and are summarized as follows:

- Large area VCSELs have shown better noise immunity than LEDs. It is recorded that large area VCSELs perform at low modal noise (relative intensity noise (RIN) < -120 dB/Hz) and low bit error rate ($<10^{-13}$) in a 4-m-long graded-index multimode fiber link at 1 GHz modulation speed [61].
- Modulation of single-mode 980-nm VCSEL up to 10 Gbits/s has demonstrated a bit error rate of $<10^{-11}$ after transmitting over 500-m-long graded-index multimode fiber (with core diameter 50 μm) [62].
- The transmission experiment over a 250-m-long graded-index multimode fiber using 850 nm oxide aperture VCSEL as the light emitting source has demonstrated a stable performance over a temperature range between −20 and 100°C [63]. The corresponding bit error rate is less than 10^{-11} at a transmission rate of 2.5 Gbits/s. The temperature-dependent power penalty for the 250-m multimode fiber (50 μm core diameter) transmission is measured to be 0.009 dB/K [63]. This experiment has indicated the reliable performance of oxide aperture VCSELs over a wide temperature range.
- The use of oxide aperture 850-nm VCSEL can boost the transmission data rate over a graded-index multimode fiber link to 12.5 Gbits/s [64].
- The data transmission experiment using 980-nm InGaAs single-mode VCSEL and butt-coupled graded-index plastic optical fiber (155 μm core diameter) has demonstrated a transmission rate of 7 Gbits/s over a distance of 80 m [65]. In addition, easy butt coupling between VCSEL and fiber with lateral alignment tolerances of 80 μm make such modules very attractive for LAN solutions such as gigabit Ethernet.

It can be concluded that VCSELs have great potential to be applied in short-distance fiber links with a Gbit/s modulation speed. It is noted that VCSELs are more than adequate to replace the LEDs in short-distance optical fiber links. This is because VCSELs exhibit a much wider modulation bandwidth, longer transmission distance over a graded-index multimode fiber, and lower power consumption over a required temperature range.

In long-haul high-speed optical fiber communication systems using single-mode silica-based fiber, long-wavelength distributed feedback (DFB) lasers are commonly used as the light emitting sources. This is because minimum absorption loss and low dispersion are the requirement for transmission over a long distance. On the other hand, long-wavelength VCSELs can provide stable single-mode operation with adequate power at low bias voltage over a range of temperature. In addition, high modulation speed and low production cost allow VCSELs to be applied in high-volume data communication applications such as gigabit Ethernet, 10-Gbit Ethernet, and the fiber channel. In fact, the merits of VCSELs over DFB lasers in long-haul high-speed optical communication systems are summarized as follows:

- In transmission experiments without optical amplification, it has been demonstrated that 1.33- and 1.5-μm single-mode VCSELs can operate at 2.5 Gbit/s and transmitting error-free through a standard single-mode fiber of >50 km long [66, 67].
- The other advantage of VCSELs over DFB lasers is low power consumption. The combination of oxide aperture VCSEL and the corresponding driver with low power operating at 2.5 Gbits/s using 0.5-μm CMOS technology has demonstrated a 7-mW power consumption for an optoelectronic transmitter. This transmitter module has launched digital signals over a graded-index multimode fiber, and the signal is detected at the other end of the fiber. The main reason for performing this experiment is to demonstrate that this CMOS driver can drive VCSEL, which works in exactly the same way as the other driver. In fact, the power dissipated in this CMOS driver is 20 times less than that of a 3.6-V electronic positive emitter-coupled logic transmit circuit and 4 times less than a 3.6-V electronic low-voltage differential signal transmit circuit [68].

From the preceding discussion, it is believed that VCSELs will sooner or later replace DFB lasers as the light emitting sources of high-speed long-haul optical fiber communication systems, provided the future development of VCSELs can improve its high-power performance and maintain single-mode operation.

In optical disk readout systems such as CD and DVD, the low-cost Fabry–Perot semiconductor laser (i.e., facet emitting laser) is usually used as the optical source. A separate external photodetector is also used in the readout system to monitor the light reflected from the optical disk. However, the optical beam emitting from the Fabry–Perot laser facets is highly asymmetric so that the allowable information density of the optical disk has to be maximized using precise design of optical lenses. Therefore, it is desired to replace Fabry–Perot lasers by VCSELs to reduce the production cost and size of the optical pickup head. This is possible because the light emitted from VCSELs is a highly symmetric circular beam and a photodetector can be integrated monolithically with VCSELs so that a low-cost and compact size optical pickup head can be realized. Figure 1.2 shows a proposed VCSEL with an intracavity QW absorber to realize the integrated optical disk readout head [69]. In this case, the CW optical beam emitting from the VCSEL is tightly focused onto the optical disk, and the reflected beam directly penetrates into the VCSEL cavity. The reflected signal from the optical disk is accurately measured by using the intracavity absorber, which is a photodetector under reverse bias. The major advantages of integrating VCSELs with an intracavity absorber as the optical pickup head are

- The circular output beam of VCSELs maximizes the data storage density of the optical disk.
- The single longitudinal mode behavior of VCSELs eliminates the intermodal noise that can be found in Fabry–Perot lasers.
- The possibility of constructing 2D arrays, which can create novel optical pickup head for parallel readout to increase the information bandwidth of the readout systems.

Hence, it is believed that the use of VCSELs in optical pickup systems will reduce the production cost but increase the storage density of the optical disk

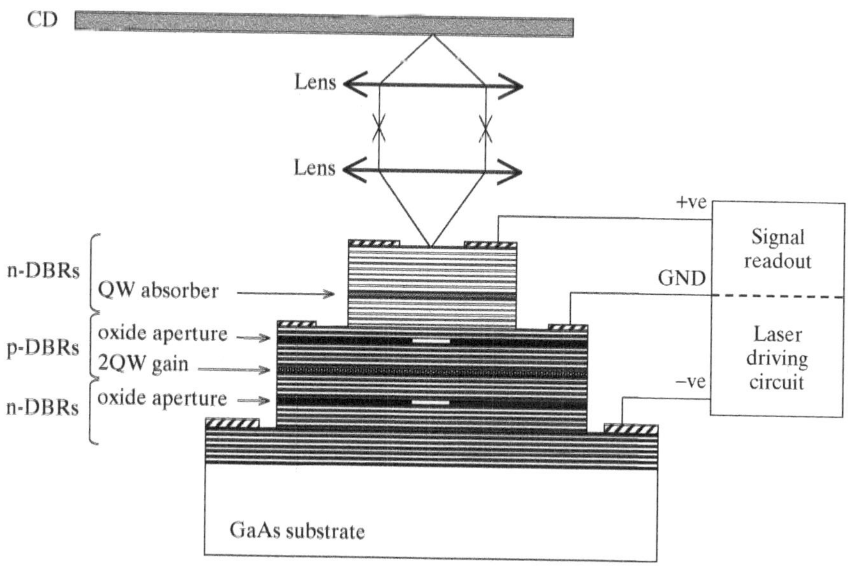

Figure 1.2 Schematic of a compact optical disk readout head using a VCSEL integrated with an intracavity absorber (after Ref. 69).

even though the VCSEL has the same lasing wavelength as the Fabry–Perot laser used in the original system.

The possibilities of using VCSELs in optical interconnections to eliminate bottlenecks in electronic connections have been studied extensively. For example, the applications of free-space optical interconnections to relieve bottlenecks associated with electrical board-to-board and chip-to-chip interconnection as well as to alleviate electromagnetic crosstalk and signal distortion [70] have been investigated. On the other hand, optical interconnections with fiber array or polymer waveguide as the transmission medium are expected to construct large bandwidth switching networks and to implement massively parallel processors [71]. There are several requirements for transmitters to realize 2D optical interconnection.

- The transmitters should operate at low power in order to avoid heating effects.
- The transmitters should have low driving voltage and current to integrate with CMOS drivers to achieve compact size and very low power consumption.
- The transmitters should have large modulation bandwidth at low power rate in order to realize large data throughput.
- Low production cost of optical interconnections is the demanding factor to compete with the electrical one.

In fact, advances in VCSEL technology have yielded transmitters that can fulfill these requirements. The most important feature of VCSEL technology is the monolithic fabrication of 2D VCSEL array, which can be fully utilized as the transmitter in the optical interconnections. It has been demonstrated that VCSELs have low threshold current, low driving voltages, high power efficiencies, and large bandwidths even at low driving conditions. Simple packaging techniques, such as flip-chip bonding for free-space interconnections, to packaging 2D arrays of hybrid VCSEL transmitters and photodetectors significantly reduce the production cost of optical interconnections [72]. Therefore, an array of VCSELs with a photodetector is the most suitable candidate to realize optical interconnections over free space as well as fiber array or waveguide.

The most important applications of parallel free-space interconnections are the replacement of electrical interconnects in chip-level interconnections of high-performance digital computers. Figure 1.3 shows the schematic of a parallel free-space interconnection that can be realized by integrating the microlens array with VCSEL array and photodetector array. It is expected that this configuration of optical interconnection has low power consumption and high data transmission speed over the electrical interconnection for a working distance of several millimeters. A highly uniform 8×8 array of GaAs/AlGaAs VCSELs (8 μm oxide aperture diameter) has been fabricated. The VCSEL array has an average threshold current of 0.72 ± 0.02 mA, an average output power of 2.05 ± 0.03 mW at 8 mA, an average voltage of 1.44 ± 0.01 V, and an average power conversion efficiency of 14.3%. In addition, the wavelength across the array is measured

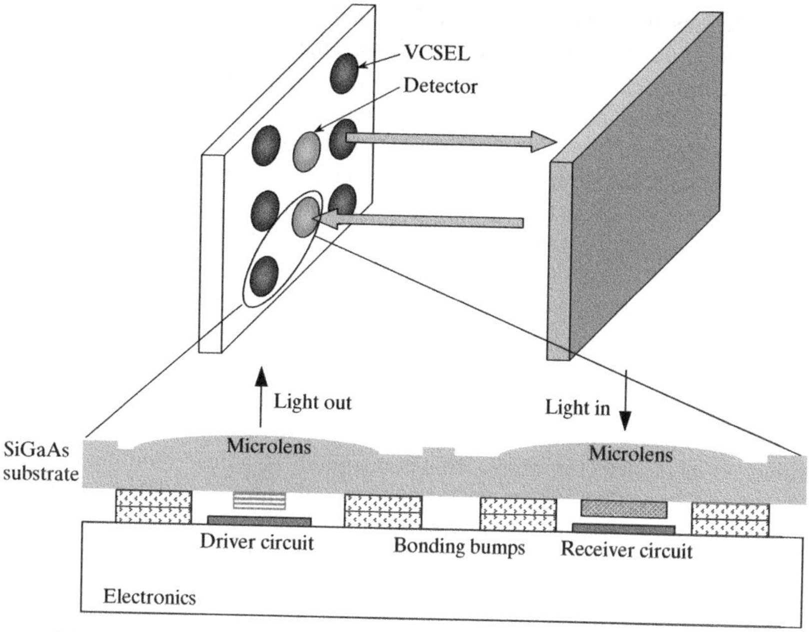

Figure 1.3 Schematic of a parallel free-space interconnection using VCSEL array and photodetector array integrated with microlens array (after Ref. 73).

to be 697 ± 0.35 nm. The maximum total power consumption is $\sim$73 mW at threshold, which is low when compared to the electrical interconnection [73]. An 8×8 array of refractive microlenses, which is fabricated in fused silica using thermal reflow of photoresist followed by mask erosion, is used to produce an $f/6.69$ beam from the array of VCSELs to be detected by the array of photodetectors. The pitch between the individual devices is 250 μm, and the whole VCSEL array occupies an area of 2.8×2.8 mm^2. This parallel optical interconnection can easily obtain a high transmission speed up to $\sim$10 Gbits/s with low power consumption. However, the minimum size of the VCSEL array is limited by optical crosstalk. One possible approach to minimize optical crosstalk in parallel optical interconnections is to use monolithic arrays of wavelength-graded VCSELs and wavelength-selective resonance-enhanced photodetectors (i.e., which are used for wavelength-division-multiplexed optical interconnect applications). These arrays can minimize optical crosstalk between neighboring channels in a parallel free-space optical interconnect by rejecting the light from neighboring wavelength-differentiated channels [74]. However, the challenge of realizing these parallel free-space interconnections is to achieve wavelength-graded arrays as optical sources and a wavelength-selective photodetector in a controllable and reproducible manner.

The development of large-capacity asynchronous transfer mode (ATM) switches and computer networks (i.e., LANs), which require high-speed interconnections between modules, is degraded by the slow traffic capacity of the electrical broad-to-broad interconnections. Therefore, it is desired to replace the existing electrical with the optical interconnections in order to increase the parallel data transfer rate of the entire system. The straightforward method to transfer data between broads is to utilize 1D or 2D VCSEL arrays and multimode fiber ribbons with the arrangement of one data channel per fiber [71]. Another cost-effective approach is to make use of the extra wide bandwidth of optical fiber by wavelength-division multiplexing (WDM) to transmit several channels within one fiber [75, 76]. Obviously, this WDM method significantly expands the transmission capacity of the optical interconnections and significantly reduces the use of optical fibers. Figure 1.4 shows a proposed low-cost WDM optical data link via a single multimode fiber. The transmitter is a monolithically integrated multiple-wavelength VCSEL array emitting wavelength-encoded optical signals, and the receiver is a channel-matched wavelength selective narrowband photodetector array. This transmitter is a rear emitting photonic integrated emitter (PIE) arrays designed to couple into a 62.5-μm multimode fiber. The PIE array consists of eight rear emitting VCSELs arranged in a pielike configuration within a 60-μm-diameter circular area. Reasons for using the rear emitting VCSEL PIE scheme are

- Large emitting area for each array element to couple effectively into a multimode fiber

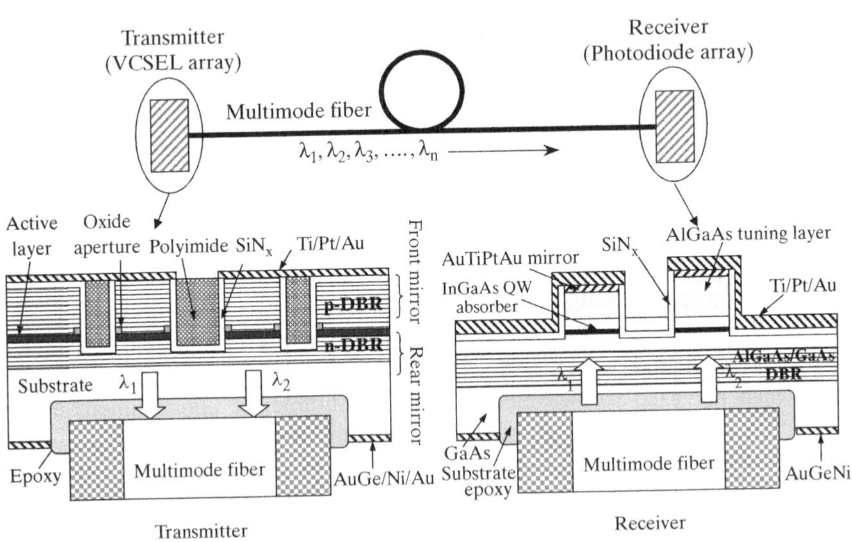

Figure 1.4 (a) A WDM optical data link transmits over a multimode optical fiber; (b) schematic of a monolithically integrated multiple wavelength VCSEL (transmitter) and a channel-matched wavelength-selective narrowband photodetector array (receiver). (After Ref. 75.)

- Compatible with flip-flop bonding scheme and provide sufficient heatsinking to reduce thermal crosstalk between channels
- Manufacturable planar processing technique that can be used to facilitate the further packaging of these VCSEL PIE arrays

Therefore, it is possible to fabricate a low-cost WDM system using PIE arrays. The Al and In composition profiles of the rear emitting VCSEL PIE array are shown in Figure 1.5. The VCSEL has AlGaAs/GaAs DBRs and three InGaAs/GaAs QWs grown by MBE. The lasing wavelength of the individual VCSEL channels can be adjusted by performing anodic oxidation on the GaAs phase tuning layer prior to growth of the rest of the top p-DBR mirror. The

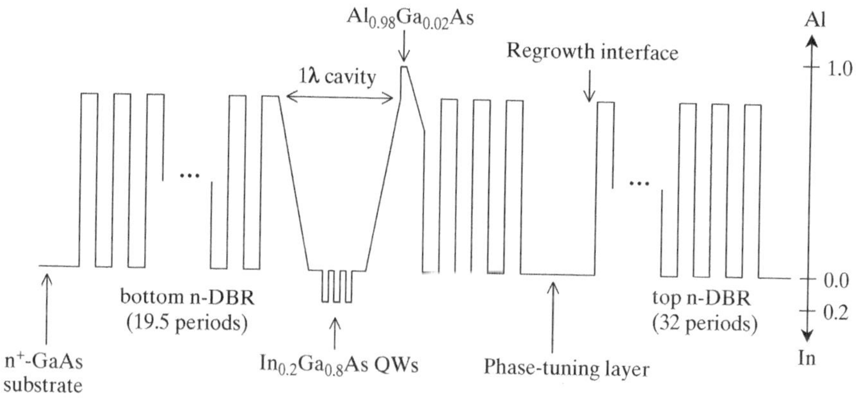

Figure 1.5 The Al and In composition profiles of the monolithically integrated multiple-wavelength VCSEL; the lasing wavelength of VCSEL section is determined by the phase detuning layer (after Ref. 75).

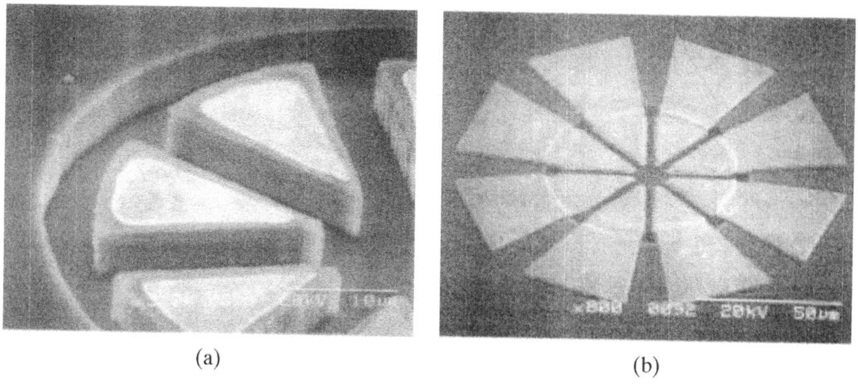

Figure 1.6 (a) SEM diagram of a part of the VCSEL PIE array after the p-contact process; (b) SEM diagram of a finished bottom emitting VCSEL PIE array. (After Ref. 75.)

GaAs phase tuning layer is the fourth GaAs quarter-wave layer above the active region, but this is grown with a thickness of 0.91λ for mode tuning, where λ is the Bragg wavelength of the DBR mirror. Figure 1.6a shows a scanning electron micrograph (SEM) of a part of the VCSEL PIE array after the p-contact process and Figure 1.6b shows a SEM of a finished bottom emitting VCSEL PIE array. The light/current characteristics of all channels of the VCSEL PIE array under CW operation at room temperature is shown in Figure 1.7a. It is noted that most of the devices lase at around 1.7 mA with a bias voltage of 2.4 V. Channels 1–6 exhibit similar electrical and spectral characteristics when the output power is below 5 mW. Figure 1.7b shows the measured lasing wavelength and threshold current of all the channels. The calculated lasing wavelength (solid line)

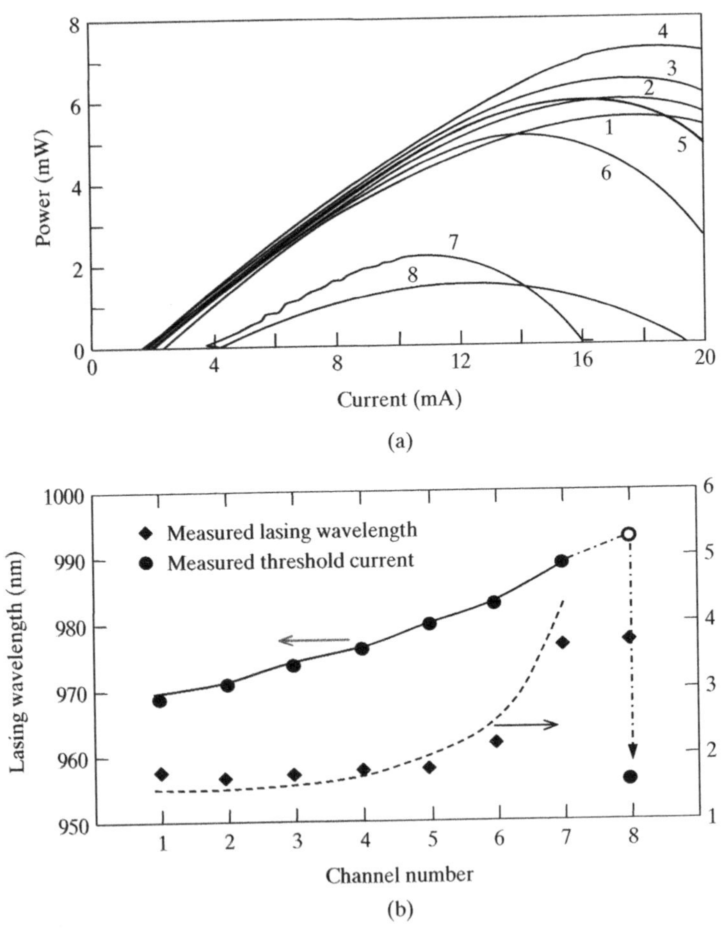

Figure 1.7 (a) Light/current curves of the VCSEL PIE array under continuous-wave operation at room temperature; (b) variation of lasing wavelength and threshold current of each channel of the VCSEL PIE array. (After Ref. 75.)

and threshold current (dotted line) are also plotted in the figure for comparison. From the figure, the achieved lasing wavelength span is 32.9 nm and the average channel spacing for channels 1–6 is around 3.0 nm. This VCSEL PIE array has undergone data transmission experiment through a 300-m-long graded-index multimode fiber (with a 62.5 μm core diameter) using an optical–electrical converter as the receiver. The results show that these devices are capable of transmitting data up to 1.5 Gbits/s per channel with bit error rate $<10^{-13}$.

In order to fulfill a complete WDM system, a pie-shaped photodetector array is also proposed for detecting the transmitted signal. The pie-shaped device is arranged within a 60-μm-diameter circle, just like the VCSEL PIE arrays. Therefore, when the fiberoptics are aligned and slightly defocused to the center of the pie array, all eight elements will be equally illuminated with all input wavelengths, but each of them will pick up only the specific channel that lies within its absorption band. The "pie"-shape wavelength-division multiplexing (WDM) photodetector arrays operate near 980 nm using a InGaAs/AlGaAs/GaAs system. The eight detectors are narrowband resonant cavity Schottky photodiodes with the Schottky Au metal as the back mirror and the AlGaAs DBR as the front mirror. Cavity modes for the individual devices can be adjusted by performing anodic oxidation on the i-AlGaAs phase tuning layers. Figure 1.4 also shows the schematic cross section of a bottom-illuminated multiple-channel resonant cavity Schottky photodetector array with the Burrus-type packaging configuration. This 8-channel WDM system has potential application for the next generation of optical interconnections but the performance of the transmitter requires further modification in order to achieve uniform threshold conditions and wavelength span for all VCSELs. In addition, further development of the photodetector array with sufficient sensitivity and bandwidth at a particular wavelength is required to realize this 8-channel WDM system. This type of transmitter and receiver design will lead to the future construction of a multiwavelength optical data link with a single dual-core multimode fiber and an integrated laser–detector array for bidirectional parallel data transfer broad-to-broad communication.

Another application of monolithic array VCSELs is the realization of functional optical interconnections [77]. In this case, optical functional devices or "smart" pixels (i.e., building block of the functional optical interconnections) are obtained by integrating VCSELs with the photodetector and transistor in the same wafer. The smart pixel features an electrophotonic structure at the device level with the advantage of electrons and photons, resulting in performance improvements such as low power consumption and functionality enhancement. The most popular usage of smart pixels is in optical information processing, which performs switching, latching, and regeneration operations and can be constructed by the basic elements of optical logic gates [78, 79]. Figure 1.8 shows a VCSEL-based smart pixel [71]. The smart pixel consists of a VCSEL thyristor and a double-cavity phototransistor (HPT). Both sections are monolithically integrated onto a GaAs substrate. The VCSEL thyristor section consists of a single vertical cavity with a p-n-p-n thyristor structure. It offers laser, switching, and latching functions. The HPT section consists of a double cavity with an n-p-n HPT

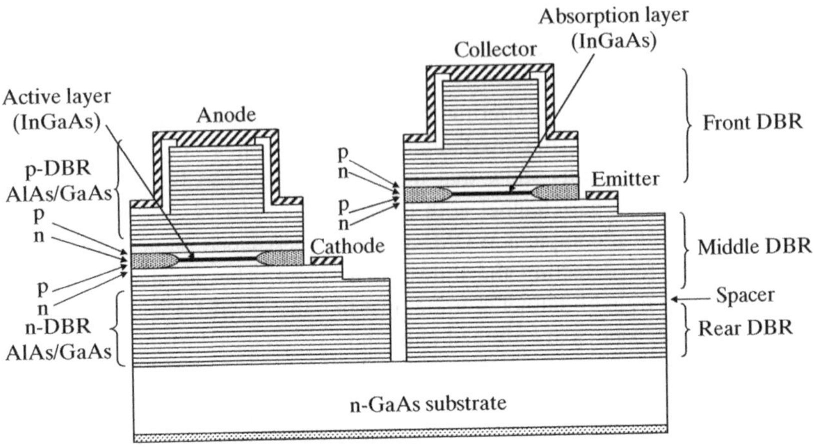

Figure 1.8 Schematic of a VCSEL-based smart pixel (after Ref. 71).

structure. It offers detection and amplification functions. The VCSEL thyristor can be switched electrically or optically. When switched on, it functions as a VCSEL modulated by the light incident on the HPT when they are connected in series. This smart pixel allows bidirectional or cascadable optical interconnections with sufficient sensitivity and bandwidth. In addition, these functions can avoid electrically induced noise in the detector section. It is expected that the two-dimensional arrays of this smart pixel is the key component for the next generation of optical switching, optical signal processing, and optical computing applications.

1.4 COMMERCIALIZATION OF VCSELs

Since the late 1990s, the explosive growth of the usage of email and the World Wide Web has driven the widespread deployment of computer networks. The popularity of the Internet has dramatically increased the amount of interchange and share of information (i.e., graphical, audio, and video data) and has outgrown the communication infrastructure beyond its current bandwidth capacities (i.e., using LEDs as transmitters). Therefore, the use of high-speed VCSELs as the replacement for LEDs is the key technology to enable the next generation of low-cost, reliable, and high-speed optical interconnections. As a result, most major semiconductor laser manufacturers have focused on the development of high-speed but low-cost VCSELs. The main challenge of VCSEL manufacturers is the capability to produce a large volume of low-cost and reliable lasers. In fact, the combination of VCSEL technology with packaging and manufacturing technologies (which enable high volume, high quality, and low production cost) will lead to the next generation of optical products. It is believed that the cost-effective deployment of higher-speed interconnections using VCSELs such as 10 Gbit/s and beyond will hinge on the widespread availability of robust fiberoptic interconnection technologies [80].

Mass production of short-wavelength VCSELs was begun in 1997 by optoelectronics manufacturers worldwide (***Honeywell, Mitel, CSEM, Emcore, Cielo,*** etc). Reliable VCSELs, which have lower production cost and higher modulation rates than the LEDs, immediately dominated the market of many high-speed LAN applications such as gigabit Ethernet as the preferred light sources. The most important advantage of VCSELs is the simplification of wafer testing procedures, which significantly reduces the required production cost. Other advantages of VCSELs over facet emitting lasers are narrow beam divergence, low current requirement, stable single-mode operation, and easy polarization control. As a result, a billion-dollar market is emerging due to the mass production of short-wavelength VCSELs as more and more businesses upgrade their premise LEDs networks to gigabit speeds. The U.S. consultancy ElectronicCast predicted that the global consumption of VCSEL-based transceivers is growing so fast that it is expected to reach U.S. $6 billion by 2008. However, fiber dispersion and high absorption at 850 and 980 nm limit the transmission rate and distance of the high-speed LANs using present short-wavelength VCSEL technology. Therefore, the future emerging higher-capacity networks, like 10-Gbit Ethernet demanding long-wavelength light sources for most interbuilding applications, will bring another billion-dollar business opportunity in manufacturing of long-wavelength VCSELs. This is a tough challenge, but developers are making progress and looking to push VCSELs all the way to the metropolitan area networks (MANs) using 1330-nm VCSEL technology. Present single-mode transceivers that can be found in the market using 1300-nm single-mode diode laser (facet emitting devices) can transmit signals over a distance of 40 km and at a gigabit-range transmission rate over a single-mode fiber. Therefore, the replacement of long wavelength single-mode facet emitting lasers in MANs requires delivering high-speed (>10-Gbit/s) 1330-nm VCSELs for a better offer.

Research in short-wavelength VCSELs began in the late 1970s, and now there are more than a dozen manufacturers producing 850- and 980-nm VCSELs, including single-diode packages, arrays, and datacommunication systems. ***Honeywell*** is one of the leading U.S. manufacturers making short-wavelength VCSELs. In April 2001, the VCSEL Products Division of Honeywell announced the full production release of its oxide aperture VCSEL product line, a set of high-performance 850-nm VCSELs packaged for high-speed (>1-Gbit/s) data communications. This product has improved high-speed operating characteristics necessary for many new datacom/telecom (datacommunication/telecommunication) applications. The cornerstone of their success in this VCSEL product is the attention to detail with respect to the reliability of VCSEL products. In 1997 ***Emcore*** Corporation, which provides integrated compound semiconductor solutions, announced the purchase of ***MicroOptical Device***, Inc., another U.S. manufacturer of microlasers and optical systems. This acquisition is expected to strengthen the development of VCSEL technology, and now Emcore is in mass production of 2.5-Gbit/s 1×12 VCSEL array, 3.125-Gbit/s 1×4 VCSEL array, and 2.5 and 10-Gbit/s chips, all at 850 nm. Its VCSEL arrays and VCSELs are designed for use in parallel fiberoptic and high-speed communication systems. ***PicoLight*** is another U.S.-based company, who claims that its parallel transmitter/receiver pair 12 channels, each at ≤2.5 Gbits/s or with an aggregate bandwidth of 20 Gbit/s at 850 nm, supplies the unique pluggable parallel interconnect that can be found on the market. This high-speed data rate of parallel interconnect products is reached using PicoLight's patented and proprietary VCSEL array technology. ***Agilent Technologies*** (formerly Hewlett-Packard Semiconductor) focuses on the fast-growing gigabit ethernet market and has developed the HRBR series (family of 850-nm VCSELs) to compete with the other system manufacturers. The HRBR products are made only for its own datacom systems such as the 1000 base-SX multimode fiber links of ≤550 m as well as the 2.125-Gbit/s low-voltage transceiver for fiber channel in SFF pluggable package. ***LaserMate*** Corporation also has mass production of ion implant and oxide aperture VCSELs, both of maximum modulation speed of 1.25 and 2.5 Gbits/s, respectively. In addition to this chip-on-die design, the company has manufactured 1D (1 × 3 array) as well as 2D (2 × 2–16 × 16 arrays) VCSELs for the applications in 850-nm fiber channel transceivers. It is a shame that ***Lucent Technologies***, the so-called "optoelectronics giant" in United States with the backup of Bell Laboratories, has no product in VCSELs or related systems.

Apart from the United States–based VCSEL manufacturers that have been mentioned in the preceding paragraphs, there are at least five manufacturers from Canada, Switzerland, Germany, and Taiwan, who have the potential to compete with the existing billion-dollar business. ***Mitel*** is a Canadian manufacturer of 850-nm VCSELs, selling fiber coupled single emitters and arrays of 1 × 4 and 1 × 12 VCSELs for the short-haul datacom market. ***Centre Suisse d'Electronique et de Microtechnique* (CSEM)** is a Switzerland-based VCSEL provider manufacturing 1 × 10 and 8 × 8 VCSEL arrays at 850 and 960-nm for datacom and gas sensing applications. The company has also fabricated individually addressable 16 × 16 VCSEL arrays emitting at 850-nm. ***Infineon Technologies*** (formerly Siemens Semiconductor) is a German manufacturer selling their multimode transceivers with oxide aperture VCSELs at 980 nm and data speeds of ≤2.125 Gbits/s. Their modules are designed for a broad range of private telecom systems, such as LANs, wide area networks (WANs) and gigabit Ethernet applications. ***Helix AG*** is another company from Switzerland manufacturing a 12-channel 1.25-Gbit/s chipset, a 4-channel 2.5/3.135-Gbit/s chipset, and a 12-channel 2.5/3.135-Gbit/s chipset using VCSEL arrays emitting at 850 nm. ***Truelight*** is the only manufacturer from an Asian country–Taiwan–mass producing 780–980-nm VCSELs in the format of chips as well as chip-in-die and at an attractive price.

The above mentioned optoelectronics manufacturers have mass-production capacity in short-wavelength VCSELs. However, none of them have any future development plans for the long-wavelength VCSELs. In fact, the highly desired 10-Gbit Ethernet is the driving force in the future development of long-wavelength VCSELs in the applications of long-distance (>60-km) and high-modulation-rate (>10-Gbit/s) datacommunication. ***Gore***

(U.S. manufacturer famous for the breathable waterproof fabric GORE-TEX) and ***Cielo Communications***, Inc. have already demonstrated their technologies to manufacture 1300-nm VCSELs. At OFC 99, Gore announced that 1310-nm single-mode version of VCSELs were available for sampling. However, up to the time of writing, its long-wavelength VCSELs have been unavailable. In 2000, Cielo Communications, partnered with Sandia National Laboratories for a technology breakthrough, demonstrated the industry's first electrically pumped VCSEL for 1300-nm applications. In addition, Cielo Communications has also demonstrated industry's first 2.5-Gbit/s VCSEL-based 1310-nm parallel array link at OFC 2000. Cielo is working to bring this long-wavelength VCSEL technology to market. CSEM is also attempting to develop long-wavelength VCSELs using AlGaAsSb mirrors. ***Avalon Photonis*** is a startup company who is working very hard to manufacture 10-Gbit/s 2D VCSEL arrays emitting at 1550 nm, but their devices are still not ready for mass production. Another startup company from Canada, ***Coretek***, which was acquired by Nortel Network in 2000, can also deliver 1550-nm tunable VCSELs at high output power of ~20 mW. This high-power tunable VCSEL is optically pumped so that the corresponding production cost is more expensive than the electrically pumped tunable VCSEL. ***Bandwidth9*** is the first manufacturer producing a VCSEL-based, long-wavelength tunable optical transmitter capable of direct modulation up to 2.5 Gbit/s. Its module, the MetroFlex "tunable integrated optical transmitter module," integrates all necessary drivers, control application-specific integrated circuitry (ASIC), modulation, and wavelength locking elements necessary to incorporate the tunable VCSEL subsystem within optical equipment. Applications of this tunable integrated optical transmitter module can be wavelength-on-demand capabilities in reconfigurable optical add-drop modules (OADMs), wavelength routers, and optical cross-connects as well as dense wavelength division multiplexing (DWDM) MAN systems and spares. Because the unique technology of manufacturing long-wavelength tunable VCSELs, products from Bandwidth9 have essentially no competition as most of the VCSEL manufacturers only concentrate on the development of high-speed VCSELs for the applications in high-speed LANs. It is believed that the future industrial standard of VCSELs is an even higher modulation rate (>10 Gbits/s), longer link lengths (>60 km), and lower power consumption (<10 mW) and the further development of 1300/1550-nm VCSEL technology in mass production is the key to the success of a billion-dollar business.

1.5 ELECTRICAL AND OPTICAL CONFIGURATIONS OF VCSELs

The motto of laser designers in realizing high-efficiency low-threshold VCSELs is "high," "small," and "well confined." The "high" means high surface reflectivity (>99.9%), the "small" means small active volume, and the "well confined" means effective optical and electrical confinement inside the small active volume. It is obvious that high surface reflectivity and small active volume will lead to ultra-low-threshold current. Well optical and electrical confinement will enhance the electrical-to-optical conversion efficiency or the wallplug efficiency of the devices. In fact, it is a challenge to realize "high," "small," and "well confined" with the available fabrication technologies. In the following paragraphs, the optoelectrical performance of the most recently designed VCSELs is reviewed.

The first issue to consider in the design of VCSELs is defining the operating wavelength. The choice of the material systems for the active region, and DBRs defines the lasing wavelength of VCSELs under construction. For example, the implementation of 850-nm VCSELs requires the use of a GaAs-based material system as the active region. In addition, the longitudinal cavity mode should match with the gain peak of the active region in order to maximize the wallplug conversion efficiency. The mirrors should have a maximum reflectivity at the operating wavelength with the appropriate number of dielectric layers to generate reflectivity greater than 99.9%. (See also Chapter 2 for the design of DBRs.) It is noted that the mirrors can be dielectric multilayered mirrors or DBRs with material system lattice-matched to the active region (i.e., GaAlAs/AlAs DBRs are lattice-matched with the GaAs material system). It is preferred to grow devices monotonically using lattice-matched active layer and DBRs as this approach will reduce fabrication cost and enhance the reliability of the devices. However, it may be difficult to realize all-monolithic growth in long-wavelength VCSELs using the InP-based system. This is because although InGaAsP/InP DBR is lattice-matched with InP based systems, the difference in refractive index between InGaAsP and InP is small, thus requiring more than 40 pairs of such epitaxial layers to obtain high reflectivity, leading to impractical mass production and difficulty in quality control. In addition, the electrical and thermal resistivities will increase with the number of epitaxial layers, and this is not desired in VCSELs. Therefore, it is very important to select an appropriate material system, that can allow the commercialization of VCSELs more easily. GaInNAs/GaAs QWs may be a better choice to realize long-wavelength VCSELs using GaAs-based DBRs. However, further improvement of the most recent processing technology of GaInNAs/GaAs QWs may be required for the realization of mass production and reliability. Finally, the cavity length (not including the DBRs) of VCSELs should be carefully selected to maximize the longitudinal overlapping between the standing-wave peak and the active layer for maximum wallplug efficiency. This is also discussed in Chapter 2. The enhancement of spontaneous emission in VCSELs may be required for the further reduction of threshold current. The design requirement of the enhancement of spontaneous emission with different cavity lengths can be found in Chapter 8.

Apart from the longitudinal confinement of optical fields requiring consideration, the transverse confinement of optical fields and electrical current also needs investigation in the design of VCSELs. Basically, the transverse optical field can be confined inside the VCSELs using gain guiding, index guiding, or antiguiding mechanisms, which is the most popular design that can be found in the most recent literature. In the early development of VCSELs, fabrication of a laser with a gain-guiding mechanism is based on a simple fabrication technique that is compatible with the existing technologies of facet emitting lasers [80].

A gain-guiding mechanism can be realized by forming a circular metal contact close to the active layer. The injection carrier concentration defines the gain region to confine the transverse mode. Figure 1.9 shows the schematic of a gain-guided VCSEL, which is similar to a metal-stripe facet emitting lasers. Using this approach, a gain-guided 850-nm VCSEL has demonstrated room-temperature CW operation [80]. The major advantage of this structure is that it is easy to fabricate but the transverse confinement of optical field and injection current density is weak so that the corresponding threshold current is high. The threshold current can be improved if the current leakage is minimized along the transverse direction. This can be easily achieved by ion implantation into the p DBR (but avoid damaging the active layer) to increase the electrical resistivity [82]. Figure 1.10 shows the schematic of an ion implanted VCSEL. As is shown, the ion-implanted region is defined selectively to control the flow of the injection current into the active layer. However, this configuration of ion-implanted region has no control on the diffusion of carrier concentration along the transverse direction of the active layer. It is possible to apply ion implantation into the active layer, but this will increase the optical absorption loss (i.e., due to scattering of ions) of the device. At high-power operation, higher-order transverse modes can be excited because of the influence of thermal lensing and spatial hole burning of carrier concentration, which are the undesired characteristics of gain-guided VCSELs. The other problem is the electrical resistivity of the DBRs, which may increase the heat generation inside the laser cavity. The only attraction of this structure is planar configuration, which guarantees the simplification in fabrication process and packaging so that low production costs can be maintained. Therefore, many manufacturers (Honeywell, etc.) manufactured products using ion implantation technology in their early-state development of VCSEL. The ion implantation technique has also been applied to fabricate long-wavelength VCSELs [42, 83].

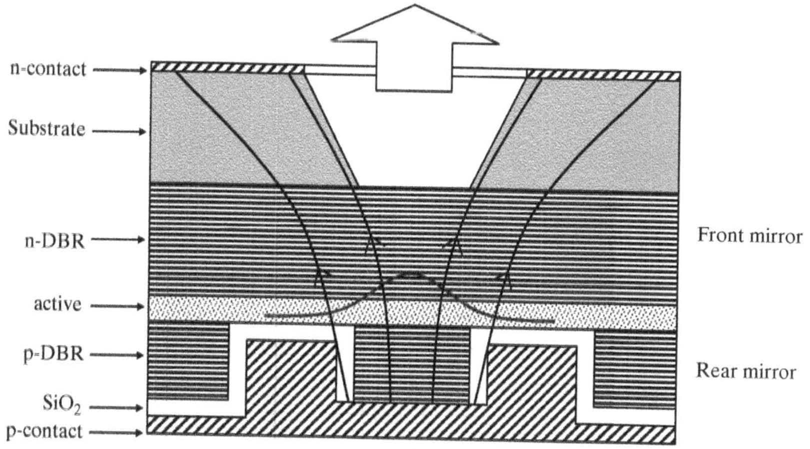

Figure 1.9 Schematic diagram of a gain-guided VCSEL with circular electrode to confine injection carrier concentration into the active layer.

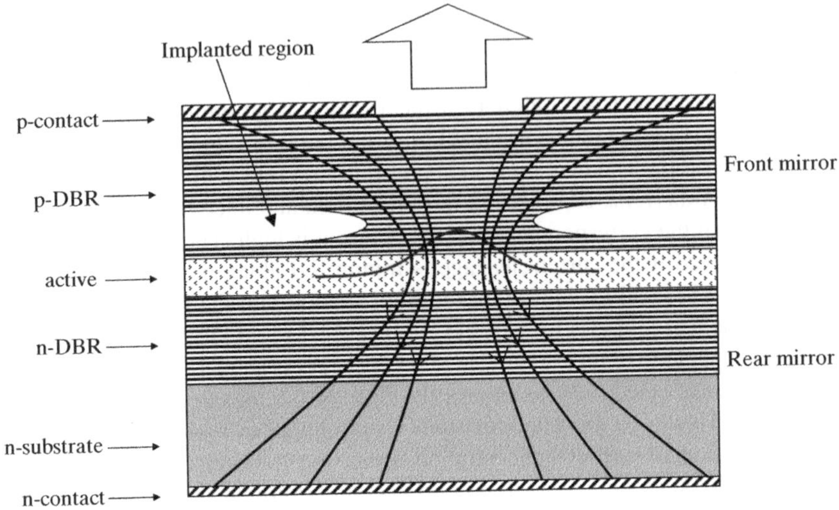

Figure 1.10 Schematic diagram of a gain-guided VCSEL with ion implantation regions to confine injection carrier concentration into the active region.

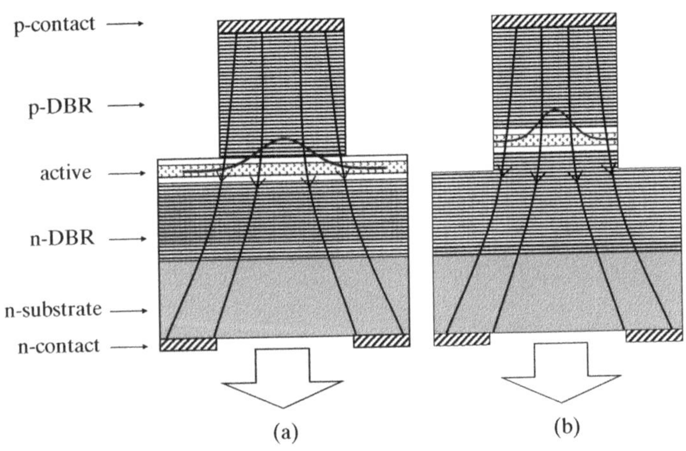

Figure 1.11 Schematic diagrams of airposted VCSELs (index-guided structure); active layer below (a) and within (b) the airposted column.

Index-guided VCSELs have better transverse confinement of optical field than do the gain-guided devices. Several types of index-guided VCSELs can be found in the literature such as airposted, etched mesa, regrowth buried heterostructure, and oxide aperture, which have different mechanisms to confine the optical field and injection current as well as different fabrication procedure complexities. Figure 1.11 shows the two possible configurations of airposted VCSELs [84, 85]. The p-DBR section of airposted VCSELs, whose edge is in direct contact with the air, and stands alone on the base material and the active layer, is sandwiched between the n- and p-type DBRs. The large difference in refractive index between the semiconductor and the air provided strong transverse confinement of the optical field. However, diffraction and scattering losses increase with the decrease (increase) of the active layer's diameter (sidewall roughness). In addition, single-mode operation is not stable in airposted structures, especially at high injection levels, due to the large difference in refractive index between the core and air regions. It is obvious that strong transverse confinement of injection carrier concentration can be obtained in this structure, but the surface recombination of carrier concentration can be enormous. Furthermore, low electrical and thermal conductivities of the airposted VCSELs can have serious problems in highly biased voltage and insufficient heatsinking. An alternative way to avoid potential dropping across the DBRs is to use an intracavity structure [86]. Figure 1.12 shows a schematic of an intracavity VCSEL. Additional layers, the p- and n-type layers, are inserted between the n and p DBRs on either surface of the active region to provide an electrical path for the injection current to reach the active region. A current constriction is also formed on top of the active layer to force the current to flow into the optical mode. This current blocking layer can be realized by ion implantation, diffusion, or a simple selective wet etch. Current crowding may occur near the periphery of the device where the optical mode is weak so that the influence of spatial hole burning of carrier concentration can be multiplied. A resistive layer between the conductive current distribution layer and the active layer can be introduced to minimize the influence of current crowding. This current layer limits the influence of current crowding by diffusion of the injected current from the center of the active layer. This device is expected to have low biasing voltage and high single-mode output power. Figure 1.13 shows a schematic of a buried heterostructure VCSEL, which is another type of index-guided structure [87]. As it is shown, the entire core region of VCSEL

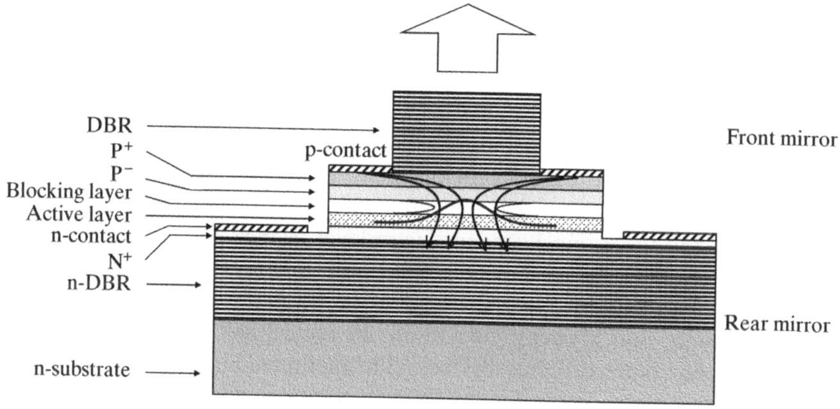

Figure 1.12 Schematic diagram of an airposted VCSEL (index-guided structure) with intracavity design.

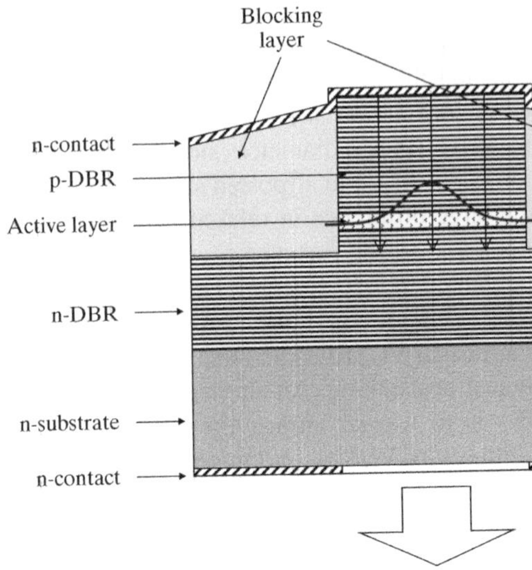

Figure 1.13 Schematic diagram of a buried heterostructure VCSEL (index-guided structure) with injection carrier concentration strongly confined uniformly inside the active region.

is immersed inside a current blocking region with careful selection of refractive index to cut off higher transverse modes. This blocking region provides enough transverse confinement of optical field and injection current, which improves the single-mode operation at high injection level. The blocking region can be polyimide or an epitaxial regrowth of p-n junction to block the diffusion of injection current outside the core region when the laser is forward-biased [88]. This structure has demonstrated a stable single-mode operation at high power and reliable performance. However, the fabrication process, which requires etching and regrowth of the blocking region, may not be suitable for mass production, and the production cost of such a device can be high. It is known that oxide aperture VCSELs, the only VCSEL with index-guided structure, have been massively produced and delivered to the market. Figure 1.14 shows a schematic of a VCSEL with double-oxide apertures. A VCSEL with single-oxide aperture is similar to that shown in Figure 1.14 but there is no oxide aperture below the active layer [89]. It is noted that the refractive index of the oxide aperture is lower than that of the original semiconductor layer. Hence, the induced effective index difference between the cavity and the surrounding region containing the oxide layer can be controlled through the thickness of the oxide layer and its position relative to the optical cavity. As a result, the optical field along the transverse direction can be confined tightly by the oxide apertures. Furthermore, the oxide apertures are insulating layers forcing the injection current through the aperture, which enhance the wallplug efficiency. Therefore, oxide aperture VCSELs exhibit extremely low threshold current [26] and high wallplug efficiency [24]. It must

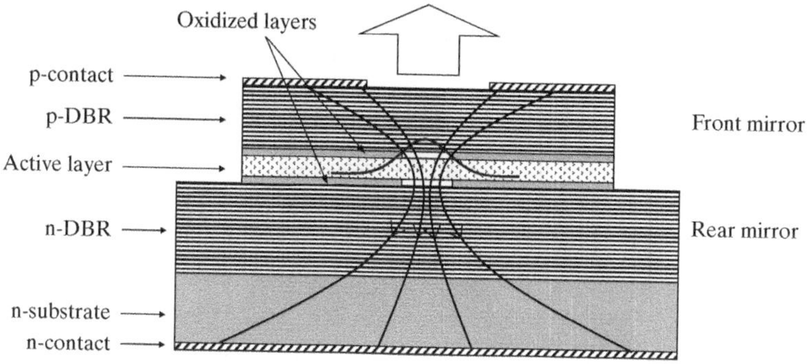

Figure 1.14 Schematic diagram of an index-guided VCSEL with double oxide apertures.

be noted that the oxidization procedure for the oxide aperture is compatible with the monolithic growth of GaAs-based material systems so that the production cost of oxide aperture VCSELs can compete with that of the ion-implanted devices. Therefore, oxide aperture VCSEL is the only index-guided structure that can be commercialized and released to the market several years after the commercialization of ion-implanted devices. As indicated in Figure 1.14, the problem of high electrical resistivity of n and p DBRs can deteriorate the performance of oxide aperture VCSELs. However, this problem can be easily solved by using the design of intracavity contacted cavity. This can be realized by removing the blocking layer as shown in Figure 1.12 by an oxide aperture or sandwiching the active layer by two oxide apertures [90]. In this case, the biased voltage of the VCSELs as well as power consumption at threshold can be further reduced.

It must be noted that oxide aperture VCSELs are now the most promising devices for replacing LEDs used in LAN and MAN networks because of their low power consumption, high modulation speed, and low production cost. However, there are two problems related to the design of oxide aperture VCSELs. The first one is the excitation of high-order transverse modes due to the high refractive index difference between the oxide and original semiconductor layer. Using the tapered waveguide design as shown in Figure 1.15 can suppress the excitation of high-order transverse modes [91]. This is because the tapered region acts as a spatial filter to remove any high-order transverse modes generated inside the active layer so that single-mode operation can be maintained in oxide aperture VCSELs at a wide range of injection levels. On the other hand, the maximum output power of VCSELs is directly related to the surface area of the active region that amplified the transverse modes. Because of the small nature of the oxide aperture, the maximum output power of oxide aperture VCSELs is limited to tens of milliwatts. It is possible to increase the diameter of the oxide aperture for higher output power, but the excitation of high-order transverse modes is inevitable even with the tapered waveguide design. In addition, the inclusion of the tapered waveguide in oxide aperture VCSELs is not preferred because of the increase in the complexity of the fabrication process as well as high production

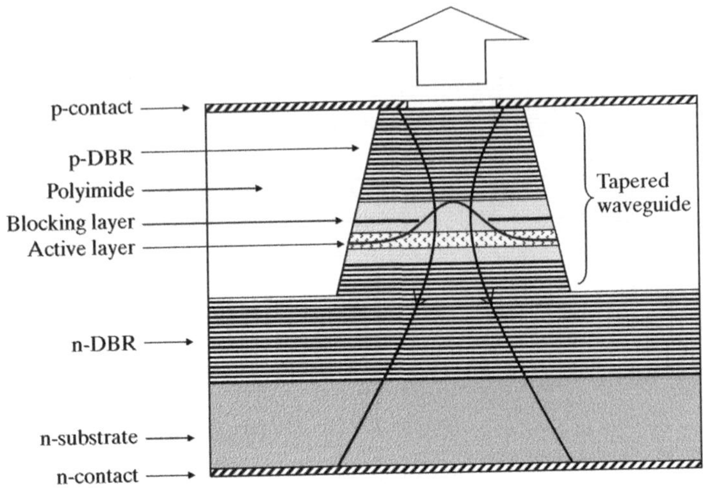

Figure 1.15 Schematic diagram of a tapered waveguide buried heterostructure VCSEL (index-guided structure).

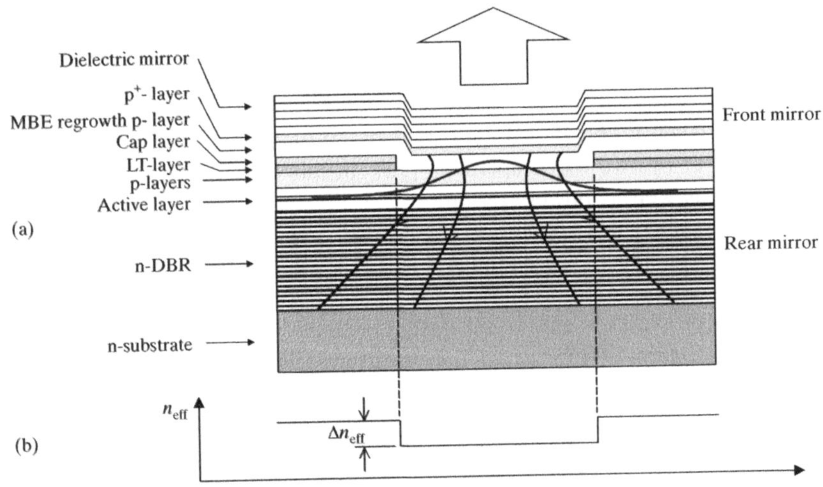

Figure 1.16 (a) Schematic diagram of an antiguided VCSEL; (b) the corresponding profile of effective refractive index n_{eff} and change of effective refractive index Δn_{eff} along the transverse direction of the laser cavity.

costs. Therefore, VCSELs with oxide apertures are not an appropriate design for high-power operation.

It may be possible to realize high-power single-mode VCSELs using an antiguided structure [92]. Figure 1.16a shows one possible configuration of antiguided VCSELs that can be found in the literature. A thick low-temperature (LT) layer with enhanced resistivity is grown on top of the p layer to confine the injected carrier concentration. An undoped cap layer is grown on top of the LT layer using molecular beam epitaxy (MBE). Hence, a circular recess can be formed on top of the cap layer using photolithography. The corresponding effective refractive index along the transverse direction of the active region is also shown in Figure 1.16b. It is noted that the antiguided structure is experienced by the transverse modes along the active layer as the LT and cap layers increase the effective refractive index of the aperture. In this structure, the aperture diameter can be as large as 6 μm to suppress the excitation of high-order transverse modes at high injection level. However, the only drawback of this structure is the inevitable high-threshold current due to strong radiation loss.

1.6 COMPUTATION TECHNIQUES IN DESIGN AND ANALYSIS OF VCSELs

In the early development of low threshold VCSELs, the required cavity size and reflectivity of mirrors were calculated by laser design engineers using simple rate equation models [93]. This is because the fabrication technologies of Fabry–Perot structures with metallic mirrors (i.e., gold deposition on semiconductor materials) used as the reflectors can avoid calculating the longitudinal standing wave inside the laser cavity so that the use of simple rate equation models is sufficient. The most recent development of VCSELs with DBR and QW active layers requires the exact calculation of the longitudinal standing-wave profile to maximize the corresponding wallplug efficiency. The scattering matrix method has been applied to analyze the longitudinal optical field distribution. The 1D beam propagation method (BPM) incorporated with scattering matrix can also be applied to investigate the transient response of optical fields inside the laser cavity [94]. The application of the scattering matrix method and 1D BPM on the analysis of optical fields inside the cavity of VCSELs is discussed in Chapter 2. On the other hand, the transverse component of optical fields can also be solved using the decoupled wave equation [95]. This approximation is acceptable if the transverse component of the optical fields has a uniform wavefront. If a VCSEL has a weakly guiding mechanism along the transverse direction, linearly polarized (LP) modes can also be applied to approximate the transverse fields. The method to study the transverse field using LP modes is given in Chapter 3.

Simple 1D BPM cannot be applied to calculate the off-axis longitudinal propagation of optical fields, which is usually observed in oxide aperture VCSELs (i.e., diffraction of optical fields from the small oxide aperture). Two possible numerical techniques—wide-angle BPM [96] and the full 3D vectorial method [96, 98]—can be applied to solve the resonant condition of off-axis longitudinal propagation of optical fields. Wide-angle BPM solves the 3D scalar wave equation with the off-axis longitudinal propagation of optical fields taken into consideration through the Padé approximant operators. In addition, the diffraction of light at the boundaries of the semiconductor layers can be evaluated using Bessel transform [96]. The merits of wide-angle BPM is simple to implement and has high computational speed. However, the accuracy of calculation is dependent on the configuration of the laser cavity and the length of the propagation step. The 3D vectorial method, which has been applied to analyze the 3D distribution of optical field inside an oxide aperture VCSEL, solves the full wave equation using the finite element method (FEM) or coupled mode theory [96,98]. The advantage of the 3D vectorial method is the exact calculation of wave equation without using any approximation. However, the major drawback of this method is the requirement of extensive computational effort. Figure 1.17 indicates the level of complexities in solving the optical field distribution inside a VCSEL. These approaches to solving the 3D profiles of the optical modes are discussed further in Chapter 3.

Calculation and analysis of optical field profiles inside the cavity of VCSELs can be easily done using the commercialization of computer-aided design (CAD) tools. For example, Beam**PROP** and *Full*WAVE (products of ***Rsoft, Inc.***) are the commercialized CAD tools for the design of integrated fiber optical devices and circuits that can be used to design circular optical waveguide of VCSELs. Beam**PROP** is capable of analyzing 2D and 3D simulation of cylindrical optical waveguide using the finite difference BPM scheme, which has a very fast performance and efficient algorithm. *Full*WAVE employs the finite-difference time-domain method for the simulation, which can be incorporated with Beam**PROP**

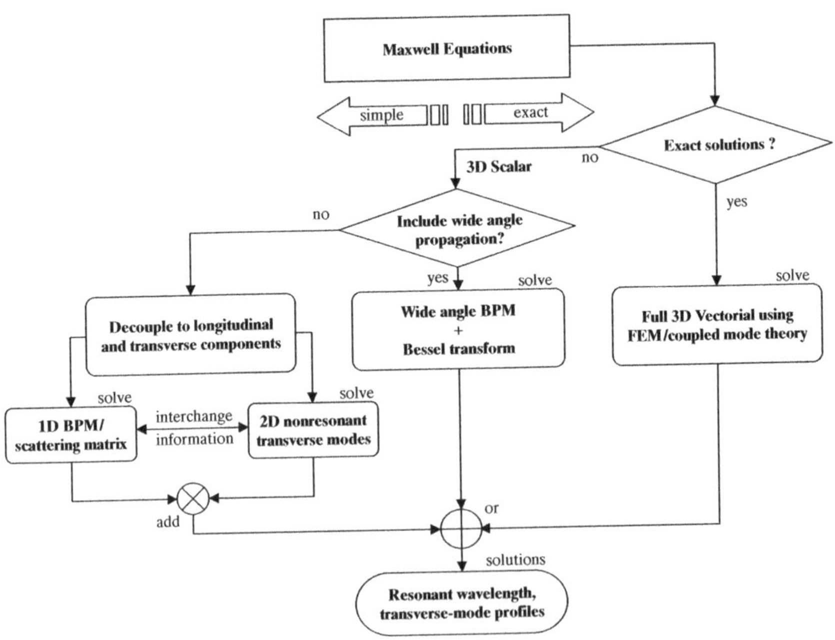

Figure 1.17 A flowchart illustrating different approaches in solving the wave equation of VCSELs.

to further improve their calculation abilities. Hence, these simulators are capable of calculating the modal propagation constants and fields under the influence of wide-angle propagation and multilayer reflection. OPTI*WAVE* (a product of ***OPTIWAVE Corp.***) is another software that can calculate optical field distribution inside a cylindrical waveguide. The most important feature of these commercial CAD tools is that they are easy to use and knowledge of different computational techniques is not required to run them.

In the design of VCSELs under CW operation, the influence of self-heating should not be avoided. Self-heating is due to the high thermal resistance of DBRs, which prohibits the dissipation of heat from the active layer to the heatsink. Understanding the self-heating effect will help improve the steady-state performance of VCSELs. The theoretical study of self-heating requires information about the distribution of heatsources and thermal resistance. The use of effective heatsource and effective thermal resistance is the simplest approach to estimate the temperature change inside the laser cavity [99]. The effective heatsource is assumed to be located at the center of the active layer, and its magnitude is roughly equal to the difference between input electrical power and output optical power. The effective thermal resistance can be deduced by adding all the semiconductor layers together. Hence, the temperature change with the increase of injection current can be easily calculated. Simple self-consistent analysis of temperature change inside the laser cavity can be performed if rate equations of carrier concentration and photon density are also factored into the calculation [100]. If calculation of detailed 3D heat distribution inside the laser cavity is required, the 3D heat equation has to be solved with the exact location of heatsources taken into consideration. FEM and the finite difference method (FDM) are the two most popular techniques used to solve the 3D heat equation numerically [99, 100]. The definition of heatsources and the numerical technique to solve the 3D heat equation are discussed in Chapter 5. It is understood that solving the heat equation from scratch can be a headache. However, there is a lot of commercial software such as ***ANSYS***, a well-developed finite element solver, which can be used to solve the 3D heat equation. Even the laser design engineers have no knowledge of the finite element technique. The graphical user interface (GUI) of ANSYS is very friendly to users, and this solver is easily used and fast enough to analyze 3D heat distribution of VCSELs.

In order to locate the heatsources, the distribution of voltage and current as well as the position of nonradiative recombination sources inside the laser cavity has to be determined. The voltage distribution inside the laser cavity can be deduced from Poisson's equation using either FEM or FDM as the solver. Boundary conditions at the interface between semiconductor layers and QW junction have to be defined in the calculation [100]. A more detailed analysis may include the continuity equation to estimate the charge accumulated at the interface between two semiconductor layers [101]. It must be noted that the potential across the QW active layer is dependent on the injection current profile, carrier recombination, and heat distribution. Therefore, development of a comprehensive model to calculate the 3D distribution of electric potential, carrier concentration, heat, and the optical field in a self-consistent manner is required [100]. Figure 1.18 illustrates a comprehensive model of VCSELs with electric potential, carrier concentration, heat, and optical field to be evaluated self-consistently. A detailed description of the development of comprehensive models of VCSELs is also given in Chapter 6. Commercial CAD tools such as PICS3D (developed by ***CrossLight Software***) have also been developed to investigate VCSELs in a self-consistent manner. PICS3D is a 3D comprehensive numerical solver of semiconductor lasers, including VCSELs. This 3D simulator offers rigorous and self-consistent treatment on thermal, electrical, and optical processes by solving the appropriate partial differential equations. Poisson's equation is used to relate the electrical potential to spatial charge distribution. The current continuity equations describe the macroscopic carrier transport mechanisms, the heat equation calculates the temperature distribution, and the wave equations in conjunction with the multilayer optics theory deduce the 3D wave profile inside the laser cavity. However, the corresponding GUI of PICS3D is less user-friendly than ANSYS, which hinders the laser design engineers with less experience starting the simulation. CrossLight is one of the known providers for a complete and detailed analysis of VCSELs that can be found on the market. Nowadays, the sophisticated commercial laser simulators enable laser design engineers, even without a knowledgeable background in laser physics, to analyze and understand their devices in great detail and gain insight into performance limitations.

The transient response of VCSELs can also be solved by the comprehensive model under time-domain calculation [102,103]. PICS3D uses a similar

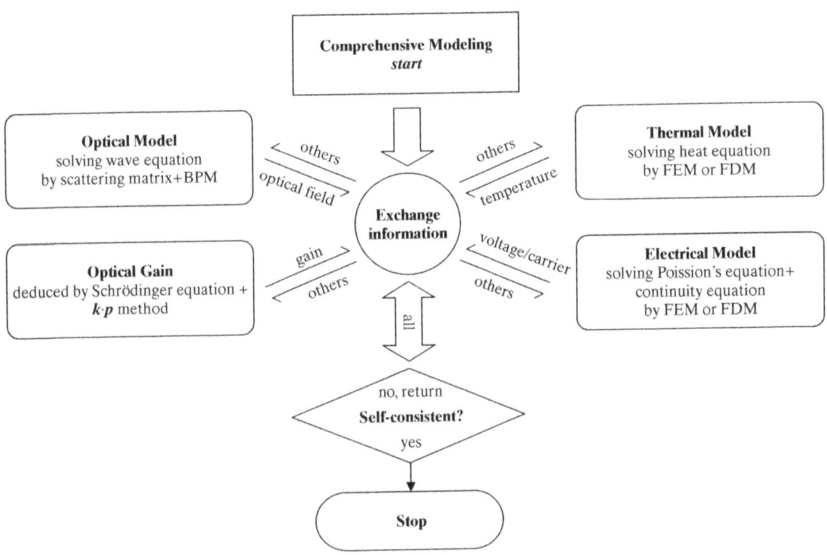

Figure 1.18 A flowchart illustrating the comprehensive self-consistent calculation of optical field, temperature, potential, and carrier concentration of VCSELs.

approach to analyze the small-signal as well as the large-signal dynamic response of VCSELs. However, the simulator for the dynamic 3D modeling of VCSELs requires extensive computational power and enormous running time for a single set of data. Therefore, simple dynamic models of VCSELs for particular applications such as AM, FM, and second-harmonic distortion (SHD) response analysis have been developed using simple rate equations [104]. However, only limited information on transverse optical confinement and self-heating effect has been introduced into the calculation. These assumptions are satisfied as the distribution of transverse modes and heat remain mostly unchanged at the DC-biased level and under small-signal modulation [104]. Furthermore, the analysis of diffraction loss can be simplified by using an effective index approach [105]. However, for the analysis of the transient response of multiple-transverse modes, parameters of the transverse confinement structure should be considered. Hence, the rate equations have to be solved simultaneously with the decoupled wave equation for the transverse modes [97]. For more detailed analysis of VCSELs, wide-angle BPM under finite difference time-domain calculation can deduce the transient response of the 3D optical field inside the laser cavity [96]. On the other hand, for VCSELs operating at subpicosecond timeframe, the carrier transport mechanism has significant influence on the dynamic response of VCSELs. The carrier energy balance equation representing the nonisothermal carrier transport inside the QW active layer should be factored into the calculation [106]. The modeling of the dynamic behavior of VCSELs with different degrees of complexity and different nonlinear effects is discussed in Chapter 7.

1.7 OPPORTUNITIES AND CHALLENGES IN FUTURE DESIGN OF VCSELs

The upcoming opportunity of long-wavelength VCSELs is to replace the short-wavelength VCSELs in LANs to further improve the transmission speed of the computer networks. Furthermore, because of the high intrinsic modulation speed of VCSELs, it is desired to replace the DFB lasers in MAN networks as well as long-haul high-speed optical fiber communication systems. Therefore, it is believed that the long-wavelength VCSELs will bring another billion-dollar business in the near future to the laser manufacturers. Challenges for the laser design engineers are to develop long wavelength VCSELs operating at high output power (>10 mW) and single transverse mode (side mode suppression ratio >40 dB). As discussed in Section 1.2, different fabrication technologies have been developed to realize 1300/1550-nm VCSELs, but only some fabrication technologies can be adopted in manufacturing low-cost mass-production long-wavelength VCSELs. For example, monolithic growth of VCSELs using lattice-matched material systems is a sensible choice to reduce the fabrication complexity so that low production cost and reliability can be maintained. Using GaInNAs/GaAs QWs with GaAs-based DBRs and substrate seems to be a promising solution to long-wavelength VCSELs but the technology to fabricate GaInNAs/GaAs QWs is not mature yet. Therefore, the first challenge is to design an appropriate fabrication technology for the manufacture of long-wavelength VCSELs. The maximum output power of VCSELs is scaled with the aperture area of the active layer. It has been shown that VCSELs with an aperture diameter of 16 μm have maximum output power of 40 mW [27], which is sufficient for the applications in MAN networks. However, high-order transverse modes, which significantly deteriorate the performance of VCSELs in single-mode optical fiber systems, can be excited in such an aperture size. Therefore, the second challenge is to design an appropriate structure of VCSELs for high-power single-mode operation. High-order transverse modes can be suppressed in index- or gain-guided structures by restraining the aperture size to its cutoff condition of higher-order modes, but the maximum output power is then limited. A tapered waveguide structure has been proposed to maintain the fundamental transverse-mode operation in index-guided VCSELs, but the maximum aperture size is limited to less than 10 μm [91]. On the other hand, an antiguided structure is used to fabricate VCSELs for improving stable single transverse-mode operation at high power [92, 107]. One of these possible configurations of antiguided VCSELs has been illustrated in Figure 1.16. Antiguided structure provides strong radiation loss for the higher-order transverse modes so that a stable fundamental mode operation can be maintained for devices with wide aperture. In addition, the influence of carrier spatial hole burning and thermal lensing can be minimized by a large index step ($\Delta n_{\mathrm{eff}} > 0.05$). Antiguided VCSELs have been fabricated either by surrounding a low-index core region by regrowth of a high-index material or by creating a low-index core region by shifting the cavity resonance (toward longer wavelength) outside the core [92]. These devices display promising results; single-mode operation up to 5–15 times its threshold for diameters as large as 16 μm aperture. However, the maximum output power has been limited to <2 mW because of the relatively large radiation loss incurred for the fundamental mode, which is the inherent characteristic of antiguided structure. The poor confinement of injection carrier concentration due to its large aperture size has also limited the wallplug efficiency as well as the threshold current of these devices. Phase-locked arrays of VCSELs may offer an alternative method to achieve higher-power diffraction-limited beams, but this may not be suitable for application in optical fiber communication systems [108].

In order to reduce the edge radiation losses for the fundamental mode of an antiguided structure, antiresonant reflecting optical waveguides (ARROWs) have been employed for lateral waveguiding in facet emitting lasers [109]. Figure 1.19a shows the lateral distribution of effective refractive index of an ARROW structure with a low-index core region, which is surrounded by a pair of reflector regions of thickness s (a multiple of quarter-wave thickness). The reflector regions are designed to be antiresonant for the fundamental lateral mode. Higher-order lateral modes, which do not meet the antiresonant condition, suffer radiation losses greater than that those the fundamental mode. ARROW, which is the favored creation of fundamental mode, has better modal discrimination than does the antiguided structure. In addition, ARROW permits a relatively large spot size (>10 μm diameter) with large built-in lateral effective index step so

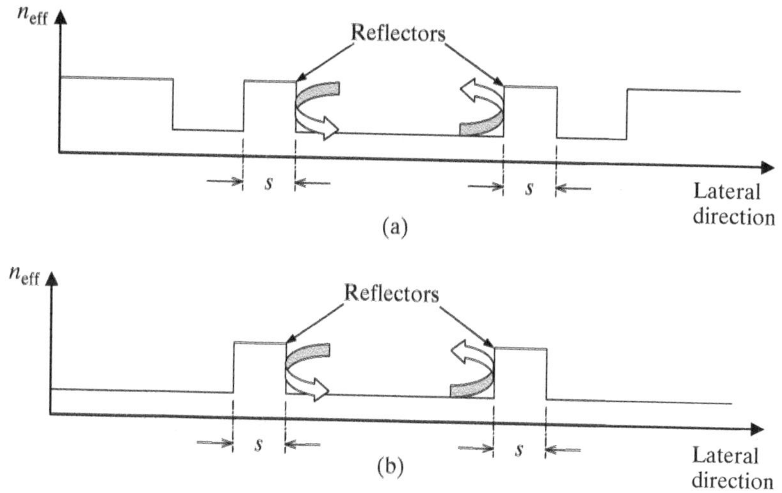

Figure 1.19 Lateral distribution of effective refractive index of facet emitting lasers with (a) ARROW structure and (b) S-ARROW structure.

that is suitable for high-power single-mode operation. A simplified version of the ARROW structure (S-ARROW), which is designed by taking out the low-index outer-reflector regions, has been found theoretically to be less sensitive to gain spatial hole burning than the conventional ARROW structure (see Fig. 1.19b). Similar to the S-ARROW facet emitting lasers, a VCSEL with an S-ARROW is proposed to reduce antiguiding loss of the fundamental transverse mode so that single-mode operation at high output power can be obtained. Figure 1.20 shows the VCSEL incorporating an S-ARROW structure, along with the corresponding profile of transverse effective refractive index. However, the introduction of S-ARROW structure shifts the cavity resonant mode to the longer wavelength. The increase in the resonant wavelength is equivalent to an increase in effective refractive index. The shift of resonant wavelength $\Delta\lambda$ can be approximated by $\Delta n_{eff}/n_{eff} \approx \Delta\lambda/\lambda_R$, where λ_R is the lasing wavelength. For a given Δn_{eff}, the width of the reflecting ring s, which satisfies the antiresonance condition, is equal to $m\lambda_1/4$, where m is an odd integer, and λ_1 is the transverse wavelength in the high-index region. It must be noted that the modal loss of an S-ARROW VCSEL is a function of reflector width, which should be selected to obtain the lowest loss region for the fundamental transverse mode. Large discrimination of higher-order modes should also be maintained at the selected width of reflector. The fabrication of the proposed S-ARROW 980-nm VCSEL involves a two-step low-pressure metal oxide chemical vapor deposition (MOCVD) growth at 700°C [109]. After the first growth, involving the layers up to the spacer layers, the ring reflector is realized by selective chemical etching with an appropriate photoresist pattern. The second growth involves the regrowth of the top p-type DBR and p^+ contact layer. Proton implantation is then used to confine the current to the central low-index

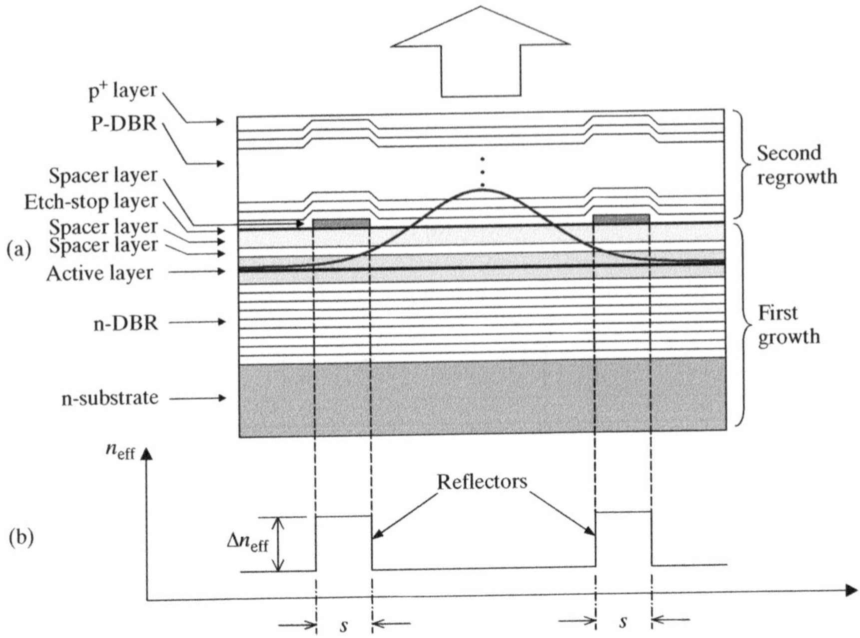

Figure 1.20 (a) Schematic diagram of an S-ARROW VCSEL; (b) the corresponding profile of effective refractive index n_{eff} and change of effective refractive index Δn_{eff} along the transverse direction of the laser cavity.

core region. Finally, metal liftoff is used to open a window in the metal contact for light output. S-ARROW VCSEL with 12-μm-diameter core region, $\Delta n_{eff} = 0.1$, and $s = 5\lambda_1/4$ demonstrates the threshold current of 4.5-mA and single-mode operation up to ~1 mW CW operation at 3 times its threshold. In this device, the output power is limited by the self-heating effect due to the unoptimized p DBR. In addition, the misalignment of gain peak ($\lambda \sim 950$ nm) with the resonant wavelength ($\lambda \sim 930$ nm) also leads to the limitation of output power. If the self-heating effect of p-DBR can be reduced and appropriate alignment of gain and resonant cavity wavelength is selected, this device has considerable potential to increase output power. It is believed that with optimization of Δn_{eff} and s, single-transverse-mode operation at high power (i.e., >10 mW from 12-μm-diameter aperture) can be obtained. This S-ARROW structure can be applied to realize long-wavelength VCSELs, but the electrical confinement of this structure needs to be improved to enhance the wallplug efficiency. The fabrication of S-ARROW VCSELs, which requires regrowth of p DBR after etching of reflectors, may not be suitable for mass production because of its relatively complicated fabrication processing (i.e., when compared with oxide aperture structure). Therefore, the challenge of realizing high-power single-mode VCSELs is to realize novel S-ARROW-type long-wavelength VCSELs with improved confinement of injection carrier concentration and simplified fabrication process.

1.8 SUMMARY

In this chapter, the most recent research and development of VCSELs have been described. It is shown that the successful demonstration of monolithic growth GaAs-based VCSELs with DBRs allows the mass production of low-cost and reliable short-wavelength lasers for applications in short-distance optical fiber communication systems such as LANs. On the other hand, different approaches to realize long-wavelength VCSELs are discussed. The difficulties in fabricating monolithic growth 1300/1550-nm VCSELs are also explained. In order to fabricate low-cost and reliable long-wavelength VCSELs, the development of new lattice-matched material systems (e.g., GaInNAs QWs or InGaAs QDs as active layer and GaAs-based material as DBRs) is required. The possibilities of obtaining visible wavelength (e.g., red, green, blue) VCSELs are also studied. Red VCSELs have been successfully fabricated using InGaP/InGaAlP QWs, but the development of green and blue VCSELs is still in progress.

The commercialization of VCSELs is also discussed in this chapter. It is noted that the replacement of LEDs by high-speed 850-nm VCSELs can significantly improve the transmission rate of the existing short-distance optical computer networks such as LANs. In addition, the transmission speed of the plastic-fiber-based communication systems can be further improved by using red VCSELs. On the other hand, optical pickup systems using VCSELs integrated with an intracavity absorber can significantly reduce the complexity as well as the production cost of CD and DVD systems. Therefore, the emergent VCSEL technologies can minimize the production cost as well as improve the performance of the available commercial products. Furthermore, the development of parallel optical interconnects and optical information processing using VCSELs is explained. It is proposed to use arrays of VCSEL and a photodetector integrated with microlens array for the realization of low-power consumption and high-data-transmission-speed optical interconnect. In addition, low-cost WDM optical data link is fabricated using monolithically integrated multiple-channel VCSEL array and photodetector. The possibility of using VCSELs as the 2D arrays of smart pixels to realize the next generation of optical switching, optical signal processing, and optical computing applications is also predicted.

There are more than 14 optoelectronics manufacturers worldwide who have massively produced VCSELs and their related products. There are about 10 United States–based manufacturers delivering short- wavelength VCSELs using ion implantation and oxide-confined technologies. Products such as VCSEL arrays and related fiber modules are also available from these companies. Other manufacturers from Canada, Switzerland, Germany, and Taiwan also offer similar products to customers worldwide. It is believed that the 1300-nm VCSELs will be the products of the next generation. These types of long-wavelength VCSELs offer high modulation bandwidth (>10 Gbit/s) and low-cost production, so they are the ideal candidates for applications in long-distance fiber computer networks such as metropolitan area networks. It is also believed that the U.S. manufacturers will predominate the future markets of long wavelength VCSELs over the other manufacturers worldwide. This is because Cielo and Core have demonstrated the possibility of mass production of 1300-nm VCSELs, and Bandwidth9 has products of 1550-nm tunable VCSELs.

The advance in fabrication technologies has significantly improved the performance of VCSELs such as low threshold, high output power, and better wallplug efficiency. In this chapter, the progress in the design of VCSELs for better optical and electrical confinement has been discussed. It is shown that the ion-implanted VCSEL (i.e., gain guiding) is the simplest structure that can be fabricated and has the lowest production cost but the corresponding wall plug efficiency is limited because of the weak electrical confinement. On the other hand, oxide aperture VCSEL (i.e., index guiding) has demonstrated the lowest threshold current and the highest wall plug efficiency, but the small aperture size limits the maximum output power of the device. Antiguided structures have been applied to realize VCSEL with single transverse mode, but the maximum output power is deteriorated by the radiation loss.

The current advancements in theoretical studies of VCSELs are briefly reviewed in this chapter. In the early investigation of VCSELs, rate equation analysis was used to identify the requirement for low-threshold operation. Detailed analysis of VCSELs requires solving the wave equation for each optical cavity using the scattering matrix method, 1D and wide-angle BPM, or the full 3D vectorial method. Apart from the calculation of optical field distribution inside the laser cavity, the distribution of heat, electrical potential, and carrier concentration, should also be evaluated. Therefore, a comprehensive self-consistent calculation technique is proposed to solve the profile of temperature, electrical potential, carrier concentration and optical field simultaneously. This method is able to calculate the steady state as well as dynamic behavior of VCSELs. In addition, more detailed analysis of VCSELs requires the consideration of carrier transport mechanism inside the QWs and carrier energy balance equations. The existing CAD tools for analysis VCSELs are also discussed in this chapter.

Finally, the challenges to realize high-power (>10-mW) long-wavelength VCSELs with stable single mode are identified. First, it is not certain which fabrication technologies are promising enough to produce low-cost and highly reliable devices, but it is certain that monolithic growth of VCSELs is a suitable choice to simplify the fabrication process. Second, a novel structure for single transverse mode operation at high power (>10 mW) should be designed for VCSELs in order to satisfy the requirement of high-speed long-haul single-mode fiber communication systems. It is believed that the ARROW-type structure may be a possible design to realize single transverse mode operation at high power. However, this laser configuration has weak electrical confinement and the fabrication requires etching and secondary growth, which may increase the production cost of the devices. Therefore, a new ARROW-type structure should be designed to eliminate the imperfection of the existing design.

REFERENCES

1. G. J. Fokken, W. L. Walters, L. E. Mattson, and B. K. Gilbert, "Low cost, multi-GHz electrical packaging for serial optoelectronic links utilizing vertical cavity surface emitting lasers," *IEEE Trans. Adv. Pack.*, Vol. 23, no. 1, 42–54, 2000.
2. J. A. Hudging, S. F. Lim, G. S. Li, W. Yuen, K. Y. Lau, and C. J. Chang-Hasnain, "Compact, integrated optical disk readout head using a novel bistable vertical cavity surface emitting lasers," *IEEE Photon. Technol. Lett.*, Vol. 11, no. 2, pp. 245–247, 1999.
3. J. Geske, V. Jayaraman, T. Goodwin, M. Culick, M. MacDougal, T. Goodmough, D. Welch, and J. Bower, "2.5 Gb/s transmission over 50 km with a 1.3 μm vertical cavity surface emitting laser," *IEEE Photon. Technol. Lett.*, Vol. 12, no. 12, pp. 1707–1709, 2000.
4. H. Kosaka, "Smart integration and packaging of 2D VCSEL's of high speed parallel links," *IEEE Select. Topics Quantum Electron.*, Vol. 5, no. 2, pp. 184–192, 1999.
5. K. Kasahara, "VSTEP based smart pixels," *IEEE J. Quantum Electron.*, Vol. 29, no. 2, pp. 757–768, 1993.
6. K. Iga, "Surface emitting laser–its birth and generation of new optoelectronics field," *IEEE J Select. Topics Quantum Electron.*, Vol. 6, no. 6, pp. 1201–1215, 2000.
7. H. Soda, K. Iga, C. Kitahara, and Y. Suematsu, "GaInAsP/InP surface emitting injection lasers," *Jpn. J. Appl. Phys.*, Vol 18., pp. 2329–2330, 1979.
8. K. Iga, S. Ishikawa, S. Ohkouchi, and T. Nishimura, "Room temperature pulsed oscillation of GaAlAs/GaAs surface emitting injection laser," *Appl. Phys. Lett.*, Vol. 45, pp. 348–350, 1984.
9. F. Koyama, S. Kinoshita, and K. Iga, "Room temperature continuous wave vertical cavity surface emitting laser and high-power 2D laser arrays," in *Tech. Digest, Conf. Lasers and Electro-Optics*, paper FC1, pp. 380–381, 1989.
10. K. Iga, S. Kinoshita, and F. Koyama, "Microcavity GaAlAs/GaAs surface emitting laser with $I_{th} = 6$ mA," *Electron. Lett.*, Vol. 23, no. 3, pp. 134–136, 1987.
11. F. Koyama, S. Kinoshita, and K. Iga, 'Room temperature continuous wave lasing characteristics of GaAs vertical cavity surface emitting laser," *Appl. Phys. Lett.*, Vol. 55, no. 3, pp. 221–222, 1989.
12. A. Chailervanitkul, S. Kinoshita, and K. Iga, "GaInAsP-InP surface emitting laser ($\lambda = 1.2$ μm, 77 K) with heteromultilayer bragg reflector," *Electron. Lett.*, Vol. 21, pp. 303–304, 1985.
13. S. A. Chalmers, K. L. Lear, and K. P. Killeen, "Low resistance wavelength reproducible p-type (Al,Ga)As distributed Bragg reflectors grown by molecular beam epitaxy," *Appl. Phys. Lett.*, Vol. 62, pp. 1585–1587, 1993.
14. J. Jewell, S. McCall, Y. Lee, A. Scherer, A. Scherer, A. Gossard, and J. English, "Lasing characteristics of GaAs micro-resonators," *Appl. Phys. Lett.*, Vol. 54, pp. 1400–1402, 1988.
15. M. A. Matin, A. F. Jezierski, S. A. Bashar, D. E. Lacklison, T. M. Benson, T. S. Cheng, J. S. Roberts, T. E. Sale, J. W. Orton, C. T. Foxon, and A. A. Rezazadeh, "Optically transparent indium-tin-oxide (ITO) ohmic contacts in the fabrication of

vertical cavity surface emitting lasers," *Electron. Lett.*, Vol. 30, no. 4, pp. 318–320, 1994.

16. Y. Lee, J. Jewell, A. Scherer, S. McCall, J Harbison, and L. Florez, "Room temperature continuous wave vertical cavity surface emitting laser single quantum well microlaser diode," *Electron. Lett.*, Vol. 25, pp. 1377–1378, 1989.
17. J. Jewell, A. Scherer, S. McCall, Y. Lee, S. Walker, J. Harbison, and L. Florez, "Low threshold electrical pumped vertical cavity surface emitting microlaser," *Electron. Lett.*, Vol. 25, pp. 1123–1124, 1989.
18. M. Sugimoto, H. Kosaka, K. Kurhara, I. Ogura, T. Numai, and K. Kasahara, "Very low threshold current density in vertical cavity surface emitting laser diodes with periodically doped distributed bragg reflectors," *Electron. Lett.*, Vol. 28, no. 4, pp. 385–387, 1992.
19. R. S. Geels, S. W. Corzine, J. W. Scott, D. B. Young, and L. A. Coldren, "Low threshold planarized vertical cavity surface emitting lasers," *IEEE Photon. Technol. Lett.*, Vol. 2, pp. 234–236, 1990.
20. R. S. Geels and L. A. Coldren, "Submilliampere threshold vertical cavity laser diode," *Appl. Phys. Lett.*, Vol. 57, pp. 1605–1607, 1990.
21. M. R. McDaniel, D. L. Huffaker, and D. G. Deppe, "Hybrid dielectric/metal reflector for low threshold vertical cavity surface emitting lasers," *Electron. Lett.*, Vol. 33, no. 20, pp. 1740–1741, 1997.
22. J. M. Dallesasse, N. Holonyak, A. R. Sugg, T. A. Richard, and N. El-Zein, "Hydrolyzed oxidation of $Al_xGa_{1-x}As$-AlAs-GaAs quantum well heterostructures and superlattices," *Appl. Phys. Lett.*, Vol. 57, pp. 2844–2846, 1990.
23. D. L. Huffaker, D. G. Deppe, K. Kumar, and T. J. Rogers, "Native oxide defined ring contact for low threshold vertical cavity lasers," *Appl. Phys. Lett.*, Vol. 65, pp. 97–99, 1994.
24. P. W. Evans, J. J. Wierer, and N. Holonyak, "$Al_xGa_{1-x}As$ native oxide based distributed bragg reflectors for vertical cavity surface emitting lasers," *J. Appl. Phys.*, Vol. 84, no. 10, pp. 5436–5440, 1998.
25. L. A. Graham, D. L. Huffaker, and D. G. Deppe, "Spontaneous lifetime control in a native oxide aperture microcavity," *Appl. Phys. Lett.*, Vol 74, no. 17, pp. 2408–2410, 1999.
26. G. M. Yang, M. MacDougal, and P. D. Dupkus, "Ultralow threshold current vertical cavity surface emitting laser obtained with selective oxidation," *Electron. Lett.*, Vol. 31, pp. 886–888, 1995.
27. W. Schmid, D. Wiedenmann, M. Grabber, R. Jager, R. Michalzik, and K. J. Ebeling, "CW operation of a diode cascade InGaAs quantum well VCSEL," *Electron. Lett.*, Vol. 34, no. 6, pp. 553–555, 1998.
28. M. Miller, M. Grabberr, R. Jager, and K. J. Ebeling, "High power VCSEL arrays for emission in Watt regime at room temperature," *IEEE Photon. Technol. Lett.*, Vol. 13, no. 3, pp. 173–175, 2001.
29. H. Martinsson, J. A. Vukusic, and A. Larsson, "Single mode power dependence on surface relief size for mode stabilized oxide confined vertical cavity surface emitting lasers," *IEEE Photon. Technol. Lett.*, Vol. 12, no. 9, pp. 1129–1131, 2000.
30. T. H. Oh, M. R. McDaniel, D. L. Huffaker, and D. G. Deppe, "Guiding and antiguiding effects in epitaxially regrown vertical cavity surface emitting lasers," *Appl. Phys. Lett.*, Vol. 72, no. 22, pp. 2782–2784, 1998.

31. R. W. Herrick and P. M. Petroff, "Gradual degradation in 850 nm vertical cavity surface emitting lasers," *IEEE Quantum Electron.*, Vol. 34, no. 10, pp. 1963–1969, 1998.

32. P. Salet, F. Gaborit, Ph. Pagnod-Rossiaux, A. Plais, E. Derouin, J, Pasquier, and J. Jacquet, "Room temperature pulsed operating of 1.3 μm vertical cavity lasers including bottom InGaAsP/InP multilayer bragg mirrors," *Electron. Lett.*, Vol. 33, no. 24, pp. 2048–2049, 1997.

33. I. F.L. Dias, B. Nabet, A. Kohl, J. ,L, Benchimol, and J. C. Harmand, "Electrical and optical characteristics of n-type doped distributed Bragg mirrors on InP," *IEEE Photon. Technol. Lett.*, Vol. 10, no. 6, pp. 763–765, 1998.

34. J. Boucart, C. Starck, F. Gaborit, A. Plais, N. Bouche, E. Derouin, L. Goldstein, C. Fortin, D. Carpentier, P. Salet, F. Brillouet, and J. Jacquet, "1 mW CW-RT Monolithic VCSEL at 1.55 μm," *IEEE Photon. Technol. Lett.*, Vol. 11, no. 5, pp. 629–631, 1999.

35. T. Baba, Y. Yogo, K. Suzuki, F. Koyama, and K. Iga, "Near room temperature continuous wave lasing characteristics of GaInAsP/InP surface emitting laser," *Electron. Lett.*, Vol. 29, no. 10, pp. 913–014, 1993.

36. H. W. Song, D. S. Song, L. Y. Han, C. K. Kim, H. Y. Ryu, and Y. H. Lee, "1.5 μm vertical cavity surface emitting lasers utilizing low loss Al_xO_y/GaAs mirrors," *Electron. Lett.*, Vol. 35, no. 4, pp. 296–297, 1999.

37. P. O. Vaccaro, H. Ohnishi, and K. Fujita, "Lateral junction vertical cavity surface emitting laser grown by molecular beam epitaxy on a GaAs (311) A oriented substrate," *Appl. Phys. Lett.*, Vol. 74, no. 25, pp. 3854–3856, 1999.

38. N. M. Margalit, K. A. Black, Y. J. Chiu, E. R. Hegblom, P. Abraham, M. Anzlowar, J. E. Bowers, J. F. Hu, and K. Streubel, "Top emitting double fused 1.5 μm vertical cavity lasers," *Electron. Lett.*, Vol. 34, no. 3, pp. 285–287, 1998.

39. Y. Qian, Z. H. Zhu, Y. H. Lo, H. Q. Hou, M. C. Wang, and W. Lin, "1.3 μm vertical cavity surface emitting lasers with double bonded GaAs/AlAs Bragg mirrors," *IEEE Photon. Technol. Lett.*, Vol. 9, no. 1, pp. 8–10, 1997.

40. S. Rapp, F. Salomonsson, J. Bentell, I. Sagnes, H. Moussa, C. Meriadec, R. Raj, K. Streubel, and M. Hammer, "Near room temperature continuous wave operation of electrical pumped 1.55 μm vertical cavity lasers with InGaAsP/InP bottom mirror," *Electron. Lett.*, Vol. 35, no. 1, pp. 49–50, 1999.

41. H. Gebretsadik, K. Kamath, K. K. Linder, X. Zhang, P. Bhattacharya, C. Caneau, and R. Bhat, "Growth of high quality GaAs/AlAs Bragg mirrors on patterned InP based quantum well mesa structures," *Appl. Phys. Lett.*, Vol. 71, no. 5, pp 581–583, 1997.

42. J. Boucart, C. Starck, F. Gaborit, A. Plais, N. Bouche, E. Derouin, L. Goldstein, C. Fortin, D. Carpentier, P. Salet, F. Brillouet, and J. Jacquet, "1 mW CW-RT monolithic VCSEL at 1.55 μm," *IEEE Photon. Lett.*, Vol. 11, pp. 629–631, no. 6, 1999.

43. C. Kazmierski, J. P. Debray, R. Madani, I. Sagnes, A. Ougazzaden, N. Bouadma, J. Etrillard, F. Alexandre, and M. Quillec, "+55°C pulse lasing at 1.56 μm of all monolithic InGaAlAs/InP vertical cavity lasers," *Electron. Lett.*, Vol. 35, no. 10, pp. 811–812, 1999.

44. J. K. Kim, E. Hall, O. Sjolund, G. Almuneau, and L. A. Coldren, "Room temperature, electrically pumped multiple active region VCSELs with high differential efficiency at 1.55 μm," *Electron. Lett.*, Vol. 35, no. 13, pp. 1084–1085, 1999.

45. E. Hall, G. Almuneau, J. K. Kim, O. Sjolund, H. Kroemer, and L. A. Coldren, "Electrically pumped, single-epitaxial VCSELs at 1.55 mm Sb-based mirrors," *Electron. Lett.*, Vol. 35, no. 16, pp. 1337–1338, 1999.

46. E. Hall, G. Almuneau, J. K. Kim, O. Sjolund, H. Kroemer, and L. A. Coldren, "Selectively etched undercut apertures in AlAsSb-based VCSELs," *IEEE Photon. Lett.*, Vol. 13, no. 2, pp. 97–99, 2001.

47. M. C. Larson, M. Kondow, T. Kitatani, K. Nakahara, H. Inoue,. and K. Uomi, "GaInNAs/GaAs long wavelength vertical cavity surface emitting laser diodes," *IEEE Photon. Technol. Lett.*, Vol. 10, no. 2, pp. 188–190, 1998.

48. A. Wagner, C. Ellmers, F. Hohnsdorf, J. Koch, C. Agert, S. Leu, M. Hofmann, W. Stolz, and W. W. Ruhle, "GaInNAs/GaAs vertical cavity surface emitting laser with ultra-broad temperature operation range," *Appl. Phys. Lett.*, Vol. 76, no. 3, pp. 271–273, 2000.

49. T. Kageyama, T. Miyamoto, S. Makino, Y. Ikenaga, N. Nishiyama, A. Matsutani, F. Koyama, and K. Iga, "Room temperature continuous wave operation of GaIn-NAs/GaAs VCSELs grown by chemical beam epitaxy with output power exceeding 1 mW," *Electron. Lett.*, Vol. 37, no. 4, pp 225–226, 2001

50. M. Yamada, T. Anan, K. Kurhara, K. Nishi, K. Tokutome, A. Kamei, and S. Sugou, "Room temperature low-threshold CW operation of 1.23 μm GaAsSb VCSELs on GaAs substrates," *Electron. Lett.*, Vol. 36, no. 7, 637–638, 2000.

51. J. A. Lott, N. N. Ledentsov, V. M. Ustinov, N. A. Ustinov, M. A. Maleev, A. E. Zhukov, A. R. Kovsh, M. V. Maximov, B. V. Volovik, Zh. I. Alferov, and D. Bimberg, "InAs-InGaAs quantum dot VCSELs on GaAs substrates emitting at 1.3 μm," *Electron. Lett.*, Vol. 36, no. 16, pp. 1384–1385, 2000.

52. M. H. Crawford, K. D. Choquette, R. J. Hickman, and K. M. Geib, "Performance of selectively oxidized AlGaInP based visible VCSELs," *OSA Trends Optics Photon. Series*, Vol. 15, pp. 104–105, 1998.

53. C. Carlsson, H. Martinsson, J. Vukusic, J. Halonen, and A. Larsson, "Nonlinear distortion and dynamic range of red (670 nm) oxide confined VCSELs," *IEEE Photon. Technol. Lett.*, Vol. 13, no. 4, pp. 358–360, 2001.

54. H. Jeon, V. Kozlov, P. Kelkar, A. V. Nurmikko, C. C. Chu, D. C. Grillo, J. Han, G. C. Hua, and R. L. Gunshor, "Room temperature optically pumped blue-green vertical cavity surface emitting lasers," *Appl. Phys. Lett.*, Vol. 67, no. 12, pp. 1668–1670, 1995.

55. K. Iga, "Possibility of Green/blue/UV surface emitting lasers," in *Int. Symp. Blue Laser Light Emitting Diodes*, March., Th-11, pp. 263–266, 1996.

56. S. Nakamura, M. Senoh, S. Nagahama, N. Iwasa, T. Yamada, T. Matsushita, H. Kiyoku, Y. Sugimoto, T. Kozaki, H. Umemoto, M. Sano, and K. Chocho, "High power, long lifetime InGaN/GaN//AlGaN based laser diodes grown on pure GaN substrates," *Jpn. J. Appl. Phys.*, Vol. 37, pp. L309–L312, 1998.

57. I. L. Krestnikov, W. V. Lundin, A. V. Sakharov, V. A. Semenov, A. S. Usikov, A. F. Tsatul'nikov, and Zh. I. Alferov, "Room temperature photopumped InGaN/GaN/AlGaN vertical cavity surface emitting laser," *Appl. Phys. Lett.*, Vol. 75, no. 9, pp. 1192–1194, 1999.

58. Y. K. Song, H. Zhou, M. Diagne, I. Ozden, A. Vertikov, C. Certer-Coman, R. S. Kem, F. A. Kish, and M. R. Krames, "A vertical cavity light emitting InGaN

quantum well heterostructure," *Appl. Phys. Lett.*, Vol. 74, no. 23, pp. 3441–3443, 1999.

59. Y. K. Song, H. Zhou, M. Diagne, A. V. Nurmikko, R. P. Schneider, Jr., C. P. Kuo, M. R. Krames, R. S. Kern, C. Carter-Coman, and F. A. Kish, "A quasicontinuous wave, optically pumped violet vertical cavity surface emitting lasers," *Appl. Phys. Lett.*, Vol. 76, no. 13, pp. 1662–1664, 2000.
60. T. Shirasawa, N. Mochida, A. Inoue, T. Honda, T. Sakaguchi, F. Koyama, and K. Iga, "Interface control of GaN/AlGaN quantum well structures in MOCVD growth," *J. Cryst. Growth*, pp. 124–127, 1998.
61. K. H. Hahn, M. R. Tan, and S. Y. Wang, "Intensity noise of large area vertical cavity surface emitting lasers in multimode optical fiber links," *Electron. Lett.*, Vol. 30, no. 2, pp. 139–140, 1994.
62. U. Fiedler, B. Moller, G. Reiner, P. Schnitzer, and K. J. Ebeling, "10 Gbit/s data transmission using top emitting VCSELs with high side-mode suppression," *Electron. Lett.*, Vol. 31, no. 19, pp. 1664–1665, 1995.
63. F. Mederer, M. Kicherer, R. Jager, H. Unold, and K. J. Ebeling, "2.5 Gbit/s data transmission with singlemode GaAs VCSELs over −20 to 100°C temperature range," *Electron. Lett.*, Vol. 36, no. 3, pp. 236–237, 2000.
64. F. Mederer, C. Jung, R. Jager, M. Kicherer, R. Michalzik, P. Schnitzer, D. Wiedenmann, and K. J. Ebeling, "12.5 Gbit/s data rate fiber transmission using single mode selectively oxidized GaAs VCSEL at $\lambda = 850$ nm," *Proc. LEOS Annual Meeting*, San Franciso, US, Vol. 2, pp. 697–698, 1999.
65. F. Mederer, R. Jager, P. Schnitzer, H. Unold, M. Kicherer, K. J. Ebeling, M. Nartomi, and R. Yoshida, "Multi-Gb/s graded index POF data link with butt-coupled single mode InGaAs VCSEL," *IEEE Photon. Technol. Lett.*, Vol. 12, no. 2, pp. 199–201, 2000.
66. J. Geske, V. Jayaraman, T. Goodwin, M. Culick, M. MacDougal, T. Goodmough, D. Welch, and J. E. Bowers, "2.5 Gb/s transmission over 50 km with a 1.3 μm vertical cavity surface emitting laser," *IEEE Photon. Technol. Lett.*, Vol. 12, no. 12, pp. 1707–1709, 2000.
67. S. Zhang, N. Margalit, T. Reynolds, and J. Bowers, "1.65-μm vertical-cavity surface-emitting laser transmission at 2.5 Gb/s," *Photon. Technol. Lett.*, vol. 9, pp. 374–376, no. 3, 1997.
68. B. Madhavan and A. F.J. Levi, "Low power 2,5 Gbit/s VCSEL driver in 0.5 μm CMOS technology," *Electron. Lett.*, Vol. 34, no. 2, pp. 178–179, 1998.
69. J. A. Hudgings, S. F. Lim, G. S. Li, W. Yuen, K. Y. Lau, and C. J. Chang-Hasnain, "Compact, integrated optical disk readout head using a novel bistable vertical cavity surface emitting laser," *IEEE Photon. Technol. Lett.*, Vol. 11, no. 2, pp. 245–247, 1999.
70. F. A.P. Tooley, "Challenges in optically interconnecting electronics," *IEEE Select. Topics Quantum Electron.*, Vol. 2, pp. 3–13, 1996.
71. H. Kosaka, "Smart integration and packaging of 2D VCSELs for high speed parallel links," *IEEE Select. Topics Quantum Electron.*, Vol. 5, no. 2, pp. 184–192, 1999.
72. O. Sjolund, D. A. Louderback, E. R. Hegblom, J. Ko, and L. A. Coldren, "Monolithic integration of substrate input/output resonant photodetectors and vertical cavity lasers," *IEEE J. Quantum Electron.*, Vol. 35, no. 7, pp. 1015–1023, 1999.

73. S. Eitel, S. J. Fancey, H. P. Gauggel, K. H. Gulden, W. Bachtold, and M. R. Taghizadeh, "Highly uniform vertical cavity surface emitting laser integrated with microlens arrays," *IEEE Photon. Technol. Lett.*, Vol. 12, no. 5, pp. 459–461, 2000.
74. S. Q. Luong, G. G. Ortiz, Y. Zhou, J. Lu, C. P. Hains, J. Cheng, H. Q. Hou, and G. A. Vawter, "Monolithic wavelength-graded VCSEL and resonance enhanced photodetector arrays for parallel optical interconnects," *IEEE Photon. Technol. Lett.*, Vol. 10, no. 5, pp. 642–644, 1998.
75. S. Y. Hu, J. Ko, E. R. Hegblom, and L. A. Coldren, "Multimode WDM optical data links with monolithically integrated multiple channel VCSEL and photodetector arrays," *IEEE J. Quantum Electron.*, Vol. 34, no. 8, pp. 1403–1414, 1998.
76. Y. Zhou, J. Cheng, and A. A. Allerman, "High speed wavelength-division multiplexing and demultiplexing using monolithic quasi-planar VCSEL and resonant photodetector arrays with strained InGaAs quantum wells," *IEEE Photon. Technol. Lett.*, Vol. 12, no. 2, pp. 122–124, 2000.
77. K. Kasahara, "VSTEP based smart pixels," *IEEE J. Quantum Electron.*, Vol. 29, no. 2, pp. 757–768, 1993.
78. J. I. Song, Y. H. Lee, J. Y, Yoo, J. H. Shin, A. Scherer, and R. E. Leibenguth, "Monolithic arrays of surface emitting laser NOR logic devices," *IEEE Photon. Technol. Lett.*, Vol. 5, no. 8, pp. 902–904, 1993.
79. L. J. Irakliotis, P. J. Stanko, C. W. Wilmsen, and P. A. Mitkas, "Demonstration of a VCSEL based optoelectronic look-up table," *IEEE Photon. Lett.*, Vol. 9, no. 4, pp. 502–504, 1997.
80. "VCSELs meet demands for datacom bandwidth," *FiberSystems* Int., Feb./May issue, pp. 39–42, 2000.
81. K. Tai, R. J. Fischer, C. W. Seabury, N. A. Olsson, T. C.D. Huo, Y. Ota, and A. Y. Cho, "Room temperature continuous wave vertical cavity surface emitting lasers," *Appl. Phys. Lett.*, Vol. 55, no. 24, pp. 2473–2475, 1989.
82. Y. J. Yang, T. G. Dziura, T. Bardin, S. C. Wang, and R. Fernandez, "Continuous wave single transverse mode vertical cavity surface emitting lasers fabricated by Helium implantation and zinc diffusion," *Electron. Lett.*, Vol. 28, no. 3, pp. 274–275, 1992.
83. Y. Qian, Z. H. Zhu, Y. H. Lo, D. L. Huffaker, D. G. Deppe, H. Q. Hou, B. E. Hammons, W. Lin, and Y. K. Tu, "Low-threshold proton implanted 1.3 μm vertical cavity top surface emitting lasers with dielectric and wafer bonded GaAs-AlAs bragg mirrors," *IEEE Photon. Technol. Lett.*, Vol. 9, no. 7, pp. 866–868, 1997.
84. H. Saito, K. Nishi, I. Ogura, S. Sugou, and Y. Sugimoto, "Room temperature lasing operation of a quantum dot vertical cavity surface emitting lasers," *Appl. Phys. Lett.*, Vol. 69, no. 21, pp. 3140–3142, 1996.
85. P. Blixt, D. I. Babic, N. M. Margalit, K. Streubel, and J. E. Bowers, "Multimode fiber transmission using room temperature double fused 1.54 μm vertical cavity lasers," *IEEE Photon. Technol. Lett.*, Vol. 8, no. 11, pp. 1564–1567, 1996.
86. J. W. Scott, B. J. Thibeault, D. B. Young, L. A. Coldren, and F. H. Peters, "High efficiency submilliamp vertical cavity lasers with intracavity contacts," *IEEE Photon. Technol. Lett.*, Vol. 6, no. 6, pp. 678–680, 1994.
87. Y. Kohama, Y. Ohiso, K. Tateno, and T. Kurokawa, "0.85 μm vertical cavity surface emitting laser diode arrays grown on p-type GaAs substrate," *IEEE Photon. Technol. Lett.*, Vol. 9, no. 3, pp. 280–282, 1997.

88. K. D. Choquette, M. Hong, R. S. Freund, J. P. Mannaerts, R. C. Wetzel, and R. E. Leibenguth, "Vertical cavity surface emitting laser diodes fabricated by In Situ dry etching and molecular beam epitaxial regrowth," *IEEE Photon. Technol. Lett.*, Vol. 5, no. 3, pp. 284–287, 1993.
89. T. H. Oh, D. L. Huffaker, and D. G. Deppe, "Comparison of vertical cavity surface emitting lasers with half-wave cavity confined by single- or double oxide apertures," *IEEE Photon. Technol. Lett.*, Vol. 9, no. 7, pp. 875–877, 1997.
90. M. H. MacDougal, J. Geske, C. K. Lin, A. E. Bond, and P. D. Dupkus, "Low resistance intracavity contacted oxide-aperture VCSEL's," *IEEE Photon. Technol. Lett.*, Vol. 10, no. 1, pp. 9–11, 1997.
91. P. Schnitzer, M. Grabberr, R. Jager, F. Mederer, R. Michalzik, D. Wiedenmann, and K. J. Ebeling, "GaAs VCSEL's at $\lambda = 780$ and 835 nm for short distance 2.5 Gb/s plastic optical fiber data links," *IEEE Photon. Technol. Lett.*, Vol. 11, no. 7, pp. 767–769, 1999.
92. T. -H. Oh, O. B. Shchekin, and D. G. Deppe, "Single-mode operation in an antiguided vertical-cavity surface-emitting laser using a low-temperature grown AlGaAs dielectric aperture," *IEEE Photon. Technol. Lett.*, Vol. 10, no. 8, pp. 1064–1066, 1998.
93. K. Iga, F. Koyama, and S. Kinoshita, "Surface emitting semiconductor lasers," *IEEE J. Quantum Electron.*, Vol. 24, no. 9, pp. 1845–1855, 1988.
94. H. A. Haus, *Waves and Fields in Optoelectronics*, Prentice-Hall, Engewood Cliffs, NJ, 1984.
95. M. J. Adams, *An Introduction to Optical Waveguide*, Wiley, New York, 1981.
96. J. F.P. Seurin and S. L. Chuang, "Discrete bessel transform and beam propagation method for modeling of vertical cavity surface emitting lasers," *J. Appl. Phys.*, Vol. 82, no. 5, pp. 2007–2016, 1997.
97. D. Burak and R. Binder, "Cold cavity vectorial eigenmodes of VCSEL's," *IEEE J. Quantum Electron.*, Vol. 33, no. 7, pp. 1205–1215, 1997.
98. M. J. Noble, J. A. Lott, and J. P. Loehr, "Quasi-exact optical analysis of oxide-apertured microcavity VCSEL's using vector finite elements," *IEEE J. Quantum Electron.*, Vol. 34, no. 12, pp. 2327–2339, 1998.
99. M. Osinski and W. Nakwaski, "Thermal effects in vertical cavity surface emitting lasers," in *Current Trends in Vertical Cavity Surface Emitting Lasers*, T. P. Lee, ed., Selected Topics in Electronics and Systems, Vol. 3, World Scientific, Singapore, 1995.
100. W. M. Man and S. F. Yu, "Comprehensive modeling of diffused quantum wells vertical cavity surface emitting lasers," *IEEE J. Select. Topics Quantum Electron.*, Vol. 4, no. 4, pp. 715–722, 1998.
101. E. Zeeb and K. J. Ebeling, "Potential barriers and current-voltage characteristics of p-doped graded AlAs-GaAs heterojunctions," *J. Appl. Phys.*, Vol. 72, no. 3, pp. 993–999, 1992.
102. L. E. Thode, G. Csanak, L. L. So, T. J.T. Kwan, and M. Campbell, "Time-dependent numerical simulation of vertical cavity lasers," *SPIE Proc.*, Vol. 2146, pp. 174–184, 1994.
103. S. F. Yu, "Dynamic behavior of vertical cavity surface emitting lasers," *IEEE J. Quantum Electron.*, Vol. 32, no. 7, pp. 1168–1179, 1996.

104. S. F. Yu, W. N. Wong, P. Shum, and E. H. Li, "Theoretical analysis of modulation response and second harmonic distortion in vertical cavity surface emitting lasers," *IEEE J. Quantum Electron.*, Vol. 32, no. 12, pp. 2139–2147, 1996.

105. S. F Yu and E. H. Li, "Influence of lateral field on the relaxation oscillation frequency of semiconductor lasers," *IEEE J. Quantum Electron.*, Vol. 32, no. 1, pp. 1–3, 1996.

106. M. Elsasser, S. G. Hense, and M. Wegener, "Subpicosecond switch-off and switch-on of a semiconductor laser due to transient hot carrier effects," *Appl. Phys. Lett.*, Vol. 70, no. 7, pp. 853–855, 1997.

107. T. H. Oh, M. R. McDaniel, D. L. Huffaker, and D. G. Deppe, "Cavity induced antiguiding in a selectively oxidized vertical cavity surface emitting laser," *IEEE Photon. Technol. Lett.*, Vol. 10, no. 1, pp. 12–14, 1998.

108. D. K. Serkland, K. D. Choquette, G. H. Hadley, K. M. Geib, and A. A. Allerman, "Two element phased array of antiguided vertical cavity lasers," *Appl. Phys. Lett.*, Vol. 75, no. 24, pp. 3764–3756, 1999.

109. D. Zhou and L. J. Mawst, "Simplified antiresonant reflecting optical waveguide type vertical cavity surface emitting lasers," *Appl. Phys. Lett.*, Vol. 76, no. 13, pp. 1659–1601, 2000.

CHAPTER 2

Simple Design Consideration of Vertical Cavity Surface Emitting Lasers

In this chapter, the simple design methodology of vertical cavity surface emitting lasers (VCSELs) under the criteria of minimum threshold current, maximum electronic conversion (current : gain) ratio as well as maximum wallplug (electrical-to-optical) efficiency are described. The corresponding design equations for VCSELs with uniform and periodic gain structures are derived for the investigation. Hence, the detailed structure of lasers can be determined for optimal performance at and above threshold operation.

2.1 INTRODUCTION

Vertical cavity surface emitting lasers (VCSELs) have optical cavities orthogonal to those of conventional facet emitting lasers [1–3]. This simple arrangement in the orientation of cavity significantly improves the output performance and fabrication flexibility of semiconductor lasers [4]. The main advantages of VCSELs over conventional facet emitting lasers are

- VCSELs emit optical beams with low divergence and a circular symmetric profile because of their wide emission surface. As a result, the coupling efficiency to optical fiber and other optical components can be improved effectively [5].
- VCSELs exhibit extremely high relaxation oscillation frequency (>70 GHz) [6] because of their short photon lifetimes. Consequently, high modulation bandwidth can be achieved.
- VCSELs facilitate wafer-scale fabrication and testing by allowing fully monolithic processes because of their vertical orientation. Therefore, the production cost and procedures for quality inspection can be reduced enormously [4,5].

Analysis and Design of Vertical Cavity Surface Emitting Lasers By Siu Fung Yu
ISBN 0-471-39124-7

Hence, VCSELs are considered as the key components in optical fiber communications, optical interconnection systems, and optical parallel processing systems.

Because of the difference between VCSELs and conventional facet emitting lasers in cavity orientation, the design consideration of facet emitting lasers may not be applied to the analysis of VCSELs. For example, the requirement of high longitudinal side-mode suppression is the major concern in the design of facet emitting lasers [7,8] but is neglected in VCSELs [4] because of the latter's extremely short cavity length. Furthermore, in facet emitting lasers, maximum wallplug (electrical-to-optical) efficiency is achieved by enhancing the transverse confinement factor (i.e., overlap between transverse field profile and optical gain) [7,8], but that is realized in VCSELs with an optimal longitudinal confinement factor (i.e., overlap between longitudinal standing wave and optical gain) [9,10]. In addition, diffraction loss, self-heating, and high reflectivity (>0.95), which are the unique characteristics of VCSELs [11–16], need to be taken into consideration in the design of high-performance VCSELs. By contrast, diffraction loss and self-heating are usually ignored in the design of facet emitting laser [7,8]. So, it can be concluded that the design criteria of VCSELs are quite different from those of facet emitting lasers.

In this chapter, the design methodology of VCSELs for optimal electrical and optical performance is discussed. The design equations for VCSELs with uniform and periodic gain structure are derived to optimize the corresponding threshold current density and differential quantum efficiency. In addition, the use of the quantum-well (QWs) active layer to enhance the steady-state performance of VCSELs is studied. It is shown that the design criteria of VCSELs are different from those of facet emitting lasers such as the requirement of extremely high reflectivity. Hence, equations for analysis and design of high-reflectivity multilayered mirror are also given. Furthermore, the abovementioned threshold characteristics of VCSELs are optimized by analyzing the corresponding wallplug efficiency with the parasitic resistance and leakage current factored into the analysis.

2.2 DESIGN EQUATIONS OF VCSELs WITH A UNIFORM GAIN STRUCTURE

In the design of semiconductor lasers with high internal quantum efficiency (i.e., $\sim$1), threshold current density and differential quantum efficiency are two parameters that must be optimized in the first place. In the following paragraphs, the equations of threshold current density and differential quantum efficiency for VCSELs are derived. In addition, the methods to minimize the threshold current density as well as to maximize the differential quantum efficiency are discussed.

2.2.1 Threshold Conditions of VCSELs with a Uniform Gain Structure

Figure 2.1 shows the schematic of VCSELs with a uniform gain structure (UGS) used in the following analysis [11,15]. An active layer is sandwiched between

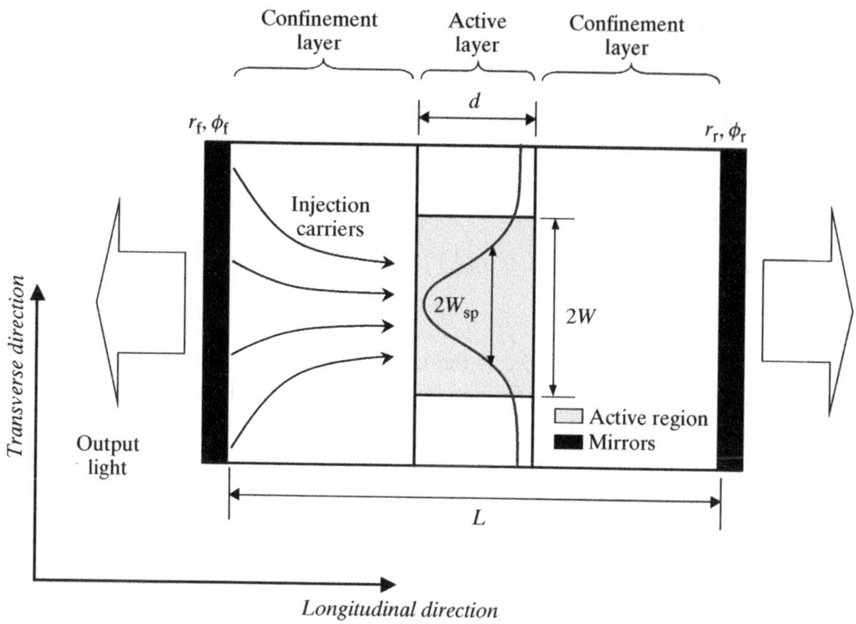

Figure 2.1 Schematic of VCSELs with uniform gain structure.

two longitudinal confinement layers (p- and n-doped), forming a Fabry–Perot resonator. Mirrors on the surface of the confinement layers provide optical feedback for the standing wave, which is amplified inside the active region, along the longitudinal direction. Stimulated emission is achieved by injecting the carrier concentration into the active region through the surface of the confinement layers. Light is emitted longitudinally from the surface of the confinement layers but is confined transversely inside the laser cavity. In the diagram, it is assumed that the total cavity length of the VCSEL is L, the thickness of the active layer is d, the radius of the active region is W, and r_f and r_r (ϕ_f and ϕ_r) are the field reflectivity (phase) of the front and rear mirrors, respectively.

In order to achieve resonant oscillation in VCSELs, the optical signal inside the resonator should obtain minimum amplification to compensate the total cavity losses so that no input signal is required to sustain the oscillation. Hence, the threshold amplitude condition is the minimum amplification of the optical signal to overcome the total optical loss, which includes diffraction loss, absorption, and scattering losses of the laser cavity as well as light output from the surfaces. In addition, the phase condition of the optical field inside the laser cavity should be a multiple of 2π in order to complete the loop phase requirement. Therefore, the threshold condition of VCSELs is similar to that of the facet emitting lasers except for the following two factors: (1) The nonuniform distribution of optical gain, absorption, and scattering losses along the longitudinal direction and (2) the excitation of diffraction loss due to the small cavity size of VCSELs [15].

The threshold conditions of VCSELs can be deduced using the approach of round-trip requirement. The optical field propagating along the longitudinal direction will be amplified inside the active region due to the stimulated emission but absorbed in the confinement layers. Part of the light will be reflected back into the laser cavity by the reflectors on the surface of the confinement layers, and the remaining optical field will emit to the surrounding area through the mirrors. Hence the round-trip condition can be written as [11,15]

$$\begin{aligned}1 = r_f r_r \exp((g_{th} - \alpha_{diff})d - \alpha_{in}L) \cdot \exp(-j2k_0(n_{ac}d + n_{conf}(L-d)) \\ - j(\phi_f + \phi_r))'\end{aligned} \tag{2.1}$$

where $j = \sqrt{-1}$, $k_0 = 2\pi/\lambda_R$, λ_R is the lasing wavelength, and n_{ac} and n_{conf} are the effective refractive index of the active and confinement layers, respectively. It can be shown from the real part of (2.1), the threshold gain, g_{th}, of the VCSELs is given by

$$\Gamma_z g_{th} = \alpha_{in} + \Gamma_z \alpha_{diff} + L^{-1} \log\left(\frac{1}{R}\right) \tag{2.2}$$

where $R = r_f r_r$, Γ_z $(= d/L)$ is the longitudinal confinement factor which is defined as the portion of light confined within the active layer so that $\Gamma_z g_{th}$ can be explained as the effective threshold gain of VCSELs. As $\Gamma_z < 1$, the required optical gain in VCSELs to achieve threshold conditions is higher than that of facet emitting lasers.

The first term on the right-hand side (RHS) of (2.2), α_{in}, represents the total internal loss of the laser cavity and is given by

$$\alpha_{in} = \Gamma_z \alpha_{ac} + (1 - \Gamma_z)\alpha_{conf}, \tag{2.3}$$

where α_{ac} and α_{conf} are the scattering and absorption loss of the active and confinement layers, respectively. As the values of α_{ac} and α_{conf} are dependent on the use of materials, α_{in} can also be considered as a material parameter, but it varies with the longitudinal confinement structure of VCSELs. If doping is applied to the active and confinement layers, the corresponding absorption loss inside the active and confinement layers can be approximated by [18]

$$\alpha_{ac}, \alpha_{conf} = 3 \times 10^{-18} n + 7 \times 10^{-18} p \quad (\mathrm{cm}^{-1}), \tag{2.4}$$

where n and p are the doping concentrations. When these doping concentrations on the confinement layer have magnitude 3.3×10^{18} cm^{-3} (1.4×10^{18} cm^{-3}), the value of α_{conf} is found to be approximately 10 cm^{-1}. The second term on the RHS of (2.2), α_{diff}, represents the diffraction loss. If we assume the distribution optical field in the transverse direction has a Gaussian function with a spot size of radius W_{sp}, the diffraction loss α_{diff} can be calculated by [15], [17]

$$\alpha_{diff} = -\frac{1}{d} \log\left(2 \cdot \left(2 + 3\left(\frac{L-d}{4k_{eff}W_{sp}^2}\right)^2 + \left(\frac{L-d}{4k_{eff}W_{sp}^2}\right)^4\right)^{-1}\right), \tag{2.5}$$

where k_{eff} ($= n_{eff} \cdot k_0$) is the effective propagation constant and n_{eff} is the effective refractive index of the laser cavity (i.e., including longitudinal confinement and active layers). From (2.5), it is noted that α_{diff} increases with the reduction of W_{sp} and d. If $L = 7$ μm, $W_{sp} = 0.5$ μm, $d = 1$ μm, $\lambda_R \sim 0.87$ μm, and $k_{eff} = n_{eff} \cdot k_0$ (i.e., $n_{eff} \sim 4$) is substituted into (2.5), it can be shown that the value of α_{diff} (i.e., $\sim 10^4$ cm^{-1}) is much higher than that of α_{in} (i.e., ~ 15 cm^{-1}) [13]. In facet emitting lasers, because of the large cavity size, the magnitude of α_{diff} is negligible. Hence, the influence of diffraction loss is one of the unique characteristics in VCSELs. The third term on the RHS of (2.2), L^{-1} $\log(1/R)$, represents the optical loss from the front and rear mirrors of the laser. If L is equal to 7 μm, a large value of R (i.e., >0.9) should be selected so that the magnitude of $L^{-1} \log(1/R)$ is close to α_{in}.

2.2.2 Threshold Current Density and Differential Quantum Efficiency

In the early state of the development of VCSELs, bulk semiconductor materials such as GaAs, InGaAs, and InGaAsP are commonly used for the active region [11–15]. In the following paragraphs, the design consideration of VCSELs with a bulk active region is discussed. In addition, the corresponding expressions of threshold current density and differential quantum efficiency are derived. These investigations will extend to study the VCSELs using QWs material as the active region.

For an active region made of bulk materials, the corresponding threshold optical gain g_{th} at peak gain wavelength can be expressed in terms of injection carrier concentration at threshold N_{th}, which is given by [8]

$$g_{th} = g_N(N_{th} - N_t), \tag{2.6}$$

where g_N is the differential gain coefficient and N_t is carrier concentration at transparency. This is a linear approximation on the optical gain with the variation of injection carrier concentration. For bulk materials such as GaAs, the values of g_N and N_t are approximately 2.5×10^{-16} cm^2 and 1.3×10^{18} cm^{-3}, respectively, at room temperature [7,8]. From the rate equation of carrier concentration, the threshold current density J_{th}, of the laser can be related to N_{th} as

$$\frac{J_{th}}{qd} = \frac{N_{th}}{\tau_{eff}}, \tag{2.7}$$

where q is the electron charge and τ_{eff} is the effective carrier lifetime. τ_{eff} can be approximated as a function of N_{th}

$$\frac{1}{\tau_{eff}} = \frac{1}{\tau_n} + B_{sp}N_{th} + C_{Aug}N_{th}^2 \approx B_{eff}N_{th}, \tag{2.8}$$

where τ_n is the carrier lifetime, B_{sp} is the bimolecular radiative recombination coefficient, and C_{Aug} is the nonradiative Auger recombination coefficient. These coefficients represent the recombination of electrons and holes inside a semiconductor material, and the recombination process can be radiative or nonradiative. It must be noted that in some materials such as GaAs, the parameters B_{sp} and C_{Aug} can be ignored in (2.8), but these parameters should be included in (2.8) for materials such as InGaAsP and InP. A detailed description of these recombination coefficients can be found in the book by Agrawal and Dutta [8]. The preceding expression of total recombination rate can be simplified using the effective recombination constant, B_{eff} ($\sim 1.5 \times 10^{-10}$ cm^3/s for GaAs) [7,8]. Substituting (2.3) into (2.2) and using the relationship as given in (2.8), the expression of J_{th} can be written as [11,15]

$$J_{th} = J_s d^2 \left(\alpha_{in} + \Gamma_z \alpha_{diff} + L^{-1} \log \left(\frac{1}{R} \right) + \Gamma_z g_N N_t \right)^2, \tag{2.9}$$

where $J_s = qB_{eff}/g_N^2 \Gamma_z^2 d$ ($\sim 7 \times 10^3$ kA/cm^2 for $d = 3$ μm and $L = 7$ μm) and the term $g_N N_t$ given in (2.9) can be interpreted as the residual absorption loss of the optical gain. The main difference between the expression of J_{th} for VCSELs and that for facet emitting lasers is the presence of α_{diff}. In addition, the value of L used in (2.9) is relatively short when compared with that of facet emitting lasers (i.e., $L \sim 400$ μm). Hence, the selected value of R should be greater than 0.9 in order to minimize the threshold current density in VCSELs. For facet emitting lasers using cleaved facets as the reflection mirrors, the corresponding power reflectivity is ~0.3.

Figure 2.2 plots the calculated threshold current density J_{th} of GaAs VCSEL against the active layer thickness d for the product of front and rear mirror reflectivity R, which varies between 0.8 and 0.99. The solid line represents $\alpha_{diff} = 0$, and the dashed line represents what is obtained from (2.5). The laser parameters used in the preceding calculation are $L = 7$ μm, $W = 1.5$ μm, and $\lambda_R = 0.87$ μm. In addition, the corresponding material parameters at room temperature are $\alpha_{ac} = 30$ cm^{-1}, $\alpha_{conf} = 10$ cm^{-1}, $g_N = 2.5 \times 10^{-16}$ cm^2, $N_t = 1.3 \times 10^{18}$ cm^{-3}, and $B_{eff} = 1.5 \times 10^{-10}$ cm/s. It is observed that the presence of α_{diff} increases the value of J_{th} especially for devices with small d (i.e., <2 μm) and large values of $R(= r_f \cdot r_r)$ (i.e., >0.99). This is because the residual absorption loss in (2.9) has a magnitude that roughly equals 400 cm^{-1}, which is less than that of α_{diff} (i.e., >500 cm^{-1}) for $d < 2$ μm. In addition, a minimum J_{th} occurs for the case $R = 0.99$ because the residual absorption loss dominates the diffraction loss and total cavity losses. Figure 2.3 plots the variation of threshold current density (solid line) and threshold current (dashed line) with the diameter of active region. It is assumed that $d = 3$ μm and $W_{sp} = W$ in the calculation with the threshold current defined as $I_{th} = J_{th} \cdot \pi \cdot W^2$. It is observed that the threshold current density decreases with the increase in the diameter of the active region and maintains uniform for $2W > 2$ μm.

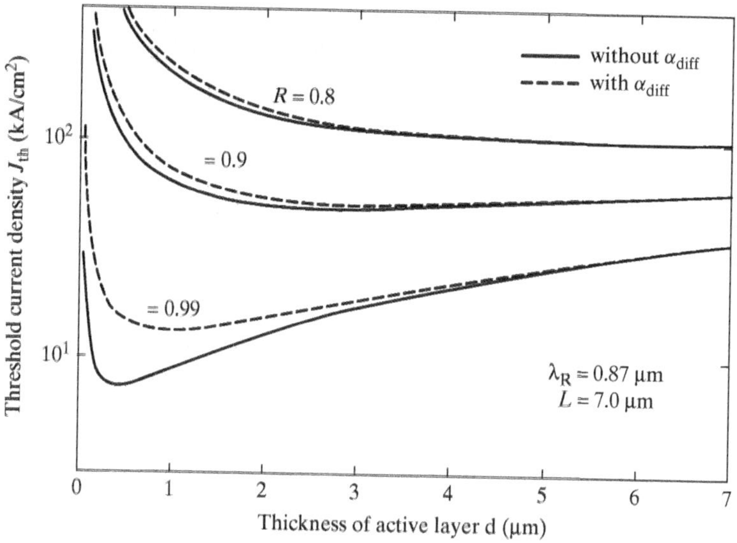

Figure 2.2 Threshold current density J_{th} versus active layer thickness d for different values of reflectivity R. Solid lines and dashed lines represent $\alpha_{diff} = 0$ and α_{diff} calculated from (2.5), respectively.

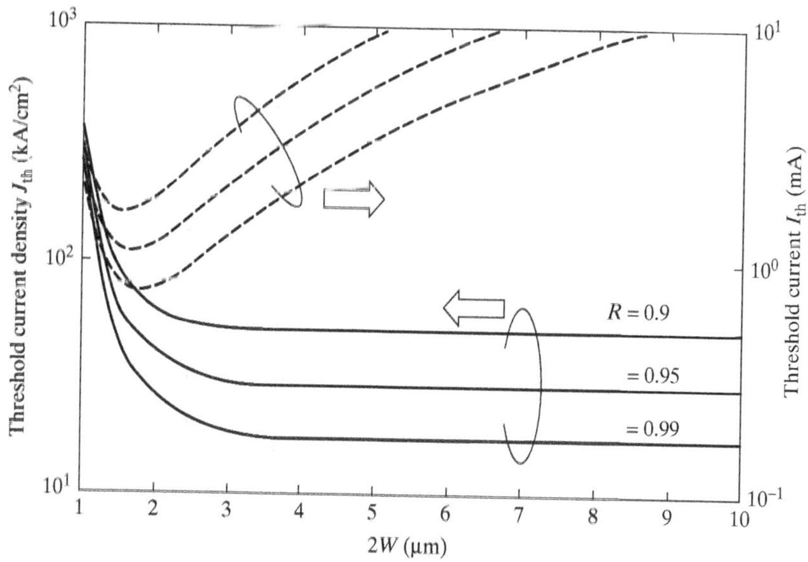

Figure 2.3 Threshold current density J_{th} (solid lines) and threshold current I_{th} (dashed lines) against the diameter of active region $2W$ for VCSELs (after Ref. 15).

The differential quantum efficiency η_d is defined as the ratio between the output optical loss to the total cavity loss of the laser, which is written as [11,15]

$$\eta_d = \eta_i \frac{\log(1/R)}{\alpha_{in} L + \alpha_{diff}\, d + \log(1/R)}, \tag{2.10}$$

where η_i is the internal quantum efficiency. For ideal lasers $\eta_d \sim 1$, which implies $\alpha_{in} = \alpha_{diff} = 0$ as no light will be absorbed inside the laser cavity. However, α_{in} and α_{diff} will always be greater than zero, so careful selection of R as well as α_{in} and α_{diff} is required in order to maximize the value of η_d. Figure 2.4 shows the variation of η_d with R for a VCSEL with $\alpha_{diff} \sim 0$ (i.e., the active region is large enough to ignore diffraction loss) and $\eta_i \sim 1$ are assumed in the calculation. The solid and dashed lines represent $\alpha_{in} = 15\ \text{cm}^{-1}$ and $50\ \text{cm}^{-1}$, respectively. Two types of mirrors are considered in the calculation: the dielectric mirror (i.e., without absorption loss) and the Au-coated mirror (i.e., with absorption loss). It is observed in both cases that η_d decreases monotonically with increase of R but dielectric mirror has better η_d than Au-coated mirror due to the presence of absorption loss. In addition, the absorption loss in Au-coated mirror reduces its reflectivity R to ~0.94 for η_d equals zero. The offset in R can be approximated by $\alpha_{Au} \cdot L$ where α_{Au}(~$85\ \text{cm}^{-1}$) is the equivalent absorption loss of Au-coated mirror. For dielectric mirrors, it is noted that $R \sim 0.95$ will be an appropriate

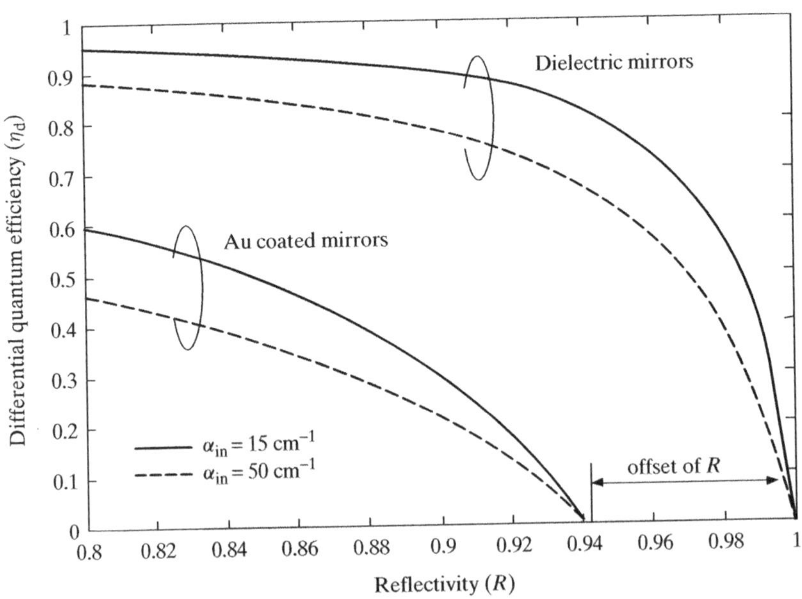

Figure 2.4 Differential quantum efficiency η_d versus reflectivity R for dielectric multi-layered and Au-coated mirrors; solid lines and dashed lines represent $\alpha_{in} = 15\ \text{cm}^{-1}$ and $\alpha_{in} = 50\ \text{cm}^{-1}$, respectively (after Ref. 29).

choice as the optimal threshold current density as well as reasonable differential quantum efficiency (~0.8) can be achieved.

As the conversion efficiency between optical and electrical signals is high in semiconductor lasers, the output power of VCSELs can be assumed to be a linear function with the injection current above threshold. Hence, the light/current characteristics of VCSELs can be roughly expressed as

$$P = A_{\mathrm{eff}}(J - J_{\mathrm{th}}) \cdot h\nu_{\mathrm{R}} \cdot \frac{\eta_{\mathrm{d}}}{q}, \tag{2.11}$$

where P (in watts) is the output power from the surfaces of the laser cavity, A_{eff} is the effective area, J is the injection current density, h is Planck's constant, and ν_{R} is the lasing frequency. The term $(J - J_{\mathrm{th}})$ is linearly proportional to the output power of light, and η_{d} factors in the total cavity losses. However, expression (2.11) is valid only for the devices operating near the threshold. In high injection conditions, the light/current curves may exhibit sublinear behavior due to the nonlinear characteristics of VCSELs such as the spatial hole burning of carrier concentration, self-heating effects, as well as gain saturation. These nonlinear characteristics of VCSELs are discussed in other chapters of this book.

2.2.3 Side-Mode and Thermal Characteristics of VCSELs

The round-trip phase condition of the longitudinal mode of wavelength, which equals λ_R, can be obtained from the imaginary part of (2.1) given by

$$\phi_{\mathrm{f}} + 2k_0(n_{\mathrm{ac}}d + n_{\mathrm{conf}}(L - d)) + \phi_{\mathrm{r}} = 2\pi m, \tag{2.12}$$

where m is an integer. ϕ_{f} and ϕ_{r} are usually assumed to be zero for Fabry–Perot-type facet emitting lasers due to the use of cleaved facets as the mirrors. Therefore, resonant wavelength is affected only by the refractive index of laser cavity. However, the phase of dielectric multilayered mirrors or distributed Bragg reflectors (DBRs), which are commonly used in VCSELs, is also dependent on its operating wavelength. Hence, the resonant wavelength of VCSELs is dependent on the refractive index of the laser cavity as well as the phase of mirrors. From (2.12), if the phase of the mirror changes by 10%, the resonant cavity wavelength will be linearly shifted by 15 and 50 Å for the center wavelength of 0.87 and 1.55 μm, respectively.

A simple expression for the mode spacing $\Delta\lambda$ of VCSELs can be deduced from (2.12) by assuming that the phase of multilayered mirrors is independent of the resonant-cavity wavelength, which gives

$$\Delta\lambda = \frac{\lambda_{\mathrm{R}}^2}{2n_{\mathrm{ac}}d + 2n_{\mathrm{conf}}(L - d)} \approx \frac{\lambda_{\mathrm{R}}^2}{2n_{\mathrm{eff}}L}, \tag{2.13}$$

where n_{eff} is the effective refractive index and λ_{R} is the lasing wavelength of the main mode. If $L = 7$ μm and $n_{\mathrm{eff}} = 4$, $\Delta\lambda$ is found to be 135 and 430 Å for lasers with $\lambda_R = 0.87$ and 1.55 μm, respectively [2]. This implies that mode spacing is much wider than that of facet emitting lasers (i.e., $\Delta\lambda$ is usually about $5 \sim 10$ Å for facet emitting lasers). Therefore, single-longitudinal-mode operation is favorable in VCSELs as the bandwidth of the optical gain spectrum is in the same order of magnitude as $\Delta\lambda$.

The gain difference and resultant mode suppression ratio of the neighboring modes can be evaluated by assuming that the gain profile of the bulk material $g(\lambda)$ is a parabolic function of wavelength, λ, that is [2]

$$g(\lambda) = g_N(N - N_t)\left(1 - \left(\frac{\lambda - \lambda_R}{\Delta\Lambda}\right)^2\right), \tag{2.14}$$

where $\Delta\Lambda$ (~400 Å for GaAs material) is the bandwidth of the optical gain spectrum and N is the carrier concentration. The side-mode suppression ratio P_0/P_1 can be deduced from a standard multimode rate equation analysis [7,8], and the corresponding expression is given by

$$\frac{P_0}{P_1} = \frac{1}{\beta_s} \cdot \frac{\Delta g}{g_{th}} \left(\frac{J}{J_{th}} - 1\right), \tag{2.15}$$

where Δg is the gain difference between the fundamental P_0 and first-order P_1 longitudinal modes, J is the injection current density, and J_{th} is the threshold current density. The parameter β_s is the spontaneous emission factor, which is given by [19]

$$\beta_s = \frac{\Gamma_z K \lambda_R^4}{4\pi^2 n_{eff}^3 V_a \Delta\lambda_s}, \tag{2.16}$$

where $\Delta\lambda_s$ (~30 Å) is the full wave at half-maximum (FWHM) of the spontaneous emission spectrum, K (~1) is Petermann's astigmatism factor, and V_a ($= \pi W^2 d$) is the volume of the active region [7]. β_s represents the amount of spontaneous emission coupled to the longitudinal modes of VCSELs. If $d =$ 3 μm, $W = 5$ μm, and $L = 7$ μm, it can be shown that β_s is in the order of 10^{-6} at $\lambda_R = 0.87$ μm. In addition, the corresponding side-mode suppression ratio of >40 dB can be achieved when the laser is biased at $1.5J_{th}$. Therefore, the influence of longitudinal side mode in VCSELs can be ignored in the calculation due to wide $\Delta\lambda$ and high side-mode suppression ratio.

If VCSELs are designed for continuous-wave (CW) operation, it is necessary to take into account the influence of self-heating. This is because the thermal resistance of VCSELs is much higher than that of conventional facet emitting lasers due to the former's small cavity size as well as the low doping concentration on the p- and n-confinement layers. In addition, the possibility of using heatsink to reduce the corresponding thermal resistivity in VCSELs is limited by the orientation of the active region. If the laser is mounted p side down, the thermal resistance of VCSELs increases with the thickness of the p-confinement layer. For VCSELs as shown in Figure 2.1, there exist five thermal sources: n-confinement layer, active layer, p-confinement layer, mirror on the p-confinement layer, and p-contact layer, with which the thermal resistance is obtained. If the total thermal resistance of VCSELs is represented by an effective value, a simple expression for the temperature change (i.e., relative to the temperature of heatsink) due to the self-heating effect ΔT can be deduced from the thermal rate Equation [20] and is given by

$$\Delta T = \kappa_{\mathrm{T}}^{-1} R_{\mathrm{TH}}(P_{\mathrm{IV}} - P), \tag{2.17}$$

where κ_{T} (W cm^{-1} K^{-1}) is the effective thermal conductivity, P_{IV} (watts) is the total input electrical power, and R_{TH} (cm^{-1}) is the normalized effective thermal resistance. It is noted that R_{TH} increases with the reduction of cavity size and is dependent only on the size of VCSELs. Using (2.17), ΔT of GaAs VCSELs biased at threshold is estimated to be 13 K, where $R_{\mathrm{TH}} = 500$ cm^{-1}, $\kappa_{\mathrm{T}} = 0.55$ W cm^{-1} K^{-1}, and $P_{\mathrm{IV}} = I_{\mathrm{th}} \cdot V_{\mathrm{J}}$ ($V_{\mathrm{J}} \sim 1.4$ V is the junction voltage of GaAs and $I_{\mathrm{th}} = 10$ mA) are assumed in the calculation. Hence, the operating temperature of VCSELs at and above threshold can be estimated and the influence of self-heating effects can be taken into account for the design of VCSELs.

2.2.4 UGS VCSELs with Quantum-Well Active Layer

From (2.9), the threshold current density of VCSELs seems to be higher than that of facet emitting lasers because of the former's small cavity size. However, this threshold current density of VCSELs can be minimized if the value of differential gain g_N can be increased. In fact, enormous enhancement in g_N can be achieved by replacing the bulk active layer with QW material, but the drawback is the reduction of the longitudinal confinement factor Γ_z with which the threshold current density is increased. In fact, using multiple quantum wells (MQWs) as the active layer of VCSELs can solve the problem of small Γ_z provided the high absorption loss inside the barriers and separate-confinement heterostructure (SCH) layers of MQWs can be avoided [21]. Therefore, there is a tradeoff between the optimal threshold current density and the dimension of MQWs.

If QW material is used as the active layer, it is more appropriate to replace the linear approximation on the optical gain and injection carrier concentration by the logarithmic gain–carrier density relation, as shown below [22–24]

$$g_{\mathrm{th}} = a_N \log\left(\frac{N_{\mathrm{th}}}{N_{\mathrm{t}}}\right), \tag{2.18}$$

where a_N is the gain coefficient and N_{t} is the carrier concentration at transparency. This equation represents the variation of optical gain at peak gain wavelength with injection carrier concentration. For GaAs/AlGaAs QWs with well width of 100 Å, a_N and N_{t} have values of $\sim$2000 cm^{-1} and $\sim 1.5 \times 10^{18}$ cm^{-3}, respectively, at room temperature [23,24]. From (2.18), the differential gain at threshold can be written as

$$g_N = \frac{a_N}{N_{\mathrm{th}}}, \tag{2.19}$$

where $g_N \sim 1 \times 10^{-15}$ cm^2 with $N_{\text{th}} \sim 2 \times 10^{18}$ cm^{-3} is assumed in the calculation. Hence, the magnitude of differential gain of QWs materials is about 10 times that of bulk materials.

Figure 2.5 shows the schematic bandgap energy of a MQW active layer used in the following analysis. If the length of VCSELs is much longer than the well width, the corresponding optical confinement of QWs Γ_w can be approximated by

$$\Gamma_{\text{w}} \approx \frac{n_{\text{w}} L_{\text{w}}}{L}, \tag{2.20}$$

where n_{w} and L_{w} are the number and the thickness of the QWs, respectively. As $n_w = 4$, it can be shown that $\Gamma_{\text{w}} \sim 2 \times 10^{-2}$ if L_{w} and L are assumed equal to 100 Å and 2 μm, respectively. The total cavity loss α_{in} is given by

$$\alpha_{\text{in}} = \Gamma_{\text{w}}\alpha_{\text{ac}} + \Gamma_{\text{b}}\alpha_{\text{b}} + \Gamma_{\text{SCH}}\alpha_{\text{SCH}} + \alpha_{\text{conf}}\frac{L-d}{L}, \tag{2.21}$$

where α_{b} and α_{SCH} are the absorption loss inside the barrier and SCH layers, respectively. Γ_{b} can be approximated by $\Gamma_{\text{b}} \approx (n_{\text{w}} - 1) \cdot L_{\text{b}}/L$, and L_{b} is the thickness of the barriers of MQWs. Γ_{SCH} is defined as $\Gamma_{\text{SCH}} = d/L - \Gamma_{\text{w}} - \Gamma_{\text{b}}$, and d is the total thickness of the active region, including MQW and SCH layers. It is noted that the length of VCSELs with QW active regions is shorter than that with a bulk active region.

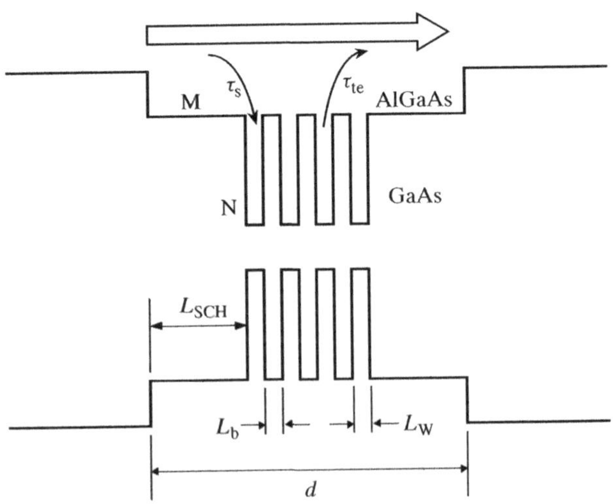

Figure 2.5 Schematic diagram of a MQW with a separate confinement heterostructure.

The relation between threshold current density J_{th} and threshold carrier concentration of QWs N_{th} can be obtained from the rate equation of carrier concentrations inside the SCH (M) and QWs (N) layers as shown below [25]:

$$\frac{\partial M}{\partial t} = \frac{J}{q n_w L_{SCH}} - \frac{M}{\tau_s} + \frac{L_w}{L_{SCH}} \frac{N}{\tau_{te}}, \tag{2.22a}$$

$$\frac{\partial N}{\partial t} = \frac{L_{SCH}}{L_w} \frac{M}{\tau_s} - \frac{\tau_n + \tau_{te}}{\tau_n \tau_{te}} N - v_g g(N) S, \tag{2.22b}$$

where L_{SCH} is the thickness of the SCH layers, τ_s (~200 ps) is the ambipolar carrier transport time across the SCH layer, τ_n (~1 ns) is the carrier lifetime in the QWs, τ_{te} (~10 ps for 100 Å QW) is the thermionic emission lifetime from the quantum wells, g is the optical gain, and v_g is the group velocity. In the rate equations, the terms with τ_{te} represent the loss of carriers from the QW and the gain by the SCH layer. However, the terms with τ_s have opposite meanings. In addition, the finite capture time of the QW can be considered in the analysis through τ_s. The corresponding transport characteristic of M and N between the SCH and QW layers is also illustrated in Figure 2.5. Hence, with the assumption that the photon density is equal to zero (i.e., $S = 0$) at threshold in (2.22), the expression of J_{th} can be deduced to be

$$\frac{J_{th}}{q n_w L_{SCH}} = \frac{1}{\tau_n} \frac{L_w}{L_{SCH}} N_{th}, \tag{2.23}$$

where $N \to N_{th}$ and $J \to J_{th}$ are used. If $\tau_n = (B_{eff} N_{th})^{-1}$ is assumed, the threshold current density J_{th} can be written as

$$J_{th} - J_s \exp\left(\frac{2}{a_N \Gamma_w}\left\{\alpha_{in} + \frac{1}{L}\left(\alpha_{diff} d + \log\left(\frac{1}{R}\right)\right)\right\}\right), \tag{2.24}$$

where $J_s = q n_w L_w B_{eff} N_t^2$ ($\sim n_w \times 55$ A/cm^2). Figure 2.6 shows the variation of threshold current density with the number of QWs used as the active layer for different values of R. In the calculation, the MQW active layer consists of GaAs QWs ($L_w = 100$ Å) separated by AlGaAs barriers ($L_b = 100$ Å) and AlGaAs SCH layers. The other parameters at room temperature used in the calculation are $\alpha_b = \alpha_{SCH} = \alpha_{ac} = 50$ cm^{-1}, $\alpha_{conf} = 10$ cm^{-1}, $B_{eff} \sim 1.5 \times 10^{-10}$ cm^3/s, $d = 0.5$ μm, $L = 2$ μm, $a_N = 2000$ cm^{-1}, and $N_t \sim 1.5 \times 10^{18}$ cm^{-3}. It is shown that the value of J_{th} reduces with the increase in n_w but remains unchanged for n_w greater than 3, 4, and 8 for $R = 0.995$, 0.99, and 0.98, respectively. It must be noted that the approximation of Γ_w by (2.20) may be invalid for large values of n_w, so the optimal values of n_w selected for both J_{th} and n_w are minimum values.

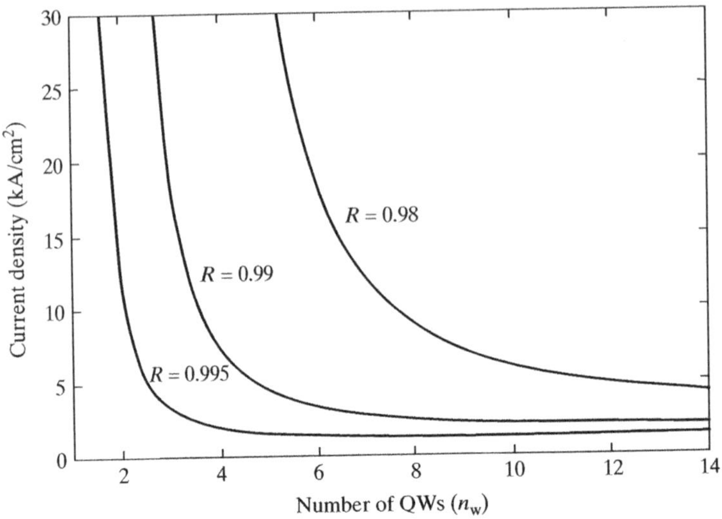

Figure 2.6 The dependence of threshold current density J_{th} on the number of quantum wells n_w with different values of reflectivity R.

2.3 DESIGN EQUATIONS OF VCSELs WITH A PERIODIC GAIN STRUCTURE

One disadvantage of using UGS in VCSELs is the poor overlap between the gain medium and the peak of the standing-wave pattern of the lasing mode. In order to improve the performance of VCSELs, the corresponding electronic structure has to be optimized for maximum the overlap between gain medium and the peak of the standing-wave pattern [9,10,26].

Figure 2.7 illustrates a VCSEL with periodic gain structure (PGS) [27]. The purpose of using PGS is to increase the overlap between the gain medium and the peak of standing-wave pattern within the cavity so that the wallplug efficiency can be maximized. It is shown that the position of the very thin active regions is aligned with the peak of the standing-wave pattern (i.e., set by the FP resonator) at a selected lasing wavelength. This can be achieved by separating the passive spacer layers with the thickness of $\lambda_R/2n$, where n is the refractive index of the spacer layer at λ_R. The other advantages of using PGS are

1. It is difficult to have longitudinal spatial hole burning because the gain medium coincides with the peaks of the standing-wave pattern.
2. The gain medium is anisotropic and wavelength-selective so that single longitudinal mode operation is favorable.
3. High external quantum efficiency as the amplified spontaneous emission is eliminated at the node of the standing-wave pattern. It is expected that the threshold current density of VCSELs with PGS can be reduced by a factor of 2 when compared to that with UGS [27].

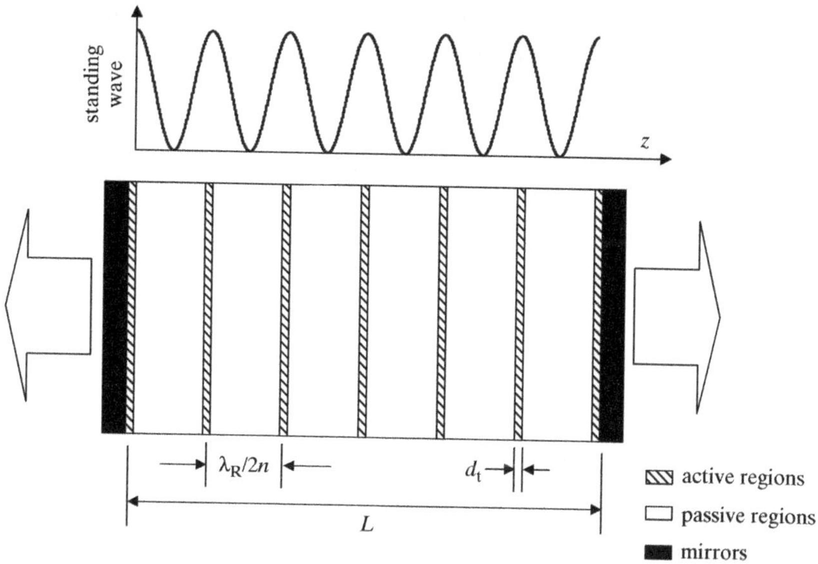

Figure 2.7 Schematic of laser cavity structures with periodic gain structure where the gain has been divided into segments placed along the electric field maxima in the cavity.

In the following sections, the design procedures for the optimal performance of VCSELs with PGS are presented.

2.3.1 Relative Confinement Factor

Because of the similarity between the longitudinal configuration of UGS and PGS, it is believed that the threshold conditions of UGS VCSELs deduced in Section 2.2 can be applied to PGS, except the longitudinal optical confinement factor Γ_z has to be replaced. In order to derive Γ_z for GPS, it is assumed that the thickness of each active layer of a VCSEL equals d_t so that the total thickness of all active segments is d and the length of the resonant cavity equals L.

For a mirror with high reflectivity, the standing-wave pattern of the optical field $E(z)$ can be approximated by [27]

$$E(z) = E_0 \cos(k_{\text{eff}} z), \tag{2.25}$$

where E_0 is a constant. Hence, the longitudinal confinement factor is given by

$$\Gamma_z = \int_{\text{active}} E^2(z)dz \bigg/ \int_L E^2(z)dz. \tag{2.26}$$

For PGS, the integral of the optical field is considered over one period with the multiplication of the top integral by the number of active segments in the cavity and that of the bottom integral by the number of half-wavelengths in the entire cavity length. Hence, Γ_z can be written as

$$\Gamma_z = \frac{\dfrac{d}{d_t}\displaystyle\int_{d_t} \cos^2(k_{\text{eff}}z)dz}{\dfrac{L}{\lambda_R/2n_{\text{eff}}}\displaystyle\int_{\lambda_R/2} \cos^2(k_{\text{eff}}z)dz}. \tag{2.27}$$

After integration, (2.27) gives

$$\Gamma_z = \frac{d}{L}\left\{1 + \frac{\sin\left(\pi\left(\dfrac{d_t}{\lambda_R/2n_{\text{eff}}}\right)\right)}{\pi\left(\dfrac{d_t}{\lambda_R/2n_{\text{eff}}}\right)}\right\} \equiv \frac{d}{L}\Gamma_r, \tag{2.28}$$

where Γ_r is defined as the relative confinement factor. It is noted that if $d_t = \lambda_R/2n_{\text{eff}}$, $\Gamma_r = 1$, which is equal to Γ_z of the UGS. However, if $d_t \to 0$, then $\Gamma_r \to 2$. Figure 2.8 shows the plot of Γ_r versus the fill factor d/L. The threshold gain of VCSELs given in (2.2) can be reduced by a factor of 2 (by substituting $\Gamma_z = \Gamma_r \cdot d/L$) if the continuous active material is replaced by very thin active layer segments, which are positioned along the peak of the standing wave, with a total thickness of d. In addition to the reduction in the threshold gain, PGS also has the advantage of increased mode selectivity. For the resonant FP mode, $\Gamma_r \approx 2$, whereas for modes off resonance, $\Gamma_r \approx 1$. The threshold gain is then seen to be twice as low for the resonant mode if the reflectivity of mirrors is

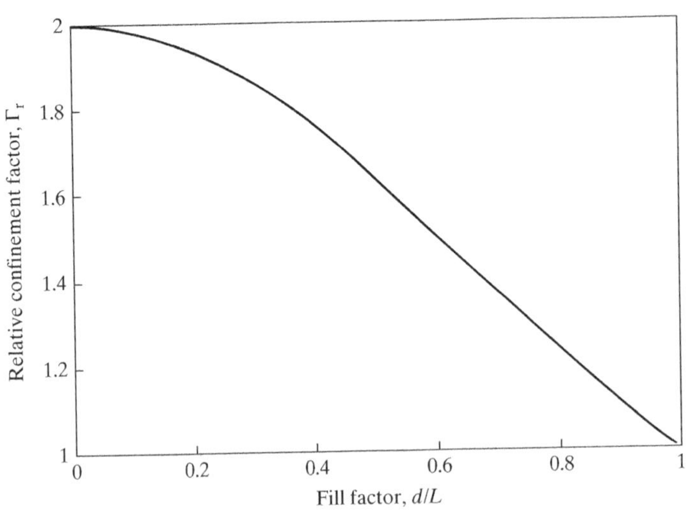

Figure 2.8 The relative confinement factor Γ_r as a function of the fill factor d/L (after Ref. 29).

independent of the operating wavelength. Furthermore, the phase condition can be approximated by (2.12) with the assumption that d is the summation of all active segments of thickness d_t, provided the difference in refractive index between the active and passive segments is small.

It must be noted that if the difference in refractive index between the active and passive segments is large, the above mentioned approach for calculating Γ_z may not be correct. This is because the PGS may produce a constructive or destructive interface dependent on the separation displacement of the active and passive segments due to their index discontinuities. The standing-wave pattern of the optical field may not be what is expected. In order to calculate the electric field distribution inside the PGS, methods such as the transmission matrix approach should be applied to take into account the strong periodic perturbation in both the index of refraction and the material gain to analyze the threshold condition of VCSELs with PGS.

2.3.2 Threshold Current Density and Differential Quantum Efficiency

In the analysis of VCSELs in Section 2.2, minimum threshold current density is recognized as a basic requirement to optimize the performance of VCSELs at threshold, especially when diffraction loss is taken into consideration. However, investigations have ignored the nonlinear relationship between the optical gain and injection current density so that the electronic conversion ratio (i.e., injection current-to optical gain conversion or $J_{\text{th}}-g_{\text{th}}$) may not be maximized for the optimal threshold current density [18,28]. In fact, it is required to minimize the threshold current density and maximize the electronic conversion ratio in the design of VCSELs at threshold. The following paragraphs show that both optimal requirements can be obtained with suitable selection of cavity length and optical gain.

Now, the expression of threshold current density is derived with electronic conversion ratio taken into consideration. From (2.24), the threshold current density J_{th} (A/cm^2) for a given threshold gain can be expressed as the multiplication of a nominal current density J_{nom} (A/cm^3), with total thickness of active region d [27]

$$J_{\text{th}} = J_{\text{nom}} d, \tag{2.29}$$

where $J_{\text{nom}} = qB_{\text{eff}} N_{\text{th}}^2$. Substituting (2.2) and (2.28) into (2.29) to replace d by cavity loss and threshold gain gives

$$J_{\text{th}} = \frac{1}{\Gamma_r}\left(\frac{J_{\text{nom}}}{g_{\text{th}}}\right)\left(\frac{\log(1/R)}{\eta_m}\right)\left(1 - \frac{\alpha_{\text{in}}}{\Gamma_z g_{\text{th}}}\right)^{-1}, \tag{2.30}$$

and

$$\eta_m = \frac{\log(1/R)}{\alpha_{\text{diff}} d + \log(1/R)}. \tag{2.31}$$

In these equations, $g_{\rm th}/J_{\rm nom}$ and $\eta_{\rm m}$ represent the electronic conversion ratio and output power efficiency, respectively. It must be noted that the electronic conversion ratio can be optimized provided that the value of $g_{\rm th}/J_{\rm nom}$ is maximized. Furthermore, the corresponding differential quantum efficiency $\eta_{\rm d}$ can be written as

$$\eta_{\rm d} = \eta_{\rm i} \frac{\log(1/R)}{\alpha_{\rm in} L + \log(1/R)} \equiv \eta_{\rm m} \left(1 - \frac{\alpha_{\rm in}}{\Gamma_z g_{\rm th}}\right). \tag{2.32}$$

The first step in optimizing the threshold current density of VCSELs is to maximize the quantity $g_{\rm th}/J_{\rm nom}$. Figure 2.9 plots the material gain and electronic conversion ratio against the nominal current density. In the calculation, the peak gain wavelength of GaAs bulk material at room temperature is approximated by (2.6) with $g_N = 3 \times 10^{-16}$ cm^2 and $N_{\rm t} \sim 1.5 \times 10^{18}$ cm^{-3}. The values of g_N and $N_{\rm t}$ are selected according to the assumption that the influence of self-heating is negligible at threshold. However, if VCSELs have high thermal resistivity so that the magnitude of ΔT obtained from (2.17) is large, different values of g_N and $N_{\rm t}$ should be used to take into account the influence of the self-heating effect on the material gain.

Figure 2.9 shows that the maximum value of $g_{\rm th}/J_{\rm nom}$ occurs when the carrier concentration is roughly twice the value of the transparency carrier concentration (i.e., $N \sim 2N_{\rm t}$). For the numbers quoted above, the corresponding material gain has a magnitude of 375 cm^{-1}. However, the peak of the curve is quite broad for the bulk material anywhere in the range of 350–500 cm^{-1}. Therefore, the value of threshold gain can be chosen on the high side of the optimal efficiency

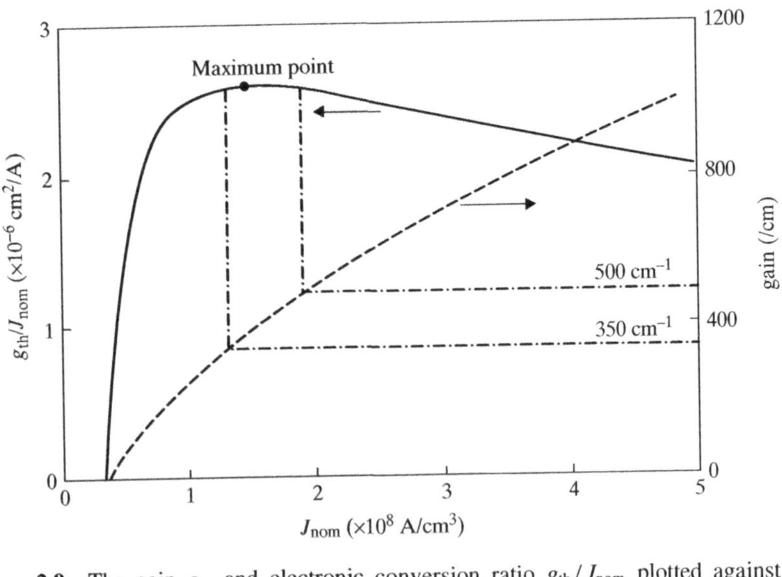

Figure 2.9 The gain $g_{\rm th}$ and electronic conversion ratio $g_{\rm th}/J_{\rm nom}$ plotted against the nominal current density $J_{\rm nom}$ (after Ref. 29).

point for the interests of minimizing the influence of total internal loss (i.e., $g_{th} > 0.1\alpha_{in}$). Therefore, the threshold gain of 500 cm^{-1} should be used for the design of VCSELs.

From (2.30) and (2.32), the optimal threshold current density and differential quantum efficiency of UGS and PGS VCSELs can be deduced for different values of fill factor. Figure 2.10 plots the normalized threshold current density J_{th}/J_N against the fill factor for both PGS and UGS (i.e., $\Gamma_r = 1$) VCSELs, where J_N is defined as $J_N = (J_{nom}/g_{th}\Gamma_z)\log(1/R)/\eta_m$. The dashed and solid lines represent the laser with UGS and PGS, respectively. It is observed that for the PGS with $\alpha_{in} = 0$, J_{th}/J_N equals $1/\Gamma_r$ and for fill factors approaching zero, the value of J_{th}/J_N approaches 0.5. On the other hand, for the UGS with $\alpha_{in} = 0$, $J_{th}/J_N = 1$ for the entire fill factor. Therefore, PGS can provide a reduction of 2 in threshold current density in the limit of small fill factors.

The threshold current densities for both UGS and PGS VCSELs with $\alpha_{in} =$ 20 cm^{-1} are higher than those with $\alpha_{in} = 0$ cm^{-1} because of the additional loss of the passive regions. In addition, for both structures with $\alpha_{in} = 20$ cm^{-1}, the corresponding threshold current densities increase rapidly with reduction of the fill factor. This is because the standing wave receives more absorption loss when the fill factor is reduced. It is shown in Figure 2.10 that for the case of UGS, the threshold current density increases monotonically with reduction of the fill factor. Hence, the optimal fill factor of UGS is 1.0 and the corresponding value of J_{th}/J_N equals 1.1. However, for the case of PGS, two competing effects affect the corre sponding threshold current density. Decrease of the fill factor from 1.0 enhances the value of the relative confinement factor, which overwhelms the influence

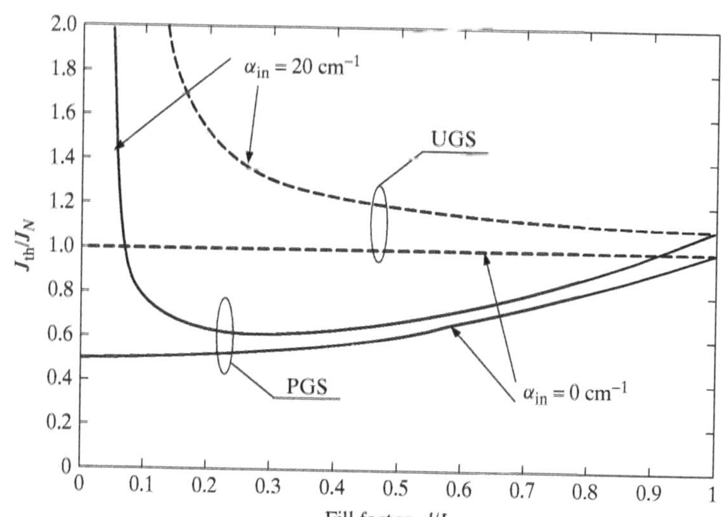

Figure 2.10 Comparison of the normalized threshold current density J_{th}/J_N of VCSELs with PGS (solid lines) and UGS (dashed lines) with and without total internal loss α_{in} as a function of fill factor d/L (after Ref. 29).

of total internal loss. On the other hand, for a small fill factor (i.e., <0.4), the corresponding threshold current density is dominated by the total internal loss so that the optimal fill factor appears between 0.2 and 0.3 (i.e., $J_{th}/J_N \sim 0.63$). Comparing the optimal values in both cases with and without total internal loss, it is shown that the PGS provides a threshold current improvement over UGS.

Figure 2.11 plots the dependence of differential quantum efficiency η_d on the fill factor d/L. As the reduction of fill factor increases, the total internal loss of VCSELs, η_d drops off rapidly, especially for $d/L < 0.2$. For a given optimal fill factor for minimum threshold current density, it is observed that the value η_d of PGS is larger than that of UGS for $d/L < 1$. This is due to the shorter cavity length necessary in PGS to provide the same threshold material gain. Hence, with the substitution of both optimal threshold current density and differential quantum efficiency into (2.11), it can be shown that PGS has a higher output power than UGS at the same injection current density.

To complete this analysis, we need to calculate the cavity length that will yield a threshold gain of 500 cm^{-1} for a given fill factor and mirror reflectivity. The optimal length of laser L_o, can be deduced from (2.2)

$$L_o = \frac{\log(1/R)}{\eta_m(\Gamma_z g_{th} - \alpha_{in})}. \tag{2.33}$$

The approximate expression of L for PGS can be deduced by neglecting the total internal loss and assuming $\Gamma_r = 2$. For a threshold gain of 500 cm^{-1} and an

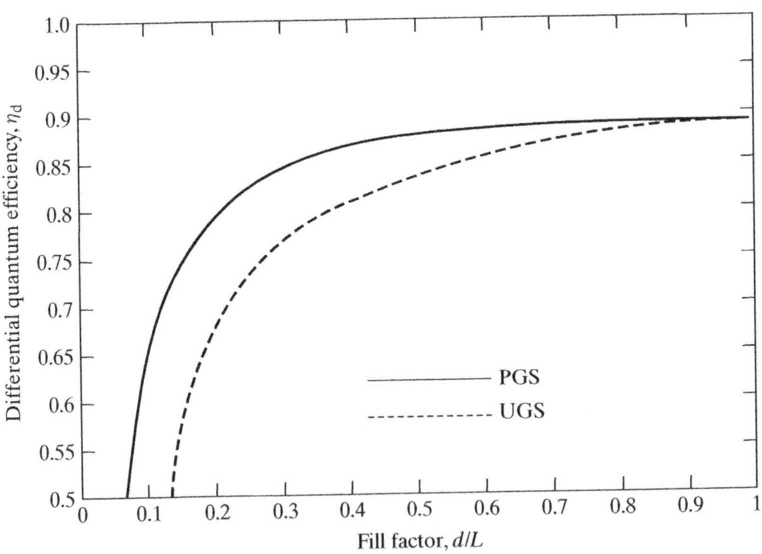

Figure 2.11 Comparison of the differential quantum efficiency η_d of VCSELs with PGS (solid lines) and UGS (dashed lines) with and without total internal loss α_{in} as a function of fill factor d/L (after Ref. 29).

optimal fill factor of 0.2–0.3, it can be shown that $L \approx 40 \times \log(1/R)$ μm. For $R = 0.95$, the optimal laser length $L_o \approx 2$ μm. The cavity length should actually be a multiple of $\lambda_R/2n$ in order to satisfy the resonance condition. In GaAs cavity, $\lambda_R/2n \sim 0.11$ μm (i.e., $\lambda_R \sim 0.87$ μm and $n \sim 4$) so that $L_o \approx 330 \times \log(1/R) \times \lambda_R/2n$. For $R = 0.95$, $L_o \sim (18 \text{ or } 19) \times \lambda_R/2n$ is the approximate optimal cavity length.

If the cavity losses or the mirror reflectivities are not known with accuracy, the exact choice of cavity length will be uncertain. Now, the choice of cavity length determines the threshold gain, but since g_{th}/J_{nom} is not a very sensitive function of gain, the exact choice of cavity length is not very critical. However, from (2.30), the threshold gain should be kept large compared to the total internal loss. Therefore, the cavity length should be underestimated if there is an uncertainty in its exact value. Another important dimension is the length of each active segment d_t. As the optimal range of the fill factor is 0.2–0.3, it corresponds to the range of d_t between 300 and 400 Å. Hence, because of the wide range of d_t that can be selected, the sensitivity to the exact choice of fill factor is low.

2.3.3 PGS VCSELs with Quantum-Well Active Layer

As explained in Section 2.2.4, in VCSELs the use of the QW active region can provide a threshold current density lower than that of using bulk materials. In order to utilize the design equations given in earlier sections to analyze the performance of the QW VCSELs with PGS, two important parameters may need to be modified for the QWs materials:

1. The first parameter is the optical gain, which has to be selected for maximum electronic conversion ratio. It must be noted that the expression of optical gain given in (2.18) is for a single QW, the optical gain of MQWs can be approximated by (2.18) with the redistribution of injection concentration inside each QWs accounted for using (2.22). Hence, the value of g_{th} for MQWs is also a function of n_w [23].
2. The second parameter is the relative confinement factor Γ_r, which needs to be adjusted to take into account the influence of QWs on the confinement of light inside the active layer. The longitudinal confinement factor Γ_z can be derived from the definition of the confinement factor

$$\Gamma_z = \frac{\displaystyle\int_{\text{thickness of QWs}} E^2(z)dz}{\displaystyle\int_L E^2(z)dz}, \tag{2.34}$$

where the numerator integrates over the total number of QWs excluding the thickness of QW barriers L_b.

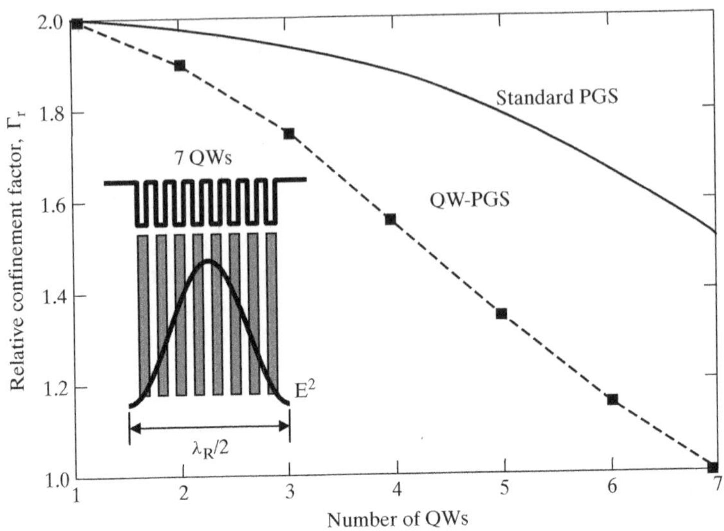

Figure 2.12 Plots of the variation of relative confinement factor Γ_r as a function of the number of quantum wells; the inset diagram illustrates the overlap between the 7 QWs and the standing-wave power (after Ref. 29).

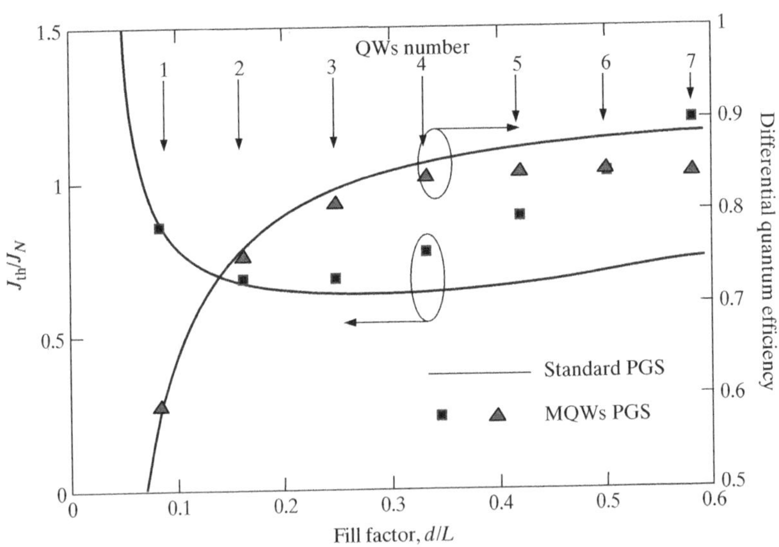

Figure 2.13 Comparison of the standard PGS with quantum-well PGS: (a) normalized threshold current density J_{th}/J_N varies with the fill factor d/L and number of quantum wells; (b) differential quantum efficiency η_d varies with the fill factor d/L and number of quantum wells (after Ref. 29).

Figure 2.12 shows the results of Γ_r calculated from various numbers of QWs in a PGS VCSEL. It can be shown that the passive higher-bandgap barriers of MQWs sacrifice the area of high optical field so that the overlap between the standing-wave pattern and material gain is reduced. This is demonstrated in the extreme case of 7 QWs ($L_w = 100$ Å and $L_b = 80$ Å), where the barriers fill up the entire half-wavelength; therefore the relative confinement factor is reduced to approximately one. Figure 2.13 plots the normalized threshold current density and differential quantum efficiency for the VCSELs with standard PGS (i.e., optical gain with no barrier assumed) and MQW PGS varying with fill factor. The optical gain used in the calculation is assumed to have an optimal value of 500 cm^{-1}. The comparison illustrates the effect of the decreased overlap when MQWs are used. For VCSELs with MQW PGS, two to three QWs provide the minimum threshold current density.

2.4 DESIGN EQUATIONS OF MULTILAYERED MIRRORS

In the preceding analysis of the threshold conditions of VCSELs, the power reflectivity of mirrors is required to be greater than 0.9 [29–31]. High reflectivity can be obtained by coating an almost perfect conductor (i.e., with large but finite conductivity) on the surface of the confinement layer. However, the high absorption from the metal coating is unavoidable; therefore, dielectric or semiconductor multilayered mirrors with high reflectivity and low absorption loss are utilized in VCSELs.

Multilayered mirrors can be realized by using a series of dielectric or semiconductor layers in which the refractive index of each subsequent layer alternates in magnitude but with equal optical thickness of a quarter-wavelength. Hence, the reflected waves interfere constructively at the design wavelength and produce a high-reflection coefficient with a phase exactly equaling zero or π. Very high peak reflection coefficients (i.e., >0.95) can be obtained even if the number of dielectric layers is small, say, 10 [32]. Multilayers of semiconductor materials, which can be fabricated by using crystal growth techniques, typically have abrupt refractive index variations across the boundaries of the layers and the material yields refractive indices in the range between 3.0 and 3.6 [29–31].

2.4.1 Numerical Calculation of Multilayered Mirrors

Transfer matrix method is generally used to analyze the amplitude and phase of the multilayered mirrors with abrupt interfaces [32]. Consider an ith dielectric layer with thickness h_i, refractive index n_i, and scattering and absorption loss α_i. Figure 2.14 shows the normal incidence of a transverse electric (TE) wave into an interface of the ith and $(i+1)$th dielectric layers. The relation between the forward traveling wave F^+ and reverse traveling waves F^- at the interface can

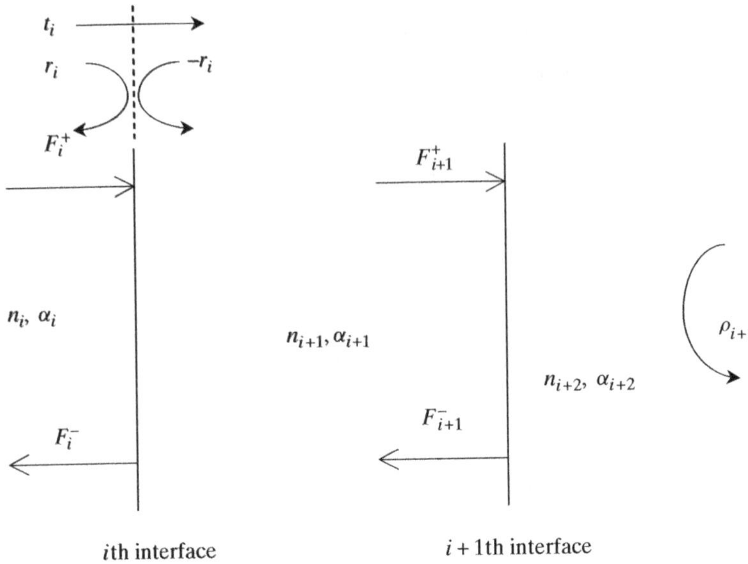

Figure 2.14 Normal incidence of a transverse electromagnetic (TEM) wave into an interface between ith and $(i+1)$th dielectric layers.

be expressed by a transmission matrix [33]

$$\begin{bmatrix} F^+ \\ F^- \end{bmatrix}_{i+1} = \begin{bmatrix} \exp(-jk_ih_i) & 0 \\ 0 & \exp(jk_ih_i) \end{bmatrix} \begin{bmatrix} \frac{1}{t_i} & -\frac{r_i}{t_i} \\ -\frac{r_i}{t_i} & \frac{1}{t_i} \end{bmatrix} \begin{bmatrix} F^+ \\ F^- \end{bmatrix}_i \equiv M_i \begin{bmatrix} F^+ \\ F^- \end{bmatrix}_i, \tag{2.35}$$

where i is an integer and the propagation constant of the ith dielectric layer k_i is defined as

$$k_i = \frac{2\pi n_i}{\lambda_R} - j\alpha_i. \tag{2.36}$$

where r_i and t_i are the field reflection and transmission coefficients, respectively, at the ith interface, which satisfy $t_i^2 + r_i^2 = 1$. The value r_i can be expressed as [32]

$$r_i = \frac{n_{i+1} - n_i}{n_{i+1} + n_i}, \tag{2.37}$$

at any abrupt ith interface. Figure 2.15 shows a multilayered mirror with m dielectric layers and $m+1$ dielectric interfaces (i.e., counts from 0 to m). Using (2.35), a matrix M_i for each ith layer of thickness h_i can be formed. Hence the relationship between the traveling waves at the mth and 0th interfaces can be

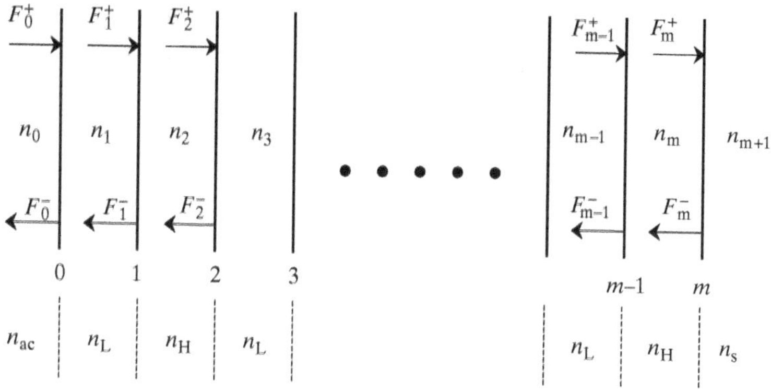

Figure 2.15 Schematic of a dielectric multilayered mirror.

expressed as

$$\begin{bmatrix} F^+ \\ F^- \end{bmatrix}_m = \prod_{i=0}^{m-1} M_i \begin{bmatrix} F^+ \\ F^- \end{bmatrix}_0 \equiv \begin{bmatrix} m_{11} & m_{12} \\ m_{21} & m_{22} \end{bmatrix} \begin{bmatrix} F^+ \\ F^- \end{bmatrix}_0, \tag{2.38}$$

so that the effective field reflectivity r_{eff}, observed from the 0th dielectric layer, can be written as

$$r_{\text{eff}} = \frac{F_0^-}{F_0^+} \equiv \frac{r_m m_{11} - m_{21}}{m_{22} - r_m m_{12}}, \tag{2.39}$$

where the boundary condition $F_m^- = r_m F_m^+$ is used and r_m is the field reflection coefficient at the mth interface. Figure 2.16 plots the variation of power reflectivity of a quarter-wavelength multilayered mirror as a function of pairs (i.e., $m/2$) with different values of α_i (=0,2,20, 200 cm^{-1}). In the calculation, it is assumed that $n_{ac} = n_s = 3.0$, $n_H = 3.52$, and $n_L = 3.0$, where the locations of n_{ac}, n_s, n_H, and n_L have been defined in Figure 2.15. It is observed that the peak reflectivity increases with the increase of pairs. However, if absorption loss is included in the calculation, the peak reflectivity saturated for $m/2$ further increases from 40. The corresponding reflection spectrum of a quarter-wavelength multilayered mirror with $m/2 = 30$ and $\alpha_i = 0$ cm^{-1} is also shown in Figure 2.16.

2.4.2 Analytic Expression of Multilayered Mirrors

As shown above, the reflectivity of the multilayered mirror at arbitrary wavelength needs to be evaluated numerically using the transmission matrix method. In fact, it is possible to formulate the reflectivity analytically, provided the thickness of all the dielectric layers is of quarter-wavelength. It can be shown that using the method of tanh substitution [see also the Appendix (Section 2A)], analytic

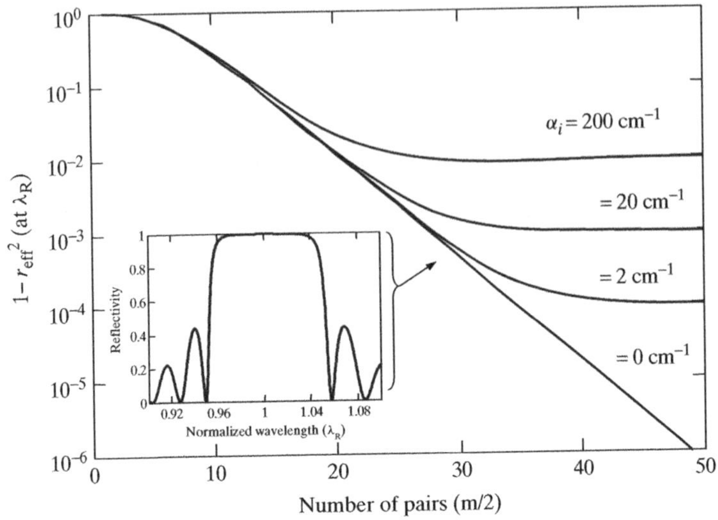

Figure 2.16 Plots of the variation of $1 - r_{\text{eff}}^2$ with the number of dielectric pairs m with different absorption coefficient α_i (=0,2,20, 200 cm^{-1}). The inset diagram illustrates the reflection spectra of a quarter-wave Bragg reflector with dielectric pairs m, which equal 30 and experience no absorption loss (after Ref. 33).

expression of r_{eff} at λ_R can be written as [34]

$$r_{\text{eff}} = \tanh\left(\sum_{i=0}^{m} |\tanh^{-1}(r_i)|\right), \tag{2.40}$$

where r_i is the field reflectivity of ith abrupt interface and is given by (2.37). By substituting (2.37) into (2.40), r_{eff} can be simplified to

$$r_{\text{eff}} = \tanh\left(\frac{1}{2}\log\left(\frac{n_{\text{ac}}}{n_s}\right) + \frac{m}{2}\log\left(\frac{n_L}{n_H}\right)\right). \tag{2.41}$$

The surprisingly simple formula for r_{eff} is due the constructive interference of the reflected wave at each interface of all multilayers (i.e., adds exactly in phase with that from every other interface). If $m = 96$, $n_s = n_{\text{ac}}$, $n_L = 3.0$, and $n_H = 3.52$, it is found that $1 - r_{\text{eff}}^2 \sim 10^{-6}$, which is close to that given in Figure 2.16. If multilayers with losses are considered in the derivation of (2.41) but under the conditions (1) $\alpha_i h_i \ll 1$, $|n_H - n_L| < 1$ and (2) $m > 90$, it can be shown that r_{eff} of the quarter-wavelength mirror at λ_R can be approximated by [33]

$$r_{\text{eff}}^2 \approx 1 - \frac{\alpha_i \lambda_R}{4\Delta n}, \tag{2.42}$$

where $\Delta n = |n_H - n_L|$ and α_i is assumed identical for all dielectric layers. If $\alpha_i = 2\ \text{cm}^{-1}$, $\lambda_R = 1\ \mu\text{m}$, and $\Delta n = 0.52$, it can be shown that $1 - r_{\text{eff}}^2 = 9.6 \times 10^{-5}$, which is close to that given in Figure 2.16 for the case $\alpha_i = 2\ \text{cm}^{-1}$. This is also true for the cases $\alpha_i = 2\ \text{cm}^{-1}$ and $200\ \text{cm}^{-1}$.

If optical loss inside the multilayered mirrors is taken into consideration, the modified threshold condition of VCSELs has to be rewritten as

$$\Gamma_z g_{\text{th}} L = \alpha_{\text{in}} L + \alpha_{\text{diff}} d + \log\left(\frac{1}{R}\right) + 2\alpha_{\text{eff}} L_{\text{eff}}, \tag{2.43}$$

where the last term on the RHS of (2.43) represents the total absorption loss inside the multilayered mirrors. In the derivation of (2.43), it is assumed that the increase in absorption loss inside the multilayers is proportional to the total cavity loss but inversely proportional to the peak reflectivity of the dielectric mirrors. Therefore, the proportional relation between the cavity loss and power reflectivity can be utilized to approximate the threshold condition given in (2.43). In this case, the effective absorption loss of multilayer α_{eff} is defined as

$$\alpha_{\text{eff}} = \frac{n_L \alpha_L + n_H \alpha_H}{n_L + n_H}, \tag{2.44}$$

and the effective length of the Bragg reflector L_{eff}, which experiences α_{eff}, can be determined and is given by [35]

$$L_{\text{eff}} = \frac{\lambda_R}{4n_{\text{ave}}}\left(\frac{n_{\text{ac}}}{n_H - n_L}\right)\left(1 - \left(\frac{n_L}{n_s}\right)^2\left(\frac{n_L}{n_H}\right)^{m-1}\right)\left(1 - \left(\frac{n_L}{n_H}\right)^m\right)$$
$$\times\left(1 - \left(\frac{n_{\text{ac}}}{n_s}\right)^2\left(\frac{n_L}{n_H}\right)^{2m-2}\right)^{-1}, \tag{2.45}$$

where n_{ave} is the average value of n_L and n_H.

The reflectivity of multilayered mirrors with metal adding on the surface of the dielectric layer can also be evaluated analytically [32,35,36]. The reflectivity r_M and the corresponding phase shift ϕ_M due to the reflection from the metal layer are given by

$$r_M = \sqrt{\frac{(n_s - n_M)^2 + k_M^2}{(n_s + n_M)^2 + k_M^2}}, \tag{2.46a}$$

$$\phi_M = \pi - \tan^{-1}\left(\frac{2n_s k_M}{n_M^2 - n_s^2 + k_M^2}\right), \tag{2.46b}$$

where n_s is the refractive index of the dielectric layer that is just adjacent to the metal and n_M and k_M are the refractive index and extinction coefficients of the metal, respectively. It must be noted that for metal with finite conductivity, there is some effective reflection plane corresponding to a penetration depth into the metal. Therefore, the dielectric layer must be slightly less than half a wavelength thick to ensure correct phase matching of the reflection waves to the metal layer. On the other hand, for an air termination, a substrate of a quarter-wavelength should be used. However, for a perfect conductor with infinite conductivity, a half-wavelength-thick dielectric layer should be used.

2.4.3 Multilayered Mirrors with Graded Interface Layers

If the interfaces of the multilayered mirror are graded, (2.39) can still be utilized to determine the corresponding reflectivity, provided the reflection from each graded interface adds exactly in phase [37]. However, (2.37) cannot be used to calculate r_i as the interfaces are no longer abrupt. Instead, a new expression of r_i has to be developed in order to take into account the graded interfaces. This can be done by applying the coupled mode theory for the derivation of coupling constant, say, κ [32,36]. This coupling constant is proportional to the fundamental Fourier component of the index variation for a quarter-wave stack. For a mirror with abrupt interfaces (i.e., a square index profile), it can be shown that the corresponding coupling constant κ_{sq} equals $\kappa_{\text{sq}} = 2\Delta n/\lambda_{\text{R}}$, where Δn is the difference between the indices of the high and low layers. If the index profile has a trapezium shape, the corresponding coupling constant κ_{gr} relative to the square index profile can be written as

$$\kappa_{\text{gr}} = \kappa_{\text{sq}} \frac{\sin(2\pi h_{\text{w}}/h_{\text{p}})}{2\pi h_{\text{w}}/h_{\text{p}}}, \tag{2.47}$$

where h_{p} is the period of the refractive index profile and h_{w} is the separation between the low and high refractive indices. Thus, the coupling constant of multilayered mirror with graded-index profile is only slightly reduced when compared with that of square index profile, provided the dielectric layers are thin.

From the standard coupled mode theory, the field reflectivity is given by $\tanh(\kappa L)$, where L is the cavity length over which coupling occurs. The reflectivity of multilayered mirrors can also be expressed by the coupled mode theory, provided the constructive interface occurred at a optical length of quarter-wavelength inside the dielectric material (i.e., $L = \lambda/4n_{\text{ave}}$); hence the new expression of r_i is

$$r_i = \tanh\left(\frac{\kappa\lambda_{\text{R}}}{4n_{\text{ave}}}\right), \tag{2.48}$$

where $n_{\text{ave}} = (n_i + n_{i+1})/2$ is the average refractive index. If $\kappa\lambda_{\text{R}}/4n_{\text{ave}} \ll 1$, (2.48) reduces to the expression of r_i given in (2.37), since $\tanh(\kappa\lambda_{\text{R}}/4n_{\text{ave}}) \sim \kappa\lambda_{\text{R}}/4n_{\text{ave}}$ in this limit. It must be noted that the coupled mode approach is valid only if Δn ($\sim |n_i - n_{i+1}|$) is a small or weak coupling between the forward- and backward-traveling waves.

2.5 DESIGN EQUATIONS FOR VCSELs OPERATION ABOVE THRESHOLD

Design equations given in Sections 2.3 and 2.4 have been developed accounting for the diffraction loss and electronic conversion ratio so that VCSELs with minimum threshold current density can be obtained. Using these design equations, suitable values of cavity length, materials used for the active layers, and the reflectivity of front and rear multilayered mirrors can be easily calculated. However, the parameters deduced from the given design equations are valid only for VCSELs operating near threshold. This is because if VCSELs operate 2 or 3 times above the threshold, the weak confinement of the injection current causes a leakage current so that the electrical power is seriously dissipated outside the active region due to the high resistivity of multilayered mirrors (i.e., parasitic resistance) [38]. Therefore, the optical power in (2.11) P will no longer be a linear function of the injection current. In this case, the wallplug efficiency (i.e., the total conversion efficiency from electrical to optical power) has to be used to optimize the abovethreshold characteristics of VCSELs so that the influence of parasitic resistance and leakage current are taken into account. In the following paragraphs, the optimal wallplug efficiency of VCSELs with the QW active layer is derived for injection current, mirror reflectivity, and cavity length.

2.5.1 Variation of Wallplug Efficiency with Injection Current

The wallplug efficiency η_{wp} of VCSELs for analysis of overall conversion efficiency between electrical and optical power is defined as [39]

$$\eta_{\text{wp}} = \frac{P}{I^2 R_s + I V_J}, \tag{2.49}$$

where P is the output optical power, I is the injection current, R_s is the series resistance, and V_J is the voltage dropped internally in the laser junction. Above threshold, V_J increases with the optical power; however, the dependence on output power is logarithmic, and so the change in the magnitude of V_J is small and can be neglected in the determination of η_{wp}. By representing the current in terms of the threshold value as

$$I - I_{\text{th}} = x \cdot I_{\text{th}}, \tag{2.50}$$

and using (2.11), the wallplug efficiency becomes

$$\eta_{\text{wp}} = \frac{h\nu_R \cdot x \cdot \eta_d / q}{(1+x)^2 I_{\text{th}} R_s + (1+x) V_J}, \tag{2.51}$$

and the values of I_{th} and η_d are assumed to be optimal (i.e., obtained by using the selection criteria given in Sections 2.2 and 2.3). Now, the operating condition of VCSELs with maximum value of η_{wp} can be obtained from (2.51). In order to determine an optimal η_{wp}, the parameter x is chosen as the variable. This will determine an optimal operating point for a particular threshold current. Taking the derivative of (2.51) with respect to x and from $\partial\eta_{wp}/\partial x = 0$, the optimal point x_o can be found and its expression is

$$x_o = \sqrt{1 + \frac{V_J}{I_{th} R_s}}. \tag{2.52}$$

Substituting x_o into (2.51) gives the optimal value of $\eta_{wp}(x_o)$:

$$\eta_{wp}(x_o) = \frac{h\nu_R}{qV_J}\eta_d \frac{(x_o - 1)}{(x_o + 1)}. \tag{2.53}$$

Figure 2.17 plots the variation of wallplug efficiency with current for a typical GaAs MQW VCSEL using the series resistance as a parameter [39]. As we can see, $\eta_{wp}(x_o)$ increase with the reduction of R_s. For $V_J \approx 1.5$ V, $I_{th} = 10$ mA, and $R_s = 10\ \Omega$, the optimal operating point is found to be $x_o = 4$, so the injection current for $\eta_{wp}(x_o)$ is 50 mA. Hence, $\eta_{wp}(x_o)$ is found to be about 0.29 for η_d and λ_R are set to 0.5 and 0.87 μm, respectively. If $R_s \to 0$, there is no well-identified maximum in $\eta_{wp}(x_o)$ and the efficiency increases monotonically with increasing drive current. With very large currents, $\eta_{wp}(x_o)$ approaches a limiting value of

$$\eta_{wp}(x_o) = \frac{h\nu_R}{qV_J}\eta_d, \tag{2.54}$$

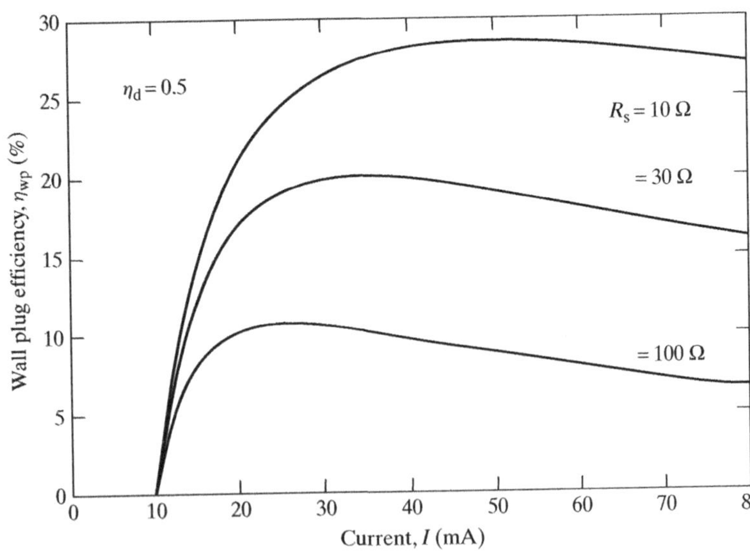

Figure 2.17 Plots of wallplug efficiency η_{wp} with the injection current I for VCSELs with different series resistance R_s.

which implies that $\eta_{wp}(x_o)$ could approach the value of η_d for large power operation. However, for VCSELs operating at high power, the dependence of V_J on the input power should be taken into account. In addition, η_d will drop due to the increase in power dissipation outside the active region, and this is the necessary phenomenon needed to be considered when VCSELs operate at high power.

If the parasitic leakage current I_p is considered in the calculation, (2.49) has to be modified into

$$\eta_{wp} = \frac{P}{(I + I_p)^2 R_s + I V_J}, \tag{2.55}$$

where I_p can be treated as a constant value. This is because the parasitic resistance is assumed to be independent of the applied voltage. With this modification, the optimal value of x, namely, x_o, becomes

$$x_o = \sqrt{1 + \frac{V_J}{I_{th} R_s} + \left(\frac{I_p}{I_{th}}\right)^2 + \frac{2I_p}{I_{th}}}. \tag{2.56}$$

Then the maximum value of η_{wp} is given by

$$\eta_{wp}(x_o) = \frac{h\nu}{qV_J}\eta_d \frac{(x_o - (I_P/I_{th} + 1))}{(x_o - (I_P/I_{th} - 1))}. \tag{2.57}$$

The dependence of η_{wp} on I for different values of R_s can be calculated by (2.57), and the corresponding profile of η_{wp} is similar to what is given in Figure 2.17 except for the values of x_o and $\eta_{wp}(x_o)$. A comparison between (2.52) and (2.56) indicates that x_o shifts to a higher value for the case where I_P is considered in the calculation, but the value of $\eta_{wp}(x_o)$ will be reduced. This can be shown by assuming that I_p equals 5 mA, $x_o = 4.15$, and $\eta_{wp}(x_o) = 0.28$ can be deduced.

2.5.2 Variation of Wallplug Efficiency with Reflectivity

Another issue of concern in designing VCSELs with maximized output power from a fixed injection current is the choice of overall mirror reflectivity R. It is noted the optimal value of η_{wp} will not vary with the selection of overall mirror reflectivity for a maximized output power. This can be shown by differentiating (2.49) with respect to R and set $\partial\eta_{wp}/\partial R$ to zero, which results in

$$\frac{\partial \eta_{wp}}{\partial R} = \frac{1}{P_{max}}\frac{\partial P_{max}}{\partial R} - \frac{1}{IR_s + V_J}\frac{\partial V_J}{\partial R} = 0, \tag{2.58}$$

where P_{max} is the maximized output power. Since V_J is a weak function of R, it can be neglected. The output power and wallplug efficiency can thus be maximized simultaneously with the same value of R as their derivatives with respect to R equal zero. Hence, it is possible to obtain an optimal operating condition for VCSELs to approximately and simultaneously maximize both η_{wp} and P.

Now, the influence of R on $\eta_{wp}(x_o)$ is ignored in the analysis and the remaining task is to find the relation between optimal reflectivity R_o and P_{max}. In order to determine R_o for P_{max}, (2.11) has to be written in the form below the optimization process

$$P = A_{eff}(J - J_{th}) \cdot h\nu_R \cdot \frac{\eta_e}{q}\left(1 + \frac{\alpha_{in}L}{\log(1/R)}\right)^{-1}, \tag{2.59}$$

where $\alpha_{diff} \sim 0$, $\eta_i \sim \eta_e$ are assumed in (2.11); η_e is defined as the electrical confinement factor that takes into account the influence of leakage current. For VCSELs with QW structure, η_e can be expressed as

$$\eta_e = \frac{J_{QW}}{J_{QW} + J_{SCH}}, \tag{2.60}$$

where J_{QW} and J_{SCH} are the recombination current density inside the QW and SCH regions, respectively, so that the total threshold current density J_{th} can be written as $J_{th} = J_{QW} + J_{SCH}$. In (2.24), it is obvious that J_{th} is a function of R, which implies that η_e is also dependent on R. It must be noted that although both J_{QW} and J_{SCH} are dependent on R, the rate of change of J_{SCH} with R is faster than J_{QW}.

In order to find the optimal value of R, (2.59) is differentiated with respect to R and set to zero, which gives

$$\frac{1}{I - I_{th}}\frac{\partial I_{th}}{\partial R} + \frac{1}{\eta_e}\frac{\partial \eta_e}{\partial R} - \frac{\alpha_{in}L}{(1 - R_o + \alpha_{in}L)^2} = 0, \tag{2.61}$$

where $\log(1/R) \approx -R$ is used in the derivation of (2.61). If $I - I_{th} = x \cdot I_{th}$ is substituted into (2.61) and with the assumptions that [40]

$$\frac{\partial J_{th}}{\partial R} = \frac{\partial J_{QW}}{\partial R} + \frac{\partial J_{SCH}}{\partial R}, \tag{2.62}$$

$$\frac{1}{J_{QW}}\frac{\partial J_{QW}}{\partial R} \cong \gamma\frac{1}{k_BT}\frac{\partial E_{FT}}{\partial R} \quad \text{and} \quad \frac{1}{J_{SCH}}\frac{\partial J_{SCH}}{\partial R} = \frac{1}{2k_BT}\frac{\partial E_{FT}}{\partial R}, \tag{2.63}$$

gives

$$\frac{1}{k_BT}\frac{\partial E_{FT}}{\partial R}\left\{\frac{(\eta_e - 1)(1 + x^{-1})}{2} + \gamma(1 - \eta_e(1 + x^{-1}))\right\} = \frac{\alpha_{in}L}{(1 - R_o + \alpha_{in}L)^2}, \tag{2.64}$$

where γ is a constant and E_{FT} is the Fermi energy in the QW at lasing condition. Furthermore, it can be shown that [40]

$$\frac{\partial E_{FT}}{\partial R} = -4k_BT\sqrt{\frac{K_{QW}/L}{1 - R + \alpha_{in}L}}, \tag{2.65}$$

for which K_{QW} is defined as [41]

$$K_{QW} = \frac{L_z^2 v_g}{2\Gamma_z h D_{np} k_B T B_i}, \tag{2.66}$$

where D_{np} is the densities of states product for the QWs, h is Planck's constant, k_B is the Boltzmann constant, and B_i is the modified Einstein coefficient for the mode. Hence, substituting (2.65) into (2.64) with the assumptions that (1) large x (i.e., ≥ 4), (2) $0.992 < R < 0.999$, (3) $\gamma \sim 0.25$, and (4) $0.5 < \eta_e < 0.9$, we can then simplify R_o to

$$1 - R_o \approx \left[\left(\frac{\alpha_{in}}{\sqrt{K_{QW}}((1 + x^{-1})(1 - \eta_e) + x^{-1})} \right)^{2/3} - \alpha_{in} \right] L. \tag{2.67}$$

Now, R_o for a chosen design of cavity length L can be determined by (2.67). In Figure 2.18, R_o is plotted as a function of cavity length. The dashed lines represent the fixed reflectivities for 3, 4, and 5 pairs of Al_xO_y/GaAs dielectric multilayered mirrors (i.e., $\Delta n \sim 1.8$), which constrain the design of the cavity length [42].

In the sections above, the optimal design of VCSELs operating above threshold is deduced. This is achieved by designing the laser from the viewpoint of wallplug

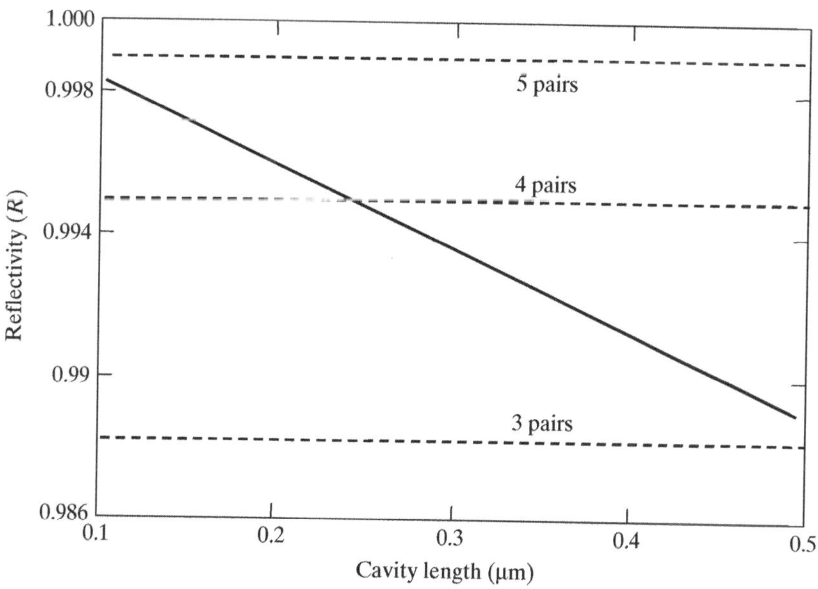

Figure 2.18 Plots of optimum reflectivity R of VCSELs with the variation of cavity length L; dashed lines represent the number of pairs (=3,4,5) of Al_xO_y/GaAs with $\Delta n \sim 1.8$ dielectric multilayered mirror (after Ref. 42).

efficiency, from a known value of series resistance, by choosing the appropriate value of x_o as the point of operation. Then, using this value of x_o to obtain the reflectivity of multilayered mirrors for a maximum output power. In the preceding derivation of design equations, the temperature of the laser cavity is assumed to be a constant. However, the power dissipated outside the active region will generate heat, which is an increasing function of injection current. It must be noted that I_{th}, K_{QW}, and η_e are functions of temperature that vary with the influence of self-heating inside the laser cavity. A simple way to take self-heating into consideration is to preset a value of temperature change for the calculation of g_{th}, I_{th}, K_{QW}, and η_e used in the design equations. Alternatively, the change of temperature can also be calculated simultaneously with the variation of g_{th}, I_{th}, K_{QW}, and η_e using (2.17).

2.6 SUMMARY

This chapter discusses the design methodology of VCSELs for optimal electrical and optical performance. Two types of VCSELs—namely, the uniform gain and periodic gain structures—are analyzed. The design equations for minimum threshold current density and maximum differential quantum efficiency are deduced under threshold condition. It is shown that the necessary conditions for optimal performance are high field reflectivity (>0.9) at the surface mirrors and low total internal loss including diffraction loss. For VCSELs with uniform gain structure, the thickness of the active layer also determines the corresponding optimal conditions. On the other hand, the electrical performance of periodic gain VCSELs is dependent on their fill factor. If the magnitude of total internal loss is small ($\sim$10 cm^{-1}), the optimal value of fill factor is around 0.2–0.3. Alternatively, MQWs can replace bulk materials as the active layer to reduce the threshold current density of VCSELs; however, the number of QWs used should be carefully selected.

It must be noted that the optical gain is a nonlinear function of injection current density; hence, the ratio between optical gain and injection current density has a maximum at some injection current density, and this ratio is referred to as the *electronic conversion ratio*. In order to maximize the performance of VCSELs at threshold, the maximum electronic conversion ratio should be used to determine the required total internal loss and reflectivity of mirrors.

The wallplug efficiency is also defined to analyze the overall conversion efficiency (i.e., conversion between electrical and optical power) of VCSELs operating above threshold. It is noted that the wallplug efficiency decreases with increase in series resistance and parasitic leakage current. This is because the power dissipated at the series resistance of the multilayered mirrors is directly proportional to the total injection current. Therefore, the wallplug efficiency of VCSELs is utilized to obtain an optimal injection current for maximum output power at abovethreshold operation.

2A APPENDIX

Substitution with tanh provides a simple way to determine the reflectivity of a quarter-wave multilayered mirror at its resonant frequency. If field reflection from a single quarter-wave layer [i.e., with ith and $(i+1)$th interfaces] is considered (see Figure 2.14), the corresponding field reflectivity ρ_{i+1} at the $(i+1)$th interface can be written as

$$\rho_{i+1} = \frac{r_{i+1} + \rho_i}{1 + r_{i+1}\rho_i}, \tag{2A.1}$$

where $\rho_i = r_i e^{-j\pi}$ is the reflection from the ith interface. Using the substitution

$$s_i = |\tanh^{-1}(r_i)| \tag{2A.2}$$

and the hyperbolic tangent formula, it can be shown that

$$|\rho_{i+1}| = \tanh(s_{i+1} + s_i). \tag{2A.3}$$

The relation for the field reflectivity of the entire multilayered mirror of mth interfaces can be derived by the application of (2A.1) and (2A.3) iteratively; hence the reflectivity from an mth interface is given by

$$|\rho_m| = \tanh\left(\sum_{i=0}^{m} \tanh^{-1}(r_i)\right) \equiv r_{\text{eff}}. \tag{2A.4}$$

This expression can be simplified by using the terms of material indices at the interfaces. Using (2A.2) gives

$$|r_i| = \frac{1 - e^{-2s_i}}{1 + e^{-2s_i}} = \frac{1 - (n_i/n_{i+1})}{1 + (n_i/n_{i+1})}, \tag{2A.5}$$

and s_i can be written as

$$s_i = -\frac{1}{2}\log\left(\frac{n_i}{n_{i+1}}\right). \tag{2A.6}$$

The field reflectivity of the entire structure is then given by

$$\rho_m = \frac{1 - b_m}{1 + b_m}, \tag{2A.7}$$

where b_m is given by

$$b_m = \prod_{i=0}^{m}\left(\frac{n_i}{n_{i+1}}\right). \tag{2A.8}$$

If the field reflectivity r_i is written as

$$|r_i| = \tanh(s_i) = \frac{1 - e^{-2s_i}}{1 + e^{-2s_i}}, \tag{2A.9}$$

then, comparing (2A.5) with (2A.9), s_i can be expressed as

$$s_i = -\frac{1}{2}\log\left(\frac{n_i}{n_{i+1}}\right). \tag{2A.10}$$

Therefore, r_{eff} is simplified to

$$r_{\text{eff}} = \tanh\left(\frac{1}{2}\log\left(\frac{n_{\text{ac}}}{n_{\text{s}}}\right) + \frac{m}{2}\log\left(\frac{n_{\text{L}}}{n_{\text{H}}}\right)\right). \tag{2A.11}$$

REFERENCES

1. Y. Suematsu, K. Iga and S. Arai, "Advanced semiconductor lasers," *Proc. of IEEE*, Vol. 80, no. 3, pp. 383–396, 1992.
2. T. P. Lee, "Recent advances in long-wavelength semiconductor lasers for optical fiber communication," *Proc. IEEE*, Vol. 79, no. 3, pp. 253–276, 1991.
3. N. Honolyak, Jr., "The semiconductor lasers: A thirty-five years perspective," *Proc. IEEE*, Vol. 85, no. 3, 1678–1693, 1997.
4. K. D. Choquette and H. Q. Hou, "Vertical cavity surface emitting lasers: Moving from research to manufacturing," *Proc. IEEE*, Vol. 85, no. 11, pp. 1730–1739, 1997.
5. W. W. Chow, K. D. Choquette, M. H. Crawford, K. L. Lear, and G. R. Haldey, "Design, fabrication and performance of infrared and visible vertical cavity surface emitting lasers," *IEEE J. Quantum Electron.*, Vol. 33, no. 10, pp. 1810–1823, 1997.
6. D. Tauber, G. Wang, R. S. Geels, J. E. Bowers, and L. A. Coldern, "Large and small signal dynamics of vertical cavity surface emitting lasers," *Appl. Phys. Lett.*, Vol. 62, pp. 325–327, 1993.
7. K. Petermann, *Laser Diode Modulation and Noise*, Kluwer, Dordrecht, 1988.
8. G. P. Agrawal and N. K. Dutta, *Long Wavelength Semiconductor Lasers*, Van Nostrand Reinhold, New York, 1986.
9. M. Y. A. Raja, S. R. J. Brueck, M. Osinski, C. F. Schaus, J. G. McInerney, T. M. Brennan, and B. E. Hammons, "Surface-emitting, multiple quantum well GaAs/AlGaAs laser with wavelength-resonant periodic gain medium," *Appl. Phys. Lett.*, Vol. 53, no. 18, pp. 1678–1680, 1988.
10. G. W. Yoffe, W. C. van der Vleuten, M. R. Leys, F. Karouta, and J. H. Wolter, "Vertical cavity surface emitting lasers with periodic gain and aluminum top contact," *Electron. Lett.*, Vol. 30, no. 10, pp. 794–795, 1994.
11. H. Soda, Y. Motegi, and K. Iga, "GaInAsP/InP surface emitting injection lasers with short cavity length," *IEEE J. Quantum Electron.*, Vol. 19, no. 6, pp. 1035–1041, 1983.
12. K. Iga, S. Kinoshita, and F. Koyama, "Microcavity GaAlAs/GaAs surface emitting laser with $I_{th} = 6$ mA," *Electron. Lett.*, Vol. 23, no. 3, pp. 134–136, 1987.

13. S. Kinoshita and K. Iga, "Circular buried heterostructure (CBH) GaAlAs/GaAs surface emitting lasers," *IEEE J. Quantum Electron.*, Vol. 19, no. 6, pp. 1035–1041, 1983.

14. T. Sakaguchi, F. Koyama, and K. Iga, "Vertical cavity surface emitting laser with an AlGaAs/AlAs Bragg reflector," *Electron. Lett.*, Vol. 24, no. 15, pp. 928–929, 1988.

15. K. Iga, F. Koyama, and S. Kinoshita, "Surface emitting semiconductor lasers," *IEEE J. Quantum Electron.*, Vol. 24, no. 9, pp. 1845–1854, 1988.

16. M. Oshikiri, H. Kawaski, F. Koyama, and K. Iga, "Reflectivity dependence of threshold current in GaInAsP/InP surface emitting laser," *IEEE Photon. Technol. Lett.*, Vol. 1, no. 1, pp. 11–13, 1989.

17. K. Moriki, H. Nakahara, K. Hattori, and K. Iga, "Single transverse mode condition of surface emitting injection lasers," *Trans. IEICE Japan*, Vol. J70-C, pp. 501–509, April. 1987.

18. G. H. B. Thompson, *Physics of Semiconductor Laser Devices*, Wiley, Chichester, UK, 1980.

19. G. Shtengel, H. Temkin, T. Uchida, M. Kim, P. Brusenbach, and C. Parsons, "Spontaneous emission factor and its scaling in vertical cavity surface emitting surface emitting lasers," *Appl. Phys. Lett.*, Vol. 64, no. 9, pp. 1062–1064, 1994.

20. N. Bewtra, D. A. Suda, G. L. Tan, F. Chatenoud, and J. M. Mu, "Modeling of quantum well lasers with electro-opto-thermal interaction," *IEEE J. Select. Topics Quantum Electron.*, Vol. 1, no. 2, pp. 331–340, 1995.

21. K. Uomi, S. J. B. Yoo, A. Scherer, R. Bhat, N. C. Andreadakis, C. E. Zah, M. A. Koza, and T. P. Lee, "Low threshold, room temperature pulsed operation of 1.5 μm vertical cavity surface emitting lasers with an optimized multi-quantum well active layer," *IEEE Photon. Technol. Lett.*, Vol. 6, no. 3, pp. 317–319, 1994.

22. Y. Arakawa and A. Yariv, "Theory of gain, modulation response and spectral linewidth in AlGaAs QW lasers," *IEEE J. Quantum Electron.*, Vol. 21, no. 10, pp. 1666–1674, 1985.

23. P. S. Zory, ed., *Quantum Well Lasers*, Academic Press, London, 1993.

24. T. A. DeTemple and C. M. Herzinger, "On the semiconductor laser logarithmic gain-current density relation," *IEEE J. Quantum Electron.*, Vol. 29, no. 5, pp. 1246–1252, 1993.

25. R. Nagarajan and J. E. Bower, "Effects of carrier transport on injection efficiency and wavelength chirping in quantum-well lasers," *IEEE J. Quantum Electron.*, Vol. 29, no. 6, pp. 1601–1608, 1993.

26. J. Behrend, J. F. Carlin, A. V. Sirbu, C. A. Berseth, A. Rudra, and E. Kapon, "InAsP/InGaAsP periodic gain structure for 1.5 μm vertical cavity surface emitting laser application," *J. Crystal Growth*, Vol. 188, pp. 295–299, 1998.

27. S. W. Corzine, R. S. Geels, J. W. Scott, R. H. Yan, and L. A. Coldren, "Design of Fabry-Perot surface emitting lasers with a periodic gain structure," *IEEE J. Quantum Electron.*, Vol. 25, no. 6, pp. 1513–1524, 1989.

28. P. McIlroy, A. Kurobe, and Y. Uematsu, "Analysis and application of theoretical gain curves to the design of multiquantum-well lasers," *IEEE J. Quantum Electron.*, Vol. 21, pp. 1958–1963, 1985.

29. R. S. Geels, S. W. Corzine, and L. A. Coldren, "InGaAs vertical cavity surface emitting lasers," *IEEE J. Quantum Electron.*, Vol. 27, no. 6, pp. 1359–1367, 1991.

30. H. Shoji, K. Otsubo, T. Fujii, and H. Ishikawa, "Calculated performances of 1.3 μm vertical cavity surface emitting lasers on InGaAs ternary substrates," *IEEE J. Quantum Electron.*, Vol. 33, no. 2, pp. 238–244, 1997.
31. M. H. MacDougal, P. D. Dapkus, A. E. Bond, C. K. Lin, and J. Geske, "Design and Fabrication of VCSELs with Al_xO_y—GaAs DBR's," *IEEE J. Select. Topics Quantum Electron.*, Vol. 3, no. 3, pp. 905–915, 1997.
32. Y. Yeh, *Optical Waves in Layered Media*, Wiley, New York, 1988.
33. T. E. Sale, "Cavity and reflector design for vertical cavity surface emitting lasers," *IEE Proc. Optoelectron.*, Vol. 142, no. 1, pp. 37–43, 1995.
34. S. W. Corzine, R. H. Yan, and L. A. Coldren, "A tanh substitution technique for the analysis of abrupt and graded interface multiplier dielectric stacks," *IEEE J. Quantum Electron.*, Vol. 27, no. 9, pp. 2086–2090, 1991.
35. D. I. Babic and S. W. Corzine, "Analytical expression for the reflection delay, penetration depth, and absorptance of quarter-wave dielectric mirrors," *IEEE J. Quantum Electron.*, Vol. 28, no. 2, pp. 514–524, 1992.
36. H. A. Haus, *Waves and Fields in Optoelectronics*, Prentice-Hall, Englewood Cliffs, NJ, 1984.
37. R. E. Sherriff, S. A. Feld, and J. P. Loehr, "Comparison of exact and approximate optical designs for graded-interface distributed Bragg reflectors," *IEEE J. Select. Topics Quantum Electron.*, Vol. 5, no. 3, pp. 582–589, 1999.
38. R. Jäger, M. Grabher, C. Jung, R. Michalzik, G. Reiner, B. Weigl, and K. J. Ebeling, "57% wallplug efficiency oxide-confined 850 nm wavelength GaAs VCSELs," *Electron. Lett.*, Vol. 33, no. 4, pp. 330–331, 1997.
39. G. W. Taylor and Q. Yang, "Optimization of the operation point of a vertical cavity surface emitting laser," *IEEE J. Quantum Electron.*, Vol. 32, no. 8, pp. 1441–1449, 1996.
40. G. W. Taylor and P. R. Claisse, "Transport solutions for the SCH quantum well laser diode," *IEEE J. Quantum Electron.*, Vol. 31, no. 12, pp. 2133–2141, 1995.
41. G. W. Taylor, "Theory of operation of the quantum well injection laser without k selection," *J. Appl. Phys.*, Vol. 70, no. 5, p. 2508, 1991.
42. H. Gebretsadik, O. Qasaimeh, H. Jiang, P. Bhattacharya, C. Caneau, and R. Bhat, "Design and realization of a 1.55 μm patterned vertical cavity surface emitting laser with lattice-mismatched mirror layers," *J. Lightwave Technol.*, Vol. 17, no. 12, pp. 2595–2604, 1999.

CHAPTER 3

Modal Characteristics of Vertical Cavity Surface Emitting Lasers

Models with various degrees of accuracy are presented in this chapter to investigate the transverse mode characteristics of VCSELs. Their merits and limitations in the study of VCSELs with different guiding structures are also discussed in detail.

3.1 INTRODUCTION

VCSELs with transverse confinement structures including those with properties such as airpost, oxide aperture, ion implanted or buried heterostructure are commonly found in literature. For VCSELs with airpost (etched mesa) configuration [1,2], a large index step between air and semiconductor provides strong transverse confinement of optical fields. VCSELs with oxide aperture configuration [3,4] have a guiding mechanism similar to that of airpost configuration. This is because the corresponding transverse fields are confined transversely by the index step of oxide aperture. Therefore, these two VCSEL confinement structures are considered as index-guided structures. For VCSELs with ion-implanted structure [5,6], the guiding mechanism of transverse modes is decided mainly by the optical gain/loss difference between the core and cladding regions of the active layer. The confinement structure is thus referred to as *gain-guided structure*. On the other hand, the guiding mechanism of buried heterostructure VCSELs with passive antiguiding region [7,8] is similar to that of the gain-guided structure except for the defocusing characteristic of the index profile. As a result, this confinement structure of VCSELs is denoted as an antiguided structure. In summary, the transverse confinement structures of VCSELs can be classified as (1) index-guiding, (2) gain-guiding, and (3) anti-index-guiding. In order to investigate the modal characteristics of VCSELs, these transverse confinement structures have to be included in the calculation.

In Chapter 2, the possibility of excitation of higher-order transverse modes is ignored in considering the design of VCSELs. It must be noted that higher-order

Analysis and Design of Vertical Cavity Surface Emitting Lasers By Siu Fung Yu
ISBN 0-471-39124-7

transverse modes can be supported in VCSELs with large cavity size. However, for the design of high-power single-mode operation, gain- or anti-index-guided VCSELs with a large cavity size are usually chosen. This is because although higher-order transverse modes are supported in VCSELs with a large cavity size, a single-mode operation can be maintained by large gain margin between fundamental and higher-order transverse modes [9]. Therefore, the modal characteristics of VCSELs with different cavity sizes and confinement structures should be studied. Furthermore, the influence of injection current density, stimulated emission, as well as thermal lensing can alter the optical gain and refractive index profiles of the transverse confinement structures so that the single transverse-mode operation may deteriorate [10–12]. These nonlinear characteristics of VCSELs should also be investigated.

This chapter is organized as follows:

1. A full set of transverse modes supported by a cylindrical dielectric waveguide is deduced. The possibility of simplifying this set of modes using the linearly polarized mode approximation is also discussed. Hence, the modal behavior of VCSELs is investigated using the linearly polarized mode approximation.
2. The origin of diffraction loss in VCSELs is explained. The mode-matching technique, which is based on the plane-wave approximation, is presented to estimate the diffraction loss.
3. The profile of transverse modes under resonant conditions is studied. A simple, two-dimensional beam propagation method is presented to study the resonant behavior of the transverse mode. The off-axis propagation of transverse modes (i.e., diffraction loss) is also considered using the scattering matrix and Bessel transform. This technique is capable of analyzing most of the VCSEL confinement structures including those with tapered waveguide geometry.
4. Finally, a general mode expansion technique, solving Maxwell's equation for vectorial solutions, is described for a rigorous study of the field distribution inside VCSEL cavities. This is achieved by the influence of mode coupling and radiation fields on the calculation of resonant wavelength. Hence, the shift of resonant wavelength and cavity loss due to the variation of cavity size are also investigated.

3.2 BEHAVIOR OF TRANSVERSE MODES OF VCSELs

Transverse mode characteristics of VCSELs can be analyzed by the cylindrical dielectric waveguide theory [13–15], because VCSELs with cylindrical geometry have transverse confinement structures similar to that of the optical fiber. In the following paragraphs, the transverse-mode characteristics of VCSELs are investigated using a full set of optical modes, including transverse electric (TE), transverse magnetic (TM), and their hybrid modes (i.e., *HE* and *EH* modes) obtained from a cylindrical dielectric waveguide. Furthermore, linearly polarization (LP) modes are proposed to replace the complete set of transverse-modes in a weakly guiding cylindrical waveguide. In this case, the LP mode approximation significantly simplifies the study of modal characteristics of VCSELs. Therefore, the LP mode approximation is used to analyze the transverse-mode characteristics in VCSELs with gain- and index-guided structures. In addition, the influence of nonuniform distribution of optical gain and refractive index on the distribution of LP modes is studied. It must be noted that the round-trip requirement inside the laser cavity is not considered in this section.

3.2.1 Solution to Maxwell's Equations in a Dielectric Cylindrical Waveguide

Consider a cylindrical waveguide, as shown in Figure 3.1, which consists of a homogeneous core of radius W and refractive index n_1, surrounded by an infinite uniform cladding of index n_2. The reason for assuming an infinitely thick cladding is that the guided modes in the core have exponentially decaying fields outside the core and do not reach the outer boundary of the cladding. Hence, the optical fields vary harmonically in the core region of refractive index n_1 and decay exponentially in the cladding region n_2. If the electromagnetic fields have a time-dependent factor $\exp(j\omega t)$, the corresponding Maxwell equation can be reduced to

$$\nabla \times \mathbf{E} = -j\omega\mu_0\mathbf{H} \qquad \text{and} \qquad \nabla \times H = j\omega\varepsilon_0 n^2\mathbf{E}, \tag{3.1}$$

where

$$n = \begin{cases} n_1 & r \le W \\ n_2 & r > W \end{cases}, \tag{3.2}$$

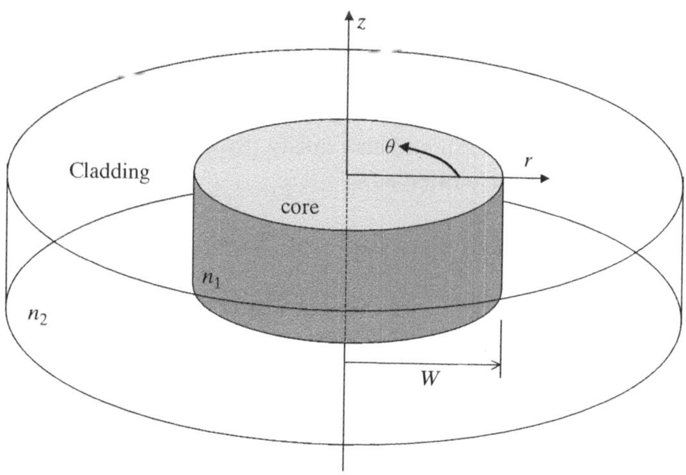

Figure 3.1 Schematic of a dielectric cylindrical waveguide.

$\omega(=ck_0)$ is the field frequency, $k_0 = 2\pi/\lambda_R$, λ_R is the wavelength, c is the velocity of light in free space, and ε_0 and μ_0 are the electric and magnetic field permeabilities, respectively. In cylindrical coordinates, the electromagnetic field components are expressed as $\mathbf{E} = (E_r, E_\theta, E_z)$ and $\mathbf{H} = (H_r, H_\theta, H_z)$. If the electromagnetic fields propagating along the longitudinal direction are assumed to have a plane-wave format, that is, $\mathbf{E}(r,\theta)\cdot\exp(-j\beta z)$ and $\mathbf{H}(r,\theta)\cdot\exp(-j\beta z)$, where β is the longitudinal propagation coefficient, Maxwell's Equation (3.1) can be simplified to a scalar wave equation as

$$\frac{\partial^2\Phi(r,\theta)}{\partial r^2} + \frac{1}{r}\frac{\partial\Phi(r,\theta)}{\partial r} + \frac{1}{r^2}\frac{\partial^2\Phi(r,\theta)}{\partial\theta^2} + \frac{q^2}{W^2}\Phi(r,\theta) = 0, \tag{3.3}$$

where $\Phi(r,\theta)$ is either $E_z(r,\theta)$ or $H_z(r,\theta)$ and

$$q^2 = \begin{cases} u^2 = W^2(k_0^2 n_1^2 - \beta^2) & r \le W \\ w^2 = W^2(\beta^2 - k_0^2 n_2^2) & r > W \end{cases}. \tag{3.4}$$

Again using (3.1), the remaining components of the electromagnetic fields, $E_r(r,\theta)$, $E_\theta(r,\theta)$, $H_r(r,\theta)$ and $H_\theta(r,\theta)$ can be written in terms of $E_z(r,\theta)$ and $H_z(r,\theta)$:

$$E_r(r,\theta) = -\frac{W^2}{q^2}\left(-\frac{j\omega\mu_0}{r}\frac{\partial H_z(r,\theta)}{\partial\theta} - j\beta\frac{\partial E_z(r,\theta)}{\partial r}\right), \tag{3.5a}$$

$$E_\theta(r,\theta) = -\frac{W^2}{q^2}\left(j\omega\mu_0\frac{\partial H_z(r,\theta)}{\partial r} - \frac{j\beta}{r}\frac{\partial E_z(r,\theta)}{\partial\theta}\right), \tag{3.5b}$$

$$H_r(r,\theta) = \frac{W^2}{q^2}\left(-\frac{jn^2\varepsilon_0\omega}{r}\frac{\partial E_z(r,\theta)}{\partial\theta} + j\beta\frac{\partial H_z(r,\theta)}{\partial r}\right), \tag{3.5c}$$

$$H_\theta(r,\theta) = \frac{W^2}{q^2}\left(jn^2\varepsilon_0\omega\frac{\partial E_z(r,\theta)}{\partial r} + \frac{j\beta}{r}\frac{\partial H_z(r,\theta)}{\partial\theta}\right). \tag{3.5d}$$

If the dielectric cylindrical waveguide has a perfectly circular symmetric structure, the wave Equation (3.3) should also satisfy periodic solutions in the angular coordinate θ, which can be written as

$$\Phi(r,\theta) = \Phi(r)\cdot\exp(\pm j\nu\theta), \tag{3.6}$$

where $\pm$ represents the possible selection of azimuthal dependence terms. This implies that the transverse modes have a twofold degeneracy in some azimuthal order $\nu(\neq 0)$.

Substituting (3.6) into (3.3), the wave equation can be reduced to

$$\frac{\partial^2\Phi(r)}{\partial r^2} + \frac{1}{r}\frac{\partial\Phi(r)}{\partial r} + \left(\frac{q^2}{W^2} - \frac{\nu^2}{r^2}\right)\Phi(r) = 0, \tag{3.7}$$

and the corresponding general solutions of $E_z(r)$ and $H_z(r)$ are given by

$$E_z(r) = \begin{cases} \dfrac{AJ_\nu(ur/W)}{J_\nu(u)} & r \leq W \\ \dfrac{AK_\nu(wr/W)}{K_\nu(w)} & r > W \end{cases}, \tag{3.8a}$$

$$H_z(r) = \begin{cases} \dfrac{BJ_\nu(ur/W)}{J_\nu(u)} & r \leq W \\ \dfrac{BK_\nu(wr/W)}{K_\nu(w)} & r > W \end{cases}, \tag{3.8b}$$

where A and B are constants to be determined by the boundary conditions. J_ν and K_ν are the Bessel and Hankel functions, respectively, of order ν. Hence, using (3.8), the corresponding expressions of $E_r(r)$, $E_\theta(r)$, $H_r(r)$, and $H_\theta(r)$ can also be obtained and are given in the Appendix, Section 3A.1. The electric fields E_z and H_z given in (3.8) can be utilized to classify the transverse modes in VCSELs. It must be noted that any guided modes with component $E_z = 0$ are defined as the *transverse electric* (TE) modes. Likewise, guided modes with component $H_z = 0$ are also referred to as the *transverse magnetic* (TM) modes. It can be shown that for dielectric cylindrical waveguide, only the cylindrical symmetric (i.e., $\nu = 0$) modes are of TE and TM types. Other modes of noncylindrical symmetry having both electric and magnetic axial field components (i.e., both E_z and $H_z \neq 0$) are called the *hybrid modes*. Hence, TE, TM, and their hybrid modes are considered to form a full set of transverse modes in VCSELs.

The full set of transverse modes can be obtained using the eigenvalue equation deduced from (3.8). The electromagnetic fields have to satisfy the boundary conditions where $r = W$ and hence evaluate the ratio of the constants A and B and find the eigenvalue equation so that β can also be calculated for the transverse modes. At the core–cladding interface, $r = W$, the boundary conditions require that the tangential components of the electric and magnetic fields be continuous. In fact, the components $E_z(r)$ and $H_z(r)$ given in (3.8) have been deliberately set for the continuous requirement at $r = W$. Hence, the remaining components for the continuous requirement at $r = W$ are $E_\theta(r)$ and $H_\theta(r)$ as they are also the tangential components of the electric and magnetic fields. The equation for the continuous requirement of $E_\theta(r)$ at $r = W$ is given by

$$A\nu\beta = -j\omega\mu_0 B\left(\frac{u^2w^2}{u^2+w^2}\right)\left(\frac{J_\nu'(u)}{uJ_\nu(u)} + \frac{K_\nu'(w)}{wK_\nu(w)}\right). \tag{3.9}$$

In addition, the corresponding equation for the continuous requirement of $H_\theta(r)$ at $r = W$ is given by

$$B\nu\beta = -j\omega\varepsilon_0 A\left(\frac{1}{u^2} + \frac{1}{w^2}\right)\left(n_1^2\frac{J_\nu'(u)}{uJ_\nu(u)} + n_2^2\frac{K_\nu'(w)}{wK_\nu(w)}\right). \tag{3.10}$$

The well-known eigenvalue equation for the transverse modes of a dielectric cylindrical waveguide can be obtained by eliminating A/B from (3.9) and (3.10) and is given by

$$\left(\frac{J_\nu'(u)}{uJ_\nu(u)}+\frac{K_\nu'(w)}{wK_\nu(w)}\right)\left(\frac{J_\nu'(u)}{uJ_\nu(u)}+(1-2\Delta)\frac{K_\nu'(w)}{wK_\nu(w)}\right)$$
$$=\left(\frac{\beta\nu}{k_0n_1}\right)^2\left(\frac{1}{u^2}+\frac{1}{w^2}\right)^2, \tag{3.11}$$

where $n_2^2/n_1^2=1-2\Delta$ is assumed in the derivation above. This eigenvalue equation can be used to study the properties of transverse modes inside a dielectric cylindrical waveguide.

Consider the special case $\nu=0$ when the RHS of (3.11) vanishes. From (3.9), it is obvious that in order to keep A [or $E_z(r)$] finite, the chosen value of B should be zero [i.e., $H_z(r)=0$], with which the special case of TM modes have to be identified. In this case, the square brackets of (3.9) should not equal zero, and hence from (3.11) the eigenvalue equation for the TM modes is

$$\left(\frac{J_1(u)}{uJ_0(u)}+(1-2\Delta)\frac{K_1(w)}{wK_0(w)}\right)=0, \tag{3.12}$$

where the relations $J_0'(u)=-J_1(u)$ and $K_0'(w)=-K_1(w)$ are used. Similarly, when $\nu=0$ and $A=0$ is chosen, the case of TE modes is obtained. Hence, it follows from (3.11) that the TE eigenvalue equation is given by

$$\left(\frac{J_1(u)}{uJ_0(u)}+\frac{K_1(w)}{wK_0(w)}\right)=0. \tag{3.13}$$

Therefore for the special case $\nu=0$ (i.e., cylindrical symmetric), the dielectric cylindrical waveguide supports modes that may be designated as TM_{0m} and TE_{0m}, where m is a radial mode number. Using (3.12) and (3.13), the corresponding propagation coefficient β of TM_{0m} and TE_{0m} modes can also be determined.

For other transverse modes $\nu\geq 1$, the corresponding eigenvalues have to be deduced from the rather complicated transcendental Equation (3.11). In this case, neither E_z nor H_z equals zero, as the corresponding transverse modes are hybrid modes, which include both E_z and H_z components. The analysis of hybrid modes using (3.11) can be simplified if Δ is of the order of 1% or less. Hence, (3.11) can be reduced to

$$\left(\frac{J_\nu'(u)}{uJ_\nu(u)}+\frac{K_\nu'(w)}{wK_\nu(w)}\right)=\pm\nu\left(\frac{1}{u^2}+\frac{1}{w^2}\right). \tag{3.14}$$

Using the properties of Bessel and Hankel functions, that is

$$J_\nu'(u)=\mp\frac{\nu J_\nu(u)}{u}\pm J_{\nu\mp1}(u)\qquad\text{and}\qquad K_\nu'(w)=\mp\frac{\nu K_\nu(w)}{w}-K_{\nu\mp1}(w), \tag{3.15}$$

it can be shown that

$$\frac{J_{\nu+1}(u)}{uJ_\nu(u)} + \frac{K_{\nu+1}(w)}{wK_\nu(w)} = 0, \tag{3.16}$$

$$\frac{J_{\nu-1}(u)}{uJ_\nu(u)} - \frac{K_{\nu-1}(w)}{wK_\nu(w)} = 0. \tag{3.17}$$

The eigenvalues corresponding to (3.16) are termed *EH modes*, and those corresponding to (3.17) are called *HE modes*. The reason for this terminology is related to the ratio of axial components of electric and magnetic fields, E_z–H_z or the ratio between A and B. Therefore, for the case $\nu \neq 0$, the dielectric cylindrical waveguide support modes may be designated as $EH_{\nu m}$ and $HE_{\nu m}$ modes. Using (3.16) and (3.17), the corresponding propagation coefficient, β, of $EH_{\nu m}$ and $HE_{\nu m}$ modes can also be determined. It must be noted that the $EH_{\nu m}$ and $HE_{\nu m}$ modes are defined under the condition that $\Delta \sim 1\%$. For a cylindrical waveguide with large Δ, the transverse modes can also be designated as $EH_{\nu m}$ and $HE_{\nu m}$ but (3.11) should be used instead of (3.16) and (3.17). Furthermore, each of the hybrid modes (i.e., $EH_{\nu m}$ and $HE_{\nu m}$ with $\nu > 0$) consists of two orthogonally polarizations to take into account the influence of $+\nu$ and $-\nu$ so that the transverse modes have twofold degeneracy.

Now, a full set of transverse modes, including its periodic solutions in the angular coordinate θ, are deduced from a perfectly cylindrical dielectric waveguide. In the following section, it is shown that the hybrid modes can be degenerated under the condition of weak transverse guidance (i.e., $\Delta \sim 0$) so that a new set of transverse modes can be redefined.

3.2.2 Linearly Polarized Modes in a Dielectric Cylindrical Waveguide

In the preceding section, $\mathrm{TE}_{\nu m}$, $\mathrm{TM}_{\nu m}$, $EH_{\nu m}$, and $HE_{\nu m}$ form a full set of transverse modes for a cylindrical dielectric waveguide. In fact, this set of modes can be transformed into a simpler set of transverse modes under the condition of weak guidance (i.e., $\Delta \sim 0$), which is referred to as *linearly polarized* (LP) modes. In the following analysis, the eigenvalue equation of LP modes is derived from the eigenvalue equations of the hybrid modes using (3.16) and (3.17). In addition, the transverse distribution of LP modes is evaluated to indicate the advantages of LP approximation.

By setting $l = \nu + 1$ in (3.16) for EH modes and $l = \nu - 1$ in (3.17) for HE modes, a new eigenvalue equation of LP_{lm} mode can be obtained and is given by

$$\frac{J_l(u)}{uJ_{l\pm1}(u)} \mp \frac{K_l(w)}{wK_{l\pm1}(w)} = 0, \tag{3.18}$$

where (3.18) holds for all $l \geq 0$. Furthermore, in the limit that $\Delta \ll 1$ and hence $n_1 \cong n_2$, (3.18) is reduced further to

$$\frac{J_l(u)}{uJ_{l-1}(u)} + \frac{K_l(w)}{wK_{l-1}(w)} = 0. \tag{3.19}$$

This equation leads to the approximated solutions for *EH*, TE, TM, and *HE* modes, from which the LP modes are formed. For example, if $l = 0$, and using the fact that $J_{-1} = -J_1$ and $K_{-1} = K_1$, (3.19) can be transformed into

$$\frac{J_0(u)}{uJ_1(u)} - \frac{K_0(w)}{wK_1(w)} = 0, \tag{3.20}$$

which is identical to the eigenvalue equation of HE_{11} mode. Hence, the fundamental mode of the weakly guided circular waveguide can be designated as LP_{01} mode. In addition, the eigenvalue equations of TM and TE modes given in (3.12) and (3.13), respectively, in the limit $\Delta \to 0$ can be recovered from (3.19). Hence, for $l = 1$, the first-order mode of the weak guided circular waveguide is designated as LP_{11} mode, which degenerates from HE_{21}, TM_{01}, as well as TE_{01} modes. Table 3.1 shows the degeneracy of the LP modes. It is clearly shown that each LP_{lm} mode $l \neq 1$ is equivalent to the $HE_{l+1,m}$ and $HE_{l-1,m}$ modes, while each LP_{1m} mode is equivalent to HE_{2m}, TM_{0m}, and TM_{0m}.

Because of the degeneration of eigenvalues in *HE* and *EH* modes, the transverse distribution of LP_{lm} mode has to be reconstructed by linearly combining $HE_{l+1,m}$ and $EH_{l-1,m}$ mode (i.e., $HE_{l+1,m} + EH_{l-1,m}$ and $HE_{l+1,m} - EH_{l-1,m}$). Both have the possibility of $\cos(l\theta)$ and $\sin(l\theta)$ polarization dependence. In fact, it can be shown that one of the two orthogonal degenerated LP_{lm} modes, in Cartesian coordinates, can be expressed as follows:

Electric component of LP_{lm}:

$$(E_x)_{HE_{l+1,m}} + (E_x)_{HE_{l-1,m}} = 0, \tag{3.21a}$$

TABLE 3.1 Degeneracies of the LP Modes

Cutoff Condition	u_c	LP Mode	Equivalent Modes	Number of Supported Modes[a]
$J_1(u_c) = 0$	0.000	LP_{01}	HE_{11}	2
$J_0(u_c) = 0$	2.405	LP_{11}	HE_{21}; TE_{01}, TM_{01}	6
$J_1(u_c) = 0$	3.832	LP_{02} (LP_{21})	HE_{12} (EH_{11}, HE_{31})	12
$J_2(u_c) = 0$	5.136	LP_{31}	EH_{21}, HE_{41}	16
$J_0(u_c) = 0$	5.520	LP_{12}	HE_{22}; TE_{02}, TM_{02}	20
$J_3(u_c) = 0$	6.380	LP_{41}	EH_{31}, HE_{51}	24
$J_1(u_c) = 0$	7.016	LP_{03} (LP_{22})	HE_{13} (EH_{12}, HE_{32})	30
$J_4(u_c) = 0$	7.588	LP_{51}	EH_{41}, HE_{61}	34
$J_2(u_c) = 0$	8.417	LP_{32}	EH_{22}, HE_{42}	38
$J_0(u_c) = 0$	8.654	LP_{13}	HE_{23}; TE_{03}, TM_{03}	42
$J_5(u_c) = 0$	8.771	LP_{61}	EH_{51}, HE_{71}	46

[a] Including the two polarization states.

$$(E_y)_{HE_{l+1,m}} + (E_y)_{HE_{l-1,m}} = \begin{cases} -2AWk_0n_1 \dfrac{J_l(ur/W)}{J_l(u)} \begin{Bmatrix} j\sin(l\theta) \\ \cos(l\theta) \end{Bmatrix} & r \le W \\ -2AWk_0n_1 \dfrac{K_l(wr/W)}{K_l(w)} \begin{Bmatrix} j\sin(l\theta) \\ \cos(l\theta) \end{Bmatrix} & r > W \end{cases}, \tag{3.21b}$$

$$(E_z)_{HE_{l+1,m}} + (E_z)_{HE_{l-1,m}} \ll (E_y)_{HE_{l+1,m}} + (E_y)_{HE_{l-1,m}}. \tag{3.21c}$$

Magnetic component of LP_{lm}:

$$(H_x)_{HE_{l+1,m}} + (H_x)_{HE_{l-1,m}} = \begin{cases} 2AWk_0n_1^2\sqrt{\dfrac{\varepsilon_0}{\mu_0}} \dfrac{J_l(ur/W)}{J_l(u)} \begin{Bmatrix} j\sin(l\theta) \\ \cos(l\theta) \end{Bmatrix} & r \le W \\ 2AWk_0n_1^2\sqrt{\dfrac{\varepsilon_0}{\mu_0}} \dfrac{K_l(wr/W)}{K_l(w)} \begin{Bmatrix} j\sin(l\theta) \\ \cos(l\theta) \end{Bmatrix} & r > W \end{cases}, \tag{3.22a}$$

$$(H_y)_{HE_{l+1,m}} + (H_y)_{HE_{l-1,m}} = 0, \tag{3.22b}$$

$$(H_z)_{HE_{l+1,m}} + (H_z)_{HE_{l-1,m}} \ll (H_x)_{HE_{l+1,m}} + (H_x)_{HE_{l-1,m}}. \tag{3.22c}$$

The other set of two orthogonally degenerated LP_{lm} modes are identical to (3.21) and (3.22) but rotated by 90° from the first set (i.e., $E_y = H_x = 0$). Hence, there is always a twofold degeneracy of polarization of the LP modes. Furthermore, there is a twofold degeneracy arising from the choice of trigonometric functions ($l \neq 0$), which has been discussed in Section 3.2.1. Hence, all the LP modes with $l \neq 0$ are of fourfold degeneration, and those with $l = 0$ are of twofold degeneration. The preceding derivation of (3.21) and (3.22) can be found in Section 3A.2.

The benefit of using this LP mode is that only one electric component and one magnetic field component are significant in LP mode (i.e., $E_y \gg E_z$, $E_x = 0$; then $H_x \gg H_z$, $H_y = 0$). The **E** vector can be chosen to lie along any arbitrary radical axis with the **H** vector along a perpendicular radical axis (i.e., in the directions of x and y as selected above). Having made this choice, there will always be a second independent mode with **E** and **H** orthogonal to the first pair. Hence, two possible polarization states are resolved from the LP modes. In addition, only a single **E** vector, with which each mode is readily visualized and drawn, need be determined. This can be shown by integrating the corresponding Poynting vector for the LP modes to find the power distribution inside the core and cladding region, that is given by

$$P_{\text{core}} = \frac{A^2}{2}(2Wk_0n_1)^2 n_1\sqrt{\frac{\varepsilon_0}{\mu_0}} \int_0^W \int_0^{2\pi} r\frac{J_l^2(ur/W)}{J_l^2(u)} \begin{Bmatrix} \sin^2(l\theta) \\ \cos^2(l\theta) \end{Bmatrix} d\theta\, dr, \tag{3.23a}$$

$$P_{\text{clad}} = \frac{A^2}{2}(2Wk_0n_1)^2 n_1\sqrt{\frac{\varepsilon_0}{\mu_0}} \int_W^\infty \int_0^{2\pi} r\frac{K_l^2(wr/W)}{K_l^2(w)} \begin{Bmatrix} \sin^2(l\theta) \\ \cos^2(l\theta) \end{Bmatrix} d\theta\, dr, \tag{3.23b}$$

which is much simpler than the full set of hybrid transverse modes. Figure 3.2 shows the power distribution of two LP modes, the LP_{01} and LP_{11} modes. Note

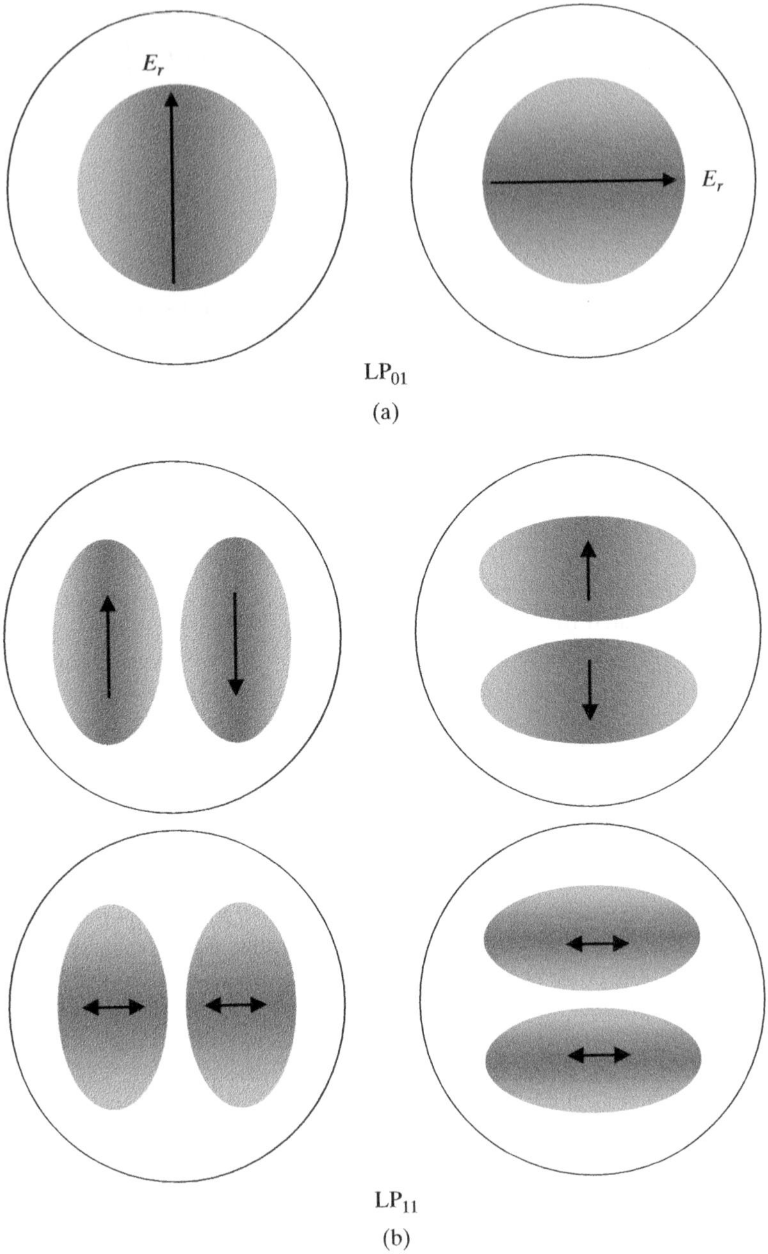

Figure 3.2 (a) Field distribution of the LP_{01} mode; (b) field distribution of the LP_{11} mode (both LP modes have two polarization states).

that, since each of the two possible polarization directions can be coupled with either a $\cos(l\theta)$ or $\sin(l\theta)$ azimuthal dependence, four discrete mode patterns are obtained for a single LP_{lm} mode for $l > 0$. However, only two discrete mode patterns are obtained for a single LP_{0m} mode.

The cutoff condition $w = 0$ (or $u = u_c$) indicates that the total number of transverse modes can be supported inside the dielectric cylindrical waveguide. From (3.19), the cutoff condition for the LP_{lm} modes is given by

$$J_{l-1}(u_c) = 0. \tag{3.24}$$

The corresponding cutoff conditions for the first 11 LP modes are given in Table 3.1. The total number of transverse modes propagating at the cutoff of the next highest LP mode, including their twofold polarization degeneracy, is also given in the table. It is noted that the increase in W (i.e., increase in u_c) increases the total number of transverse modes supported inside the dielectric cylindrical waveguide.

3.2.3 Effective Gain and Wavelength of the Transverse Modes in VCSELs

In the preceding section, the cutoff condition determines the total number of guided modes inside the dielectric cylindrical waveguide. However, the excitation of transverse modes in VCSELs is also dependent on their relative modal gain, which implies that not all the supported transverse modes will be excited under lasing condition. In the following paragraphs, the transverse-mode characteristics of VCSELs with gain- [4,5] and index-guided [6,7] structures are analyzed using the LP mode approximation. The corresponding modal gain and oscillation wavelength of each LP mode are deduced at threshold. It can be shown that gain-guided VCSELs have more effective gain difference than do index-guided lasers.

In the analysis of gain-guided VCSELs, it is assumed that the optical gain/loss and refractive index are uniformly distributed inside the core and cladding regions of the active layer. Therefore, the complex permittivity, ε_1 and ε_2, inside the core and cladding regions, respectively, are given by [9]

$$\sqrt{\varepsilon_1} = n_1 + \frac{jg}{2k_0} \quad \text{and} \quad \sqrt{\varepsilon_2} = n_2, \tag{3.25}$$

where g is the optical gain inside the core region. In gain-guided structure, it is noted that $n_1 = n_2 \equiv n$ and $n \ll g/2k_0$ so that the transverse modes are weakly guided inside the laser. Hence, the corresponding modal characteristics can be analyzed by LP mode approximation. From Section 3.2.2, the field strength of LP modes Ψ can be written in forms of Bessel and Hankel functions

$$\Psi = A_l \exp(-j\beta z) \begin{cases} \dfrac{J_l(ur/W)}{J_l(u)} & r \le W \\ \dfrac{K_l(wr/W)}{K_l(w)} & r > W \end{cases}, \tag{3.26}$$

where Ψ can be either E_y or H_x and A_l is the electric strength at $r = W$. The azimuthal variation of Ψ has been ignored in the calculation for a simple reason. The parameters u and w given in (3.26) have the following expressions

$$u^2 = W^2(k_0^2\varepsilon_1 - \beta^2) \quad \text{and} \quad w^2 = W^2(\beta^2 - k_0^2\varepsilon_2), \tag{3.27}$$

where $\beta \equiv k_0 n_{\text{eff}} + j g_{\text{eff}}/2$ and n_{eff} and g_{eff} are defined as the effective refractive index and optical gain, respectively, experienced by the transverse modes inside the active layer. Again, from Section 3.2.2, the corresponding eigenvalue equation of transverse modes is given by

$$\frac{J_l(u)}{u J_{l-1}(u)} + \frac{K_l(w)}{w K_{l-1}(w)} = 0, \tag{3.28}$$

with which the value of β can be solved numerically.

The modal properties of the gain-guided lasers can be investigated by the universal design curves (i.e., normalized propagation constant b vs. normalized frequency, v), where b and v are defined as [9]:

$$b = \frac{w^2}{w^2 + u^2} \approx \frac{g_{\text{eff}}}{g} - j\frac{2k_0 \Delta n_{\text{eff}}}{g}, \tag{3.29a}$$

$$v = \sqrt{u^2 + w^2} \approx W(1 + j)\sqrt{\frac{n \cdot g \cdot k_0}{2}}, \tag{3.29b}$$

where $\Delta n_{\text{eff}} = n_{\text{eff}} - n$ and the terms of orders g^2 and Δn_{eff}^2 have been neglected in the derivation of (3.29). As the real and imaginary parts of v are approximately the same, only the real part of v is considered in the following investigation. From (3.29), it is noted that the real part of b [Re(b)], which is normalized to $0 \leq \text{Re}(b) \leq 1$, represents the amount of gain experienced by the modes. On the other hand, the imaginary part of b [Im(b)], which is greater than zero, represents the deviation of the modal effective index from the material refractive index (i.e., Δn_{eff}). Figures 3.3a and 3.3b plot Re(b) versus Re(v) and Im(b) versus Re(v) (i.e., solid lines), respectively, for some LP modes. In Figure 3.3a, Re(b) goes to zero at some value of Re(v), which is the cutoff condition of the LP modes. On the other hand, Im(b) maintains a finite value at the cutoff condition (see also Fig. 3.3b). Hence, the influence of transverse confinement on the modal characteristics of gain-guided VCSELs is roughly understood from (3.29) through Figure 3.3. However, the threshold condition of the LP modes cannot be deduced from (3.29) as the universal design curves factor in only the influence of transverse effects, but the round-trip condition of the LP modes is neglected in the derivation.

From (2.2), the effective optical gain g_{eff} of VCSELs operating at threshold (i.e., round-trip condition) can be written as

$$g_{\text{eff}} = \Gamma_z^{-1} \cdot \left(\alpha_{\text{in}} + L^{-1} \log\left(\frac{1}{R}\right)\right), \tag{3.30}$$

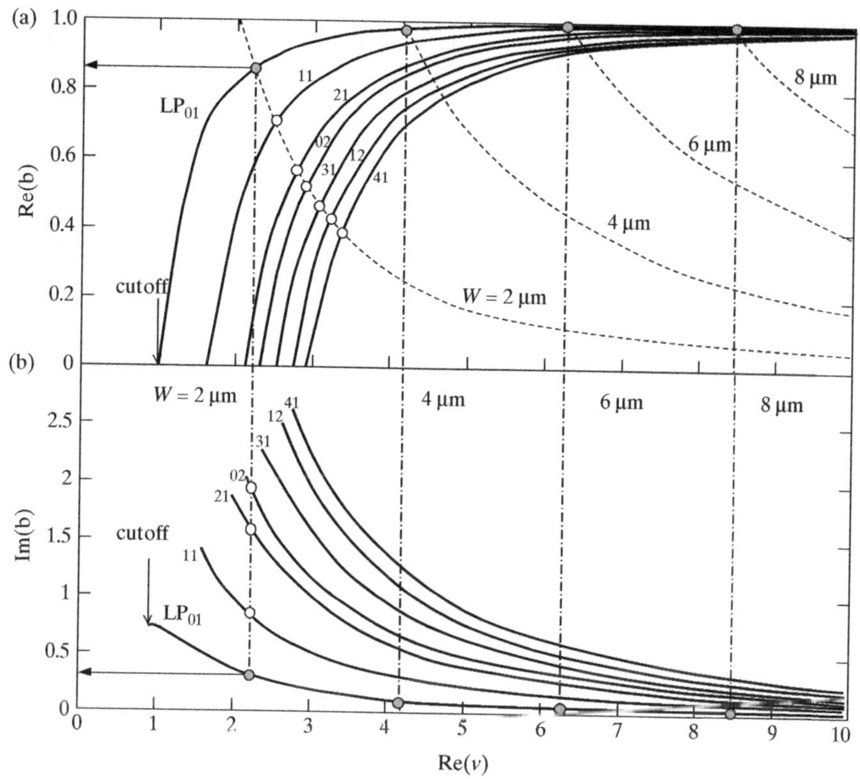

Figure 3.3 Diagrams of (a) Re{B} versus Re{ν} and (b) Im{B} versus Re{ν} for the first seven LP_{lm} modes guided by a step gain profile (solid lines). The dashed lines in (a) are the plot of threshold condition using (3.32) with $\lambda_R = 1$ μm, $n = 3.5$, $g = 1000$ cm^{-1}, and $W = 2, 4, 6$, and 8 μm. (After Ref. 9).

where L is the cavity length of VCSELs, Γ_z is the confinement factor in the longitudinal direction, α_{in} is the total internal loss, R is the power reflectivity of the combined mirrors, and the influence of diffraction loss is ignored. In (3.30), g_{eff} represents the total amount of optical gain received by the transverse mode from the core region of VCSELs. If the transverse modes are completely enclosed inside the active region, $g_{\text{eff}} = g$ can be written where g is the optical gain of active region. However, if the optical field is weakly guided along the transverse direction, the amount of optical gain received by the transverse modes will be reduced. Therefore, a transverse confinement factor Γ_t should be introduced to take into account the reduction of optical gain

$$\Gamma_t g = g_{\text{eff}}, \tag{3.31}$$

where $\Gamma_t \leq 1$. Comparing (3.31) and (3.29a), it can be shown that $\Gamma_t \equiv \text{Re}(b)$. Now, the threshold condition of VCSELs with gain-guided structure can be written as

$$\mathrm{Re}(b) = \frac{1}{\mathrm{Re}^2(v)} \frac{\pi n W^2}{\lambda_R} g_{\mathrm{eff}}, \tag{3.32}$$

where $k_0 = 2\pi/\lambda_R$ is used in the derivation of (3.32). Using (3.32), the variation of $\mathrm{Re}(b)$ with $\mathrm{Re}(v)$ with different values of W can be calculated, and the result is given in Figure 3.3a in terms of the dashed lines. In the calculation, it is assumed that $n = 3.5$, $\lambda_R = 1$ μm, $g = 1000$ cm^{-1}, and W varying between 2 and 8 μm. Therefore, the threshold condition of VCSELs can be obtained from the intersection points of the solid and dashed curves. This is because the intersection points give the corresponding modal gain received by the LP modes (e.g., the arrows indicates the value of the modal gain of LP_{01} mode for VCSEL with $W = 2$ μm). It is also shown in the figure that LP_{01} mode receives the highest modal gain, more than the other LP modes. This implies that the LP_{01} mode has the lowest threshold at this laser configuration so that the other LP modes suppress at threshold. On the other hand, the increase in W in turn increases Γ_t due to the higher $\mathrm{Re}(b)$ values so that the modal gain received by the higher-order modes is increased. Figure 3.4 shows the effective gain difference between the fundamental (LP_{01}) and first-order (LP_{11}) modes obtained from Figure 3.3a. It is shown that the effective gain difference increases with the reduction of W. This is because the profile of LP_{01} mode has a better

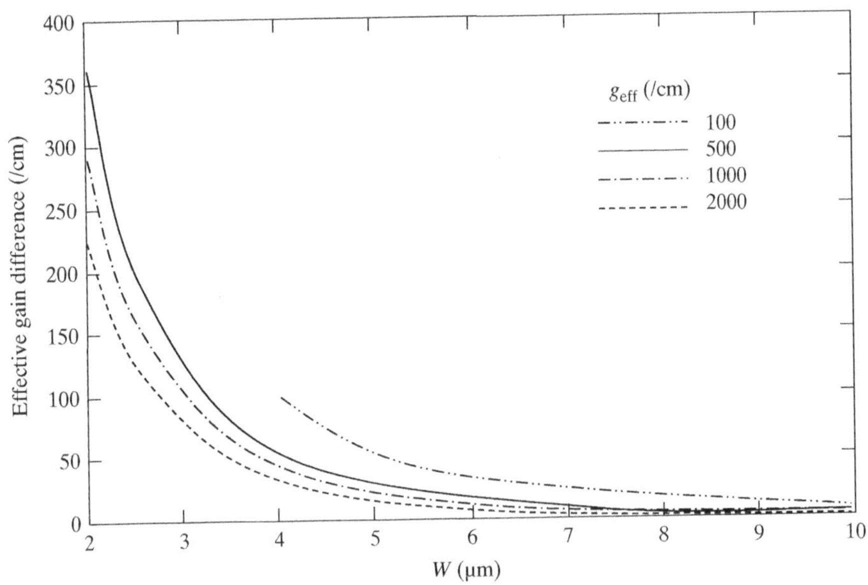

Figure 3.4 Difference in effective modal gain between LP_{01} and LP_{11} modes as a function of W for $\lambda_R = 1$ μm and $n = 3.5$, and with various g of the gain-guided lasers (after Ref. 9).

overlapping with the core region than that of the LP_{11} mode, especially for small W. Furthermore, the effective gain difference increases with the decrease in g_{eff}, which implies that the effective gain difference can be enhanced by the reduction of total internal loss.

The corresponding effective index difference of LP_{lm} mode at threshold ($\Delta n_{eff,lm}$), can be obtained from the intersection points between solid curves and vertical dashed–dotted lines as shown in Figure 3.3b (e.g., the arrows indicate the value of the effective index difference of LP_{01} mode with $W = 2\ \mu m$). It is noted that the phase and oscillation wavelength of the transverse modes vary with $\Delta n_{eff,lm}$. Under the requirement of phase condition (2.12), the shift of wavelength of LP_{lm} mode, $\Delta\lambda_{lm}$, from λ_R can be deduced from $\Delta n_{eff,lm}$ and is given by [9]

$$\Delta\lambda_{lm} = \lambda_R \Delta n_{eff,lm} \frac{d}{n_{eff} d + n_{conf}(L - d)}, \tag{3.33}$$

where n_{conf} is the refractive index of confinement layer (see Fig. 2.1). It is shown that $\Delta n_{eff,lm}$ of the transverse modes is nearly constant at different d values. This implies that the transverse modes have almost the same spacing. In addition, the mode spacing is reduced with the increase in W. It is noted that higher-order transverse modes have resonance wavelengths shorter than that of the fundamental mode.

Now consider VCSELs with index-guided structure and the corresponding refractive indices of core and cladding regions are n_1 and n_2, respectively. In this case, the guiding mechanism of the transverse modes is less dependent on the optical gain of the core region. Under this assumption, the eigenvalue equation can be obtained and is similar to (3.28) except for the following parameters:

$$u^2 = k_0^2\ W^2(n_1^2 - n_{eff}^2) \quad \text{and} \quad w^2 = k_0^2\ W^2(n_{eff}^2 - n_2^2). \tag{3.34}$$

In addition, the corresponding normalized propagation constant b versus normalized frequency v are defined as

$$b = \frac{n_{eff}^2 - n_2^2}{n_1^2 - n_2^2}, \tag{3.35a}$$

$$v = k_0 W \sqrt{n_1^2 - n_2^2}. \tag{3.35b}$$

The transverse confinement factor in an index-guided structure can also be expressed as [9]

$$\Gamma_t = \frac{\int_0^W |\Psi|^2 r\, dr}{\int_0^\infty |\Psi|^2 r\, dr}. \tag{3.36}$$

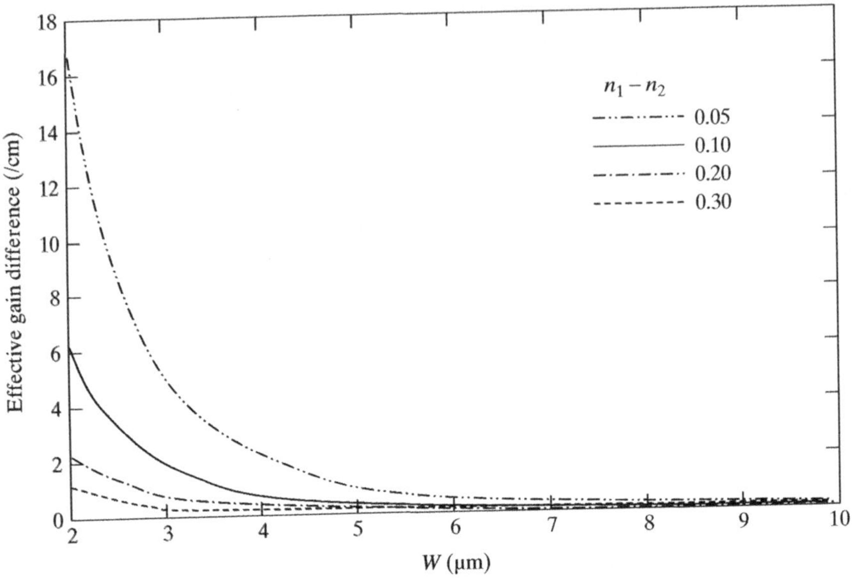

Figure 3.5 Difference in effective modal gain between LP_{01} and LP_{11} modes as a function of W for $\lambda_R = 1\ \mu m$ and $g = 1000\ cm^{-1}$; the curves are drawn for several values of $n_1 - n_2$ (after Ref. 9).

Hence, the effective gain difference between transverse modes can be calculated by multiplying the material gain g by the difference of Γ_t for the corresponding transverse modes. Figure 3.5 shows the effective gain difference between LP_{01} and LP_{11} modes as a function of W for several refractive index steps, $n_1 - n_2$. In the calculation, $g = 1000\ cm^{-1}$ is used and the influence of the gain-guiding mechanism is neglected. A comparison between Figures 3.4 and 3.5 shows that the modal gain difference of the gain-guided structure is much better than that of the index-guided structure. Therefore, the gain-guided structure should provide better side-mode suppression and narrow spectral envelopes than the index-guided structure of comparable W [3]. Finally, the corresponding shift of wavelength of LP_{lm} mode ($\Delta\lambda_{lm}$), from λ_R can be obtained from (3.33) and is given by

$$\Delta\lambda_{lm} = \lambda_R(n_1 - n_2)(1 - b_{lm})\frac{d}{n_{\text{eff}}\, d + n_{\text{conf}}(L - d)}. \tag{3.37}$$

It is shown that the transverse mode spacing of the index-guided structure is similar to that of the gain-guided structure.

In the preceding analysis of VCSELs with gain- and index-guided structures, the optical gain and refractive index are assumed uniformly distributed along the core and cladding regions of the active layer. However, the influence of spatial hole burning [10,11] and thermal lensing [12] modifies the distribution of refractive index and optical gain inside the active layer of VCSELs. Hence, the modal gain and transverse distribution of each LP mode may be affected significantly. In the following paragraphs, a multilayer approximation is described in an analysis of the transverse distribution of LP modes with the consideration of nonuniform distribution of refractive index and optical gain inside the active layer of VCSELs.

3.2.4 Influence of Nonuniform Distribution of Carrier Concentration and Temperature on the Transverse-Mode Characteristics of VCSELs

Due to the influence of spatial hole burning [10,11] and thermal lensing [12], the profiles of refractive index and optical gain inside the active layer of VCSELs are nonuniformly distributed. In order to analyze the nonuniform distribution of refractive index and optical gain, the active layer of VCSELs is subdivided into $2n+1$ small equal segments along the transverse direction and each segment has uniform permittivity in the azimuthal direction as illustrated in Figure 3.6. The weakly guided transverse modes inside the active layer of VCSELs can be calculated using the LP_{lm} mode approximation given in Section 3.2.3 but with a slight modification. If Ψ is assumed to be the electric field of a LP_{lm} mode, Ψ over each segment of the active layer can be expressed in terms of Bessel functions J_l, Y_l, and Hankel functions K_l, I_l of order l as

$$\Psi = \exp(-j\beta z)\begin{cases} A_1 J_l\left(\dfrac{u_1 r}{W}\right) & 0 \le r < r_1 \\ A_2 J_l\left(\dfrac{u_2 r}{W}\right) + B_2 Y_l\left(\dfrac{u_2 r}{W}\right) & r_1 \le r < r_2 \\ \vdots & \\ A_n J_l\left(\dfrac{u_n r}{W}\right) + B_n Y_l\left(\dfrac{u_n r}{W}\right) & r_{n-1} \le r < r_n \\ A_{n+1} K_l\left(\dfrac{w_{n+1} r}{W}\right) + B_{n+1} I_l\left(\dfrac{w_{n+1} r}{W}\right) & r_n \le r < r_{n+1} \\ \vdots & \\ A_{2n} K_l\left(\dfrac{w_{2n} r}{W}\right) + B_{2n} I_l\left(\dfrac{w_{2n} r}{W}\right) & r_{2n-1} \le r < r_{2n} \\ A_{2n+1} K_l\left(\dfrac{w_{2n+1} r}{W}\right) & r \ge 2\,W \end{cases}, \quad (3.38)$$

where $r_k = k \cdot \Delta r$, k is an integer, and $\Delta r = W/n$. The parameters u_k and w_k are defined by

$$u_k^2 = W^2(\varepsilon_k k_0^2 - \beta^2) \quad \text{and} \quad w_k^2 = W^2(\beta^2 - \varepsilon_k k_0^2), \quad (3.39)$$

where ε_k is the relative permittivity at the kth segment of the active layer and β is the longitudinal propagation coefficient to be determined by the eigenvalue equation. It must be noted that ε_k is a complex number, and the real and imaginary

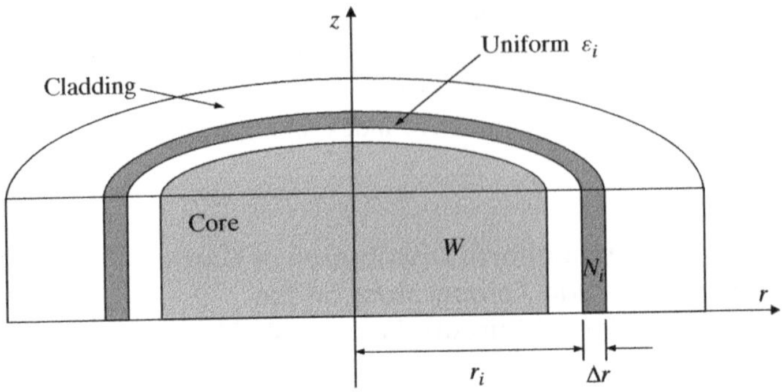

Figure 3.6 Piecewise description of the nonuniform distribution of permittivity along the azimuthal direction of the active layer of VCSEL.

parts of ε_k represent the refractive index and gain of the segment of the active layer, respectively.

The corresponding eigenvalue equation of (3.38) can be obtained by matching Ψ and its derivative at the boundary of the segments (i.e., $r_1, r_2, \ldots, r_{2n}$). Hence, a set of linear homogeneous equations for the complex coefficient A_k and B_k (for $k = 1, 2, .., 2n + 1$) can be deduced. For example

$$\begin{bmatrix} A_k \\ B_k \end{bmatrix} = \begin{bmatrix} J_l\left(\dfrac{u_k r_k}{W}\right) & K_l\left(\dfrac{w_k r_k}{W}\right) \\ u_k J_l'\left(\dfrac{u_k r_k}{W}\right) & w_k K_l'\left(\dfrac{w_k r_k}{W}\right) \end{bmatrix}^{-1} \begin{bmatrix} J_l\left(\dfrac{u_{k+1} r_k}{W}\right) & K_l\left(\dfrac{w_{k+1} r_k}{W}\right) \\ u_{k+1} J_l'\left(\dfrac{u_{k+1} r_k}{W}\right) & w_{k+1} K_l'\left(\dfrac{w_{k+1} r_k}{W}\right) \end{bmatrix} \times \begin{bmatrix} A_{k+1} \\ B_{k+1} \end{bmatrix}. \tag{3.40}$$

Hence, it can be shown that

$$\begin{bmatrix} m_{11} & m_{12} \\ m_{21} & m_{22} \end{bmatrix} \begin{bmatrix} A_1 \\ A_{2n+1} \end{bmatrix} = \begin{bmatrix} 0 \\ 0 \end{bmatrix}, \tag{3.41}$$

where m_{11}, m_{12}, m_{21}, and m_{22} can be evaluated from (3.38) using the approach similar to the derivation of (3.40). Therefore, the condition for nontrivial solutions of (3.41) determines the complex propagation coefficient β and the field profile of Ψ.

The influence of nonuniform distribution of carrier concentration and temperature on the LP modes can also be factored into the model proposed above. This can be done if the corresponding variation of refractive index and optical gain are taken into account in the relative permittivity ε as

$$\Delta\varepsilon_k = 2\sqrt{\varepsilon_k}\Delta n_k + \frac{j\sqrt{\varepsilon_k}g_k}{k_0}, \tag{3.42}$$

where g_k and Δn_k are the optical gain and change in refractive index, respectively, inside the kth segment of the active layer. The optical gain g_k at the kth segment is approximated by

$$g_k = g_N(T_k)\{N_k - N_t(T_k)\} - \alpha_{m,k}, \tag{3.43}$$

where $\alpha_{m,k}$ is the effective material losses (including free-carrier absorption loss and scattering loss) of the kth segment. The differential gain coefficient g_N and the transparency carrier concentration N_t are both functions of temperature. The change in refractive index Δn_k, due to the variation of carrier concentration ΔN_k and temperature ΔT_k at the kth segment, can be approximated by [11]

$$\Delta n_k = \frac{\partial n}{\partial N}\Delta N_k + \frac{\partial n}{\partial T}\Delta T_k, \tag{3.44}$$

where $\Delta N_k = N_k - N_t$, $\partial n/\partial N \sim -5 \times 10^{-22}\ \text{cm}^3$, and $\partial n/\partial T \sim 5 \times 10^{-4}\ \text{K}^{-1}$ for GaAs materials [16]. Now, the remaining task is to calculate the carrier concentration profile and temperature distribution over the active layer. Using the approach given by Dutta [17] for VCSELs with circular metal contact of radius W and with the assumption that the current is uniformly injected, the carrier concentration distribution near threshold can be approximated as

$$N_k = \begin{cases} \dfrac{JL_n^2(1 - WK_1(W/L_n)I_0(r_k/L_n)/L_n)}{qD_nd} & 0 \le r_k \le W \\[2ex] \dfrac{JL_nWI_1(W/L_n)K_0(r_k/L_n)}{qD_nd} & r_k > W \end{cases}, \tag{3.45}$$

where J is the injection current density; q is the electron charge; d is the active layer thickness; D_n and L_n are the diffusion coefficient and diffusion length, respectively, of the carrier concentration; and I_0 and K_0 are the modified Bessel and Hankel functions, respectively, of the zeroth order. For the laser operating near threshold, ΔT_k can be approximated by (2.17), that is, $\Delta T_k = \kappa_T^{-1} R_{TH} P_{IV,k}$. The electrical power $P_{IV,k}$, generated at the kth segment can be expressed in terms of N_k and is given by

$$P_{IV,k} = \frac{N_k q d}{\tau_n} A_{\text{eff}} V_J, \tag{3.46}$$

where A_{eff} is the effective area. It must be noted that the calculation of carrier concentration profile and distribution of temperature proposed above are of a first-order approximation, which is true only for a device operating near or at threshold. For the abovethreshold calculation, a self-consistent technique is required to simultaneously solve optical field, carrier concentration, and temperature profiles, discussed in other chapters.

3.3 CALCULATION OF DIFFRACTION LOSS IN VCSELs USING MODE-MATCHING TECHNIQUE

In Section 3.2, the longitudinal propagation of transverse modes inside a cylindrical dielectric waveguide is deduced under the assumption of plane-wave approximation [i.e., $\exp(-j\beta z)$]. However, the application of plane-wave approximation on the longitudinal propagation of transverse modes may not be appropriate. This is because the transverse modes are only partially confined inside the active layer so that the off-axis propagation of transverse modes is unavoidable inside the multilayer mirrors, especially for VCSELs with a small cavity size (i.e., in order of λ_R). Therefore, the origin of diffraction loss can be explained as the scattering of light from the core region of the active layer. As a result, the off-axis propagation of optical field reduces the reflectivity of the multilayered mirror so that the total cavity loss is increased. Hence, this phenomenon of transverse modes is referred to as *diffraction loss*. In order to get more accurate results of the analysis of the transverse modes, the influence of diffraction loss should be considered.

Simple approximation methods for the off-axis propagation of transverse modes have been utilized to calculate diffraction loss in VCSELs. For example, the effective index method [18], the position-dependent beam propagation method [19], or the intracavity lens counteracting diffraction approximation [20] have been developed to analyze the diffraction loss in VCSELs with AlAs oxide aperture (i.e., strong index-guided structure). In fact, the development of these approximation methods was based on an intuitive interpretation of laser cavity. For a more accurate approach, off-axis propagation of transverse modes have to be evaluated directly using methods such as vector plane wave decomposition [21,22], mode-matching technique [23–25] or modified beam propagation method [26–28] to calculate diffraction loss. In the following paragraphs, the self-consistent mode matching technique is described to calculate the reflectivity of multilayered mirrors.

Figure 3.7 shows the schematic of a multilayered mirror used in the calculation. It is assumed that the multilayered mirror and a GaAs active layer are all grown on a GaAs substrate. The multilayered mirror has a circular symmetry with pairs of GaAs–AlAs layers of half-wavelength each. A refractive index step is applied to the core–cladding region of the active layer so that a VCSEL with index-guided structure is obtained. The calculation of total fields inside the laser cavity can be simplified by assuming that the reflected field at $z = 0$ is the sum of unperturbed incident field and fields reflected from the multilayered mirror. However, other cavity modes such as leaky guiding and radiation modes are ignored in the consideration. Therefore, the LP modes (i.e., plane-wave approximation) given in Section 3.2.2 can be used to describe the total fields if they are well confined inside the core region of the active layer. Furthermore, the corresponding eigenvalue equation for the LP modes can be simplified by using the concept of effective width [29]. This can be done by approximating a penetration

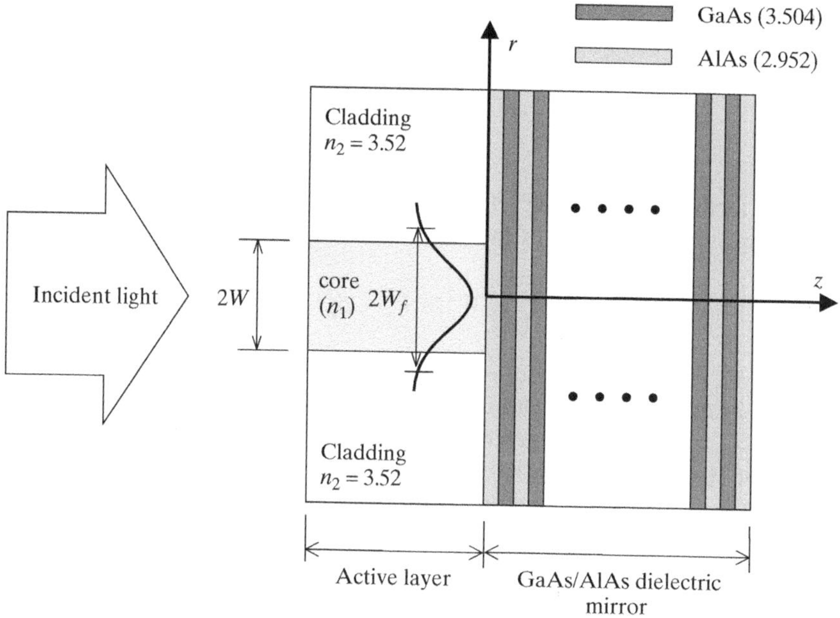

Figure 3.7 Schematic of the multilayered mirror used to calculate the reflectivity under the influence of diffraction.

depth W_f of the LP modes into the cladding region as [23,24]

$$W_f = W + \frac{1}{\sqrt{\beta_f^2 - n_2^2 k_0^2}}, \tag{3.47}$$

where β_f is the propagation constant of the incident mode inside the active layer. If the LP modes are the confined transverse modes inside the active layer, the incident field Ψ_f with the above-defined boundary conditions can be written as

$$\Psi_f(r, z) = J_1\left(\frac{u_f r}{W_f}\right) e^{-j\beta_f z} \qquad \text{for } z \leq 0 \tag{3.48}$$

where u_f is determined by the boundary condition $J_1(u_f) = 0$ and the propagation constant β_f is given by

$$u_f^2 = W_f^2(n_1^2 k_0^2 - \beta_f^2). \tag{3.49}$$

The off-axis propagation of the reflection field Ψ_r due to the influence of diffraction loss of the multilayered mirrors, is approximated by a set of LP modes (i.e., a set of plane waves), which is given by

$$\Psi_r(r,z) = A_f J_1\left(\frac{u_f r}{W_f}\right) e^{j\beta_f z} + \sum_{\substack{p=1\\ p\neq f}}^{\infty} A_p J_1\left(\frac{u_p r}{W_p}\right) e^{j\beta_p z} \qquad \text{for } z \geq 0, \quad (3.50)$$

where Ψ_r includes propagating (β_p real) and evanescent (β_p imaginary) modes. The propagation constant β_p of the pth mode is obtained from

$$u_p^2 = W_p^2(n_1^2 k_o^2 - \beta_p^2). \qquad (3.51)$$

The term A_p can be interpreted as the modal reflection coefficient, u_p is determined by $J_1(u_p) = 0$ for the pth mode, and W_p can be obtained from (3.47) by replacing β_f with β_p. In the preceding derivation, the optical fields Ψ_f and Ψ_r are assumed propagating paraxially (i.e., $\partial^2/\partial z^2 = 0$) inside the laser cavity.

In order to calculate the magnitude of the reflected fields, the modal reflection coefficient A_p is solved by mode matching with the spectrum response of the multilayered mirror. This can be done by using the Bessel transform $\hat{\Psi}(s)$ of an electric field $\Psi(r)$ which is defined by

$$\hat{\Psi}(s)e^{-j\beta z} = \int_0^{\infty} \Psi(r,z)\cdot J_1(sr)\cdot r\,dr, \qquad (3.52)$$

where β is the corresponding propagation coefficient. Hence, the transformation of the total fields $\Psi_f + \Psi_r$ at $z = 0$ is given by

$$\int_0^{\infty} [\Psi_f(r,z) + \Psi_r(r,z)]\cdot J_1(sr)\cdot r\,dr$$
$$= \hat{\Psi}_f(s)e^{-j\beta_f z} + A_f\hat{\Psi}_f(s)e^{j\beta_f z} + \sum_{\substack{p=1\\ p\neq f}} A_p\hat{\Psi}_p(s)e^{j\beta_p z}, \qquad (3.53)$$

where $\hat{\Psi}_p(s)$ is the first-order Bessel transform of the pth mode. It is noted that the spectrum response of the multilayered mirror can be represented by a uniform medium having a single response function $\beta_{\mathrm{D}}(s)$ [30], so that for $z \geq 0$ the spectrum response of the multilayered mirror can be written as

$$\int_0^{\infty} \Psi_{\mathrm{D}}(r,z)\cdot J_1(sr)\cdot r\,dr = \hat{\Psi}_{\mathrm{D}}(s)\cdot e^{j\beta_{\mathrm{D}} z}, \qquad (3.54)$$

where $\Psi_{\mathrm{D}}(r,z)$ and β_{D} can be considered as the transmitted field and propagation coefficient, respectively, of the LP modes inside the multilayered mirror. $\hat{\Psi}_{\mathrm{D}}(s)$ can be determined by the boundary condition at the interface between active layer and mirror. See Section 7A.3 for the derivation of β_{D} from a multilayered mirror.

If the electric component of incident field is perpendicular to the plane of incidence or parallel to the interface between the active layer and multilayered mirror, the incident and reflected fields are referred to as *TE-polarized*. Hence the corresponding boundary conditions are given by

$$\Psi_f(r,z) + \Psi_r(r,z)\,|_{z=0} = \Psi_{\rm D}(r,z)|_{z=0}\,, \tag{3.55a}$$

$$\frac{\partial}{\partial z}\Psi_f(r,z) + \Psi_r(r,z)\,|_{z=0} = \frac{\partial}{\partial z}\Psi_{\rm D}(r,z)\bigg|_{z=0}\,. \tag{3.55b}$$

Using (3.53), (3.54), (3.55), and the orthonormalization properties of $\hat{\Psi}$

$$\int_0^\infty \hat{\Psi}_p(s)\cdot\hat{\Psi}_q^*(s)\,ds = \begin{cases}1 & p=q\\ 0 & p\neq q\end{cases}, \tag{3.56}$$

the magnitude A_f of the fundamental mode (i.e., LP_{01} mode) can be deduced and is given by

$$A_f = \int_0^\infty \left(\frac{\beta_f-\beta_{\rm D}}{\beta_f+\beta_{\rm D}}\right)\hat{\Psi}_f\hat{\Psi}_f^*\,ds - \sum_{\substack{p=1\\ p\neq f}}^{\infty} A_p \int_0^\infty \left(\frac{\beta_p+\beta_{\rm D}}{\beta_f+\beta_{\rm D}}\right)\hat{\Psi}_p\hat{\Psi}_f^*\,ds. \tag{3.57}$$

Therefore, A_p given in (3.57) can be written as

$$A_p = \int_0^\infty \left\{\frac{\beta_0-\beta_{\rm D}}{\beta_p+\beta_{\rm D}} - A_f\frac{\beta_f+\beta_{\rm D}}{\beta_p+\beta_{\rm D}}\right\}\hat{\Psi}_f\hat{\Psi}_p^*\,ds$$
$$- \sum_{\substack{q=1\\ q\neq f,p}}^{\infty} A_q \int_0^\infty \left(\frac{\beta_q+\beta_{\rm D}}{\beta_p+\beta_{\rm D}}\right)\hat{\Psi}_q\hat{\Psi}_p^*\,ds. \tag{3.58}$$

Equations (3.57) and (3.58) form a system of linear equations, which can be solved iteratively to yield A_f and A_p. As the magnitude of the incident field Ψ_f is set to 1, A_f can be interpreted as the magnitude of the reflection wave so that the total loss, α, for the fundamental mode in VCSELs, can be written as

$$\alpha = \alpha_{\rm in} + L_{\rm eff}^{-1}\log\left(\frac{1}{|A_f|^2}\right), \tag{3.59}$$

where $L_{\rm eff}$ is the effective cavity length. The second term on the RHS of (3.59) represents the sum of cavity loss and diffraction loss inside the DBR. Total loss due to higher-order LP mode can also be expressed in a similar manner [23,24].

In the preceding paragraphs, the model is designed to calculate the diffraction loss of VCSELs with an index-guided structure. However, diffraction loss of VCSELs with other transverse confinement structures such as gain- or antiguided structures can also be estimated by using the same technique. However, W_f given in (3.47) has to be replaced by [31]

$$W_f = W + \frac{1}{\mathrm{Re}(\beta)}, \tag{3.60}$$

where β is the complex propagation coefficient of the VCSELs. For VCSELs with pure gain-guided structure, it was shown in Section 3.2.3 that $\mathrm{Re}(\beta) = g_{\mathrm{eff}}/2$. This assumption is reasonable because highly effective gain reduces the penetration depth of the transverse modes into the cladding region.

Figure 3.8 shows the variation of cavity loss with W due to LP_{01} mode of VCSELs. It is assumed that the VCSELs operating at $\lambda_R = 0.85$ μm with weakly index-guiding (anti-index-guiding) waveguide, which has $n_1 - n_2 = 0.0225$ ($n_1 - n_2 = -0.0225 + j0.012$) and the complex refractive index is uniformly distributed along the core and cladding regions of the active layer [31]. In addition, $L_{\mathrm{eff}} = 2$ μm and $\alpha_{\mathrm{in}} = 0\ \mathrm{cm}^{-1}$ is set in the calculation. As expected, the cavity loss is negligible for $W > 2$ μm but increases dramatically for $W < 1$ μm. From these results, it is noted that the diffraction loss has a significant influence on the modal gain of the transverse modes, especially for VCSELs with a small cavity size. Furthermore, it can be shown that higher-order LP modes have higher diffraction loss than does the LP_{01} mode. This is because higher-order LP modes are only supported in dielectric waveguide with large W so that the influence of diffraction loss on LP_{01} mode is minimized. On the other

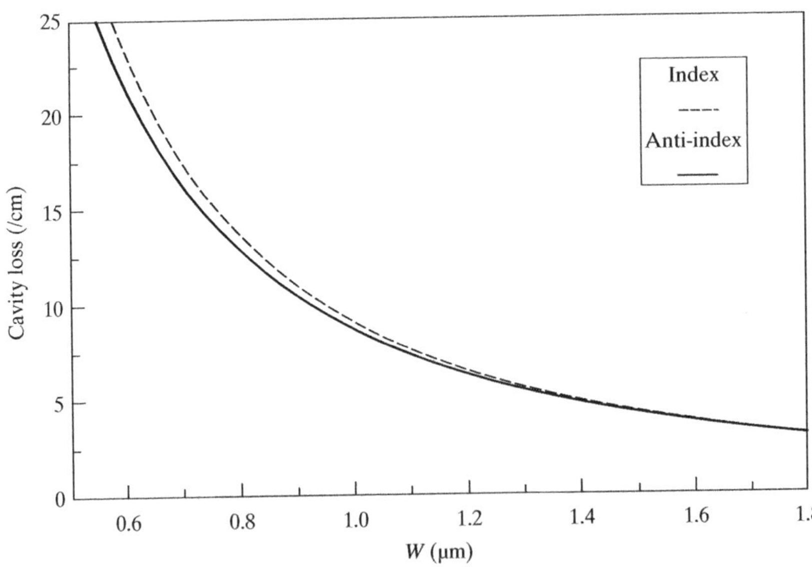

Figure 3.8 Variation of cavity loss α with core radius W for device of index-guided (dotted line) and anti-index-guided (solid line) structures (after Ref. 31).

hand, when W is small, only the LP_{01} mode is supported and there is no higher-order LP mode to be affected by diffraction loss. Therefore, the diffraction loss may be used to enhance the suppression of side mode in VCSELs with large W.

3.4 RESONANT TRANSVERSE-MODE CHARACTERISTICS IN VCSELs

In Section 3.3, the transverse modes are evaluated for different transverse confinement structures. However, for VCSELs in resonant conditions, the transverse modes should also fulfill the round-trip requirement. In this case, the longitudinal propagation coefficient and modal profile of the transverse modes can be modified by the longitudinal structure of the laser cavity. For VCSELs with single-layer reflectors such as those given in Section 3.2.2, the round-trip requirement can be determined using the effective index method [18]. However, if multilayered mirrors are used as the reflectors, effective index approximation may fail to predict the round-trip requirement, especially in the presence of diffraction loss. Hence, different approaches should be adopted to evaluate the transverse fields inside the multilayered mirrors. In the following paragraphs, the influence of the round-trip condition on the transverse modes is studied through VCSELs with tapered waveguide geometry. A two-dimensional beam propagation model is described to solve the resonant conditions of transverse modes. In addition, multilayered mirrors are taken into calculation through the scattering matrix method. The off-axis propagation of transverse fields at the interface of layers is also investigated using Bessel transform so that diffraction loss can be determined accurately.

3.4.1 Calculation of Resonant Transverse Modes in VCSELs with Tapered Waveguide Using Beam Propagation Method

In Section 3.2.3, it is shown that high side-mode suppression can be achieved in gain-guided VCSELs with a small cavity size such as the proton-implanted small-area VCSELs [5]. In fact, the modal gain difference can be further enhanced if the cylindrical cavity of VCSELs is tapered along the longitudinal direction [32]. In this section, VCSELs with tapered waveguide structure are analyzed using a two-dimensional beam propagation method. It is shown that the transverse-mode behavior of VCSELs also depends on the round-trip conditions.

A schematic diagram of VCSELs with tapered waveguides is shown in Figure 3.9a. Alternating semiconductor layers form the n- and p-type distributed Bragg reflectors (DBRs) in which the spacer and active layers are sandwiched. All the layers are tapered and buried inside a passive region, which forms the cladding region of the lasers. Figure 3.9b shows the model of tapered waveguide VCSELs used for the numerical analysis. It is assumed that the refractive index of the core region, including the DBRs, active and spacer regions, is of value n_1. The corresponding refractive index of the cladding region is n_2. The effective cavity length of the tapered waveguide is represented by L_{eff}, which is determined

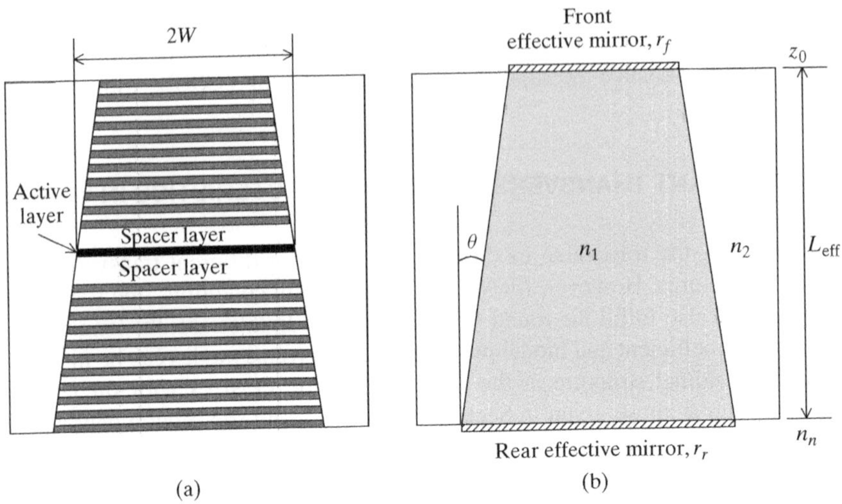

Figure 3.9 (a) Schematics diagram of tapered waveguide VCSELs and (b) its two-dimensional approximated model for calculation.

by the energy-penetrated depth of the electromagnetic wave into the mirror [33]. A planar mirror replaces the reflectivity of the multilayered mirrors with the same reflectivity.

In order to analyze the tapered waveguide geometry, the Maxwell equation (3.1) is simplified by substituting the longitudinal propagation of electromagnetic fields having a plane wave format, that is, $\mathbf{E}(r,z)\cdot\exp(\pm j\beta z)$ and $\mathbf{H}(r,z)\cdot\exp(\pm j\beta z)$ and ignoring the azimuthal dependence of the fields. Hence, the corresponding paraxial approximated two-dimensional wave equation can be written as follows [25,26,34]:

$$\pm\frac{\partial\psi^{\pm}(r,z)}{\partial z}=\frac{j}{2\beta}\left[\frac{1}{r}\frac{\partial}{\partial r}\left(r\frac{\partial}{\partial r}\right)+\Delta\varepsilon(r,z)k_0^2\right]\psi^{\pm}(r,z), \tag{3.61}$$

where $\Psi^{+}(r,z)$ and $\Psi^{-}(r,z)$ are the slowly varying envelope of the forward and reverse traveling waves, respectively, along the longitudinal direction z; $\beta(=n_{\text{eff}}k_0)$ is the longitudinal propagation coefficient; n_{eff} is the effective refractive index of the dielectric waveguide; and

$$\Delta\varepsilon=n_2\begin{cases}2\left(n_1(r,z)-n_2+\Delta n(N)+\dfrac{j(g(N)-\alpha_{\text{m}})}{k_0}\right) & \text{active layer}\\ 2\left(n_1(r,z)-n_2-\dfrac{j\alpha_{\text{m}}}{k_0}\right) & \text{elsewhere}\end{cases}. \tag{3.62}$$

In Equation (3.62), $n_1(r,z)$ is the refractive index distribution of the core region, α_m is the effective material losses (including free-carrier absorption and scattering losses) of the dielectric materials, $g(N)$ is the optical gain, and $\Delta n(N)$ is the change of refractive index with the function of carrier concentration N. In the derivation of (3.61), the Fresnel approximation (i.e., $|\partial^2\Psi/\partial z^2| \ll |2j\beta \cdot \partial\Psi/\partial z|$) has been used. Furthermore, it is assumed that the transverse modes have no azimuthal variation and Ψ depends only on the transverse position r. Hence, there is no preferred axis of symmetry in the circular cross section and the transverse field Ψ can be two orthogonally degenerated LP_{lm} modes.

Direct integration of (3.61) between two adjacent points $z = z_i$ and z_{i+1} gives the corresponding change in phase and magnitude of $\Psi^{\pm}$

$$\Psi^{\pm}(r, z_i \pm \Delta z) = \Psi^{\pm}(r, z_i)\exp(\pm[H]\Delta z), \tag{3.63}$$

where $\Delta z = z_{i+1} - z_i$ and

$$H = \frac{j}{2\beta}\left[\frac{1}{r}\frac{\partial}{\partial r}\left(r\frac{\partial}{\partial r}\right) + \Delta\varepsilon(r, z)k_0^2\right]. \tag{3.64}$$

Furthermore, the output power is related to the boundary conditions between the front and rear effective mirrors, which are given by

$$\Psi^{-}(r, z_0) = r_f\Psi^{+}(r, z_0) \quad \text{and} \quad \Psi^{+}(r, z_n) = r_r\Psi^{-}(r, z_n), \tag{3.65}$$

where z_0 and z_n are the position of the field reflectivities r_f and r_r, respectively, of the effective mirror. Hence, the resonant conditions of the transverse modes can be evaluated by solving (3.63) and (3.65) simultaneously. The numerical approximation in solving (3.63) can be found in Section 7A.4.

The calculated modal loss as a function of $2W$ is shown in Figure 3.10. The solid lines represent the case of VCSELs with uniform cylindrical waveguides; the dashed lines, VCSELs with tapered waveguide (tapered by $\theta = 34°$). It is assumed in the calculation that $n_1 = 3.2$, $n_2 = 3.425$, $\lambda_R = 0.92$ μm, $r_f = r_r = 0.996$ [32]. For VCSELs with uniform waveguide, the modal loss difference between the fundamental and first-order modes is about 160 cm^{-1}. This magnitude is significant compared with the total cavity loss of VCSELs, and single-mode emission is possible. For VCSELs with tapered waveguides, it is observed that more losses are introduced to the higher order modes. This is because tapered VCSELs have an additional beam size-limiting aperture so that the modal loss difference is improved over the uniform waveguide lasers. Thus, it can be concluded that the tapered waveguide geometry has a higher modal selection property than the uniform one and the influence of the round-trip condition should be considered in deriving transverse modes.

3.4.2 Consideration of Multilayered Mirrors Using Scattering Matrix

In the preceding analysis of VCSELs with tapered waveguide, the DBR is replaced by a uniform medium. In fact, the longitudinal propagation of transverse fields along the DBRs can also be determined by applying the scattering

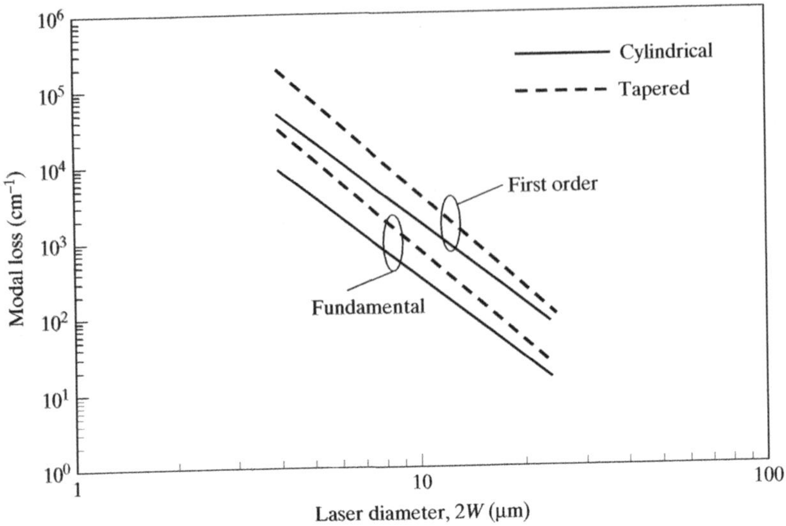

Figure 3.10 The calculated modal losses as a function of $2W$ for a tapered waveguide VCSEL (after Ref. 32).

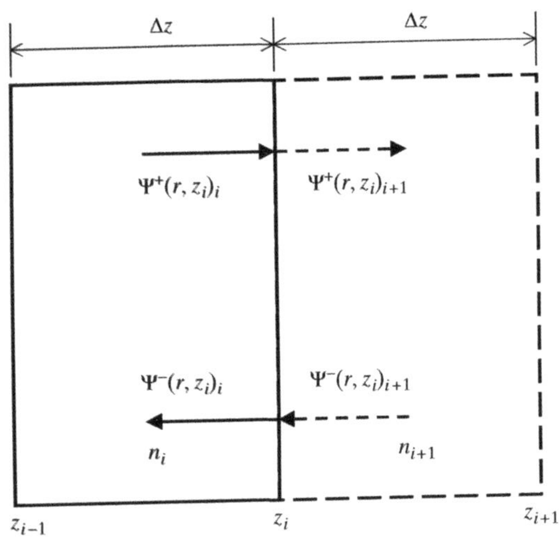

Figure 3.11 Reflection and transmission of optical fields at the interface of two adjacent layers.

matrix at the interfaces between semiconductor layers [30]. If the refractive index is assumed uniform within each layer, the scattering matrix can be applied to calculate the reflection and transmission of transverse modes at the interfaces. Figure 3.11 shows the transverse fields of TE polarized at the boundary between two adjacent layers, ith and $(i+1)$th at positions z_i and z_{i+1}, respectively. Hence, the traveling fields between two adjacent layers are given by [34]

$$\begin{bmatrix} \Psi^+(r, z_i) \\ \Psi^-(r, z_i) \end{bmatrix}_{i+1} = \frac{1}{t_i(r)} \begin{bmatrix} 1 & r_i(r) \\ r_i(r) & 1 \end{bmatrix} \begin{bmatrix} \Psi^+(r, z_i) \\ \Psi^-(r, z_i) \end{bmatrix}_i, \tag{3.66}$$

where $\Psi^{\pm}(r, z_i)_i$ and $\Psi^{\pm}(r, z_i)_{i+1}$ are the traveling waves at the interface of ith and $(i+1)$th layers, respectively; t_i and r_i represent the radical variation of transmission and reflection coefficients and can be expressed as

$$t_i(r) = \frac{2n_{i+1}(r)}{n_i(r) + n_{i+1}(r)}, \tag{3.67a}$$

$$r_i(r) = \frac{n_{i+1}(r) - n_i(r)}{n_{i+1}(r) + n_i(r)}, \tag{3.67b}$$

where n_i and n_{i+1} are the refractive indices inside the ith and $(i+1)$th layers, respectively. Hence, the propagation of transverse modes over the DBR can be calculated by a combination of the beam propagation method and the scattering matrix. If the boundary conditions at the front and rear surfaces of the DBRs are applied, the corresponding round-trip conditions of the transverse modes can be deduced.

Now, the dependence of transverse modes on the DBRs and the transverse confinement structure of VCSELs are obtained under the resonant conditions using the beam propagation method and scattering matrix. This is because the nonuniform distribution of refractive index and optical gain inside the active layer is factored in through the phase terms [i.e., $(1/r)(\partial/\partial r)[r(\partial/\partial r)]$ and $\Delta\varepsilon$]. In addition, the propagation of optical fields inside the DBRs is taken into calculation. Hence, the wavevectors along the transverse and longitudinal directions can be obtained simultaneously in the resonant conditions. However, one drawback of using scattering matrix for the boundary condition is the ignorance of off-axis propagation of traveling fields at the semiconductor interfaces.

3.4.3 Calculation of Off-Axis Propagation of Traveling Waves Using Discrete Bessel Transform

In Section 3.3, Bessel transform is used to analyze the off-axis propagation of traveling fields between the interface of the active layer and DBR. The motivation for using this approach is to avoid solving the off-axis propagation of traveling fields in the spatial domain. In fact, the traveling fields can be transformed in the Bessel domain, so the corresponding off-axis propagation of reflected fields could be determined from the spectral response function of the DBRs. Now, the method of Bessel transform is again applied to modify the scattering matrix given in Section 3.4.2. If $\hat{\Psi}_f(s)$ is defined as the Bessel transform of the incident electric field $\Psi_f(r)$, which is given by [30]

$$\hat{\Psi}_f(s) = \int_0^\infty \Psi_f(r) \cdot J_0(sr) \cdot r\,dr, \tag{3.68}$$

where s is the transverse wavevector and J_0 is the Bessel function of the first kind of zeroth order and Ψ_f is assumed to be independent of z. The reflected and transmitted fields in the spatial domain, $\Psi_r(r)$ and $\Psi_t(r)$, respectively, at an interface can be expressed in terms of $\hat{\Psi}_f$, which are given by [28]

$$\Psi_r(r) = \int_0^\infty r_{2p}(s) \cdot \hat{\Psi}_f(s) \cdot J_0(sr) \cdot s\,ds, \tag{3.69a}$$

$$\Psi_t(r) = \int_0^\infty t_{2p}(s) \cdot \hat{\Psi}_f(s) \cdot J_0(sr) \cdot s\,ds, \tag{3.69b}$$

where $r_{2p}(s)$ and $t_{2p}(s)$ are the plane-wave reflection and transmission coefficients in the Bessel domain, respectively, at the ith interface. In the definition of forward and reverse Bessel transform above, it is assumed only the fundamental LP mode is supported by the laser cavity. However, higher-order LP modes (say, l mode) can also be accounted for by replacing J_0 to J_l. The off-axis propagation of traveling fields at the ith interface is implicitly introduced into (3.69) through the method of Bessel transform.

If the Bessel operations given above are written in terms of the Bessel transform operator B_0, which is assumed to be symmetric and unitary, (3.69) can be transformed into

$$\Psi_r(r) = R'_{2p}\Psi_f(r) \qquad \text{and} \qquad \Psi_t(r) = T'_{2p}\Psi_f(r), \tag{3.70}$$

where $R'_{2p} = B_0^{-1}R_{2p}B_0$ and $T'_{2p} = B_0^{-1}T_{2p}B_0$. Operators R_{2p} and T_{2p} are obtained from the linearization of r_{2p} and t_{2p}, respectively. Hence, operators R'_{2p} and T'_{2p} simply represent R_{2p} and T_{2p}, respectively, on a different basis defined by the operator B_0 and can thus be seen as generalized field reflection and transmission coefficients. Now, the remaining task is to express the scattering matrix given Section 3.4.2 in terms of R'_{2p} and T'_{2p}. Please note that if the off-diagonal elements of R'_{2p} and T'_{2p} are zero, there will be no off-axis propagation of fields at the interfaces of the multilayered structure.

In order to apply Bessel transform into the beam propagation method, (3.69) has to be discretized and written in a matrix formation as shown below [28]:

$$\underline{\Psi}_r = [B_0]^{-1} \cdot [R_{2p}] \cdot [B_0] \cdot \underline{\Psi}_f, \tag{3.71a}$$

$$\underline{\Psi}_t = [B_0]^{-1} \cdot [T_{2p}] \cdot [B_0] \cdot \underline{\Psi}_f, \tag{3.71b}$$

where $\underline{\Psi}_r$, $\underline{\Psi}_t$, and $\underline{\Psi}_f$ are the corresponding traveling fields expressed in the format of a $(2n+1)\times 1$ column vector. $[B_0]$ and $[B_0]^{-1}$ are the $(2n+1)\times(2n+1)$ matrix of the Bessel function operators. The elements of $[B_0]$ and $[B_0]^{-1}$, b_{pq} and $\tilde{b}_{pq}$, can be written as

$$b_{p,q} = \frac{2}{W^2}\frac{J_0(\zeta_p\zeta_q/\zeta_{2n+1})}{J_1^2(\zeta_q)}, \tag{3.72a}$$

$$\tilde{b}_{p,q} = \frac{2}{S^2}\frac{J_0(\zeta_p\zeta_q/\zeta_{2n+1})}{J_1^2(\zeta_q)}, \tag{3.72b}$$

where p and $q(=1,2,\ldots,2n+1)$ are integers, $2n+1$ is the total number of grid points along the transverse direction, W is the radius of the laser cavity, ζ_p is the pth zero of J_0 [i.e., $J_0(\zeta_p)=0$], and $S=\zeta_{2n+1}/W$. For more information on discrete Bessel transform, see Section 3A.5. Hence, $[R'_{2p}]$ and $[T'_{2p}]$ can be determined with the knowledge of the reflection and transmission matrices $[R_{2p}]$ and $[T_{2p}]$ at the interface, which are defined as

$$[R_{2p}] = \begin{bmatrix} r_{2p}(s_1) & & 0 \\ & \ddots & \\ 0 & & r_{2p}(s_{2n+1}) \end{bmatrix}, \tag{3.73a}$$

$$[T_{2p}] = \begin{bmatrix} t_{2p}(s_1) & & 0 \\ & \ddots & \\ 0 & & t_{21p}(s_{2n+1}) \end{bmatrix}, \tag{3.73b}$$

where $s_k - \zeta_k/W$ for $k=1,2,\ldots,2n+1$. If the interface is assumed to be TE polarized, the elements of the reflection and transmission matrices of the incident transverse fields, r_{2p} and t_{2p}, respectively, can be expressed as

$$r_{2p}(s_k) = \frac{\beta_{i+1}(s_k)-\beta_i(s_k)}{\beta_{i+1}(s_k)+\beta_{i+1}(s_k)}, \tag{3.74a}$$

$$t_{2p}(s_k) = \frac{2\beta_{i+1}(s_k)}{\beta_{i+1}(s_k)+\beta_{i+1}(s_k)}, \tag{3.74b}$$

where $\beta_i(s_k)=\sqrt{n_i^2k_0^2-s_k^2}$ and $\beta_{i+1}(s_k)=\sqrt{n_{i+1}^2k_0^2-s_k^2}$. Obviously, this approach can include the effects of off-axis light propagation at the interface.

If normal incident is assumed in the calculation (i.e., $s_k=0$), $[R'_{2p}]$ and $[T'_{2p}]$ can be simplified to

$$[R'_{2p}] = \frac{n_{i+1}-n_i}{n_{i+1}+n_i}[I], \tag{3.75a}$$

$$[T'_{2p}] = \frac{2n_{i+1}}{n_{i+1}+n_i}[I], \tag{3.75b}$$

where the identity matrix $[I]$ has a dimension of $(2n + 1) \times (2n + 1)$. The beam propagation method with Bessel transform described in this section is a general technique for analyzing the transverse mode characteristics of VCSELs with different confinement structures such as index- or gain-guided structures. For VCSELs with strong index-guided structure, the influence of diffraction loss is implicitly considered.

The calculation given in Section 3.4.1 is repeated with consideration of off-axis propagation of fields inside the DBRs. In the calculation, the reflectivity of the DBRs is chosen to be 0.996. The length of the DBRs is converted from L_{eff} as used in Section 3.4.1. In addition, the corresponding effective refractive indices of the core and cladding regions are chosen to be n_1 and n_2, respectively. It can be shown that the modal loss behavior of the transverse modes obtained using the transfer matrix method and Bessel transform is similar to that given in Figure 3.10. However, the magnitude of modal loss for the tapered waveguide VCSELs increases especially for the small W ($<1\ \mu$m) of the active layer. This is because in Section 3.4.1, the diffraction loss inside the multilayered mirrors has been ignored in the calculation.

3.5 VECTORIAL RESONANT MODES OF VCSELs

In Section 3.2, linearly polarized modes are derived to approximate the transverse-mode behavior of VCSELs. However, it is based on the assumption that the transverse modes are weakly guided inside the cylindrical dielectric waveguide and that the propagation fields are of plane-wave approximation. Hence, the mode-matching technique and beam propagation method given in Sections 3.3 and 3.4 are used to study the off-axis propagation of fields inside VCSELs with an index-guided structure. Similar numerical methods based on scalar wave equations are also proposed for the same reason [35–39]. However, with these numerical approaches, it is difficult to obtain accurate resonant wavelength and threshold gain of the transverse modes in VCSELs with a small cavity and strong index contrast (i.e., oxide aperture devices). This is because the mode coupling (i.e., coupling of TE and TM modes) and radiation fields have been ignored in the analysis. Therefore, a vectorial approach should be adopted to solve the off-axis propagation of transverse fields and mode coupling simultaneously [40–43].

In the following sections, a general mode expansion technique based on the vectorial solution of Maxwell's equations is presented [40,43]. This approach gives exact results in the case of devices with high transverse index contrast such as oxide aperture lasers. Furthermore, a common-mode approximation is proposed to simplify the vectorial transform matrix for the calculation of resonant wavelength and threshold gain of the transverse modes.

3.5.1 Formulation of Vectorial Modes inside Multilayered Mirrors Using General Mode Expansion Technique

In order to use the general mode expansion technique to calculate the reflection and transmission of multilayered mirrors, the basis of the expanded fields should be selected. It is possible to select the transverse modes derived from Section 3.2.1 as the corresponding basis, but a continuum of evanescent modes must be considered [21,22]. Analysis can be simplified if the structure under study is enclosed inside a metallic cylinder so that the corresponding discrete set of eigenmodes (i.e., due to cutoff condition) can be used to describe the different field components within the cylinder for any arbitrary dielectric profile and any excitation [40]. This can be done by introducing a perfectly reflecting metallic cylindrical wall at a certain radius W_f around the entire cylindrical waveguide. It must be noted that the selected W_f has to be large enough to avoid having any influence on the phenomena under study. In fact, this approach is similar to the effective width used in the mode-matching technique as described in Section 3.3.1.

Now, the general solutions for $E_z(r)$ and $H_z(r)$ can be expressed as [40]

$$\begin{bmatrix} E_z(r) \\ H_z(r) \end{bmatrix} = \begin{bmatrix} A \\ B \end{bmatrix} \cdot J_\nu\left(\frac{ur}{W_f}\right), \tag{3.76}$$

where $u^2 = W_f^2(k_0^2n^2 - \beta^2)$, and A and B are constants to be determined. The other components of the vector fields can be obtained from (3.5) and (3.6) using $E_z(r)$ and $H_z(r)$. The wall introduces boundary conditions that $E_z(r)$ and $E_\theta(r)$ as well as the derivative of $H_z(r)$ and $H_\theta(r)$ should vanish at $r = W_f$. Hence, the corresponding eigenvalue equation to determine A, B, and β is given by

$$A \cdot B \cdot J_\nu(u) \cdot J_\nu'(u) = 0, \tag{3.77}$$

and the coupling between the electric and magnetic fields is given by

$$\frac{A}{B} = \frac{\omega\mu_0}{\nu u \beta}\frac{J_\nu'(u)}{J_\nu(u)}. \tag{3.78}$$

Because a Bessel function and its derivative never vanish at the same point, the eigenvalue equation can be expressed in two possible formats.

If $E_z(r)$ vanishes at the perfectly conducting wall, the eigenvalue equation reduces to

$$J_\nu(u) = 0. \tag{3.79}$$

For $E_\theta(r)$ to vanish at the perfectly conducting wall, $H_z(r)$ has to be identically zero (i.e., $B = 0$) but $J_\nu'(u) \neq 0$, which implies $B = 0$, and hence a TM mode is obtained. Therefore, the propagation coefficient β, obtained from (3.79), is for TM mode, and the corresponding fields are also of TM polarization. These fields can be denoted as $\mathbf{E}_z^{\mathrm{TM}}$, $\mathbf{E}_t^{\mathrm{TM}}$, and $\mathbf{H}_t^{\mathrm{TM}}$, where $\mathbf{E}_z^{\mathrm{TM}} + \mathbf{E}_t^{\mathrm{TM}} \equiv (E_z, E_r, E_\theta)^{\mathrm{TM}}$ and $\mathbf{H}_t^{\mathrm{TM}} \equiv (0, H_r, H_\theta)^{\mathrm{TM}}$.

On the other hand, the other possible eigenvalue equation is

$$J_\nu'(u) = 0. \tag{3.80}$$

It is possible for E_z and E_θ to vanish at $r = W_f$ only if $E_z = 0$ but $J_\nu(u) \neq 0$, which implies $A = 0$ and, hence a TE mode is obtained. Therefore, the propagation coefficient β, obtained from (3.80), is for TE mode, and the corresponding fields are also of TE polarization. These fields can be denoted as $\mathbf{H}_z^{\mathrm{TE}}$, $\mathbf{H}_t^{\mathrm{TE}}$, and $\mathbf{E}_t^{\mathrm{TE}}$, where $\mathbf{E}_t^{\mathrm{TE}} \equiv (0, E_r, E_\theta)^{\mathrm{TE}}$ and $\mathbf{H}_z^{\mathrm{TE}} + \mathbf{H}_t^{\mathrm{TE}} \equiv (H_z, H_r, H_\theta)^{\mathrm{TE}}$. The azimuthal dependence of both the TE and TM modes can be obtained by substituting (3.76) and (3.6) into (3.5), and the corresponding field components are shown in Table 3.2. In general, for $\nu \geq 1$, the general solution given in the table cannot be separated into two independent groups of TE and TM modes. Both have to be taken into account to satisfy the boundary conditions. The resulting hybrid mode is referred to as *HE* or *EH* mode as discussed in Section 3.2.

Consider a nonuniform layer with the refractive index dependent on the radial direction r, but independent of the azimuthal and longitudinal directions, θ and z. The fields confined inside that layer can be expressed as a series of suitable "basis modes." These modes can be the set of TE and TM modes deduced from a uniform layer with refractive index n. If the first μ modes are selected from the discrete set of eigenmodes, the "layer mode" propagating in such a layer can be written as

$$\mathbf{E}(r, \theta, z)_t = \exp(-j\beta z) \sum_{p=1}^{\mu} a_{\mathrm{TM},p} \mathbf{E}_p^{\mathrm{TM}}(r, \theta)_t + a_{\mathrm{TE},p} \mathbf{E}_p^{\mathrm{TE}}(r, \theta)_t, \tag{3.81a}$$

$$\mathbf{H}(r, \theta, z)_t = \exp(-j\beta z) \sum_{p=1}^{\mu} b_{\mathrm{TM},p} \mathbf{H}_p^{\mathrm{TM}}(r, \theta)_t + b_{\mathrm{TE},p} \mathbf{H}_p^{\mathrm{TE}}(r, \theta)_t, \tag{3.81b}$$

TABLE 3.2 Azimuthal Dependence of Transverse Electric and Magnetic Modes[a]

	TE Modes	TM Modes
E_θ	$-j\omega\mu_0 BJ_\nu'(ur/W_f)\sin(\nu\theta)W_f^2/u^2$	$-j\beta\nu AJ_\nu(ur/W_f)\sin(\nu\theta)W_f^2/ru^2$
E_r	$j\nu\omega\mu_0 BJ_\nu(ur/W_f)\cos(\nu\theta)W_f^2/ru^2$	$j\beta AJ_\nu'(ur/W_f)\cos(\nu\theta)W_f^2/u^2$
E_z	0	$AJ_\nu(ur/W_f)\cos(\nu\theta)$
H_θ	$j\nu\beta\mu_0 BJ_\nu(ur/W_f)\cos(\nu\theta)W_f^2/ru^2$	$jn^2\varepsilon_0\omega AJ_\nu'(ur/W_f)\cos(\nu\theta)W_f^2/u^2$
H_r	$j\beta BJ_\nu'(ur/W_f)\sin(\nu\theta)W_f^2/ru^2$	$j\nu n^2\varepsilon_0\omega AJ_\nu'(ur/W_f)\sin(\nu\theta)W_f^2/ru^2$
H_z	$BJ_\nu(ur/W_f)\sin(\nu\theta)$	0

[a] It must be noted that another set of solutions can be obtained substitution $\cos(\nu\theta) \to \sin(\nu\theta)$ and $\sin(\nu\theta) \to -\cos(\nu\theta)$, so the most general form of solution consists of a superposition of these two orthogonally polarized modes.

$$\mathbf{E}(r,\theta,z)_z = \exp(-j\beta z)\sum_{p=1}^{\mu} c_{\mathrm{TM},p}\mathbf{E}_p^{\mathrm{TM}}(r,\theta)_z, \tag{3.81c}$$

$$\mathbf{H}(r,\theta,z)_z = \exp(-j\beta z)\sum_{p=1}^{\mu} c_{\mathrm{TE},p}\mathbf{H}_p^{\mathrm{TE}}(r,\theta)_z, \tag{3.81d}$$

where β is the corresponding propagation coefficient inside the nonuniform layer; $a_{\mathrm{TM},p}$, $a_{\mathrm{TE},p}$, $b_{\mathrm{TM},p}$, $b_{\mathrm{TE},p}$, $c_{\mathrm{TM},p}$, and $c_{\mathrm{TE},p}$ are relative amplitudes of the basis modes. The layer mode given in (3.81) should also be the solution to Maxwell's equations in the nonuniform layer with r-dependent refractive index $n(r)$. Now, the propagation coefficient β and the relative amplitude of the basis modes, $a_{\mathrm{TM},p}$, $a_{\mathrm{TE},p}$, $b_{\mathrm{TM},p}$, $b_{\mathrm{TE},p}$, $c_{\mathrm{TM},p}$, and $c_{\mathrm{TE},p}$, can be deduced by substituting (3.81) into the Maxwell's curl equations (3.1). In addition, the orthogonality relation for the basis modes inside uniform layers has been used in the derivation

$$\int_S \mathbf{E}_p(r,\theta)_{\mathrm{t}} \times \mathbf{H}_q(r,\theta)_{\mathrm{t}} \cdot d\mathbf{S} = \delta_{pq}, \tag{3.82}$$

where p and q are the mode number. In (3.82), the surface integral runs over a complete cross section **S**, bounded by the perfect conductor at $r = W_f$ and δ_{pq} is the Krönecker delta. Hence, a $6 \times \mu$ matrix eigenvalue equation is deduced, and the detailed derivation can be found in a paper by Omar and Schunemann [44]. Using this matrix eigenvalue equation, a set of μ'th-layer modes supported by the corresponding nonuniform layer can be evaluated.

3.5.2 Formulation of Cavity Problem Using the Vectorial Transform Matrix Approach

If the laser cavity shown in Figure 3.12 is considered in the analysis, any field inside a ith layer of the mirror can be expressed by superposition of all μ'th-layer modes propagating in the positive and negative z directions. The total transverse fields are given by

$$\mathbf{E}_i^{\mathrm{tot}}(r,\theta,z)_{\mathrm{t}} = \sum_{p=1}^{\mu'} (A_{i,p}^{+}\exp(-j\beta_{i,p}z) + A_{i,p}^{-}\exp(j\beta_{i,p}z))\mathbf{E}_{i,p}(r,\theta)_{\mathrm{t}}, \tag{3.83a}$$

$$\mathbf{H}_i^{\mathrm{tot}}(r,\theta,z)_{\mathrm{t}} = \sum_{p=1}^{\mu'} (A_{i,p}^{+}\exp(-j\beta_{i,p}z) - A_{i,p}^{-}\exp(j\beta_{i,p}z))\mathbf{H}_{i,p}(r,\theta)_{\mathrm{t}}, \tag{3.83b}$$

in which $\mathbf{H}_{i,p}(r,\theta)_{\mathrm{t}}$ and $\mathbf{E}_{i,p}(r,\theta)_{\mathrm{t}}$ are the profiles of pth layer modes in the ith layer and $\beta_{i,p}$ is the corresponding propagation coefficient, which are obtained from the eigenvalue equations of (3.81). $A_{i,p}^{+}$ and $A_{i,p}^{-}$ are the amplitude of layer mode propagating in the positive and negative directions of z, respectively. It

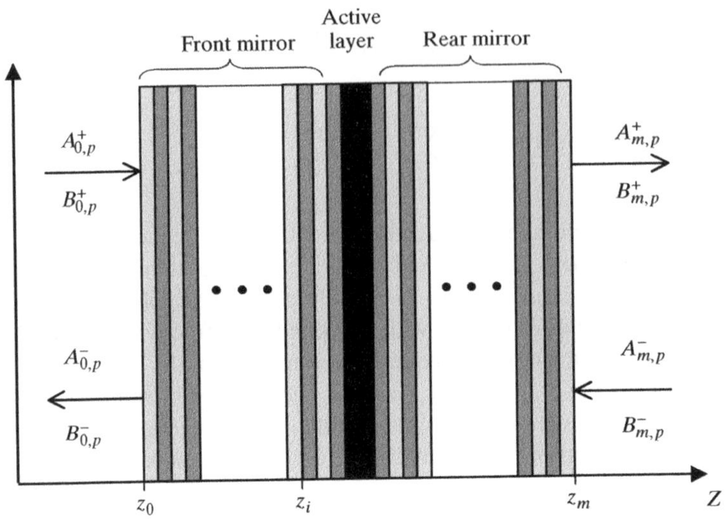

Figure 3.12 Schematic diagram of an airpost VCSEL.

must be noted that in realistic mirrors, most layers are uniform without any transverse index steps. In this case, the discrete set of basis modes given in Table 3.2 can be used to represent the total transverse fields inside the layer so that the time consuming process of solving the matrix eigenvalue equation of (3.81) can be avoided.

In order to calculate the field distribution inside the laser cavity, all the transverse field components at the boundary between any two adjacent layers have to be matched so that mode coupling and radiation losses can be accounted. For example, the reflection and transmission at an interface between two layers, say, the ith and $(i+1)$th layers, the boundary conditions are applied to the tangential field components, which can be written as

$$\mathbf{E}_{i+1}^{\text{tot}}(r,\theta,z_i)_{\text{t}} = \mathbf{E}_{i}^{\text{tot}}(r,\theta,z_i)_{\text{t}}, \tag{3.84a}$$

$$\mathbf{H}_{i+1}^{\text{tot}}(r,\theta,z_i)_{\text{t}} = \mathbf{H}_{i}^{\text{tot}}(r,\theta,z_i)_{\text{t}}, \tag{3.84b}$$

where the boundary is positioned at $z = z_i$. Again, using the orthogonality property of the layer modes, the relationship between the coefficients $A_{i,p}^{\pm}$ and $A_{i+1,p}^{\pm}$ can be obtained and is given by

$$A_{i+1,p}^{+}\exp(-j\beta_{i+1,p}z_i) + A_{i+1,p}^{-}\exp(j\beta_{i+1,p}z_i) = \sum_{q}^{\mu'}(A_{i,p}^{+}\exp(-j\beta_{i,p}z_i) + A_{i,p}^{-}\exp(j\beta_{i,p}z_i))\mathbf{K}_{p,q}, \tag{3.85a}$$

$$A_{i+1,p}^{+}\exp(-j\beta_{i+1,p}z_i) - A_{i+1,p}^{-}\exp(j\beta_{i+1,p}z_i) = \sum_{q}^{\mu'}(A_{i,p}^{+}\exp(-j\beta_{i,p}z_i) - A_{i,p}^{-}\exp(j\beta_{i,p}z_i))\mathbf{L}_{p,q}, \tag{3.85b}$$

where the following overlap integrals have been used:

$$\mathbf{K}_{p,q} = \left(\int_{\mathbf{S}} \mathbf{E}_{i+1,p}(r,\theta)_{\mathrm{t}} \times \mathbf{H}_{i+1,p}(r,\theta)_{\mathrm{t}} \cdot d\mathbf{S}\right)^{-1} \cdot \left(\int_{\mathbf{S}} \mathbf{E}_{i,q}(r,\theta)_{\mathrm{t}} \times \mathbf{H}_{i+1,p}(r,\theta)_{\mathrm{t}} \cdot d\mathbf{S}\right), \tag{3.86a}$$

$$\mathbf{L}_{p,q} = \left(\int_{\mathbf{S}} \mathbf{E}_{i+1,p}(r,\theta)_{\mathrm{t}} \times \mathbf{H}_{i+1,p}(r,\theta)_{\mathrm{t}} \cdot d\mathbf{S}\right)^{-1} \cdot \left(\int_{\mathbf{S}} \mathbf{E}_{i+1,p}(r,\theta)_{\mathrm{t}} \times \mathbf{H}_{i,q}(r,\theta)_{\mathrm{t}} \cdot d\mathbf{S}\right). \tag{3.86b}$$

Hence, addition and subtraction of (3.85a) and (3.85b) gives

$$A_{i+1,p}^{+} = \frac{1}{2}\sum_{q}^{\mu'}\{(\mathbf{K}_{p,q} + \mathbf{L}_{p,q})\exp(-j\beta_{i,p}z_i + j\beta_{i+1,p}z_i)A_{i,p}^{+} + (\mathbf{K}_{p,q} - \mathbf{L}_{p,q})\exp(j\beta_{i,p}z_i + j\beta_{i+1,p}z_i)A_{i,p}^{-}\}, \tag{3.87a}$$

$$A_{i+1,p}^{-} = \frac{1}{2}\sum_{j}^{\mu'}\{(\mathbf{K}_{p,q} - \mathbf{L}_{p,q})\exp(-j\beta_{i,p}z_i - j\beta_{i+1,p}z_i)A_{i,p}^{+} + (\mathbf{K}_{p,q} + \mathbf{L}_{p,q})\exp(j\beta_{i,p}z_i - j\beta_{i+1,p}z_k)A_{i,p}^{-}\}. \tag{3.87b}$$

Writing (3.87) in a matrix form, one obtains

$$\begin{bmatrix} A_{i+1,p}^{-} \\ A_{i+1,p}^{+} \end{bmatrix} = \overline{\mathbf{M}}_{i,p}\begin{bmatrix} A_{i,p}^{-} \\ A_{i,p}^{+} \end{bmatrix}, \tag{3.88}$$

where the elements of $\overline{\mathbf{M}}_{k,p}$ can be obtained from (3.87). Finally, multiplication of the transform matrices $\overline{\mathbf{M}}_{k,i}$ for all interfaces leads to the total transform matrix of the VCSEL structure

$$\overline{\mathbf{M}}_p = \prod_{i=0}^{m-1}\overline{\mathbf{M}}_{i,p} \equiv \begin{bmatrix} [\overline{\mathbf{M}}_p]_{11} & [\overline{\mathbf{M}}_p]_{12} \\ [\overline{\mathbf{M}}_p]_{21} & [\overline{\mathbf{M}}_p]_{22} \end{bmatrix}, \tag{3.89}$$

and using the boundary conditions at $z = z_0$ and $z = z_m$, $A^-_{i=m,p} = A^+_{i=0,p} = 0$, the eigenvalue equation for the amplitude of propagation fields can be expressed as

$$\overline{\mathbf{MM}}_p \cdot \begin{bmatrix} A^-_{0,p} \\ A^+_{m,p} \end{bmatrix} = \begin{bmatrix} 0 \\ 0 \end{bmatrix}, \tag{3.90}$$

where

$$\overline{\mathbf{MM}}_p = \begin{bmatrix} -[\overline{\mathbf{M}}_p]_{21} \cdot [\overline{\mathbf{M}}_p]_{22}^{-1} & [\overline{\mathbf{M}}_p]_{22}^{-1} \\ [\overline{\mathbf{M}}_p]_{11} & 0 \end{bmatrix} \begin{bmatrix} I & 0 \\ -[\overline{\mathbf{M}}_p]_{12} & I \end{bmatrix}^{-1}. \tag{3.91}$$

Therefore, the resonant condition for the laser cavity has the form

$$\det\{\overline{\mathbf{MM}}_p\} = 0. \tag{3.92}$$

The real and imaginary parts of (3.92) give the threshold gain and resonant wavelength, respectively, for the device. In addition, the corresponding eigenmode profile is given by the eigenvectors, $A^-_{0,p}$ and $A^+_{m,p}$.

Consider a cylindrical airpost VCSEL guided by a strong index structure and designed to be at resonance at $\lambda = 0.789$ μm, where the longitudinal distribution of semiconductor layers is similar to that shown in Figure 3.12. All layers are assumed to be a quarter of a wavelength thick. The refractive index of the active layer is 3.67, the refractive indices of DBR pairs are 2.99 and 3.44. The number of pairs for the front and rear multilayered mirrors is 24 and 28, respectively. Figure 3.13a shows the resonant wavelength λ_R of the airpost VCSELs as a function of cavity radius W. It is observed that the resonant wavelength λ_R is blueshifted relatively to the planar cavity resonant wavelength λ. In addition, both the blueshift and the spectral distance between modes increases with decreasing W (increasing mode number). Figure 3.13b shows the normalized cavity loss as a function of cavity radius W. It is shown that for large-cavity and low-mode orders, diffraction loss is minimized because of the large overlap between the layer modes. However, as the cavity radius decreases and/or the mode order increases, the mismatch between the corresponding layer modes increases so that the diffraction loss in the mirrors grows larger. Hence, diffraction loss increases with the decrease of cavity radius and the increase of mode orders. A similar conclusion is obtained from Section 3.3.

3.5.3 Simplification of the Vectorial Transform Matrix by Common-Mode Approximation

If the mismatch of the layer modes between adjacent layers is small, mode coupling and radiation fields can be omitted in the calculation of vectorial transform matrix. However, the ignorance of mode coupling and radiation fields in a new solution results in inconsistency between the new solution and (3.88). This is because the hybrid and radiation modes no longer exist in the new solution and

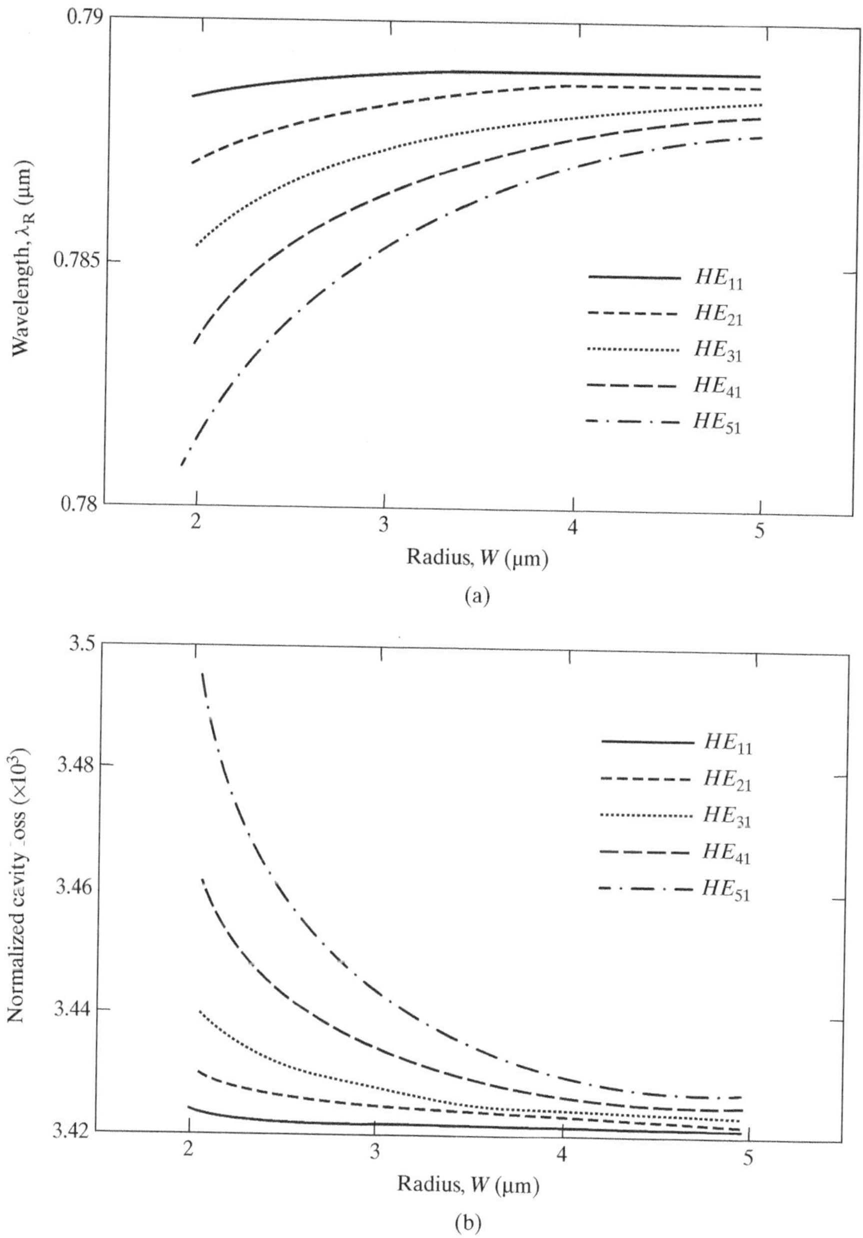

Figure 3.13 (a) Resonant wavelength λ_R and (b) normalized cavity loss αL for an airpost VCSEL as a function of cavity radius W; the results are shown for HE_{11}, HE_{21}, HE_{31}, HE_{41}, and HE_{51} hybrid eigenmodes. (After Ref. 22).

the longitudinal propagation of TE and that of TM modes have to be treated separately throughout the entire structure. In order to construct the new solutions, it is required that the amplitudes of $\mathbf{E}_T^{\mathrm{TE}}$ and $\mathbf{E}_T^{\mathrm{TM}}$ ($\mathbf{H}_T^{\mathrm{TE}}$ and $\mathbf{H}_T^{\mathrm{TM}}$) fields are proportional to $1/\beta_i$ and β_i (1 and n_i^2), respectively, at the interfaces. Hence, the continuity of polarized fields between any two adjacent layers can be maintained. It must be noted that separate treatments for TE and TM modes have also been applied to the mode-matching technique and the scattering matrix method as given in Sections 3.3 and 3.4.

Consider a cylindrical airpost VCSEL with longitudinal distribution of semiconductor layers similar to that of Figure 3.12. If there is no radial variation of refractive index for all the layers, the TE and TM modes can be expressed by the basis functions given in Table 3.2 with A and B as the corresponding amplitude of the propagation modes. As (3.77) cannot be used to determine the propagation coefficient of the new solutions, the longitudinal propagation constant $\beta_{i,p}$ of pth mode in the ith layer is approximated by

$$\beta_{i,p} = \sqrt{k_0^2 n_i^2 - k_{t,p}^2}, \tag{3.93}$$

where $i = 0, 1, \ldots, m$ and $k_{t,p}$ is the transverse propagation constant of the pth mode at the active layer. Equation (3.93) is deduced under two assumptions: (1) the relative mismatch of dielectric constants between adjacent layers is much smaller than the mismatch between layers and the surroundings, and (2) the transverse propagation constant $k_{t,p}$ of the corresponding local modes of the layers are approximately the same as those in the active layer. Hence, this approach (i.e., common value of $k_{t,p}$ over the entire semiconductor layers) is also referred to as *common-mode approximation* [21].

The transverse field components in the ith layer can be rewritten in matrix form

$$\begin{bmatrix} H_r & H_\theta & E_r & E_\theta \end{bmatrix}_p^T = \overline{\overline{M}}_{i,p} \cdot \begin{bmatrix} A_i^- & A_i^+ & B_i^- & B_i^+ \end{bmatrix}_p^T, \tag{3.94}$$

where the superscript T stands for the transpose of matrix and $\overline{\overline{M}}_{i,p}$ is a 4×4 matrix with its elements specified in Table 3.2. Matching these field components with the field in the $(i+1)$th layer leading to the transform matrix at the ith boundary gives

$$\begin{bmatrix} A_{i+1}^- \\ A_{i+1}^+ \\ B_{i+1}^- \\ B_{i+1}^+ \end{bmatrix}_p = \begin{bmatrix} \mathbf{M}_i^{\mathrm{TE}} & \overline{\mathbf{O}} \\ \overline{\mathbf{O}} & \mathbf{M}_i^{\mathrm{TM}} \end{bmatrix}_p \begin{bmatrix} A_i^- \\ A_i^+ \\ B_i^- \\ B_i^+ \end{bmatrix}_p, \tag{3.95}$$

where $\overline{\mathbf{O}}$ is a 2×2 matrix, with all elements zero. The matrices $\mathbf{M}_i^{\mathrm{TE}}$ and $\mathbf{M}_i^{\mathrm{TM}}$ are given by [21]

$$\mathbf{M}_i^{\text{TE}} = \frac{1}{2}\begin{bmatrix} \dfrac{1+1/\beta_i^{\text{R}}}{\mathbf{c}_i^-} & \dfrac{1-1/\beta_i^{\text{R}}}{\mathbf{c}_i^+} \\ \dfrac{1-1/\beta_i^{\text{R}}}{\mathbf{c}_i^+} & \dfrac{1+1/\beta_i^{\text{R}}}{\mathbf{c}_i^-} \end{bmatrix}_p, \tag{3.96a}$$

$$\mathbf{M}_i^{\text{TM}} = \frac{1}{2}\begin{bmatrix} \dfrac{(\varepsilon_i^{\text{R}}+\beta_i^{\text{R}})}{\mathbf{c}_i^-} & \dfrac{(\varepsilon_i^{\text{R}}-\beta_i^{\text{R}})}{\mathbf{c}_i^+} \\ \dfrac{(\varepsilon_i^{\text{R}}-\beta_i^{\text{R}})}{\mathbf{c}_i^+} & \dfrac{(\varepsilon_i^{\text{R}}+\beta_i^{\text{R}})}{\mathbf{c}_i^-} \end{bmatrix}_p, \tag{3.96b}$$

where $\beta_i^{\text{R}} = \beta_{i,p}/\beta_{i+1,p}$, $\varepsilon_i^{\text{R}} = (n_{i,p}/n_{i+1,p})^2$, and $\mathbf{c}_i^{\pm} = \exp[j(\beta_{i+1,p} \pm \beta_{i,p})z_i]$. Therefore, the transmitted amplitude on the right side of the structure are related to the amplitude, incident from the left side, as

$$\begin{bmatrix} A_m^- \\ A_m^+ \end{bmatrix}_p = \mathbf{M}^{\text{TE}} \begin{bmatrix} A_0^- \\ A_0^+ \end{bmatrix}_p, \tag{3.97a}$$

$$\begin{bmatrix} B_m^- \\ B_m^+ \end{bmatrix}_p = \mathbf{M}^{\text{TM}} \begin{bmatrix} B_0^- \\ B_0^+ \end{bmatrix}_p, \tag{3.97b}$$

where

$$\mathbf{M}^{\text{TE}} = \prod_{i=0}^{m-1} \mathbf{M}_i^{\text{TE}} \quad \text{and} \quad \mathbf{M}^{\text{TM}} = \prod_{i=0}^{m-1} \mathbf{M}_i^{\text{TM}}. \tag{3.98}$$

where $\mathbf{M}^{\text{TE}}$ and $\mathbf{M}^{\text{TM}}$ are the transform matrices of the VCSEL for TE and TM polarization, respectively. It should be noted that there is no coupling between TE and TM modes as hybrid modes are ignored in the calculation and the longitudinal modes are determined by their polarization components. Again, set the determinants of $\mathbf{M}^{\text{TE}}$ and $\mathbf{M}^{\text{TM}}$ to zero, and the threshold gain and resonant wavelength of the TE and TM modes can be obtained.

Figure 3.14 compares the resonant wavelength λ_{R} obtained from general mode expansion technique and common-mode approximation. It is observed that both methods provide the same predictions for the fundamental HE_{11} mode. For higher-order modes, the agreement between the two methods remains satisfactory provided that the cavity radius is greater than 2 μm. However, for VCSELs with small cavity sizes ($W \sim 1$ μm), the theory predicts that TE_{51} and TM_{51} modes have different resonant wavelengths, indicating the breakdown of common-mode approximation. Hence, the common-mode approximation may be used for the calculation of lower-order hybrid modes (i.e., HE_{11}, HE_{21}), even for relatively small cavities (i.e., $W \sim 2\lambda_{\text{R}}$). For higher-order modes and within the limit of $W \sim \lambda_{\text{R}}$, the diffraction losses are dominant in the cavity so that the common-mode approximation breaks down. In fact, this limitation is also applied to the mode-matching technique and beam propagation method as discussed in Sections 3.3 and 3.4.

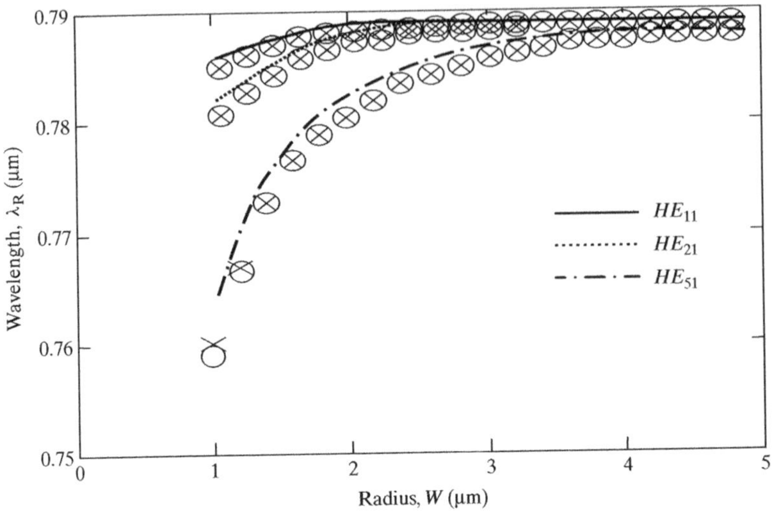

Figure 3.14 Resonant wavelength λ_R versus cavity radius W as calculated from general mode expansion technique (lines) and common-mode approximation (crosses and circles correspond to the TE and TM components of the hybrid modes, respectively) (after Ref. 21).

3.6 SUMMARY

In this chapter, the theory of cylindrical dielectric waveguide is applied to analyze the transverse-mode characteristics of VCSELs. A full set of transverse modes, including the TE, TM, and their hybrid modes, are deduced. It must be noted that because of the cylindrical symmetric of waveguide geometry, TE and TM modes can be separated only for the lowest azimuthal order (i.e., $\nu = 0$). For higher azimuthal orders, the TE and TM modes are coupled together, forming the hybrid modes usually denoted as *HE* or *EH* modes depending on the relative magnitude of mode components E_z and H_z. Under weak guidance, it can be shown that the full set of transverse modes degenerate, and a new set of cavity modes, referred to as (LP) modes, linearly polarized is formed. The main advantage of LP modes is that only one transverse electric component and one transverse magnetic field component are significant so that the corresponding expression of field profile can be simplified enormously. Hence, the LP modes are commonly used in studying the transverse-mode characteristics of VCSELs.

Diffraction loss, arising from the off-axis propagation of transverse modes, is a unique characteristic of VCSELs. A simple method, the mode-matching technique, is described to approximate the off-axis propagation of transverse modes by a set of plane waves (i.e., LP modes). In this approximation, the transverse modes are assumed TE- or TM-polarized so that the appropriate boundary conditions can be applied at the interface. This method is used to study the influence of diffraction loss on the total cavity loss of VCSELs with index- and anti-index-guided structures.

For VCSELs operating at resonant condition, the corresponding transverse modes should also fulfill the round-trip requirement. In this case, the profile of transverse modes can be modified significantly by the longitudinal distribution of the layers inside the laser cavity. Hence, a two-dimensional beam propagation method is used to solve the longitudinal propagation of transverse modes. The boundary conditions at all the interfaces are introduced into calculation by using scattering matrix. In addition, the off-axis propagation of transverse modes at the interfaces can also be factored in using Bessel transform. The modified beam propagation method has been applied to investigation of the transverse-mode characteristics of VCSELs with tapered waveguide geometry. It is shown that the effective gain difference between the fundamental and first-order transverse modes can be enhanced by the tapered design.

It must be noted that for VCSELs with small cavity sizes and strong index guidance, the influence of mode coupling and radiation fields affects the resonant wavelength and threshold gain of the transverse modes. Hence a rigorous approach, the general mode expansion technique, is used to solve the vectorial solution of Maxwell's equation. This method is realized by expressing the total fields in terms of a set of known basis functions, with which the influence of mode coupling and radiation fields can be considered. All transverse components of the total fields are matched at all the interfaces so that the round-trip condition can be fulfilled. Hence, a vectorial transform matrix is obtained to calculate the resonant wavelength and threshold gain of the transverse modes. On the other hand, the vectorial transform matrix can be simplified by using the common-mode approximation. The common-mode approximation is based on the assumption that the mode of each layer is approximately the same as if in the active layer. The TE and TM components of the hybrid mode are allowed to propagate separately in order to make it possible to determine the longitudinal mode. However, the influence of mode coupling and radiation fields has been ignored in the analysis of VCSELs using common-mode approximation.

3A APPENDIX

3A.1 Optical Field Distribution inside a Cylindrical Waveguide

The electromagnetic fields $E_r(r)$, $E_\theta(r)$, $H_r(r)$, and $H_\theta(r)$ inside a dielectric cylindrical waveguide can be calculated by substituting (3.8) into (3.5) with the general solution of the θ dependent fields expressed in terms of $\exp(j\nu\theta)$

$$E_r(r) = \begin{cases} -\dfrac{W^2}{u^2}\left[B\dfrac{\nu w\mu_0 J_\nu(ur/W)}{rJ_\nu(u)} - A\dfrac{j\beta u J_\nu'(ur/W)}{WJ_\nu(u)}\right] & r \le W \\[2ex] \dfrac{W^2}{w^2}\left[B\dfrac{\nu w\mu_0 K_\nu(wr/W)}{rK_\nu(w)} - A\dfrac{j\beta w K_\nu'(wr/W)}{WK_\nu(w)}\right] & r > W \end{cases}, \tag{3A.1a}$$

$$E_\theta(r) = \begin{cases} -\dfrac{W^2}{u^2}\left[B\dfrac{jw\mu_0 u J'_\nu(ur/W)}{WJ_\nu(u)} - A\dfrac{\beta\nu J_\nu(ur/W)}{r_a J_\nu(u)}\right] & r \le W \\ \dfrac{W^2}{w^2}\left[B\dfrac{jw\mu_0 w K'_\nu(wr/W)}{WK_\nu(w)} - A\dfrac{\beta\nu K_\nu(wr/W)}{r_a K_\nu(w)}\right] & r > W \end{cases}, \tag{3A.1b}$$

$$H_r(r) = \begin{cases} \dfrac{W^2}{u^2}\left[A\dfrac{\nu n_1^2 w\varepsilon_0 J_\nu(ur/W)}{rJ_\nu(u)} + B\dfrac{j\beta u J'_\nu(ur/W)}{WJ_\nu(u)}\right] & r \le W \\ -\dfrac{W^2}{w^2}\left[A\dfrac{\nu n_2^2\, w\varepsilon_0 K_\nu(wr/W)}{rK_\nu(w)} - B\dfrac{j\beta w K'_\nu(wr/W)}{WK_\nu(w)}\right] & r > W \end{cases}, \tag{3A.1c}$$

$$H_\theta(r) = \begin{cases} \dfrac{W^2}{u^2}\left[A\dfrac{jn_1^2 w\varepsilon_0 u J'_\nu(ur/W)}{WJ_\nu(u)} - B\dfrac{\beta\nu J_\nu(ur/W)}{rJ_\nu(u)}\right] & r \le W \\ -\dfrac{W^2}{w^2}\left[A\dfrac{jn_2^2 w\varepsilon_0 w K'_\nu(wr/W)}{WK_\nu(w)} - B\dfrac{\beta\nu K_\nu(wr/W)}{rK_\nu(w)}\right] & r > W, \end{cases} \tag{3A.1d}$$

where A and B are constants to be determined by the boundary conditions and J' and K' represent the derivative of J and K, respectively, with respect to r once.

3A.2 Linearly Polarized Modes inside a Weakly Guiding Cylindrical Dielectric Waveguide

Let the general solutions of $E_z(r, \theta)$ and $H_z(r, \theta)$ be

$$E_z(r, \theta) = \begin{cases} A\dfrac{J_\nu(ur/W)}{J_\nu(u)} \cdot g_\nu(\theta) & r \le W \\ A\dfrac{K_\nu(wr/W)}{K_\nu(w)} \cdot g_\nu(\theta) & r > W \end{cases}, \tag{3A.2a}$$

$$H_z(r, \theta) = \begin{cases} B\dfrac{J_\nu(ur/W)}{J_\nu(u)} \cdot f_\nu(\theta) & r \le W \\ B\dfrac{K_\nu(wr/W)}{K_\nu(w)} \cdot f_\nu(\theta) & r > W \end{cases}, \tag{3A.2b}$$

where

$$g_\nu(\theta) = \begin{Bmatrix} \cos(\nu\theta) \\ j\sin(\nu\theta) \end{Bmatrix} \quad \text{and} \quad f_\nu(\theta) = \begin{Bmatrix} -\sin(\nu\theta) \\ j\cos(\nu\theta) \end{Bmatrix}, \tag{3A.3}$$

with which the influence of $+\nu$ and $-\nu$ can be taken into account. From (3.9), (3.10) and (3.16), (3.17), the ratio between A and B can be deduced under the condition that $\Delta \sim 0$, and is given by

$$\frac{A}{B} = \mp \frac{j}{n_1}\sqrt{\frac{\mu_0}{\varepsilon_0}}, \tag{3A.4}$$

where the minus and plus signs stand for *EH* and *HE* modes, respectively. Substitute (3A.2) into (3.5) and with the auxiliary equations (3A.2) and (3A.4). It can be shown that the total fields of the transverse modes are given by [13]

$$E_z(r,\theta) = \begin{cases} Au\dfrac{J_v(ur/W)}{J_{v\pm1}(u)} \cdot g_v(\theta) & r \le W \\ \mp Aw\dfrac{K_v(wr/W)}{K_{v\pm1}(w)} \cdot g_v(\theta) & r > W \end{cases}, \tag{3A.5a}$$

$$H_z(r,\theta) = \begin{cases} \pm An_1\sqrt{\dfrac{\varepsilon_0}{\mu_0}}\dfrac{uJ_v(ur/W)}{J_{v\pm1}(u)} \cdot f_v(\theta) & r \le W \\ -An_1\sqrt{\dfrac{\varepsilon_0}{\mu_0}}\dfrac{wK_v(wr/W)}{K_{v\pm1}(w)} \cdot f_v(\theta) & r > W \end{cases}, \tag{3A.5b}$$

$$E_r(r,\theta) = \begin{cases} \mp jAk_0Wn_1\sqrt{\dfrac{\varepsilon_0}{\mu_0}}\dfrac{J_{v\pm1}(ur/W)}{J_{v\pm1}(u)} \cdot g_v(\theta) & r \le W \\ \mp jAk_0Wn_1\sqrt{\dfrac{\varepsilon_0}{\mu_0}}\dfrac{K_{v\pm1}(wr/W)}{K_{v\pm1}(w)} \cdot g_v(\theta) & r > W \end{cases}, \tag{3A.5c}$$

$$E_\theta(r,\theta) = \begin{cases} jAk_0Wn_1\dfrac{J_{v\pm1}(ur/W)}{J_{v\pm1}(u)} \cdot f_v(\theta) & r \le W \\ jAk_0Wn_1\dfrac{K_{v\pm1}(wr/W)}{K_{v\pm1}(w)} \cdot f_v(\theta) & r > W \end{cases}, \tag{3A.5d}$$

$$H_r(r,\theta) = \begin{cases} -jAk_0Wn_1^2\sqrt{\dfrac{\varepsilon_0}{\mu_0}}\dfrac{J_{v\pm1}(ur/W)}{J_{v\pm1}(u)} \cdot f_v(\theta) & r \le W \\ -jAk_0Wn_1^2\sqrt{\dfrac{\varepsilon_0}{\mu_0}}\dfrac{K_{v\pm1}(wr/W)}{K_{v\pm1}(w)} \cdot f_v(\theta) & r > W \end{cases}, \tag{3A.5e}$$

$$H_\theta(r,\theta) = \begin{cases} \mp Ak_0Wn_1^2\sqrt{\dfrac{\varepsilon_0}{\mu_0}}\dfrac{J_{v\pm1}(ur/W)}{J_{v\pm1}(u)} \cdot g_v(\theta) & r \le W \\ Ak_0Wn_1^2\sqrt{\dfrac{\varepsilon_0}{\mu_0}}\dfrac{K_{v\pm1}(wr/W)}{K_{v\pm1}(w)} \cdot g_v(\theta) & r > W \end{cases}, \tag{3A.5f}$$

where the normalized constants $uJ_v(u)/(2J_{v\pm1}(u))$ and $\mp wK_v(w)/(2K_{v\pm1}(w))$ have been applied to the electric fields, again the upper and lower signs apply to *EH* and *HE* modes, respectively.

In order to seek the expression of LP modes, the field components given in (3A.5) have to be transformed from polar to Cartesian coordinates with the usual transformation as shown below [13]

$$\begin{bmatrix} E_x \\ E_y \end{bmatrix} = \begin{bmatrix} \cos\theta & -\sin\theta \\ \sin\theta & \cos\theta \end{bmatrix} \begin{bmatrix} E_r \\ E_\theta \end{bmatrix}, \tag{3A.6}$$

with a similar equation for H_x, H_y in terms of H_r, H_θ. Therefore, the electric and magnetic fields expressed in Cartesian coordinates are given by

$$E_x(r,\theta) = \begin{cases} \mp jAWk_0 n_1 \dfrac{J_{\nu\pm1}(ur/W)}{J_{\nu\pm1}(u)} \cdot g_{\nu\pm1}(\theta) & r \le W \\ \mp jAWk_0 n_1 \dfrac{K_{\nu\pm1}(wr/W)}{K_{\nu\pm1}(w)} \cdot g_{\nu\pm1}(\theta) & r > W \end{cases}, \tag{3A.7a}$$

$$E_y(r,\theta) = \begin{cases} jAWk_0 n_1 \dfrac{J_{\nu\pm1}(ur/W)}{J_{\nu\pm1}(u)} \cdot f_{\nu\pm1}(\theta) & r \le W \\ jAWk_0 n_1 \dfrac{K_{\nu\pm1}(wr/W)}{K_{\nu\pm1}(w)} \cdot f_{\nu\pm1}(\theta) & r > W \end{cases}, \tag{3A.7b}$$

$$H_x(r,\theta) = \begin{cases} -jAk_0 W n_1^2 \sqrt{\dfrac{\varepsilon_0}{\mu_0}} \dfrac{J_{\nu\pm1}(ur/W)}{J_{\nu\pm1}(u)} \cdot f_{\nu\pm1}(\theta) & r \le W \\ -jAk_0 W n_1^2 \sqrt{\dfrac{\varepsilon_0}{\mu_0}} \dfrac{K_{\nu\pm1}(wr/W)}{K_{\nu\pm1}(w)} \cdot f_{\nu\pm1}(\theta) & r > W \end{cases}, \tag{3A.7c}$$

$$H_y(r,\theta) = \begin{cases} \mp jAk_0 W n_1^2 \sqrt{\dfrac{\varepsilon_0}{\mu_0}} \dfrac{J_{\nu\pm1}(ur/W)}{J_{\nu\pm1}(u)} \cdot g_{\nu\pm1}(\theta) & r \le W \\ \mp jAk_0 W n_1^2 \sqrt{\dfrac{\varepsilon_0}{\mu_0}} \dfrac{K_{\nu\pm1}(wr/W)}{K_{\nu\pm1}(w)} \cdot g_{\nu\pm1}(\theta) & r > W \end{cases}, \tag{3A.7d}$$

This set of equations is interpreted with upper and lower signs corresponding to *EH* and *HE* modes, respectively, while the trigonometric functions f and g refer to the polarization of each mode. Letting $l = \nu + 1$ for *EH* modes and $l = \nu - 1$ for HE modes, the transverse distribution of LP_{lm} mode has to be reconstructed by linearly combining $HE_{l+1,m}$ and $EH_{l-1,m}$ modes. The corresponding LP_{lm} modes are given in (3.21) and (3.22).

3A.3 Response Function of Multilayered Mirror

The response function β_D is determined by a standard matrix method, equivalent to the cascade process [30]. In each layer of the DBR mirror, the Bessel transformed field Ψ also satisfies the Helmholtz equation

$$\frac{\partial^2 \Psi}{\partial z^2} + (k_i^2 - s^2)\Psi = 0, \tag{3A.8}$$

where k_i is the wavevector at the ith layer. The transfer function is defined by the following relation:

$$-j\beta_D = \frac{1}{\Psi}\frac{\partial \Psi}{\partial z}\bigg|_{z=0}. \tag{3A.9}$$

The solution to the Helmholtz equation inside the layer can be approximated by a combination of forward- and backward-traveling plane waves as given by

$$\Psi = A_i \exp(-j\beta_i z) + B_i \exp(j\beta_i z), \tag{3A.10}$$

where β_i is the eigenvalue, which is given by

$$\beta_i = \begin{cases} \sqrt{k_i^2 - s^2} & \text{for } k_i^2 - s^2 \geq 0 \\ -j\sqrt{s^2 - k_i^2} & \text{for } k_i^2 - s^2 \leq 0 \end{cases}. \tag{3A.11}$$

A_i and B_i are the magnitudes of the corresponding plane waves. Using the continuity properties of the transformed fields as they move from layer ith into layer $i+1$, as so on, it can be shown that

$$\begin{bmatrix} A_i \\ B_i \end{bmatrix} = \begin{bmatrix} e^{-j\beta_i z_i} & e^{j\beta_i z_i} \\ -j\beta_i e^{-j\beta_i z_i} & j\beta_i e^{j\beta_i z_i} \end{bmatrix}^{-1} \begin{bmatrix} e^{-j\beta_{m+1} z_i} & e^{j\beta_{i+1} z_i} \\ -j\beta_{i+1} e^{-j\beta_{i+1} z_i} & j\beta_{i+1} e^{j\beta_{i+1} z_i} \end{bmatrix}$$
$$\times \begin{bmatrix} A_{i+1} \\ B_{i+1} \end{bmatrix} = [M_i] \begin{bmatrix} A_{l+1} \\ B_{i+1} \end{bmatrix}, \tag{3A.12}$$

where z_i is the position of the interface between layers ith and $i+1$. Hence, for mth pairs, it can be shown that

$$\begin{bmatrix} A_0 \\ B_0 \end{bmatrix} = \prod_{i=0}^{m-1} [M_i] \begin{bmatrix} A_m \\ B_m \end{bmatrix}. \tag{3A.13}$$

If an infinite air layer is added beyond the DBR mirror, a plane wave of amplitude Ψ_D moves in the positive z direction, which leads to

$$\begin{bmatrix} A_0 \\ B_0 \end{bmatrix} = \prod_{i=0}^{m} [M_i] \begin{bmatrix} \Psi_D \\ 0 \end{bmatrix} = \begin{bmatrix} m_{11} & m_{12} \\ m_{21} & m_{22} \end{bmatrix} \begin{bmatrix} \Psi_D \\ 0 \end{bmatrix}, \tag{3A.14}$$

where

$$[M_m] = \begin{bmatrix} e^{-j\beta_m z_m} & e^{j\beta_m z_m} \\ -j\beta_m e^{-j\beta_m z_m} & j\beta_m e^{j\beta_m z_m} \end{bmatrix}^{-1} \begin{bmatrix} e^{-jk_0 z_m} & e^{jk_0 z_m} \\ -jk_0 e^{-jk_0 z_m} & jk_0 e^{jk_0 z_m} \end{bmatrix}. \tag{3A.15}$$

Hence, the response function can be written as

$$\beta_D(s) = \beta_1 \frac{m_{11} - m_{21}}{m_{11} + m_{21}}. \tag{3A.16}$$

3A.4 Numerical Technique for Beam Propagation Method

The propagation equations of the traveling waves

$$\Psi^{\pm}(r, z_i \pm \Delta z) = \Psi^{\pm}(r, z_i) \cdot \exp(\pm[H]\Delta z) \tag{3A.17}$$

can be expressed in a discrete format by assuming that $\Psi^{\pm}$ is discretized as a column vector

$$\underline{\Psi}^{\pm} = \begin{bmatrix} \psi_1^{\pm} \\ \psi_2^{\pm} \\ \vdots \\ \psi_{2n+1}^{\pm} \end{bmatrix}, \tag{3A.18}$$

where $\psi_1^{\pm} \equiv \Psi^{\pm}(r = 0)$ represents the transverse variation of optical field from the middle of the waveguide ($r = 0$) to the side of the laser and at a longitudinal position z_i.

The matrix $[H]_{z,t}$ can be written as

$$[H] = [I] + [H] \cdot \Delta z + \tfrac{1}{2}[H]^2 \cdot \Delta z^2 + \tfrac{1}{6}[H]^3 \cdot \Delta z^3 + \tfrac{1}{24}[H]^4 \cdot \Delta z^4 + \cdots, \tag{3A.19}$$

where $[I]$ is an $m \times (2n + 1)$ identity matrix. $[H]$ is an $m \times (2n + 1)$ tridiagonal matrix representing the operator H [45] where

$$H = \frac{j}{2\beta}\left[\frac{1}{r}\frac{\partial}{\partial r}\left(r\frac{\partial}{\partial r}\right) + \Delta\varepsilon(r, z)k_0^2\right]. \tag{3A.20}$$

The derivatives in the operator H can be approximated by the finite difference

$$\left.\frac{\partial^2 \psi^{\pm}}{\partial r^2}\right|_k = \frac{\psi_{k+1}^{\pm} - 2\psi_k^{\pm} + \psi_{k-1}^{\pm}}{\Delta r^2}, \tag{3A.21a}$$

$$\left.\frac{1}{r}\frac{\partial \psi^{\pm}}{\partial r}\right|_k = \frac{\psi_{k+1}^{\pm} - \psi_{k-1}^{\pm}}{2(k-1)\Delta r^2}, \tag{3A.21b}$$

where k (=2 to $2n + 1$) is an integer and Δr ($\sim W/n$) is the radial separation between two discrete points so that $k = n$ implies $r = W$. Because of the symmetry of the field, the second derivative at the middle of the waveguide (i.e., $r = 0$ or $k = 1$) takes the particular form of

$$\left.\frac{\partial^2 \psi^{\pm}}{\partial r^2}\right|_{k=1} = \frac{2(\psi_2^{\pm} - \psi_1^{\pm})}{\Delta r^2}. \tag{3A.22}$$

Now, the elements of matrix $[H]$ can be written as

$$H_{1,1} = \frac{j(-4/\Delta r^2 + k_0^2\Delta\varepsilon)}{2\beta}, \tag{3A.23a}$$

$$H_{1,2} = \frac{2j}{\Delta r^2\beta}. \tag{3A.23b}$$

For $k > 1$, we obtain

$$H_{k,k\pm 1} = \frac{j(1 \pm 1/2(k-1))}{2\beta \Delta r^2}, \tag{3A.24a}$$

$$H_{k,k} = \frac{j(-2/\Delta r^2 + k_0^2 \Delta\varepsilon)}{2\beta}. \tag{3A.24b}$$

Other matrix elements are zero and $\psi^{\pm} \to 0$ for $k \to$ large.

The convergence of the beam propagation method requires small Δz and an appropriate ratio between Δz and Δr. It can be shown that the corresponding truncation error is of second-order accuracy in grid size between Δz and Δr [45]. This finite difference approach (i.e., Crank–Nicolson difference method) has a major advantage of being simple in implementation but at the expense of high truncation error, especially in the analysis of small geometry devices. Therefore, low-truncation-error finite difference equations with truncation errors up through fourth-order accuracy are derived and can be found in Hadley's paper [46]. The main advantages of these low-truncation-error finite difference equations are insensitivity of the truncation error to the grid size, low memory requirement, and fast runtime execution for the implementation in a PC-level computing power.

3A.5 Bessel Transform and Discrete Bessel Transform

The Bessel transformation of a field function $\Psi(r)$ in spatial domain r into one in Bessel domain s is defined as

$$\hat{\Psi}(s) = \int_0^{\infty} \Psi(r) \cdot J_l(sr) \cdot r\, dr, \tag{3A.25a}$$

and the corresponding inverse Bessel transform is given by

$$\Psi(r) = \int_0^{\infty} \hat{\Psi}(s) \cdot J_l(sr) \cdot s\, ds, \tag{3A.25b}$$

where J_l is the Bessel function of the first kind of lth order. If only the fundamental mode is considered in the calculation, then l can be set to zero.

Consider the transverse field Ψ and its Bessel transform $\hat{\Psi}$, which can be discretized as column vectors of $2n + 1$ elements

$$\underline{\Psi} = \begin{bmatrix} \psi_1 \\ \psi_2 \\ \vdots \\ \psi_{2n+1} \end{bmatrix} \quad \text{and} \quad \underline{\hat{\Psi}} = \begin{bmatrix} \hat{\psi}_1 \\ \hat{\psi}_2 \\ \vdots \\ \hat{\psi}_{2n+1} \end{bmatrix}, \tag{3A.26}$$

where $\psi_{k'} = \psi(r_{k'})$ and $\hat{\psi}_k = \hat{\psi}(s_k)$. $\{r_{k'}\}$ and $\{s_k\}$ are the grid points in spatial and Bessel domain, respectively, where the grid points $\{r_{k'}\}$ and $\{s_k\}$ are defined as

$$r_{k'} = \frac{\zeta_{k'}}{S} \quad \text{and} \quad s_k = \frac{\zeta_k}{W}, \tag{3A.27}$$

where $k', k = 1, 2, \ldots, 2n+1$; ζ_k is the kth order of $J_l(\zeta_k) = 0$, and the boundary condition $\Psi(W) = 0$ is imposed. In addition, $r_{2n+1} = W$ and $s_{n+1} = S$ so that $\zeta_{2n+1} = SW$. As a consequence, S is uniquely determined by $2n+1$ and W. Hence, the Bessel transform of Ψ can be approximated by

$$\hat{\Psi}(s) = \int_0^{\infty} \Psi(r) \cdot J_0(sr) \cdot r\,dr \approx \int_0^{W} \Psi(r) \cdot J_0(sr) \cdot r\,dr, \tag{3A.28}$$

where the fundamental mode is considered in the transformation. The orthogonality properties of the Bessel transform are given by [28]

$$\int_0^{W} J_0(s_p r) \cdot J_0(s_q r) \cdot r\,dr = \delta_{pq} \frac{W^2}{2} [J_0'(s_p W)]^2, \tag{3A.29a}$$

$$\int_0^{S} J_0(s r_q) \cdot J_0(s r_p) \cdot s\,ds = \delta_{qp} \frac{S^2}{2} [J_0'(S r_q)]^2. \tag{3A.29b}$$

If the off-axis propagation field Ψ is be expanded as a set of plane waves in the basis of J_0

$$\Psi(r) = \sum_{k'=1}^{2n+1} a_{k'} J_0(\alpha_{k'} r), \tag{3A.30}$$

the corresponding discrete Bessel transform is given by

$$\hat{\Psi}(s_q) = \sum_{k'=1}^{2n+1} a_{k'} \int_0^{W} J_0(\alpha_{k'} r) \cdot J_0(s_q r) \cdot r\,dr. \tag{3A.31}$$

By choosing $a_k = s_k = \zeta_k / W$ and the orthogonality relation of the Bessel functions, (3A.31) becomes

$$\hat{\Psi}(s_k) = a_k \frac{W^2}{2} [J_1'(\zeta_k)]^2. \tag{3A.32}$$

Thus, for $r = r_k = \zeta_k / S$, we obtain

$$\Psi(r_q) = \frac{2}{W^2} \sum_{k=1}^{2n+1} \frac{J_0(\zeta_k \eta_q / \zeta_{2n+1})}{[J_1(\zeta_k)]^2} \hat{\Psi}(s_k) \tag{3A.33}$$

and the corresponding Bessel transform is given by

$$\hat{\Psi}(r_q) = \frac{2}{S^2} \sum_{k'=1}^{2n+1} \frac{J_0(\zeta_{k'} \zeta_q / \zeta_{2n+1})}{[J_1(\zeta_{k'})]^2} \Psi(r_{k'}). \tag{3A.34}$$

Thus, the discrete Bessel transform matrix elements for the forward $B_0 = [b_{pq}]$ and the inverse $B_0^{-1} = [\tilde{b}_{pq}]$ transforms, respectively, are

$$b_{p,q} = \frac{2}{W^2} \frac{J_0(\zeta_p \zeta_q / \zeta_{2n+1})}{[J_1(\zeta_q)]^2}, \tag{3A.35a}$$

$$\tilde{b}_{p,q} = \frac{2}{S^2} \frac{J_0(\zeta_p \zeta_q / \zeta_{2n+1})}{[J_1(\zeta_q)]^2}, \tag{3A.35b}$$

and $[B_0^{-1}][B_0] \approx [I]$.

REFERENCES

1. P. Blixt, D. I. Babic, N. M. Margalit, K. Streubel, and J. E. Bower, "Multimode fiber transmission using room temperature double-fused 1.54 μm vertical cavity lasers," *IEEE Photon. Technol. Lett.*, Vol. 8, no. 11, pp. 1564–1566, 1996.
2. Y. Kohama, Y. Ohiso, K. Tateno, and T. Kurokawa, "0.85 mm vertical cavity surface emitting laser diode array grown on p-type GaAs substrate," *IEEE Photon. Technol. Lett.*, Vol. 9, no. 3, pp. 280–282, 1997.
3. K. L. Lear, K. D. Choquette, R. P. Schneider, Jr., S. P. Kilcoyne, and K. M. Geib, "Selectively oxidized vertical cavity surface emitting lasers with 50% power conversion efficiency," *Electron. Lett.*, Vol. 31, no. 3, pp. 208–209, 1995.
4. T. H. Oh, D. L. Huffaker, and D. G. Deppe, "Comparison of vertical cavity surface emitting lasers with half-wave cavity spacer confined by single- or double oxide apertures," *IEEE Photon. Technol. Lett.*, Vol. 9, no. 7, pp. 875–877, 1997.
5. K. Tai, Y. Lai, K. F. Huang, T. C. Huang, T. D. Lee, and C. C. Wu, "Transverse mode emission characteristics of gain-guided surface emitting lasers," *Appl. Phys. Lett.*, Vol. 63, no. 19, pp. 2624–2625, 1993.
6. J. Boucat, C. Starck, F. Gaborit, A. Plais, N. Bouche, E. Derouin, L. Goldstein, C. Fortin, D. Carpentier, P. Salet, F. Brillouet, and J. Jacquet, "1-mW CW-RT monolithic VCSEL at 1.55 μm," *IEEE Photon. Lett.*, Vol. 11, no. 6, pp. 629–631, 1999.
7. C. J. Chang-Hasnain, Y. A. Wu, G. S. Li, G. Hasnain, K. D. Choquette, C. Caneau, and L. T. Florez, "Low threshold buried heterostructure vertical cavity surface emitting lasers," *Appl. Phys. Lett.*, Vol. 63, no. 10, pp. 1307–1309, 1993.
8. Y. A. Wu and C. J. Chang-Hasnain, "Transverse mode selection with a passive antiguide region in vertical cavity surface emitting lasers," *IEEE Photon. Technol. Lett.*, Vol. 6, no. 8, pp. 924–926, 1994.
9. R. Michalzik and K. J. Ebeling, "Generalized BV diagrams for higher order transverse modes in planar vertical cavity laser diodes," *IEEE J. Quantum Electron.*, Vol. 31, no.8, pp. 1371–1379, 1995.
10. G. C. Wilson, D. M. Kuchta, J. D. Walker, and J. S. Smith, "Spatial hole burning and self-focusing in vertical cavity surface emitting lasers," *Appl. Phys. Lett.*, Vol. 64, pp. 542–544, 1994.
11. D. Vakhshoori, J. D. Wynn, G. J. Zydzik, R. E. Leibenguth, M. T. Asom, K. Kojima, and R. A. Morgan, "Top surface emitting lasers with 1.9 V threshold voltage and effect of spatial hole burning on their transverse mode operation and efficiencies," *Appl. Phys. Lett.*, Vol. 62, pp. 1448–1450, 1993.

12. M. Brunner, K. Gulden, R. Hovel, M. Moser, and M. Iiegems, "Thermal lensing effects in small oxide confined vertical cavity surface emitting lasers," *Appl. Phys. Lett.*, Vol. 76, no. 1, pp. 7–9, 1999.
13. M. J. Adams, *An Introduction to Optical Waveguide*, Wiley, New York, 1981.
14. A. W. Snyder and J. D. Love, *Optical Waveguide Theory*, Chapman & Hall, London, 1983.
15. J. E. Midwinter, *Optical Fibers for Transmission*, Wiley, New York, 1979.
16. N. K. Dutta and R. J. Nelson, "Temperature dependence of the lasing characteristics of the 1.3 mm InGaAsP-InP and GaAs-$Al_{0.26}Ga_{0.64}As$ DH lasers," *IEEE J. Quantum Electron.* Vol. 18, no.5, pp. 871–875, 1982.
17. N. K. Dutta, "Analysis of current spread, carrier diffusion and transverse mode guiding in surface emitting lasers," *J. Appl. Phys.*, Vol. 68, pp. 1961–1963, 1990.
18. G. R. Hadley, "Effective index model for vertical cavity surface emitting lasers," *Opt. Lett.*, Vol. 20, no. 13, pp. 1483–1485, 1995.
19. E. R. Hegblom, D. I. Babic, B. J. Thibeault, and L. A. Coldren, "Estimation of scattering loss in dielectrically apertured vertical cavity lasers," *Appl. Phys. Lett.*, Vol. 68, no. 13, pp. 1757–1759, 1996.
20. L. A. Coldren, B. J. Thibeault, E. R. Hegblom, G. B. Thompson, and J. W. Scott, "Dielectric apertures as intracavity lenses in vertical cavity lasers," *Appl. Phys. Lett.*, Vol. 68, no. 3, pp. 313–315, 1996.
21. D. Burak and R. Binder, "Cold cavity vectorial eigenmodes of VCSELs," *IEEE J. Quantum Electron.*, Vol. 33, no.7, pp. 1205–1215, 1997.
22. D. Burak and R. Binder, "Electromagnetic characterization of vertical cavity surface emitting lasers based on a vectorial eigenmode calculation," *Appl. Phys. Lett.*, Vol. 72, no. 8, pp. 891–893, 1998.
23. R. R. Burton, M. S. Stern, P. C. Kendall, and P. N. Robson, "VCSEL diffraction loss theory," *IEE Proc-Optoelectron.*, Vol. 142, no. 2, pp. 77–81, 1995.
24. R. R. Burton, M. S. Stern, P. C. Kendall and P. N. Robson, "Modeling of diffraction in pillar vertical cavity surface emitting lasers with embedded bragg layers," *Opt. Quantum Electron.*, Vol. 28, pp. 1677–1684, 1996.
25. D. I. Babic, Y. C. Chung, N. Dali, and J. E. Bowers, "Modal reflection of quarter-wave mirrors in vertical cavity lasers," *IEEE J. Quantum Electron.*, Vol. 29, no.6, pp. 1950–1962, 1993.
26. M. Shimizu, F. Koyama, and K. Iga, "Transverse mode analysis for surface emitting laser using beam propagation method," *IEICE Trans.*, Vol. E74, no. 10, pp. 3334–3341, 1991.
27. J. P. Zhang and K. Petermann, "Beam propagation model for vertical cavity surface emitting lasers: Threshold properties," *IEEE J. Quantum Electron.*, Vol. 30, no.7, pp. 1529–1536, 1994.
28. J. F.P. Seurin and S. L. Chuang, "Discrete bessel transform and beam propagation method for modeling of vertical cavity surface emitting lasers," *J. Appl. Phys.*, Vol. 82, no. 5, pp. 2007–2016, 1997.
29. P. W.A. Mcilroy, M. S. Stern, and P. C. Kendall, "Spectral index method for polarized modes in semiconductor rib waveguides," *J. Lightwave Technol.*, Vol. 8, no.1, pp. 113–117, 1990.
30. E. Hechit, *Optics*, 2nd ed., Addison-Wesley, Reading, MA, 1987, Chapter 9.

31. S. F. Yu, "Analysis and design of vertical cavity surface emitting lasers for self-sustained pulsation operation," *IEEE J. Quantum Electron.*, Vol. 34, no.3, pp. 497–505, 1998.

32. Y. A. Wu, G. S. Li, R. F. Nabiev, K. D. Choquette, C. Caneau, and C. J. Chang-Hasnain, "Single mode, passive anti-guide vertical cavity surface emitting lasers," *IEEE J. Select. Topics Quantum Electron.*, Vol. 1, no. 2, pp. 629–637, 1995.

33. D. I. Babic and S. W. Corzine, "Analytical expression for the reflection delay, penetration depth, and absorptance of quarter-wave dielectric mirrors," *IEEE J. Quantum Electron.*, Vol. 28, no. 2, pp. 514–524, 1992.

34. S. F. Yu and C. W. Lo, "Influence of transverse modes on the dynamic response of vertical cavity surface emitting lasers," *IEE Proc-Optoelectron.*, Vol. 143, no. 2, pp. 189–194, 1996.

35. H. K. Bissessur, F. Koyama, and K. Iga, "Modeling of oxide-confined vertical cavity surface emitting lasers," *IEEE J. Select. Topics Quantum Electron.*, Vol. 3, no. 2, pp. 344–353, 1997.

36. K. Y. Chang, J. Woodhead, and P. N. Robson, "Modal threshold analysis of dielectrix-apertured vertical cavity surface emitting lasers," *Appl. Phys. Lett.*, Vol. 72, no. 3, pp. 335–337, 1998.

37. K. Y. Chang, J. Woodhead, J. Roberts, and P. N. Robson, "A scalar variation method for resonant modes of oxide-apertured vertical cavity surface emitting lasers," *J. Appl. Phys.*, Vol. 87, no. 1, pp. 10–21, 2000.

38. Y. Z. Huang, "Effect of reflectivity at the interface of oxide layer on transverse mode control in oxide confined vertical cavity surface emitting lasers," *J. Appl. Phys.*, Vol. 83, no.7, pp. 3769–3772, 1998.

39. M. J. Noble, J. H. Shin, K. D. Choquette, J. P. Loehr, J. A. Lott, and Y. H. Lee, "Calculation and measurement of resonant-mode blueshifts in oxide-aperture VCSEL's," *IEEE Photon. Technol. Lett.*, Vol. 10, no. 4, pp. 475–477, 1998.

40. B. Demeulenaere, D. De Zutter, and R. Baets, 'Rigorous electromagnetic study of diffraction loss in VCSEL mirrors," *IEE Proc-Optoelectron.*, Vol. 143, no. 4, pp. 221–227, 1996.

41. M. J. Noble, J. P. Loehr, and J. A. Lott, "Analysis of microcavity VCSEL lasing modes using a full-vector weighted index method," *IEEE J. Quantum Electron.*, Vol. 34, no.10, pp. 1890–1903, 1998.

42. M. J. Noble, J. A. Lott, and J. P. Loehr, "Quasi-exact optical analysis of oxide-apertured microcavity VCSEL's using vector finite elements," *IEEE J. Quantum Electron.*, Vol. 34, no.12, pp. 2327–2339, 1998.

43. B. Demeulenaere, P. Bienstman, B. Dhoedt, and R. G. Baets, "Detailed study of AlAs-oxidized apertures in VCSEL cavities for optimized modal performance," *IEEE J. Quantum Electron.*, Vol. 35, no.3, pp. 358–367, 1999.

44. A. S. Omar and K. F. Schunemann, "Complex and backward-wave modes in inhomogeneously and anisotropically filled waveguides," *IEEE Trans. Microwave Theory Techl*, Vol. MTT-35, no. 3, pp. 268–275, 1987.

45. G. R. Hadley, "Low-truncation-error finite difference equations for photonics simulation I: Beam propagation," *J. Lightwave Technol.*, Vol. 16, no.1, pp. 134–141, 1998.

46. G. R. Hadley, "Low-truncation-error finite difference equations for photonics simulation II: Vertical cavity surface emitting lasers," *J. Lightwave Technol.*, Vol. 16, no.1, pp. 142–151, 1998.

CHAPTER 4

Polarization Properties of Vertical Cavity Surface Emitting Lasers

The intrinsic optical anisotropies of VCSELs are investigated. A simple rate equation model (i.e., two-level model) is derived with the consideration of in-plane gain anisotropy to analyze the polarization dynamics of VCSELs. Furthermore, a four-level model, which takes into account the phase information of two orthogonal polarizations through saturable dispersion, is deduced to analyze the polarization stability and noise of VCSELs.

4.1 INTRODUCTION

It is explained in Chapter 3 that if VCSELs have cylindrical symmetric waveguides, there exists a twofold degeneracy of the orthogonal polarizations for each transverse mode. The two orthogonal polarizations can be located at any azimuthal direction, and their transverse profiles are identical, suggesting that the two polarizations have the same modal gain and can be excited simultaneously above threshold. However, it has been shown experimentally that the unintentional anisotropy may be introduced during the fabrication of VCSELs and the optical fields are polarized relative to a particular crystallographic plane [1]. This is because the distribution of in-plane gain is dependent on the orientation of the crystallographic plane of the QW materials. In fact, it has been shown that the growth of QW epitaxial layers on the non (001) substrates gives in-plane gain anisotropy [2–5]. Furthermore, it is observed experimentally that the two orthogonal polarizations of VCSELs exhibit optical switching and optical bistability [6]. It is believed that these nonlinear characteristics of VCSELs are caused by gain saturation, which is also the intrinsic property of optical materials. In addition, the discrimination of optical gain and frequency of the two orthogonal polarizations are observed in VCSELs as a result of the electrooptically induced birefringence [7]. Hence, these intrinsic optical anisotropies have a significant

Analysis and Design of Vertical Cavity Surface Emitting Lasers By Siu Fung Yu
ISBN 0-471-39124-7

influence on the polarization properties in VCSELs and therefore need to be investigated further.

Extrinsic optical anisotropy can also be introduced into VCSELs deliberately to discriminate for optical gain and oscillation frequency between the two orthogonal polarizations. For example, VCSELs with noncircular cavity [8, 9] or tilted pillar structures [10] can enhance the in-plane gain anisotropy. The use of a polarized filter is also possible to break the symmetry of the cylindrical waveguide of VCSELs [11–13]. Another possible method is to control the in-plane gain anisotropy of the QW active layer due the fabrication of VCSELs. This can be done by the introduction of anisotropic stress/strain tensors on the QW active layer so that polarization emission is allowed only in a particular crystallographic axis. The anisotropic stress/strain tensors can be realized by using elliptically etched substrate hole [14] or the growth of epitaxial layer on the misoriented substrate [15–17]. However, these extrinsic optical anisotropies in VCSELs are trivial and will not be discussed in this chapter.

This chapter is organized as follows. The influence of gain saturation, crystal orientation, and birefringence on the intrinsic optical anisotropy of VCSELs is investigated. A simple rate equation model (i.e., two-level model) is then developed to analyze the polarization properties of VCSELs. Furthermore, the electrooptically induced birefringence due to the internal electric field inside the entire structure of VCSELs is studied. A four-level model is also derived with phase information between the two orthogonal polarizations included in the calculation so that the modal gain and oscillation frequency of the two polarizations can be evaluated simultaneously. In addition, the laser parameters such as loss anisotropy and birefringence are extracted from the measured noise spectra of VCSELs by the modified four-level model.

4.2 IN-PLANE OPTICAL GAIN IN VCSELs

In-plane gain anisotropy may be observed in VCSELs at high-power operation due to the presence of nonlinear susceptibilities. This is because the nonlinear susceptibilities induce self- and cross-saturation, which modifies the optical gain of the two orthogonal polarizations. On the other hand, in-plane gain anisotropy can also be realized in strained QWs of non (001) orientation because of their asymmetric crystal distribution over the in-plane direction. Furthermore, it can be shown that birefringence can induce in-plane gain anisotropy in VCSELs. In this section, the formations of in-plane gain anisotropy in VCSELs are analyzed: (1) the self- and cross-saturation coefficients for the two orthogonal polarizations are derived by solving the equation of motion for the density matrix in perturbation series, (2) the in-plane gain anisotropies in strained QW VCSELs for various crystal orientations are studied theoretically, and (3) the modal gain of the two orthogonal polarizations under the influence of birefringence is calculated using the effective index method.

4.2.1 Self- and Cross-Gain Saturation-Induced Optical Gain Anisotropy

If the active layer of a VCSEL is parallel to the (001) plane and the direction of field is propagated along the [001] or k direction, the direction of polarized fields in VCSELs can be at any direction over the (001) plane and is determined by anti-symmetric properties of active region or geometric shape of laser. In this study, it is assumed that the two orthogonal polarizations are in the [110] [denoted by h-mode (horizontal-mode)] and $[1\bar{1}0]$ [denoted by v-mode vertical-mode)] directions. Two Cartesian coordinates are employed to describe the direction of susceptibility tensors. The first one is fixed to the principal crystallographic axes spanned by $(\hat{x}, \hat{y}, \hat{z})$, and the second one is constructed by the axis of light propagation and the axes of the two orthogonal polarizations (h and v modes), that is, $(\hat{h}, \hat{v}, \hat{k})$, where $\hat{z} = \hat{k}$. The definitions of the directions are illustrated in Figure 4.1. Hence, the physical properties associated with the material electronic structure are calculated in the frame $(\hat{x}, \hat{y}, \hat{z})$, and the nonlinear optical coefficients given in this frame are then converted to the values in the frame $(\hat{h}, \hat{v}, \hat{k})$. The nonlinear optical coefficients in the frame $(\hat{h}, \hat{v}, \hat{k})$ are used to calculate self- and cross-saturation coefficients for the h and v modes. In the following, the vector components are labeled with subscripts (x, y, z) and (h, v, k) for the $(\hat{x}, \hat{y}, \hat{z})$ and $(\hat{h}, \hat{v}, \hat{k})$ frames, respectively.

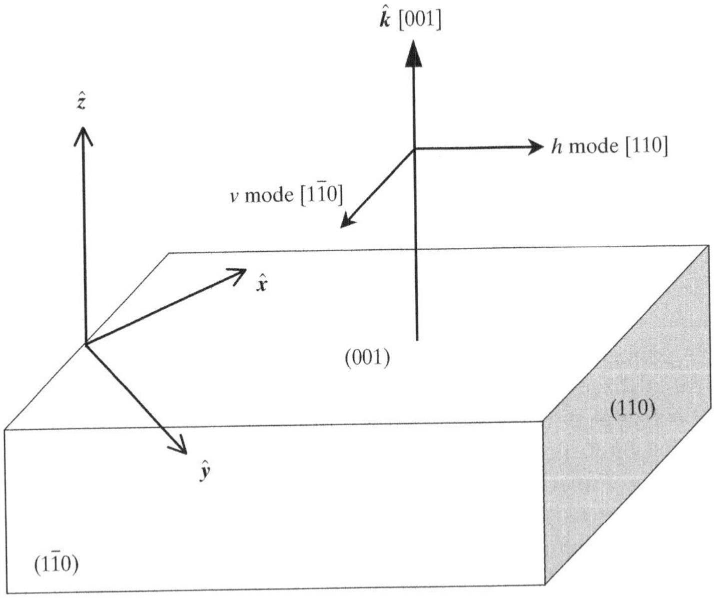

Figure 4.1 The definitions of coordinates in the present study. The principal crystal axes are spanned by $(\hat{x}, \hat{y}, \hat{z})$. The light propagates in [001] with h and v modes in [110] and $[1\bar{1}0]$, respectively.

If the induced electric polarization is considered in derivation of the wave equation, (3.1) has to be written as

$$\nabla \times \boldsymbol{E} = -\mu_{\mathrm{r}}\mu_0 \frac{\partial \boldsymbol{H}}{\partial t} \quad \text{and} \quad \nabla \times \boldsymbol{H} = \varepsilon_0 \frac{\partial \boldsymbol{E}}{\partial t} + \frac{\partial \boldsymbol{P}}{\partial t}, \tag{4.1}$$

where $\boldsymbol{P}$ is the induced electric polarization and μ_{r} is the relative permeability. Hence, the corresponding wave equation is given by

$$\nabla^2 \boldsymbol{E} - \frac{1}{c^2}\frac{\partial^2}{\partial t^2}\boldsymbol{E} = \mu_0 \frac{\partial^2}{\partial t^2}\boldsymbol{P}, \tag{4.2}$$

where c is the velocity of light in free space and $\mu_{\mathrm{r}} \sim 1$ is assumed in the derivation of the wave equation given above. The electric field propagating along the k axis can be expressed as [18]

$$\boldsymbol{E}(\boldsymbol{r}, t) = \tfrac{1}{2}\boldsymbol{E}_0(\omega; t) \cdot [\exp(-j\boldsymbol{k} \cdot \boldsymbol{r} + j\omega t) + \exp(j\boldsymbol{k} \cdot \boldsymbol{r} - j\omega t)], \tag{4.3}$$

where $\boldsymbol{E}_0(\omega; t)$ is the slowly varying time vector field at frequency ω, which is uniformly distributed over the displacement $\boldsymbol{k}$. In addition, the induced electric polarization $\boldsymbol{P}$ can be assumed to be of the form

$$\boldsymbol{P}(\boldsymbol{r}, t) = \tfrac{1}{2}\varepsilon_0 \chi \boldsymbol{E}_0(\omega; t) \cdot (\exp(-j\boldsymbol{k} \cdot \boldsymbol{r} + j\omega t) + \exp(j\boldsymbol{k} \cdot \boldsymbol{r} - j\omega t)), \tag{4.4}$$

where χ is the complex susceptibility. Substituting (4.3) and (4.4) into the wave equation and ignoring the terms $\partial^2 \boldsymbol{E}_0/\partial t^2$ gives

$$2j\omega\varepsilon_{\mathrm{g}} \frac{\partial \boldsymbol{E}_0}{\partial t} + \left(\varepsilon_{\mathrm{r}}\omega^2 - \frac{n_{\mathrm{b}}^2 \Omega^2}{c^2}\right)\boldsymbol{E}_0 = 0, \tag{4.5}$$

where it is assumed that $\varepsilon_{\mathrm{g}} \approx \varepsilon_{\mathrm{b}} + \omega/2 \cdot \partial\varepsilon_{\mathrm{b}}/\partial\omega$ is the group permittivity; ε_{b} is the background relative permittivity; $\boldsymbol{k} \cdot \hat{\boldsymbol{n}}_{\mathrm{k}} = n_{\mathrm{b}}\Omega/c$, $\hat{\boldsymbol{n}}_{\mathrm{k}}$ is the unit vector; and $n_{\mathrm{b}} = (\sqrt{\varepsilon_{\mathrm{b}}})$ is the background refractive index. Under the condition of external carrier injection, ε_{r} can be written as

$$\varepsilon_{\mathrm{r}} \approx n_{\mathrm{b}}^2 + 2n_{\mathrm{b}}\Delta n + jn_{\mathrm{b}}\left(\frac{\alpha_{\mathrm{m}} c}{\omega} + \frac{\mathrm{Im}(\chi)}{n_{\mathrm{b}}}\right), \tag{4.6}$$

where α_{m} is the free-carrier absorption and scattering losses. Because of the external carrier injection, Δn is the small change of refractive index and $\mathrm{Im}(\chi)$ is the change of optical gain. Hence (4.5) can be expressed as

$$\frac{\partial \boldsymbol{E}_0}{\partial t} = j\frac{n_{\mathrm{r}}}{n_{\mathrm{g}}}(\omega - \Omega)\boldsymbol{E}_0 + \frac{j\omega}{n_{\mathrm{g}}}\left(\Delta n_{\mathrm{b}} + j\frac{1}{2}\left(\frac{\alpha_{\mathrm{m}} c}{\omega} + \frac{\mathrm{Im}(\chi)}{n_{\mathrm{b}}}\right)\right)\boldsymbol{E}_0, \tag{4.7}$$

where $n_g = \sqrt{\varepsilon_g}$, $n_r = \sqrt{\varepsilon_r}$, and $(\omega^2 - \Omega^2) \cong 2\,\omega(\omega - \Omega)$ is used in the derivation of (4.7). It is useful to separate (4.7) into real and imaginary parts using

$$\mathbf{E}_0(t) = \mathbf{A}_0(t) \cdot \exp(-j\phi(t)), \tag{4.8}$$

where $\mathbf{A}_0$ is a vector. Therefore, the rate equation of $\mathbf{A}_0$ can be written as

$$\frac{\partial \mathbf{A}_0}{\partial t} = -\frac{1}{2}\nu_g\left(\alpha_m + \frac{\omega}{\nu_b}\cdot\frac{\mathrm{Im}(\chi)}{\varepsilon_b}\right)\mathbf{A}_0, \tag{4.9}$$

where $\nu_g(= c/n_g)$ is the group velocity and $\nu_b(= c/n_b)$ is the velocity of the transverse modes.

The susceptibility χ can be expanded in the perturbation series of the electric field up to the third order. Two optical fields are considered in the susceptibility; one is the lasing light and the other is injected light with polarization orthogonal to the lasing field. It is assumed that the lasing field is h mode, propagating in the direction k with an angular frequency ω_h, and the injected field is v mode with angular frequency ω_v. Hence, the susceptibility tensor $\chi_h(\omega_h)$, induced by the h mode, $\boldsymbol{E}_{oh}(\omega_h)$, is given by [19, 20]

$$\begin{aligned}\chi_h(\omega_h) = \chi_{hh}^{(1)}(\omega_h) &+ \chi_{hhhh}^{(3)}(\omega_h, \omega_h, \omega_h)|\boldsymbol{E}_{oh}(\omega_h;t)|^2\\ &+ \chi_{hvvh}^{(3)}(\omega_v, \omega_v, \omega_h)|\boldsymbol{E}_{ov}(\omega_v;t)|^2,\end{aligned} \tag{4.10a}$$

and the corresponding susceptibility tensor $\chi_v(\omega_v)$, induced by the v mode, $\boldsymbol{E}_{ov}(\omega_v)$, can be deduced in a similar manner and is given by

$$\begin{aligned}\chi_v(\omega_v) = \chi_{vv}^{(1)}(\omega_v) &+ \chi_{vvvv}^{(3)}(\omega_v, \omega_v, \omega_v)|\boldsymbol{E}_{ov}(\omega_v;t)|^2\\ &+ \chi_{vhhv}^{(3)}(\omega_h, \omega_h, \omega_v)|\boldsymbol{E}_{oh}(\omega_h;t)|^2,\end{aligned} \tag{4.10b}$$

where the electric fields are written as

$$\mathbf{A}_{oh}(\omega_h;t) = \hat{\boldsymbol{h}}\cdot\boldsymbol{E}_{oh}(\omega_h;t) \quad \text{and} \quad \mathbf{A}_{ov}(\omega_v;t) = \hat{\boldsymbol{v}}\cdot\boldsymbol{E}_{ov}(\omega_v;t). \tag{4.11}$$

It is more convenient to write the amplitude Equation (4.9) in terms of the photon density $S(t)$ defined using

$$S_h(t) = \frac{1}{\hbar\omega_h}\frac{\varepsilon_h}{2}|\boldsymbol{E}_{oh}(\omega_h;t)|^2 \quad \text{and} \quad S_v(t) = \frac{1}{\hbar\omega_v}\frac{\varepsilon_v}{2}|\boldsymbol{E}_{ov}(\omega_v;t)|^2, \tag{4.12}$$

where $\hbar\omega_{h,v}$ is the photon energy. Since $S_{h,v}(t)$ is proportional to $\mathbf{A}^2_{oh,ov}$, the corresponding photon rate equation of the h mode can be written as

$$\frac{\partial S_h}{\partial t} = \nu_g g_h\left(1 - \frac{\alpha_m}{g_h} - \zeta_{sh}S_h - \zeta_{chv}S_v\right)S_h, \tag{4.13a}$$

where g_h is the linear gain coefficient, ζ_{sh} is the self-saturation coefficient, and ζ_{chv} is the cross-saturation coefficient between the two orthogonal polarizations. A similar expression of a laser operating at v mode with angular frequency of ω_v and in the presence of the orthogonal polarization (i.e., h mode of frequency ω_h) is given by

$$\frac{\partial S_v}{\partial t} = \nu_g g_v \left(1 - \frac{\alpha_m}{g_v} - \zeta_{sv} S_v - \zeta_{cvh} S_h\right) S_v, \tag{4.13b}$$

and the parameters given in (4.13b) have meaning similar to that of (4.13a). It must be noted that the derivation of (4.13) has ignored the influence of longitudinal distribution of the optical field. If the profile of longitudinal field is taken into consideration, Γ_z should be multiplied to the RHS of (4.13).

It can be shown that the optical gain and saturation coefficients can be expressed in terms of imaginary parts of linear and nonlinear susceptibilities [19,20]:

$$g_\eta = \frac{-\omega_\eta \mathrm{Im}(\chi_{\eta\eta}^{(1)})}{\nu_\eta \varepsilon_\eta} \quad \text{for } \eta = v \quad \text{or } h, \tag{4.14a}$$

$$\zeta_{sv} = \frac{2\hbar\omega_v^2 \mathrm{Im}(\chi_{vvvv}^{(3)})}{g_v \nu_v \varepsilon_v^2}, \tag{4.14b}$$

$$\zeta_{sh} = \frac{2\hbar\omega_h^2 \mathrm{Im}(\chi_{hhhh}^{(3)})}{g_h \nu_h \varepsilon_h^2}, \tag{4.14c}$$

$$\zeta_{cvh} = \frac{2\hbar\omega_v\omega_h \mathrm{Im}(\chi_{vhhv}^{(3)})}{g_v \nu_v \varepsilon_v \varepsilon_h}, \tag{4.14d}$$

$$\zeta_{chv} = \frac{2\hbar\omega_h\omega_v \mathrm{Im}(\chi_{hvvh}^{(3)})}{g_h \nu_h \varepsilon_h \varepsilon_v}, \tag{4.14e}$$

where $\varepsilon_{h,v}$ and $\nu_{h,v}$ are the dielectric constant and velocity, respectively, of the laser material. Compare (4.13) with the convention expression of photon rate equation [21], the corresponding expressions of g for h and v modes are

$$g = \begin{cases} g_h(1 - \zeta_{sh} S_h - \zeta_{chv} S_v) & h \text{ mode} \\ g_v(1 - \zeta_{sv} S_v - \zeta_{cvh} S_h) & v \text{ mode} \end{cases}, \tag{4.15}$$

where $\alpha_m < (g_h, g_v)$ is assumed and the terms inside the brackets represent the gain saturation due to the two orthogonal polarizations. It is observed in (4.15) that if the two orthogonal polarizations are in resonance (i.e., $\omega_h \approx \omega_v$) so that $g_h \approx g_v$, gain anisotropy can be induced by high intensities of the two orthogonal polarizations through self- and cross-saturation coefficients. The method used to calculate the nonlinear susceptibilities of QWs is described in the following section.

4.2.2 Modeling of Self- and Cross-Gain Saturation Coefficients

General expressions for the linear and nonlinear optical susceptibility tensors can be deduced using the density matrix approach [22]. The quantum-mechanical Liouville equation for a density matrix ρ is given by

$$\frac{\partial \rho(\boldsymbol{k},t)}{\partial t} = \frac{1}{j\hbar}[H(\boldsymbol{k}), \rho(\boldsymbol{k},t)], \tag{4.16}$$

where $\boldsymbol{k}$ can also be interpreted as the wavevector of the crystallographic direction, say, [001]. The Hamiltonian operator $H(\boldsymbol{k})$ consists of three parts:

$$H = H_0 + H' + H_{\text{relax}}. \tag{4.17}$$

Where H_0 is the unperturbed Hamiltonian for electron and hole and H' takes into account the interaction of matters such as electron–photon interaction, and can be expressed as

$$H' = -\boldsymbol{M} \cdot \boldsymbol{E}(t) = -\sum_{i=x,y,z} M^i E_i(t), \tag{4.18}$$

where $\boldsymbol{M}$ is the dipole operator and H_{relax} includes the relaxation effects due to incoherent scattering process. Therefore, the general expression for the Liouville equation is given by

$$\frac{\partial \rho(\boldsymbol{k},t)}{\partial t} = \frac{1}{j\hbar}[H_0(\boldsymbol{k}) + H'(\boldsymbol{k}), \rho(\boldsymbol{k},t)] + \left(\frac{\partial \rho(\boldsymbol{k},t)}{\partial t}\right)_{\text{relax}}, \tag{4.19}$$

and

$$\left(\frac{\partial \rho(\boldsymbol{k},t)}{\partial t}\right)_{\text{relax}} = \frac{1}{j\hbar}[H_{\text{relax}}(\boldsymbol{k}), \rho(\boldsymbol{k},t) - \rho^{(0)}], \tag{4.20}$$

where $\rho^{(0)}$ is the density matrix at thermal equilibrium. It is noted that the total electron wavefunctions can be expanded in terms of $|\Psi_p\rangle$, where Ψ_p refers to the wavefunction of the pth band. If the density matrix in this basis function is written as $\langle \Psi_p|\rho(\boldsymbol{k},t)|\Psi_q\rangle \equiv \rho_{p,q}(\boldsymbol{k},t)$, the corresponding quantum mechanical Liouville equation can be expressed as

$$\frac{\partial \rho_{p,q}(\boldsymbol{k},t)}{\partial t} = \frac{1}{j\hbar}[H_0(\boldsymbol{k}) + H'(\boldsymbol{k}), \rho(k,t)]_{p,q} - (\rho_{p,q}(\boldsymbol{k},t) - \rho^{(0)}_{p,q}(\boldsymbol{k},t))\gamma_{p,q}, \tag{4.21}$$

where $\gamma_{p,q}$ is the population relaxation rate $(p = q)$ or phase relaxation rate $(p \neq q)$. Please note that this perturbation approximation is applied only to the case of weak optical fields.

In order to solve the quantum-mechanical Liouville equation, the perturbation method is again used to expand the density matrices as below:

$$\rho(\boldsymbol{k},t) = \rho^{(0)}(\boldsymbol{k},t) + \rho^{(1)}(\boldsymbol{k},t) + \rho^{(2)}(\boldsymbol{k},t) + \cdots, \tag{4.22}$$

in which diagonal components are the Fermi distributions in each band with vanishing off-diagonal components. Substituting (4.22) into (4.21) gives

$$\frac{\partial \rho_{p,q}^{(n+1)}}{\partial t} = \frac{1}{j\hbar}([H_0, \rho^{(n)}] + [H', \rho^{(n)}])_{p,q} - \rho_{p,q}^{(n+1)} \gamma_{p,q}, \tag{4.23}$$

for $n = 0, 1, 2$ and $\langle \Psi_p | \rho^{(n)} | \Psi_q \rangle = \rho_{p,q}^{(n)}$ is defined in the calculation. If it is assumed that

$$[H_0, \rho^{(n)}]_{p,q} = \langle \Psi_p | H_0 \rho^{(n)} - \rho^{(n)} H_0 | \Psi_q \rangle = \omega_{p,q} \rho_{p,q}^{(n)}, \tag{4.24}$$

where $\omega_{p,q} (= E_p - E_q/\hbar)$ is the oscillation frequency of the generated photon due to the recombination of carriers between energy levels E_p and E_q, (4.23) can be expressed in terms of basis functions as shown below:

$$\begin{aligned} \frac{\partial \rho_{p,q}^{(n+1)}}{\partial t} &= (-j\omega_{p,q} - \gamma_{p,q}) \rho_{p,q}^{(n+1)} \\ &\quad - \frac{1}{j\hbar} \sum_{p'} \sum_{i=x,y,z} (M_{p,p'}^i \rho_{p',q}^{(n)} - \rho_{p,p'}^{(n)} M_{p',q}^i) E_i(t). \end{aligned} \tag{4.25}$$

Now, if the optical field $\boldsymbol{E}(\boldsymbol{r}, t)$ and the density matrix $\rho_{p,q}^{(1)}(t)$ are written in Fourier expressions

$$\boldsymbol{E}(\boldsymbol{r}, t) = \tfrac{1}{2}(\boldsymbol{E}(\omega_v) e^{(j\boldsymbol{k}\cdot\boldsymbol{r} - j\omega_v t)} + \boldsymbol{E}(\omega_h) e^{(j\boldsymbol{k}\cdot\boldsymbol{r} - j\omega_h t)}) + \text{c.c.}, \tag{4.26}$$

$$\rho_{p,q}^{(1)}(t) \approx \rho_{p,q}^{(1)}(\omega_v) e^{-j\omega_v t} + \rho_{p,q}^{(1)}(\omega_h) e^{-j\omega_h t}, \tag{4.27}$$

(where c.c. = complex conjugate) then, by substituting (4.26), (4.27) into (4.25) and matching the time-dependent exponential terms, it can be shown that

$$\rho_{p,q}^{(1)}(\omega_\eta) = \frac{\rho_{p,p}^{(0)} - \rho_{q,q}^{(0)}}{\hbar(\omega_\eta - \omega_{p,q} - j\gamma_{p,q})} \sum_{i=x,y,z} M_{p,q}^i E_i(\omega_\eta), \tag{4.28}$$

for $\omega_\eta = \omega_v$ or ω_h. Hence, the frequency dependence of $\rho_{p,q}^{(2)}$ and $\rho_{p,q}^{(3)}$ can also be solved in a similar manner, provided the solution of $\rho_{p,q}^{(0)}$ is known.

The polarization per unit volume $\boldsymbol{P}(\omega)$ is calculated from the trace of the matrix product of the dipole moment matrix $\boldsymbol{M}$ and the density matrix ρ:

$$\boldsymbol{P}(\omega) = \frac{1}{L_\text{w}} \int \text{Tr}[\rho(\omega) \boldsymbol{M}] \frac{d\boldsymbol{k}^2}{(2\pi)^2}, \tag{4.29}$$

where L_w is the width of QW

$$\begin{aligned} Tr[\rho(\omega)\boldsymbol{M}] &= \sum_{p,q=\text{c,v}} \rho_{p,q}(\omega) M_{p,q} \\ &= \rho_{\text{c,c}} M_{\text{c,c}} + \rho_{\text{c,v}} M_{\text{v,c}} + \rho_{\text{v,c}} M_{\text{c,v}} + \rho_{\text{v,v}} M_{\text{v,v}}, \end{aligned} \tag{4.30}$$

and $M_{p,q}$ is a 3×3 matrix. The subscripts c and v represent the conduction and valence bands, respectively. Hence, the ith component of the polarization density, to the first order in the optical field, is given by

$$P_i(t) = (M^i_{\mathrm{v,c}}\rho^{(1)}_{\mathrm{c,v}}(\omega_\eta) + M^i_{\mathrm{c,v}}\rho^{(1)}_{\mathrm{v,c}}(\omega_\eta))e^{-j\omega_\eta t}, \tag{4.31}$$

and, with the definition of the component of electric susceptibility tensor of first order, $\chi^{(1)}_{i,i'}$, it is expressed as

$$P_i(t) = \sum_{i'=x,y,z} \varepsilon_0\chi^{(1)}_{i,i'}(\omega_\eta)E_{i'}(\omega_\eta)e^{-j\omega_\eta t}. \tag{4.32}$$

Compare (4.31) and (4.32) and using (4.28), the linear susceptibility tensor of the QW is given by

$$\varepsilon_0\chi^{(1)}_{i,i'}(\omega_\eta) = \sum_{\mathrm{c,v}} \frac{1}{L_{\mathrm{w}}} \int \frac{M^i_{\mathrm{v,c}}(\boldsymbol{k})M^{i'}_{\mathrm{v,c}}(\boldsymbol{k})}{\hbar(\omega_\eta - \mu_{\mathrm{c,v}}(\boldsymbol{k}))}(\rho^{(0)}_{\mathrm{c,c}} - \rho^{(0)}_{\mathrm{v,v}})\frac{d\boldsymbol{k}^2}{(2\pi)^2}, \tag{4.33}$$

where i and i' are the crystallographic axis directions (i.e., x, y, or z direction). The summation indices c and v run over conduction subbands and valence subbands, respectively. They are the lowest subbands (spin-degenerate two bands, $\mathrm{c}_\pm$) in the conduction band and the six subbands (three spin-degenerate pairs: $v_{1\pm}$, $v_{2\pm}$, and $v_{3\pm}$) in the valence band. $M^i_{v,c}$ are the dipole transition matrix elements in i direction. $\rho^{(0)}_{\mathrm{c,c}}(\rho^{(0)}_{\mathrm{v,v}})$ are the diagonal components of the lowest order density matrix in the subbands $c_\pm(v_{1\pm}, v_{2\pm}, v_{3\pm})$, which are Fermi distribution functions with specified Fermi levels and temperatures. $\hbar\mu_{\mathrm{c,v}}(\boldsymbol{k})$ is the energy difference between the conduction and valence bands:

$$\hbar\mu_{\mathrm{c,v}}(\boldsymbol{k}) = \hbar(\omega_{\mathrm{c}} - \omega_{\mathrm{v}}) - j\hbar\gamma_{\mathrm{c,v}} = (E_{\mathrm{c}}(\boldsymbol{k}) - E_{\mathrm{v}}(\boldsymbol{k})) - j\hbar\gamma_{\mathrm{c,v}}, \tag{4.34}$$

where $E_{\mathrm{c}}(E_{\mathrm{v}})$ is the energy of the electron (hole) with wavevector $\boldsymbol{k}$. Higher-order susceptibility tensor in the frame of $(\hat{\boldsymbol{x}}, \hat{\boldsymbol{y}}, \hat{\boldsymbol{z}})$ can also be deduced in a similar manner, and their expressions are given in Section 4A.1.

In the derivation above, the linear and nonlinear susceptibility tensors are given in the frame of $(\hat{\boldsymbol{x}}, \hat{\boldsymbol{y}}, \hat{\boldsymbol{z}})$ but the natural coordinate system of VCSELs is spanned by $(\hat{\boldsymbol{h}}, \hat{\boldsymbol{v}}, \hat{\boldsymbol{k}})$, which are used to express the optical gain and saturation coefficients in the rate equation model. The coordinate transformation from $(\hat{\boldsymbol{x}}, \hat{\boldsymbol{y}}, \hat{\boldsymbol{z}})$ to $(\hat{\boldsymbol{h}}, \hat{\boldsymbol{v}}, \hat{\boldsymbol{k}})$ is a simple rotation by 45° on z, which is expressed by a 3×3 matrix R [23]. For example, it can be shown that

$$\chi^{(1)}_{\eta,\eta}(\omega_\eta) = \tfrac{1}{2}(\chi^{(1)}_{x,x}(\omega_\eta) + \chi^{(1)}_{y,y}(\omega_\eta)), \tag{4.35a}$$

$$\chi^{(3)}_{\eta\eta\eta\eta}(\omega_\eta, \omega_\eta, \omega_\eta) = \tfrac{1}{2}(\chi^{(3)}_{xxxx} + \chi^{(3)}_{xxyy} + \chi^{(3)}_{xyxy} + \chi^{(3)}_{xyyx}), \tag{4.35b}$$

where $\eta = v$ or h.

Evaluation of the expression of $\chi^{(1)}$ and $\chi^{(3)}$ requires the knowledge of the dipole transition matrix elements $M_{c,v}(\boldsymbol{k})$ and the energy dispersions $E_{c,v}(\boldsymbol{k})$, which can be obtained from the band calculations. The energy dispersion of the conduction band can be approximated by the parabolic band with effective masses for the well and barrier materials [22]. The valence band states are calculated by the envelope function approximation with the 4×4 Luttinger–Kohn Hamiltonian [22]. The spin–orbit splitoff band can be ignored in the envelope function approximation, as its contribution to the gain saturation is small in the energy range of interest. However, the axial approximation has not been used because the anisotropy in the (k_x, k_y) plane is considered to be important for the polarization dependence of saturation coefficients. Detailed modeling of energy band structure of QWs can be found in section 4A.2.

Consider the numerical calculation of linear gain and saturation coefficients of a $In_{0.732}Ga_{0.268}As_{0.582}P_{0.418}/InP$ QW material at operation wavelength around 1.3 μm. The concentrations of Ga and As are chosen so that the well material is lattice-matched to InP. In the calculation, the conduction band offset and the thickness of the well are set to 0.4 and 10 nm, respectively. Inside the well (barrier), the energy bandgap, dielectric constant ε_0, and effective mass of electron are assumed to be 0.95 eV (1.35 eV), 13.3 (12.4), and 0.0573 (0.08), respectively. In addition, the Luttinger parameters of the well (barrier) are assumed to be $\gamma_1 = 7.8$ (4.95), $\gamma_2 = 2.8$ (1.65), and $\gamma_3 = 3.6$ (2.35) for the InGaAsP/InP QW. Figure 4.2 plots the gain spectra of the two orthogonal polarized modes varying with sheet density of carrier concentration at room temperature. Because of the crystal isotropy along any of the transverse crystallographic axes of the (001) plane, the linear gain of h mode $[g_h(\omega)]$, and that of v mode, $[g_v(\omega)]$, are identical. Two peaks at photon energies of 1.028 and 1.052 eV are observed from the gain spectra at sheet density of 3×10^{12} cm^{-2}. The band calculation shows that the first subband state (v_1) in the valence band is dominated by $|v; \frac{3}{2}, \pm\frac{3}{2}\rangle$ components and the second subband (v_2) by $|v; \frac{3}{2}, \pm\frac{1}{2}\rangle$ components in the vicinity of the zone center. The transition between the conduction (c) and (v_1), which is $|c; \frac{1}{2}, \pm\frac{1}{2}\rangle \rightarrow |v; \frac{3}{2}, \pm\frac{3}{2}\rangle$ dominant, leads to the first gain peak and the second peak is caused by the transition between c–v_2, which is $|c; \frac{1}{2}, \pm\frac{1}{2}\rangle \rightarrow |v; \frac{3}{2}, \pm\frac{1}{2}\rangle$. These selection rules also affect the spectra of saturation coefficients. Figure 4.3 shows the spectra of the corresponding self-saturation coefficient of the two orthogonal polarizations with different sheet density concentrations. Again because of the crystallographic isotropy, the self-saturation coefficients of the h and v modes $[\zeta_{sh}$ and $\zeta_{sv}]$, are identical for the entire spectrum. A peak is observed near photon energy of 1.02 eV, which implies that the c–v_1 transition has influence on the self-saturation coefficient.

There are two types of cross-saturation coefficients: (1) ζ_{cvh}, which represents the gain saturation of lasing light S_v caused by the injected orthogonal polarization S_h, is the saturation coefficient for two orthogonal polarizations; and (2) ζ_{chv}, which has a meaning similar to those inferred from indices. These two cross-saturation coefficients are the function of two optical frequencies, ω_v and ω_h. It is noted that the cross-saturation coefficients are maximized only when both

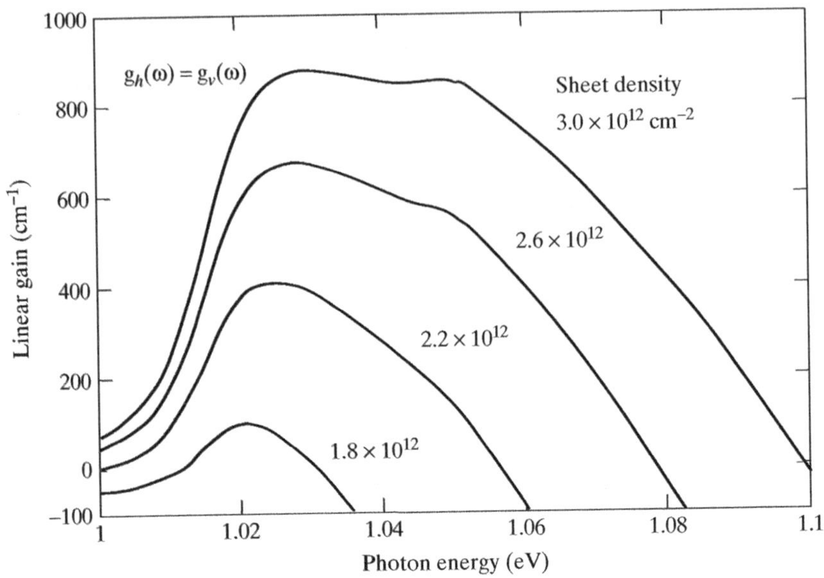

Figure 4.2 Plot of linear gain spectra of h and v modes in InGaAsP/InP quantum wells at room temperature (after Ref. 20).

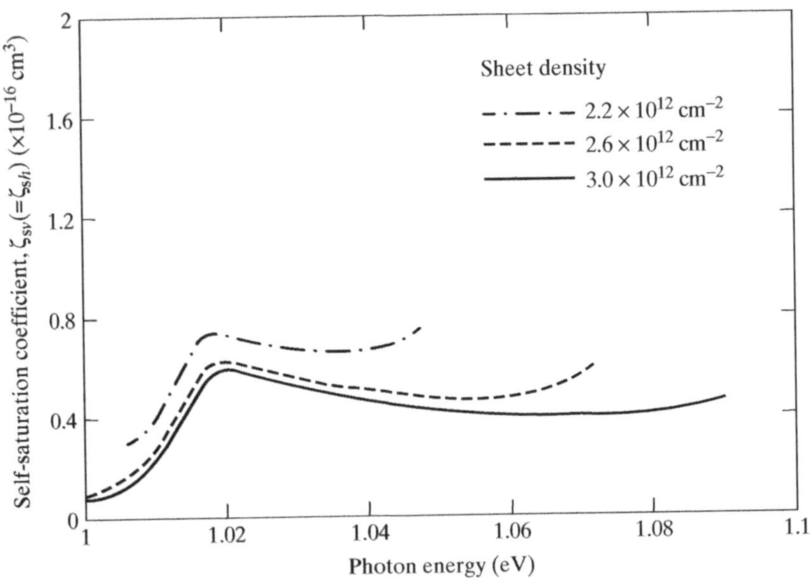

Figure 4.3 Self-saturation coefficients of h and v-modes at resonance (i.e., $\omega_h \approx \omega_v$) in InGaAsP/InP quantum wells (after Ref. 20).

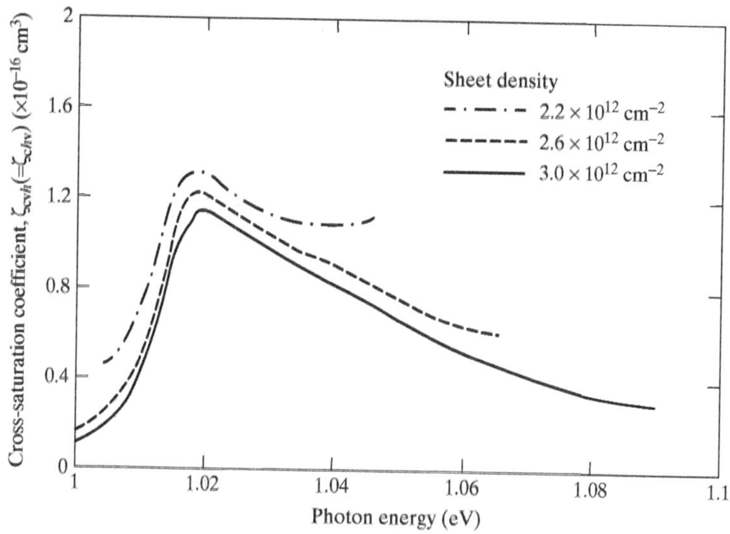

Figure 4.4 Cross-saturation coefficients of h and v-modes at resonance (i.e., $\omega_h \approx \omega_v$) in InGaAsP/InP quantum wells (after Ref. 20).

photon energies are in resonance (i.e., $\omega_v \cong \omega_h$) but vanish at the other photon energies. Furthermore, due to the isotropy of the crystal structure, $\zeta_{chv} = \zeta_{cvh}$ is expected and the corresponding spectra of $\zeta_{chv}(= \zeta_{cvh})$ at resonance is shown in Figure 4.4. It is noted that the profile of cross-saturation coefficients at resonance is dependent on the photon energies and has a peak at 1.02 eV, which indicates the domination of $c{-}v_1$ transition.

From Figures 4.3 and 4.4, it can be concluded that although the optical gains of the two orthogonal polarizations are identical, the presence of self- and cross-saturation coefficients may induce optical gain anisotropy due to the high polarization intensity at a particular resonance frequency. In addition, the in-plane orientation of the two orthogonal polarizations can be defined by the external injection of the polarized field. Hence, this formation of in-plane optical gain anisotropy can be used to control the polarization characteristics of VCSELs such as polarization switching and polarization bistability [19,20]

4.2.3 Dependence of Optical Gain on Crystal Orientation for Strained QWs

In Section 4.2.2, the in-plane gain distribution on the (001) plane of the QW active layer is found to be isotropy. This is because of the crystalline symmetry of [001]-oriented QW heterostructures. However, if the active region of strained QWs is allocated at crystal orientations other than [001], the in-plane gain anisotropy can be obtained from the dependence of strain and crystal orientation on the band structures. In analysis of GaAs/InP strained QW growth on InP substrate with various crystallographic directions, it has been found that emission polarization can be controlled and optical gains can be enhanced with non (001) orientations. The optical gain of VCSELs with orientations other than (001) has a higher value. This is because the decrease of valence band density of states (i.e., reduction of in-plane effective mass) increases the optical gain of QWs with non (001) orientation.

The orientation dependence of the optical gain can be calculated using (4.33) and the dipole moment matrix as given in Section 4A.2. Modification of the dipole moment matrix is required in order to consider the band mixing between heavy hole–light hole states and the anisotropy and nonparabolicity on the valence bands. The band mixing effect in the valence band can be treated by a multiband effective mass theory using a 4×4 Luttinger–Kohn Hamiltonian matrix, taking into account strain–orbit interaction [22]. This effective Hamiltonian matrix can be written in two parts. The first part describes the symmetry of the crystal, quadratically related to the components of the wavevector. The second part describes the strain–orbit interaction caused by the biaxial stress [22]. Hence, the effective Hamiltonian is again described in terms of a 4×4 matrix. In addition, the effective Hamiltonian for arbitrary crystal orientation is obtained by spinor rotation of the (001) effective Hamiltonian to a required direction. The conduction band is again solved by using the Kane model but accounting for the orientation dependence of band structures. In fact, detailed analysis of optical gain can be found in a paper by Niwa et al. [24].

The dependence of in-plane orientation on the optical gain of a strained InGaAs/InP QWs is shown in Figure 4.5. Five cases of QW structures–(001), (110), (112), (221), and (111) in-plane QWs–are considered in the analysis of gain anisotropy. It is observed that the (001) orientation has lower optical gain than do the other orientations and (111) has the highest optical gain. For the other cases, it is clear that the maximum gains are obtained at $\theta = 90^\circ$ for $(NN1)$ and at $\theta = 0^\circ$ for $(11N)$ $(N \geq 2)$ when θ is the polarization angle. The gain curves of $(NN1)$ orientations $(N \geq 3)$, which are not plotted for clarity, are located between the curves of (221) and (110). The higher gains in VCSELs with $(NN1)$ $(N \geq 2)$ are attributed to the smaller hole density of states compared with (001)-oriented VCSELs. This is because the decrease of hole density of states causes a reduction of asymmetry between the conduction and valence bands so that the optical gain is enhanced. In fact, the hole density of states is the smallest (i.e., optical gain is the highest) at the (111) orientation. Hence, an optimum orientation for the VCSELs exists between (111) and (110), that is, (221) in Figure 4.5 and the orientation of the polarized fields can be well defined if the non (001) orientation of QW structure, except (111), is considered as the QW plane. In this analysis, it is shown that the optical gain and direction of emission polarization can be controlled in a particular in-plane orientation due to the intrinsic optical anisotropy of the QW materials.

4.2.4 Modeling of Polarization Modes in VCSELs

In this section, a rate equation model is deduced to analyze the polarization properties of VCSELs [25]. Using this model, the influence of in-plane gain anisotropy and gain saturation on the polarization properties of VCSELs is studied. The rate

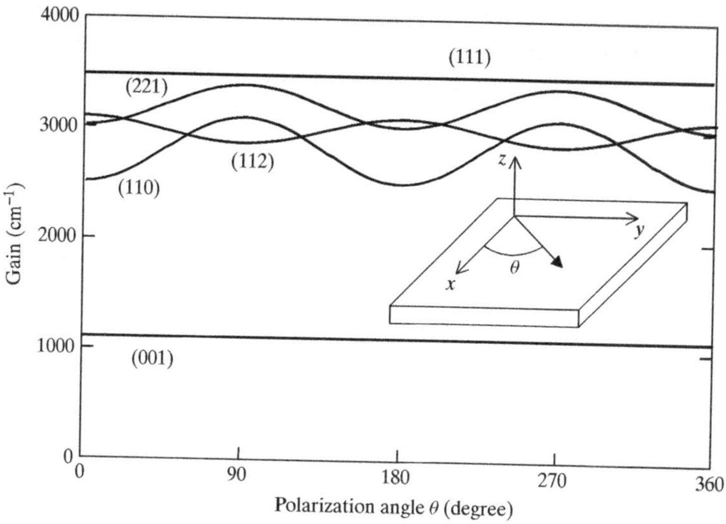

Figure 4.5 In-plane gain anisotropy for the different orientation growth of five epitaxial layers are plotted as a function of an in-plane angle θ (after Ref. 24).

equation model is also applied to investigate the influence of birefringence (i.e., refractive index anisotropy) on the modal gain of the two orthogonal polarizations [26]. It can be shown that a relatively small amount of refractive index anisotropy (~0.005%) is adequate to induce in-plane gain anisotropy to discriminate the polarization state of VCSELs.

The polarization properties of VCSELs with symmetry cylindrical waveguide structure are analyzed by studying the modal gain of the polarized transverse modes. Suppose that the transverse modes are weakly guided inside the core region of the active layer so that the LP mode approximation can be applied. Hence, polarized electric field distribution, $\Psi_{h,lm}$ and $\Psi_{v,lm}$, can be expressed as

$$\Psi_{h,lm}(r,\theta) = \Phi_{lm}(r)\sin(l\theta) \qquad \text{for } h \text{ mode}, \tag{4.36a}$$

$$\Psi_{v,lm}(r,\theta) = \Phi_{lm}(r)\cos(l\theta) \qquad \text{for } v \text{ mode}, \tag{4.36b}$$

where $\Phi_{lm}(r)$ is the radial distribution of the transverse mode of order lm, which has an expression of (3.26) with the corresponding eigenvalue equation given by (3.28). The corresponding azimuthal variation of the polarized electric field is described by the functions $\sin(l\theta)$ and $\cos(l\theta)$ for the h and v modes, respectively. These functions ensure the orthogonal orientation of the two polarized fields. Hence, the modal gain $g_{\eta,lm}$ of $\Psi_{\eta,lm}$ can be obtained by a standard perturbation analysis, expressed as [25]

$$\langle g_{\eta,lm}\rangle = \frac{\displaystyle\int_0^\infty\int_0^{2\pi} \Psi_{\eta,lm}(r,\theta)g(r,\theta)\Psi^*_{\eta,lm}(r,\theta)r\,d\theta\,dr}{\displaystyle\int_0^\infty\int_0^{2\pi} \Psi_{\eta,lm}(r,\theta)\Psi^*_{\eta,lm}(r,\theta)r\,d\theta\,dr}, \tag{4.37}$$

where the asterisk superscript (*) represents a complex conjugate, $\eta = v, h$ and $g(r, \theta)$ is the in-plane gain distribution of the active layer. Hence, the corresponding photon rate equations can be obtained by substituting (4.37) into (4.13) as deduced in Section 4.2.1, with the consideration of higher-order transverse modes and spontaneous emission. It can be shown that the corresponding photon rate equation can be written as

$$\frac{dS_{\eta,lm}}{dt} = v_g \Gamma_{z,lm} \langle g_{\eta,lm} \rangle \left(1 - \zeta_{s\eta,lm} S_{\eta,lm} - \sum_{\substack{\eta \neq \eta' \\ p,q \neq l,m}} \zeta_{c\eta\eta',pq} S_{\eta',pq} \right) S_{\eta,lm} - \frac{S_{\eta,lm}}{\tau^{\eta}_{p,lm}} + \Gamma_{z,lm} \beta^{\eta}_{s,lm} B_{\mathrm{eff}} \langle N \rangle^2, \tag{4.38}$$

where a spontaneous emission term is deliberately introduced into the photon rate equation and $\langle N \rangle$ is the average carrier concentration over the active layer. The photon lifetime τ_p and spontaneous emission factor β_s are assumed to be dependent on the profile of the transverse modes. Furthermore, in order to determine the modal properties of the polarized modes for the laser operating above threshold, it is necessary to calculate the spatial distribution of carrier concentration inside the active layer using the carrier rate equation [25]

$$\frac{\partial N(r, \theta, t)}{\partial t} = \frac{J}{qd} - \frac{N}{\tau_n} + D_n \left(\frac{1}{r} \frac{\partial}{\partial r} \left(r \frac{\partial N}{\partial r} \right) \right) + \frac{D_n}{r^2} \frac{\partial^2 N}{\partial \theta^2} - v_g \sum_{\substack{\eta = v,h \\ l,m}} g(r, \theta) |\Psi_{\eta,lm}(r, \theta)|^2 S_{\eta,lm}, \tag{4.39}$$

where τ_n is the carrier lifetime, D_n is the carrier diffusion constant, d is the thickness of the active layer, and J is the injection current density. Hence, the transverse polarized fields at or above threshold can be obtained numerically by simultaneously solving (4.38) and (4.39) using the finite difference method.

If fundamental transverse mode (i.e., LP_{01}) with two orthogonal polarizations is considered only in the analysis of the polarization properties of VCSELs, the rate equation models can be simplified to [27]

$$\frac{dS_h}{dt} = v_g \Gamma_{z,h} g_{N,h} (N - N_t)(1 - \zeta_{sh} S_h - \zeta_{chv} S_v) S_h - \frac{S_h}{\tau^h_{\mathrm{p}}} + \Gamma_{z,h} \beta^h_{\mathrm{s}} B_{\mathrm{eff}} N^2, \tag{4.40a}$$

$$\frac{dS_v}{dt} = v_g \Gamma_{z,v} g_{N,v} (N - N_t)(1 - \zeta_{sv} S_v - \zeta_{cvh} S_h) S_v - \frac{S_v}{\tau^v_p} + \Gamma_{z,v} \beta^v_{\mathrm{s}} B_{\mathrm{eff}} N^2, \tag{4.40b}$$

$$\frac{\partial N}{\partial t} = \frac{J}{qd} - \frac{N}{\tau_n} - v_g g_{N,v} (N - N_t) S_v - v_g g_{N,h} (N - N_t) S_h, \tag{4.40c}$$

where $g_{N,h}(= 3.0 \times 10^{-16}\ \text{cm}^{-2})$ and $g_{N,v}(= 2.85 \times 10^{-16}\ \text{cm}^{-2})$ are the differential gain coefficients of the h and v modes, respectively and N_t $(= 2.0 \times 10^{18}\ \text{cm}^{-3})$ is the carrier concentration at transparency. In the derivation of (4.40), it is assumed that the carrier concentration N is uniformly distributed along the active layer and the QW optical gain of is linearly approximated with the injected carrier concentration (i.e., for more appropriate calculation, the logarithmic gain–carrier density relation should be used). In addition, it is assumed that $\Gamma_{z,h} = \Gamma_{z,v}$, $\beta_s^v = \beta_s^h = 1 \times 10^{-3}$, $\tau_p^v = \tau_p^h = 2$ ps, $B_{\text{eff}} = 1.0 \times 10^{-10}\ \text{cm}^3\text{s}$, the self- and cross-saturation coefficients for both polarization modes are assumed identical ($\zeta_{sv} = \zeta_{sh} = 1.0 \times 10^{-17}$ cm^3 and $\zeta_{cvh} = \zeta_{chv} = 2.0 \times 10^{-17}\ \text{cm}^3$). It is noted that the differential gain coefficients for the h and v modes are different and the $g_{N,h}$ is assumed larger than $g_{N,v}$. In addition, the value of the cross-saturation coefficients is assumed to be larger than that of the self-saturation coefficients. These assumptions are based on the experimental observation of bistable operation of two orthogonal polarizations. In fact, to realize this bistable operation, the condition of optical gain anisotropy with $\zeta_{sv} \times \zeta_{sh} < \zeta_{cvh} \times \zeta_{chv}$ is required [27]. Hence, it can be shown that the v mode is well suppressed by more than 10 dB and only the h mode is supported for the laser biased about twice the threshold (i.e., equivalent to gain difference of 5%). However, further increase of the lasing mode intensity reduces the modal gain of the h mode through the gain saturation and mode switching is observed. Assuming that $g_{N,v} = g_{N,h}(= 3 \times 10^{-16}\ \text{cm}^{-2})$ is applied to the model and the other parameters remained unchanged, the numerical calculation has demonstrated that both orthogonal polarizations can be excited but are dependent on the initial conditions. If no randomly generated spontaneous emission noise is used on β_s, both polarizations can be excited simultaneously above threshold.

In the preceding calculations, the oscillation frequencies of the two orthogonal polarizations degenerate as both polarizations are of the same fundamental transverse-mode profile. However, it is seldom found experimentally that the two orthogonal polarizations have identical oscillation frequencies [28]. This is because the two polarizations experience birefringence (i.e., anisotropy in refractive index), which induces a split in oscillation frequency. In addition, birefringence may alternate the radial distribution of the two orthogonal polarizations and introduces in-plane gain anisotropy. In order to analyze the influence of birefringence on the polarization properties of VCSELs, the photon rate equations derived in Section 4.2.1 have to be modified. Consider that the total field $\boldsymbol{E}$ includes the two polarized states of the fundamental transverse mode (i.e., LP_{01} mode), which can be written as

$$\boldsymbol{E}(r,\theta,t) = \tfrac{1}{2}\{\Psi_v(r,\theta)E_v(t)\exp(-j\beta_v z + j\omega_v t) + \Psi_h(r,\theta)E_h(t)\exp(-j\beta_h z + j\omega_h t)\}, \tag{4.41}$$

and $\Psi_\eta(\eta = \text{v}, h)$ satisfies the Holmholtz equation as [29]

$$\frac{1}{r}\frac{\partial}{\partial r}\left(r\frac{\partial \Psi_\eta}{\partial r}\right) + \frac{1}{r^2}\frac{\partial^2 \Psi_\eta}{\partial \theta^2} + (\varepsilon_r(r)k_0^2 - \beta_{\text{eff},\eta}^2)\Psi_\eta = 0, \tag{4.42}$$

where $\beta_{\text{eff},\eta}$ is the corresponding effective propagation coefficient in the transverse direction. Substituting (4.41) into (4.5) and using (4.42), it can be shown that

$$\frac{2j\omega}{c^2}\varepsilon_{\text{g}}\frac{\partial E}{\partial t} - \frac{\omega^2}{c^2}(\langle\varepsilon_r\rangle - \langle\varepsilon_{\text{b},\eta}\rangle)E + 2\beta_\eta(\beta_{\text{eff,n}} - \beta_\eta)E = 0 \tag{4.43}$$

The term given below

$$\langle\varepsilon_r\rangle - \langle\varepsilon_{\text{b},\eta}\rangle = n_{\text{b},\eta}\begin{cases} 2\Delta n_\eta + \dfrac{jc(g_\eta - \alpha_{\text{m}})}{\omega_\eta} & \text{core} \\ \dfrac{-jc\alpha_m}{\omega_\eta} & \text{cladding} \end{cases} \tag{4.44}$$

represents the effective optical gain experienced by the polarized fields, and the term $2\beta_\eta(\beta_{\text{eff},\eta} - \beta_\eta)$ represents the detuning from the waveguide mode. Hence, the rate equation model can be simplified to

$$\frac{dS_v}{dt} = v_{\text{g}}\Gamma_{z,v}g_vS_v - \frac{S_v}{\tau_{\text{p}}^v} + \beta_{\text{s}}^v\Gamma_{z,v}B_{\text{eff}}N^2, \tag{4.45a}$$

$$\frac{dS_h}{dt} = v_{\text{g}}\Gamma_{z,h}g_hS_h - \frac{S_h}{\tau_{\text{p}}^h} + \beta_{\text{s}}^h\Gamma_{z,h}B_{\text{eff}}N^2, \tag{4.45b}$$

and the corresponding carrier rate equation is similar to (4.40c). These two photon rate equations have ignored the influence of self- and cross-gain saturation, and the calculation of g_v and g_h is different from that given in (4.40) as birefringence has been taken into calculation. Figure 4.6 shows the calculation of the normalized difference between the h and v modes (i.e., with transverse distribution of LP_{01}), that is, $\delta\Psi = \Psi^h(r)/ < \Psi^h > -\Psi^v(r)/ < \Psi^v >$. It is assumed that the corresponding background permittivity of the two orthogonal polarizations are $\varepsilon_{\text{b},h}(r < W) = 12.2514$, $\varepsilon_{\text{b},h}(r > W) = 11.56$, $\varepsilon_{\text{b},v}(r < W) = 12.25$, and $\varepsilon_{\text{b},v}(r > W) = 11.56$. The core radius W of the symmetry cylindrical waveguide of VCSELs is 3 μm. It is noted that the refractive index of the h mode in the core region is larger than that of the v mode, so the h mode has better transverse confinement than the v mode. Therefore, it is expected that the h mode has higher optical gain than the v mode and the corresponding gain difference is found to be ~ 0.04 cm^{-1}, which is sufficient to maintain a single h-mode operation above threshold [30]. Furthermore, it is expected that the birefringence separates the oscillation frequency of the two orthogonal polarizations and the lasing mode, the h mode has higher oscillation frequency. This is because $\varepsilon_{\text{b},h} > \varepsilon_{\text{b},v}$ implies $\beta_{\text{eff},h} > \beta_{\text{eff},v}$ so that the lasing mode has higher oscillation frequency than the nonlasing mode. Hence, polarization mode have higher oscillation frequency will be excited by birefringence only if VCSELs have uniform in-plane gain distribution. In the following section, the origin of birefringence in VCSELs is analyzed more rigorously.

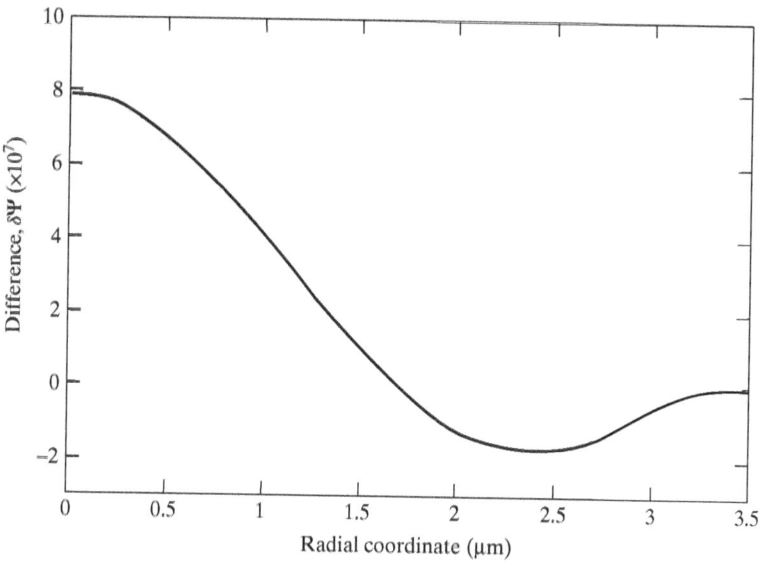

Figure 4.6 The normalized intensity difference $\delta\Psi$ versus radial coordinate (after Ref. 30).

4.3 ELECTROOPTIC EFFECT AND POLARIZATION BIREFRINGENCE IN VCSELs

For VCSELs operating above threshold, a small amount of birefringence may be experienced by the polarized fields. In fact, birefringence is produced by an inevitable internal field $\mathscr{E}$ (due to the applied voltage) arising from the electro-optic effect. This effect causes the in-plane gain anisotropy as well as the split of frequency between the two orthogonal polarizations in VCSELs. If a VCSEL is grown on a (001) substrate and an internal field $\mathscr{E}$, is pointed in the [001] direction, this orientation of the applied field will lead to a change in refractive index, which can be written as [31]

$$\Delta\left(\frac{1}{n^2}\right) = \pm r_{41}\mathscr{E} + R_{12}\mathscr{E}^2, \tag{4.46}$$

where the + and − signs apply to light polarized along the [110] and [1$\bar{1}$0] directions and r_{41} and R_{12} are the linear and quadratic electrooptic effects, respectively. Hence, the corresponding difference in refractive index between the two orthogonal polarized modes is given by

$$n_h(z) - n_v(z) = -n^3(z)r_{41}(z)\mathscr{E}(z), \tag{4.47}$$

where the subscripts h and v denote the polarized modes along the [110] and [1$\bar{1}$0] directions, respectively, and $n(z)$ is the refractive index in zero electric field. The variable z indicates the position dependence of n and r_{41} along the [001] direction so that (4.47) evaluates the local birefringence of the laser cavity.

In order to calculate the overall birefringence of the laser cavity, (4.47) has to be modified by the nonuniform distribution of optical field profile E. The optical field distribution inside the VCSEL can be obtained from [29]

$$\left(\frac{d^2}{dz^2} + \left(\frac{\omega}{c}\right)^2 [n(\omega, z) + \Delta n(\omega, z)]\right) \cdot E(z) = 0, \tag{4.48}$$

where Δn is the change in refractive index. If Δn is interpreted as the birefringence (i.e., $\Delta n = n_h - n_v$), the corresponding frequency splitting, $\Delta\omega(= \omega_v - \omega_h)$, between the two orthogonal polarizations can be estimated from (4.48). By substituting $\omega = \omega_s + \Delta\omega$ and $E = E_s + \Delta E$ into (4.48), where the subscript ('s') stands for steady state, and using the first-order perturbation method, the approximated solution to $\Delta\omega$ is given by [32]

$$\frac{\Delta\omega}{\omega_s} = -\frac{\displaystyle\int (\Delta n(z)/n(z))|E_s(z)|^2 dz}{\displaystyle\int (n_g(z)/n(z))|E_s(z)|^2 dz}, \tag{4.49}$$

where n_g is the group refractive index. Substituting (4.47) into (4.49) gives

$$\frac{\omega_v - \omega_h}{\omega_s} \approx \left\langle \frac{n^3}{n_g} r_{41} \right\rangle \langle \mathscr{E} \rangle. \tag{4.50}$$

The brackets $\langle\rangle$ denote the normalized spatial averaging over the longitudinal direction, and $\langle \mathscr{E} \rangle$ is defined as

$$\langle \mathscr{E} \rangle = \frac{\displaystyle\int \mathscr{E}(z)|E_s(z)|^2 dz}{\displaystyle\int |E_s(z)|^2 dz}. \tag{4.51}$$

As a result, the local refractive index changes induced by the internal electric field $\mathscr{E}$ have to be weighted by the optical intensity $|E_s(z)|^2$ in order to evaluate their influence on the overall birefringence and cavity resonance frequency. $|E_s(z)|^2$ can be calculated using the transfer matrix method as discussed in Chapter 2 if the laser structure is known. The approximated value of $\langle$ E$\rangle$ can also be obtained by a simple approach shown below.

It is noted that $\mathscr{E}$ is dependent on the configuration of lasers such as material composition and doping levels as well as the external bias voltage. Figure 4.7 sketches a typical VCSEL structure with multilayered mirrors. An active layer, comprising three QWs (of material type 1), is centered in a 1λ spacer layer (of material type 2), which is surrounded by multilayered mirrors composed of alternating $\lambda/4$ semiconductor layers (of material 2 and material 3). Bandgaps and refractive indices of materials 1, 2, and 3 are denoted by $E_{g1} < E_{g2} < E_{g3}$ and $n_1 > n_2 > n_3$, respectively. It is assumed that series resistance is uniformly distributed as indicated by the slope of Fermi levels shown in Figure 4.7.

There are basically three important contributions to $\langle \mathscr{E} \rangle$. These contributions arise from (1) the built-in potential across the active layer, $\langle \mathscr{E}_{\text{active}} \rangle$, (2) the series resistance in the device $\langle \mathscr{E}_{\text{series}} \rangle$, and (3) the localized fields at various heterojunctions of the multilayered mirrors $\langle \mathscr{E}_{\text{mirrors}} \rangle$. $\langle \mathscr{E}_{\text{active}} \rangle$ is due to the electrostatic potential V_{active} across the active layer. If the laser is biased at the situation that the transparency is reached and population inversion is obtained, the electric field can be assumed uniformly along the active layer [i.e., $V_{\text{active}} \sim (E_{g2} - E_{g1})/q$] so that $\langle \mathscr{E}_{\text{active}} \rangle$ can be written as

$$\langle \mathscr{E}_{\text{active}} \rangle = \frac{\int_{\text{active}} \mathscr{E}(z)|E_s(z)|^2 dz}{\int_{\text{entire}} \mathscr{E}(z)|E_s(z)|^2 dz} \approx \frac{2V_{\text{active}}}{L_{\text{eff}}}, \tag{4.52}$$

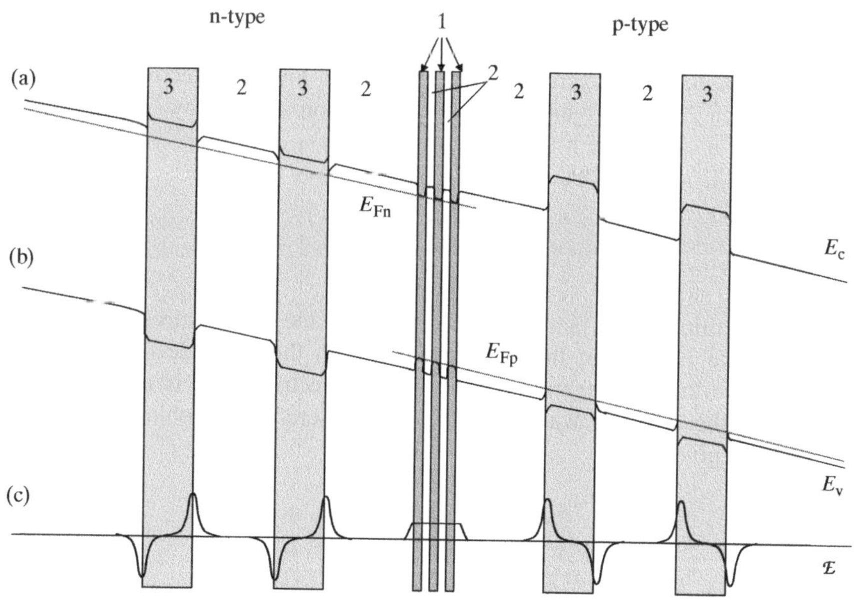

Figure 4.7 Sketch of the position dependence of some important quantities in a typical VCSEL, comprising three quantum wells of material type 1 centered in a 1λ cavity of material type 2, which is surrounded by multilayered mirrors composed of materials 2 and 3. The various curves show (a) the energy level of the conduction band, (b) the energy level of the upper valence band, and (c) the internal electric field. The dotted lines in (a) and (b) denote the electron and hole Fermi levels, respectively.

where the factor 2 in (4.52) is due to resonant FP mode as discussed in Section 2.3.1 and L_{eff} is the effective cavity length resulting from the spatial integral in the denominator of (4.52). L_{eff} can be expressed as $L_{\mathrm{eff}} \approx L_{\mathrm{spacer}} + 2L_{\mathrm{pen}}$, where L_{pen} is the penetration length of the optical field inside the multilayered mirror and can be calculated from (2.45).

Furthermore, $\langle \mathscr{E}_{\mathrm{series}} \rangle$ is due to the series resistance of the VCSEL. If the series resistance is assumed to be uniformly distributed over the multilayered mirrors, $\langle \mathscr{E}_{\mathrm{series}} \rangle$ can be expressed as

$$\langle \mathscr{E}_{\mathrm{series}} \rangle = -\frac{2fV_{\mathrm{pair}}}{L_{\mathrm{eff}}}, \tag{4.53}$$

where V_{pair} is the voltage drops across a pair of dielectric layers (i.e., low and high dielectric pairs), $f(\approx L_{\mathrm{pen}}/(\lambda_{\mathrm{R}}/2n))$ is the effective number of pairs felt by the penetration optical intensity (i.e., $\lambda_{\mathrm{R}}/2n$ is the length of a dielectric pair). The negative sign is due to the fact that the electric field is pointing toward negative z (from p to n mirror).

The last term contribution to $\langle \mathscr{E} \rangle$, namely, $\langle \mathscr{E}_{\mathrm{mirrors}} \rangle$, rises from the localized fields at various heterojunctions of the mirrors. These internal fields are similar to those appearing inside a p-n homojunction except for a difference in electron affinity between the two semiconductor layers. This can be accounted for by introducing conduction band offset, η_{CBO} (i.e., the fraction of the bandgap difference), into the Fermi level of the composing materials. Hence, if the doping levels of the semiconductor multilayered mirrors are not too low, the electrostatic potential across each 2–3 heterojunction, V_{32}, can be approximated by $(E_{g3} - E_{g2})/q$ multiplied by η_{CBO} for the n-type mirror and $1 - \eta_{\mathrm{CBO}}$ for the p-type mirror. As a result, $\langle \mathscr{E}_{\mathrm{mirrors}} \rangle$ can be deduced for (4.51), which is given by

$$\langle \mathscr{E}_{\mathrm{mirrors}} \rangle = \frac{(2f+1)V_{32}}{L_{\mathrm{eff}}}, \tag{4.54}$$

where the multiplication factor $2f + 1$ is due to the constructive interference of the positive polarity of the electric fields with the antinode of the optical standing wave but the negative polarity of the electric fields is eliminated by the node of the standing wave (see Fig. 4.8). Therefore, combining the three contributions gives

$$\begin{aligned} \langle \mathscr{E} \rangle &= \langle \mathscr{E}_{\mathrm{active}} \rangle + \langle \mathscr{E}_{\mathrm{series}} \rangle + \langle \mathscr{E}_{\mathrm{mirrors}} \rangle \\ &= \frac{2V_{\mathrm{active}}}{L_{\mathrm{eff}}} - \frac{2fV_{\mathrm{pair}}}{L_{\mathrm{eff}}} + \frac{(2f+1)V_{32}}{L_{\mathrm{eff}}}. \end{aligned} \tag{4.55}$$

Assumed that material 1 is GaAs, material 2 is $Al_{0.18}Ga_{0.82}As$, and material 3 is AlAs. The corresponding bandgap energies and refractive indices (at room temperature and at operating wavelength λ_{R}, 850 nm) are $E_{g1} = 1.42$ eV, $E_{g2} = 1.67$ eV, $E_{g1} = 2.17$ eV, and $n_1 = 3.64$, $n_2 = 3.46$, $n_3 = 2.99$, respectively. Furthermore, if $L_{\mathrm{space}} = 0.24$ μm and the total number of dielectric pairs of the

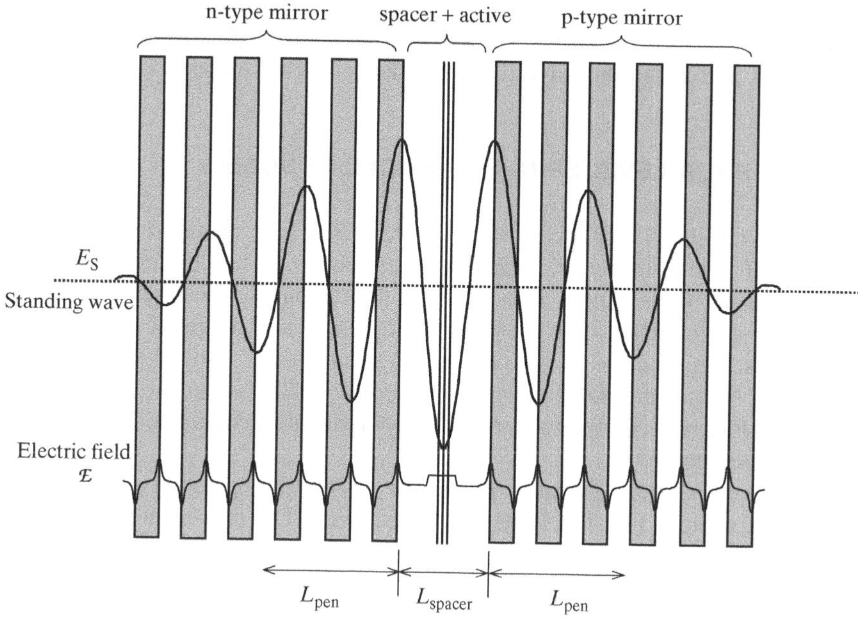

Figure 4.8 The longitudinal distribution of optical field $E_s(z)$ inside a typical VCSEL. L_{space} is the spacer length between the two multilayered mirrors, and L_{pen} is the penetration depth of the optical field inside the multilayered mirrors.

n-type and p-type mirrors is (20 + 25), it can be shown that $f = 3.4$ and $L_{pen} = 0.45$ μm. If the device is biased at 2.5 V, it can also be shown that $V_{active} \approx 0.2$ V, $V_{32} \approx 2.17$ V, and $V_{pair} \approx 0.024$ V. Using these parameters, the split of frequency between h and v modes is given by

$$\nu_h - \nu_v \approx \Delta\nu_{active} + \Delta\nu_{series} + \Delta\nu_{mirrors} \approx +1.5 - 0.6 + 14.4 \text{ GHz} = +15.3 \text{ GHz}. \tag{4.56}$$

This addition clearly shows that the internal fields at the heterojunctions of multilayered mirrors are the domination factor for the determination of birefringence through the electrooptic effect. The positive value of (4.56) indicates light polarized along the [110] axis having a higher resonance frequency than light polarized along $[1\bar{1}0]$.

These calculations have been compared with the experimental measurement. It is found that the lasing mode along the [110] direction has a resonance frequency of about +10 GHz more than the nonlasing one along the $[1\bar{1}0]$ direction for VCSELs from the 1×16 arrays but only +4 GHz more than that from the 1×8 arrays [33]. Hence, it has been proved that the theoretical prediction gives the right direction of detuning but the corresponding magnitude of detuning is overestimated. This implies that birefringence has significant influence on the split off frequency between the two orthogonal polarizations, but other intrinsic optical anisotropies in the laser cavity can also alternate the polarization properties of VCSELs and are further investigated in the following sections.

4.4 THE FOUR-LEVEL SPIN–FLIP MODEL OF VCSELs

In Section 4.2, the polarization properties of VCSELs are analyzed using a two-level model. The influence of intrinsic optical anisotropies on the stability of the two orthogonal polarizations is considered throughout the nonlinear properties of in-plane optical gain. One drawback of this two-level model is its ignorance of the phase information of the orthogonal polarizations through the saturable dispersion. Hence, the polarization stability of VCSELs under the influence of phase coupling between the two orthogonal polarizations cannot be evaluated by the two-level model.

In this section, the polarization properties of VCSELs operating at LP_{01} mode are studied using a four-level rate equation model. The four-level model takes into account the spin sublevels of the conduction and valence bands of the QW materials. Therefore, the lasing field of different polarizations associated with the transition between different spin sublevels can be included in the calculation. It is assumed that the transition of two orthogonal polarizations is coupled through the spin–flip relaxation process. Hence, the phase information of the two orthogonal polarizations can be evaluated simultaneously using the four-level model. In addition, this model allows the introduction of saturable dispersion and intrinsic optical anisotropies in the calculation. In the following paragraphs, the model for the polarization dynamics of VCSELs based on the angular momentum dependence of the conduction and valence bands of the QW semiconductor is derived. Gain anisotropy and birefringence are also incorporated into the four-level model for the calculation of the polarization stability of VCSELs.

4.4.1 Derivation of the Four-Level Model of VCSELs

The polarization of laser light is of a quantum nature and originates in the spin sublevels of the lasing transition between the conduction and valence bands of the semiconductor. For unstrained QWs with quantum confinement in the z direction, the zero-momentum electron state of the conduction band has a total angular momentum of $\boldsymbol{J}_z = \pm\frac{1}{2}$, while the zero-momentum hole state is dominated by heavy hole valence band with total angular momentum of $\boldsymbol{J}_z = \pm\frac{3}{2}$. If the QW active layer is perpendicular to the direction of laser emission z, which coincides with the quantization axis so that the electric field is in the $h-v$ plane, two circular polarized modes can be modeled by the transition between these spin sublevels of conduction and valence bands (see Fig. 4.9). Because of the quantum nature of light, dipole transitions are allowed only for $\Delta\boldsymbol{J}_z = \pm 1$ where left and right circularly polarized light corresponds to $\Delta\boldsymbol{J}_z = +1$ and $\Delta\boldsymbol{J}_z = -1$, respectively. Therefore, the allowed dipole transition from $\boldsymbol{J}_z = \frac{1}{2}$ to $\boldsymbol{J}_z = \frac{3}{2}$ is associated with

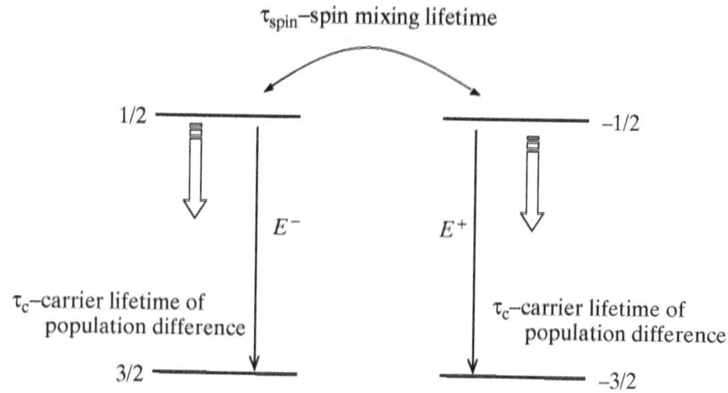

Figure 4.9 Four-level model for polarization dynamics in quantum-wells VCSELs.

the left circularly polarized light and that from $\boldsymbol{J}_z = -\frac{1}{2}$ to $\boldsymbol{J}_z = -\frac{3}{2}$ is associated with the right circularly polarized light.

The dipole transitions can be modeled using a four-level model. The vector electric field for a single transverse mode can be written as

$$\boldsymbol{E} \cong (F_h(x, y, t)\hat{\boldsymbol{h}} + F_v(x, y, t)\hat{\boldsymbol{v}}) \cdot \exp(-j\beta z + j\omega t), \tag{4.57}$$

where F_h and F_v are the slowly varying amplitudes of the linearly polarized components of the transverse mode in h and v directions. The linearly polarized fields can be transformed in terms of the left and right circularly polarized components

$$F_{\pm} = \frac{1}{\sqrt{2}}(F_v \pm jF_h), \tag{4.58}$$

where $F_{\pm}$ are associated with the transitions $\mp\frac{1}{2} \rightarrow \mp\frac{3}{2}$, respectively. The Maxwell–Bloch equations can be written as [34]

$$\frac{\partial}{\partial t}F_{\pm} = -\tau_p^{-1}F_{\pm} - jg_0^*P_{\pm} \tag{4.59}$$

$$\frac{\partial}{\partial t}P_{\pm} = -(\tau_d^{-1} + j(\omega_g - \omega))P_{\pm} + jg_0F_{\pm}(D \pm d), \tag{4.60}$$

$$\frac{\partial}{\partial t}D = -\tau_c^{-1}(D - J) + jg_0^*(F_+^*P_+ + F_-^*P_-), \tag{4.61}$$

$$\frac{\partial}{\partial t}d = -\tau_J^{-1}d + jg_0^*(F_+^*P_+ - F_-^*P_-), \tag{4.62}$$

where $P_{\pm}$ are the corresponding slowly varying amplitudes of the polarization fields, ω is the frequency of the laser cavity, ω_g is the frequency associated with the bandgap, and g_0 and J are the coupling constant and incoherent pumping parameters, respectively, associated with the injection current. In (4.59)–(4.62),

τ_p and τ_d represent the photon lifetime of the laser cavity and lifetime of the dipole polarization, respectively. The population difference D has a lifetime of τ_c, and that of d has a lifetime of τ_J; τ_J can be expressed as

$$\tau_J^{-1} = \tau_c^{-1} + 2\tau_{spin}^{-1}, \tag{4.63}$$

where τ_{spin} is the lifetime of spin–flip carriers, which takes into account the mixing of populations with the opposite value of $\boldsymbol{J}_z$. Hence, the influence of the "spin–flip relaxation process" can be factored into this model. The population differences D and d are defined as

$$D = \tfrac{1}{2}((\vec{n}_{1/2} + \vec{n}_{-1/2}) - (\vec{n}_{3/2} + \vec{n}_{-3/2})), \tag{4.64a}$$

$$d = \tfrac{1}{2}((\vec{n}_{1/2} + \vec{n}_{-3/2}) - (\vec{n}_{1/2} + \vec{n}_{3/2})), \tag{4.64b}$$

where $\vec{n}_i$ is the population of the ith spin sublevel ($i = \pm\frac{1}{2}, \pm\frac{3}{2}$), D is associated with the total population difference between the conduction and the valence bands, and d is associated with the difference in population inversions associated with right and left circularly polarized emissions.

The four-level model can be simplified if the relaxation rate of dipole polarization $1/\tau_d$ is assumed to be much faster than the time derivative of dipole polarization. Hence, (4.60) can be rewritten as

$$P_\pm = \chi_\pm F_\pm \cong \frac{g_0(\Omega + j\tau_d^{-1})}{\tau_d^{-2} + \Omega^2}(D \pm d)F_\pm, \tag{4.65}$$

where $\Omega = \omega_g - \omega$ is called the *detuning* [35]. Hence, the vector rate equations of $F_\pm$ can be simplified by substituting (4.65) into (4.59), which gives

$$\frac{\partial}{\partial t}F_\pm = -\tau_p^{-1}F_\pm + \tau_p^{-1}(1 - j\alpha_H)(g'\tau_p\tau_d^{-1}(D \pm d))F_\pm, \tag{4.66}$$

where the linewidth enhancement factor α_H, of the single-mode semiconductor laser theory [18] is defined in terms of the susceptibilities $\chi_\pm$ as

$$\alpha_H \equiv \frac{\mathrm{Re}(\chi_\pm)}{\mathrm{Im}(\chi_\pm)} = \Omega\tau_d. \tag{4.67}$$

In semiconductor materials, α_H usually has magnitude, which varies between 2 to 6. Furthermore, multiplying (4.66) by $\sqrt{2g'}$ gives

$$\frac{\partial}{\partial t}E_\pm = -\tau_p^{-1}E_\pm + \tau_p^{-1}(1 - j\alpha_H)(N \pm n)E_\pm. \tag{4.68}$$

The negative sign in front of α_H represents the reduction of refractive index due to the increase of $N \pm n$, which is consistent with the expression of the traveling-wave Equation (4.7). Again, the carrier rate Equation (4.61) can also be transformed as follows using the procedures described above

$$\frac{\partial}{\partial t}N = -\tau_c^{-1}(N-\mu) - \tau_d^{-1}(|E_+|^2 + |E_-|^2)N - \tau_d^{-1}(|E_+|^2 - |E_-|^2)n, \tag{4.69}$$

where $g'\tau_p/\tau_d$ has been multiplied to (4.61). Similarly, it can be shown that the rate equation of d is given by

$$\frac{\partial}{\partial t}n = -\tau_J^{-1}n - \tau_d^{-1}(|E_+|^2 - |E_-|^2)N - \tau_d^{-1}(|E_+|^2 + |E_-|^2)n. \tag{4.70}$$

In the preceding derivation of rate equations, the following substitutions have been used:

$$E_\pm = \sqrt{2g'}F_\pm, \qquad N = \frac{g'\tau_p}{\tau_d}D, \qquad n = \frac{g'\tau_p}{\tau_d}d, \tag{4.71a}$$

$$\alpha_H = \Omega\tau_d = (\omega_g - \omega)\tau_d, \qquad g' = \frac{|g_0|^2}{\tau_d^{-2} + \Omega^2},$$

$$J_c = \frac{\tau_d}{g'\tau_p}, \qquad \text{and} \qquad \mu = \frac{J}{J_c}. \tag{4.71b}$$

Therefore, the variable N can be interpreted as the difference between the actual carrier number and carrier number at transparency, and the variable n is the difference in the carrier numbers of the two magnetic sublevels; n contributes to the cross-saturation coupling of magnitude and phase between the left and right circularly polarized fields.

It is noted that (4.68) - (4.70) can be reduced to the familiar two-level rate equation model, provided $n = 0$ and $E_+ = E_-$ are assumed. On the other hand, if the magnitude of τ_J in (4.70) is very small, n can quickly relax to zero with the increase of time. As a result, the two modal amplitudes E_+ and E_- are coupled to a single carrier population N and the rate equations of E_+ and E_- are degenerated. This model is sometimes assumed phenomenologically for the dual-polarization semiconductor lasers (i.e., $E_+ \neq E_-$). This limit corresponds to a very fast mixing of populations with different spins in which the spin dynamics can be adiabatically eliminated. However, if τ_J takes on its maximum value given by the radiative lifetime of the carriers (i.e., when $\tau_J = \tau_d$), the right and left circularly polarized transitions can be decoupled and two sets of independent equations for $(E_+, N_+ = N + n)$ and $(E_-, N_- = N - n)$ emerge.

From experimental measurements of spin relaxation lifetime τ_{spin}, in QW materials [36, 37], it is estimated that $\tau_c \times 10^{-2} \leq \tau_{spin} \leq \tau_c$, where $\tau_c \sim 1$ ns. It is noted that the spin–flip lifetime (i.e., τ_{spin}) is on a larger timescale than the photon lifetime (i.e., $\tau_p \sim 1$ ps) of VCSELs. In (4.59)—(4.62), the lifetime of dipole polarization τ_d has the shortest timescale because $\tau_c \geq \tau_d$. There are several spin relaxation processes for electrons, and holes have been identified in semiconductors [36,38–40], including scattering by defects [41,42], exchange interactions between electrons and holes [41], and exciton–exciton exchange interactions [42], all of which contribute to τ_{spin}.

4.4.2 Gain Anisotropy and Birefringence in VCSELs

The polarization state of light emitted by a laser depends on two main facts. The first is the angular momentum of the quantum states involved in the material transitions for emission or absorption. Emission of a quantum of light with right (left) circular polarization corresponds to a transition in which the projection of the total material angular momentum on the direction of propagation changes by +1 (−1). This factor of polarization selection has been discussed in Section 4.4.1. The second factor is associated with the intrinsic optical anisotropies, which lead to a preference for a particular polarization state of the laser light. The second factor can be deliberately introduced into (4.68) through the loss anisotropy γ_a and birefringence γ_{bf}. The modified four-level model of vector rate equations is given by

$$\frac{dE_\pm}{dt} = -\tau_{\text{p}}^{-1}E_\pm + j\omega_{\text{sf}}E_\pm + \tau_{\text{p}}^{-1}(1 - j\alpha_{\text{H}})(N \pm n)E_\pm - (\gamma_{\text{a}} + j\gamma_{\text{bf}})E_\mp, \tag{4.72}$$

$$\frac{dN}{dt} = -\tau_{\text{c}}^{-1}(N - \mu) - \tau_{\text{d}}^{-1}(N + n)|E_+|^2 - \tau_{\text{d}}^{-1}(N - n)|E_-|^2, \tag{4.73}$$

$$\frac{dn}{dt} = -\tau_{\text{J}}^{-1}n - \tau_{\text{d}}^{-1}(N + n)|E_+|^2 + \tau_{\text{d}}^{-1}(N - n)|E_-|^2, \tag{4.74}$$

where $\omega_{\text{sf}}(= \alpha_{\text{H}}/\tau_{\text{p}})$ is deliberately introduced into (4.72) to normalize the complex field amplitude at threshold to zero. It is noted that the values of γ_{a} and γ_{bf} depend critically on the waveguide design of VCSELs. Because of the optical anisotropies, there are often two preferred orthogonal polarizations that coincide with the crystal axes of the laser cavity. The meaning and effect of the parameters γ_{a} and γ_{bf} are most clearly displayed when these vector rate equations are rewritten in terms of the two orthogonal polarizations, E_h and E_v, as shown below:

$$E_h = -j\frac{E_+ - E_-}{\sqrt{2}} \quad \text{and} \quad E_v = \frac{E_+ + E_-}{\sqrt{2}}. \tag{4.75}$$

Hence, the four-level model expressed in terms of E_h and E_v are given by [43]

$$\frac{dE_h}{dt} = -(\tau_{\text{p}}^{-1} - \gamma_{\text{a}})E_h + j(\tau_{\text{p}}^{-1}\alpha_{\text{H}} + \gamma_{\text{bf}})E_h + \tau_{\text{p}}^{-1}(1 - j\alpha_{\text{H}})(NE_h - jnE_v), \tag{4.76}$$

$$\frac{dE_v}{dt} = -(\tau_{\text{p}}^{-1} + \gamma_{\text{a}})E_v + j(\tau_{\text{p}}^{-1}\alpha_{\text{H}} - \gamma_{\text{bf}})E_v + \tau_{\text{p}}^{-1}(1 - j\alpha_{\text{H}})(NE_v + jnE_h), \tag{4.77}$$

$$\frac{dN}{dt} = -\tau_d^{-1}(N(1 + |E_h|^2 + |E_v|^2) - \mu + jn(E_h E_v^* - E_v E_h^*)), \tag{4.78}$$

$$\frac{dn}{dt} = -\tau_J^{-1} n - \tau_d^{-1}(n(|E_h|^2 + |E_v|^2) + jN(E_h E_v^* - E_v E_h^*)), \tag{4.79}$$

where $\tau_c = \tau_d$ is assumed in the derivation. If $\gamma_a \neq \gamma_{bf} \neq 0$, the solutions of (4.76)—(4.79) are orthogonal polarizations and their orientation is restricted to one of two specific states polarized in the h and v directions by γ_a and γ_{bf}. As the in-plane gain distribution is assumed to be uniform in (4.76)—(4.79), the optical gain and lasing frequency of the two orthogonal polarizations depend mainly on γ_a and γ_{bf}. It can be shown that γ_a leads to different thresholds for these two polarizations and that the h mode has the lower threshold when γ_a is positive (the v mode is favored if γ_a is negative). In addition, γ_{bf} leads to a frequency difference of $2\gamma_{bf}$ between the h and v modes (the v mode has lower frequency when γ_{bf} is positive). The main difference between the four-level model derived in this section and the two-level model given in Section 4.2 is that in the four-level model, the intrinsic detuning of frequency between the two orthogonal polarizations has been introduced into calculation through the presence of the linewidth enhancement factor α_H (i.e., saturable dispersion) and $n(\neq 0)$. Hence using the four-level model, the phase information of the two orthogonal polarizations can be evaluated in VCSELs.

4.4.3 Stability Analysis of Polarizations in Isotropic Medium

The stability of the two orthogonal polarizations can be analyzed by the perturbation method. This can be done by assuming that the steady-state solutions of $E_\pm$, N, and n have the form

$$E_\pm = Q_{s\pm} \exp(j(\omega_{s\pm} t \pm \theta_r) + j\theta_a), \qquad N = N_s, \qquad \text{and } n = n_s, \tag{4.80}$$

where the subscript "s" stands for steady state, θ_a is an arbitrary phase that can be ignored, and θ_r is a relative phase. For the case of isotropy (i.e., $\gamma_a = \gamma_{bf} = 0$), the two circularly polarized components degradation in amplitude and frequency can be deduced by setting the time derivative terms of (4.72)—(4.74) to zero, it can be shown that

$$Q_{\pm s} = \sqrt{\frac{\mu - 1}{2}} \qquad \text{and} \qquad \omega_{\pm s} = 0, \tag{4.81}$$

where $N_s \to 1$ and $n_s \to 0$ are assumed in the derivation of (4.81). Furthermore, the projections of the circular polarizations on the h and v directions are found to be

$$E_h = \sqrt{\mu - 1}\sin\theta_r \quad \text{and} \quad E_v = \sqrt{\mu - 1}\cos\theta_r, \tag{4.82}$$

which are the orthogonal polarizations at an arbitrary value of θ_r. Therefore, it is shown that in a steady-state condition, this solution is linearly stable for any finite value of parameters, but if $\tau_J \to 0$ (implies $|E_-|^2 - |E_+|^2 \to \infty$), it becomes marginally stable with respect to amplitude fluctuations. This means that the finite value of τ_J stabilizes the linearly polarized emission and destabilizes circularly polarized or elliptically polarized emission in an isotropic medium. In addition, the lasing frequency of the two orthogonal polarizations degenerates to zero as $n_s \to 0$ at steady state.

4.4.4 Stability Analysis of Polarizations in Birefringence

For the case $\gamma_{bf} \neq 0$ but $\gamma_a = 0$, four types of steady-state solutions can be obtained (see Fig. 4.10). Two of them have orthogonal linear polarization, which are defined as the h and v modes. For each of these modes, the circular polarized components have equal amplitudes. The other two types of solutions are elliptically polarized for which the circular polarized components have unequal amplitude [43]:

1. The solutions of the two orthogonal linear polarizations are derived (i.e., Figs. 4.10a and 4.10b). Letting $E_h = 0$, and setting the time derivative terms of

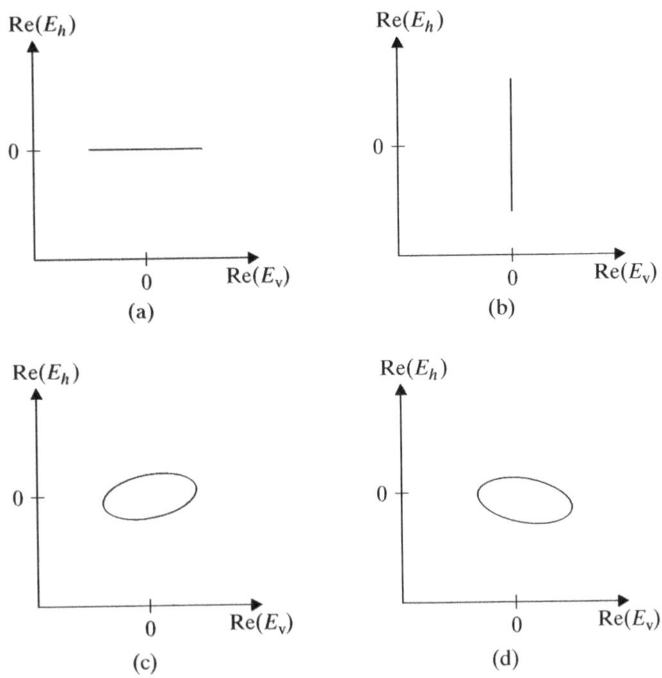

Figure 4.10 Steady-state solutions of (4.72)–(4.74): (a) v mode; (b) h mode; and (c,d) elliptically polarized (after Ref. 34).

(4.73) and (4.74) to zero, the v-mode solution can be expressed as

$$Q_{s\pm} = \sqrt{\frac{\mu - 1}{2}}, \omega_{s\pm} = -\gamma_{bf}, \quad \text{and} \quad \theta_r = 0 \tag{4.83}$$

and the corresponding two orthogonal polarizations are given by

$$E_v = \sqrt{\mu - 1}\exp(-j\gamma_{bf}t) \quad \text{and} \quad E_h = 0. \tag{4.84}$$

Again letting $E_v = 0$ and setting the time derivative terms of (4.73) and (4.74) to zero, the h-mode solution can be expressed as

$$Q_{s\pm} = \sqrt{\frac{\mu - 1}{2}}, \omega_{s\pm} = \gamma_{bf}, \quad \text{and} \quad \theta_r = \frac{\pi}{2} \tag{4.85}$$

and the corresponding orthogonal polarization are given by

$$E_v = 0 \quad \text{and} \quad E_h = \sqrt{\mu - 1}\exp(j\gamma_{bf}t). \tag{4.86}$$

The corresponding steady-state values of the total carrier population and the population difference between the sublevels with opposite value of the spin for both linearly polarized solutions are

$$N_s = 1 \quad \text{and} \quad n_s = 0. \tag{4.87}$$

Hence, it is shown that γ_{bf} leads to a frequency difference of $2\gamma_{bf}$ between the h and v modes.

2. The solutions of the two elliptical polarizations, which is similar to that of the derivation of orthogonal polarizations, are deduced (i.e., Figs 4.10c and 4.10d). This is done by setting the time derivative terms of (4.72)—(4.74) to zero and assuming that $E_+ \neq E_- \neq 0$. The two elliptically polarized solutions are given by

$$Q_{s\pm}^2 = \frac{1}{2}(\mu - N_s)\left(1 \mp \frac{N_s - 1}{n_s}\right), \tag{4.88}$$

$$\omega_{s\pm} = -\tau_p^{-1}\alpha_H \frac{(N_s - 1)^2 - n_s^2}{N_s - 1}, \tag{4.89}$$

$$\tan(2\theta_r) = -\frac{1}{\alpha_H}\frac{N_s - 1}{n_s}. \tag{4.90}$$

These two solutions are distinguished by the two values for the normalized population difference n_s, which are given by

$$n_s^2 = \frac{(\mu - N_s)(N_s - 1)N_s}{\tau_d\tau_J^{-1} + \mu - N_s}. \tag{4.91}$$

The corresponding value for N_s is obtained from

$$\gamma_{bf}^2\left(\frac{\tau_d}{\tau_J}+\mu-N_s\right)^2=\tau_p^{-2}\left(\left(\frac{\tau_d}{\tau_J}+\mu-N_s\right)-\frac{\tau_d}{\tau_J}N_s\right)$$

$$\times\left(\left(\frac{\tau_d}{\tau_J}+\mu-N_s\right)(N_s-1)+\alpha_H^2(\mu-N_s)N_s\right) \quad (4.92)$$

but is restricted to the condition that $N_s > 1$, which implies the pumping level should satisfy

$$\mu > 1+\frac{\tau_p\tau_d\gamma_{bf}}{\tau_J(\alpha_H-\tau_p\gamma_{bf})}. \quad (4.93)$$

The stability of the four types of steady-state solutions given above can be examined by writing their solutions as [43]

$$E_\pm=(Q_{s\pm}+a_\pm)\exp(j(\omega_{s\pm}t\pm\theta_r)+j\theta_a), \quad (4.94a)$$

$$N=N_s+\Delta N, \quad (4.94b)$$

$$n=N_s+\delta n, \quad (4.94c)$$

where $a_\pm$ is a complex perturbation of the field amplitude, and ΔN and δn are real perturbations of N and n, respectively. After substituting the perturbed solutions (4.94) into (4.72)—(4.74) and linearizing to the first-order in the perturbation, one obtains the following set of linear coupled differential equations for $a_\pm$, ΔN, and δn:

$$\frac{\partial a_\pm}{\partial t}=\tau_p^{-1}(1-j\alpha_H)(N_s\pm N_s-1)a_\pm+j\alpha_H\tau_p^{-1}a_\pm$$

$$+\tau_p^{-1}(1-j\alpha_H)(\Delta N\pm\delta n)Q_{s\pm}-j\gamma_{bf}a_\pm\exp(\mp j2\theta_r), \quad (4.95a)$$

$$\frac{\partial \Delta N}{\partial t}=-\tau_d^{-1}(N_s+N_s)Q_{s+}(a_++a_+^*)-\tau_d^{-1}(N_s-N_s)Q_{s-}(a_-+a_-^*)$$

$$-\tau_d^{-1}(1+Q_{s+}^2-Q_{s-}^2)\Delta N-\tau_d^{-1}(Q_{s+}^2-Q_{s-}^2)\delta n, \quad (4.95b)$$

$$\frac{\partial \delta n}{\partial t}=-\tau_d^{-1}(N_s+N_s)Q_{s+}(a_++a_+^*)+\tau_d^{-1}(N_s-N_s)Q_{s-}(a_-+a_-^*)$$

$$-\tau_d^{-1}(Q_{s+}^2-Q_{s-}^2)\Delta N-[\tau_J^{-1}+\tau_d^{-1}(Q_{s+}^2+Q_{s-}^2)]\delta n. \quad (4.95c)$$

To simplify the notation, (4.95) is written in vectorial form

$$\frac{\partial}{\partial t}\mathbf{A}=\boldsymbol{M}\mathbf{A}, \quad (4.96)$$

where $\mathbf{A}=(a_+,a_+^*,a_-,a_-^*,\Delta N,\delta n)^T$ and $\boldsymbol{M}$ is a 6×6 matrix whose coefficients can be easily derived from (4.95). The eigenvalues of $\boldsymbol{M}$ are determined by a sixth-order polynomial that has to be solved. The stability of a steady-state solution is given by the real parts of the eigenvalues λ [i.e., stable if $\mathrm{Re}(\lambda) < 0$ for all λ], while the imaginary part of λ gives a frequency characteristic of the evolution of the perturbation.

The stability of the two orthogonal polarizations can be analyzed by substituting the steady-state v mode solutions, (4.83) and (4.87) or the steady-state h-mode solutions, (4.85) and (4.87), in (4.96). The set of equations given by (4.96) can be decoupled into two independent subsets if the equations are rewritten for the variables $S = a_+ + a_-$ and $R = a_+ - a_-$ [34]. The first subset is

$$\frac{\partial S}{\partial t} = 2\tau_{\mathrm{p}}^{-1}(1 - j\alpha_{\mathrm{H}})Q\Delta N, \tag{4.97a}$$

$$\frac{\partial S^*}{\partial t} = 2\tau_{\mathrm{p}}^{-1}(1 + j\alpha_{\mathrm{H}})Q\Delta N, \tag{4.97b}$$

$$\frac{\partial \Delta N}{\partial t} = -\tau_{\mathrm{d}}^{-1}QS - \tau_{\mathrm{d}}^{-1}QS^* - \tau_{\mathrm{d}}^{-1}(1 + 2Q^2)\Delta N, \tag{4.97c}$$

where $Q = Q_{\mathrm{s}\pm}$ is assumed. These equations determine the stability of a polarization (i.e., h or v mode) with respect to the perturbation of same polarization. This subset of equations is independent of γ_{bf} and τ_{J}. The general solutions of (4.97) are

$$\begin{bmatrix} S \\ S^* \\ \Delta N \end{bmatrix} = \begin{bmatrix} S_{\mathrm{s}} \\ S_{\mathrm{s}}^* \\ \Delta N_{\mathrm{s}} \end{bmatrix} \exp\left(\frac{\lambda t}{\tau_{\mathrm{d}}}\right). \tag{4.98}$$

It can be shown that one of the three eigenvalues of (4.98) has zero value and the other two are of complex expressions but with negative real part. This means that each steady-state solution of the two orthogonal polarizations is unconditionally stable with respect to amplitude perturbation of the same polarization.

The second subset of equations is

$$\frac{\partial R}{\partial t} = 2\tau_{\mathrm{p}}^{-1}(1 - j\alpha_{\mathrm{H}})Q\delta n \pm j2\gamma_{\mathrm{bf}}R, \tag{4.99a}$$

$$\frac{\partial R^*}{\partial t} = 2\tau_{\mathrm{p}}^{-1}(1 + j\alpha_{\mathrm{H}})Q\delta n \mp j2\gamma_{\mathrm{bf}}R^*, \tag{4.99b}$$

$$\frac{\partial \delta n}{\partial t} = -\tau_{\mathrm{d}}^{-1}QR - \tau_{\mathrm{d}}^{-1}QR^* - (\tau_{\mathrm{J}}^{-1} + 2\tau_{\mathrm{d}}^{-1}Q^2)\delta n, \tag{4.99c}$$

where the upper and lower signs denote the stabilities of h and v modes, respectively, in steady state. This subset determines the stability of a polarized solution with respect to perturbations of the "orthogonal polarization." To determine the eigenvalues of (4.99), it is assumed that

$$\begin{bmatrix} R \\ R^* \\ \delta n \end{bmatrix} = \begin{bmatrix} R_{\mathrm{s}} \\ R_{\mathrm{s}}^* \\ \delta n_{\mathrm{s}} \end{bmatrix} \exp\left(\frac{\lambda t}{\tau_{\mathrm{d}}}\right), \tag{4.100}$$

and the resulting eigenequation (i.e., third-order polynomial) for λ is

$$P(\lambda) = \lambda^3 + \underbrace{\left(\frac{\tau_d}{\tau_J} + \mu - 1\right)}_{\lambda^2}\lambda^2 + \underbrace{\left(2\frac{\tau_d}{\tau_p}(\mu - 1) + 4\left(\frac{\tau_d}{\tau_J}\right)^2\right)}_{\lambda^1}\lambda$$

$$\underbrace{+ 4(\gamma_{bf}\tau_d)^2\left(\frac{\tau_d}{\tau_J} + \mu - 1\right) \pm 4\frac{\tau_d^2\gamma_{bf}}{\tau_p}\alpha_H(\mu - 1)}_{\lambda^0}, \tag{4.101}$$

where the upper and lower signs denote the stabilities of v and h modes, respectively, in steady state. If the coefficient of λ^0 of (4.101) is zero (i.e., equivalent to, say, $\gamma_{bf} = 0$), there is a zero eigenvalue associated with the arbitrariness of the polarization direction. It can also be shown that the other two complex eigenvalues of (4.101) always have a negative real part, which implies that the system is unconditionally stable. However, if the coefficient of λ^0 of (4.101) is nonzero, the zero eigenvalue becomes nonzero, which determines the stability of the system. On the other hand, if $\alpha_H = 0$ (i.e., if there is no coupling between amplitude and phase fluctuations), both solutions of the two orthogonal polarizations are always stable [i.e., the coefficients of the polynomial (4.101) are all positive as $\mu > 1$]. Hence, there exists a regime of bistability for any value of μ or γ_{bf} and no polarization switching occurs as the injection current is changed, but the nonvanishing value of α_H, together with the phase anisotropy, causes polarization switching.

Now, the stability of a particular solution for a general value of α_H in terms of two control parameters, the injection current μ and the birefringence parameter $\gamma_{bf}\tau_d$, is analyzed. This can be done by selecting a critical value of μ so that any variation from that value of μ can lead to purely exponential growth or decay of the polarized solutions. The condition for the h-polarized solution can be obtained from (4.101) by setting the coefficient of λ^0 to zero. The critical value of μ at which the stability of this solution change is given by

$$\mu_h = 1 + \frac{\tau_p\tau_d\gamma_{bf}}{\tau_J(\alpha_H - \tau_p\gamma_{bf})}. \tag{4.102}$$

Using the same augment (i.e., the coefficient of λ^0 should be less than or equal to zero for the requirement that all λ real parts be negative), it can be shown that the h-mode solution is always stable for any $\mu < \mu_h$. For the v-polarized solution, the condition for the v-polarized solution can be obtained from (4.101) by equating the product of coefficients of λ^2 and λ^1 to the coefficient of λ^0. Hence, the critical value of μ at which the stability of this solution change is given by

$$\mu_v = 1 - \tau_d\tau_J^{-1} + 2\alpha_H\tau_d\gamma_{bf}. \tag{4.103}$$

For the v-polarized solution to be stable, all real parts of λ values must be negative, which implies that $\mu > \mu_v$. Figure 4.11 shows the four different regions

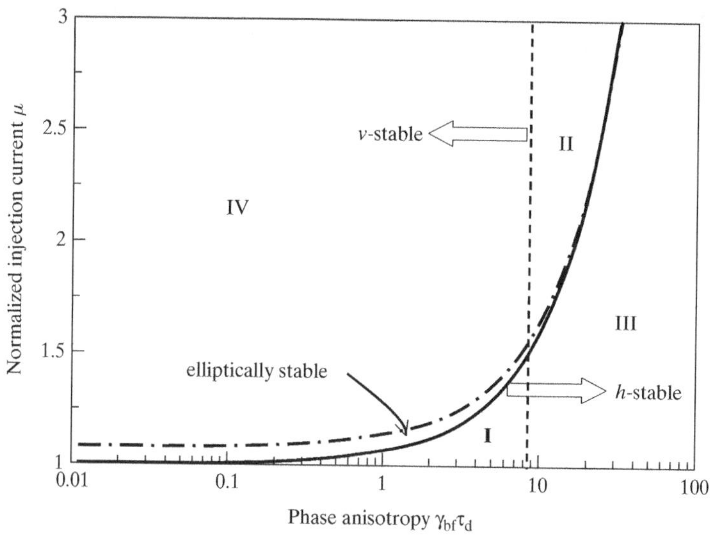

Figure 4.11 Stability diagram for the steady-state solutions. The h mode is stable below the solid curve, while the v mode is stable to the left of the dashed line. These two curves divide four zones of different stability for the two linearly polarized solutions. In I, both solutions are stable; in II, neither solution is stable; in III, only the h mode is stable; and in IV, only the v mode is stable. Elliptically polarized solutions are stable within the narrow region between the solid and dashed–dotted curves. (After Ref. 43).

with different stabilities for the linearly polarized solutions obtained from (4.102) and (4.103): region I—both orthogonal polarizations are stable; region II—both are unstable; region III—only the h-mode solution is stable; region IV—only the v-mode solution is stable. The stability diagram is a consequence of the combined effect of saturable dispersion associated with the α_H factor and spin dynamics associated with a finite value of τ_J. In the calculation above, the following parameters have been used : $\tau_p = 300$ ns, $\tau_d = 1$ ns, $\tau_J = 50$ ns, and $\alpha_H = 3.0$.

The linear stability of elliptically polarized solutions can also be examined by solving (4.96) numerically for the sixth-order polynomial eigenvalues. To find the stability of a particular elliptical polarization, the coefficients of the polynomial of the eigenequation have to be obtained numerically in order to deduce the corresponding eigenvalues. The stability is determined by looking at the real part of the eigenvalue as described previously. The condition of a stable elliptically polarized solution is also shown in Figure 4.11. The elliptically polarized solution is stable in a narrow domain of parameters bounded by the dashed and solid lines in which μ is close to but larger then μ_h.

4.4.5 Stability Analysis of Polarizations in Gain Anisotropy and Birefringence

For the case γ_{bf} and $\gamma_a \neq 0$, the h and v modes have different thresholds. This is a typical experimental situation as small amplitude anisotropies are unavoidable.

Using an approach similar to that described in the sections above, the steady-state v-mode solutions are given by

$$Q_{s\pm}^2 = \frac{1}{2}\frac{\mu - N_s}{N_s}, \qquad \omega_{s\pm} = -\gamma_{bf} - \gamma_a\alpha_H, \qquad \theta_r = 0, \tag{4.104a}$$

$$N_s = 1 + \gamma_a\tau_p \quad \text{and} \quad n_s = 0, \tag{4.104b}$$

and the steady-state h-mode solutions by:

$$Q_{s\pm}^2 = \frac{1}{2}\frac{\mu - N_s}{N_s}, \qquad \omega_{s\pm} = \gamma_{bf} + \gamma_a\alpha_H, \qquad \theta_r = 0, \tag{4.105a}$$

$$N_s = 1 - \gamma_a\tau_p \quad \text{and} \quad n_s = 0. \tag{4.105b}$$

These equations show that the two orthogonal polarizations have different thresholds and oscillation frequencies. The split in threshold current and oscillation frequency between h and v modes are $-2\gamma_a\tau_p$ and $2\gamma_{bf} + 2\gamma_a\alpha_H$, respectively.

As in Section 4.4.4, the stability of the two orthogonal polarizations can be analyzed by the following set of equations:

$$\frac{\partial R}{\partial t} = 2\tau_p^{-1}(1 - j\alpha_H)Q\delta n \pm 2(\gamma_a + j\gamma_{bf})R, \tag{4.106a}$$

$$\frac{\partial R^*}{\partial t} = 2\tau_p^{-1}(1 + j\alpha_H)Q\delta n \pm 2(\gamma_a - j\gamma_{bf})R^*, \tag{4.106b}$$

$$\frac{\partial \delta n}{\partial t} = -\tau_d^{-1}N_sQR - \tau_d^{-1}N_sQR^* - (\tau_J^{-1} + 2\tau_d^{-1}Q^2)\delta n, \tag{4.106c}$$

where the upper and lower signs are for the stabilities of h and v modes, respectively, in steady state. The eigen equation can also be deduced in a similar manner and the corresponding characteristic polynomial for the eigenvalue λ is given by

$$\begin{aligned} P(\lambda) &= \lambda^3 + \left(\frac{\tau_d}{\tau_J} + 2Q^2 \mp \gamma_a\tau_d\right)\lambda^2 \\ &+ \left((\gamma_{bf}^2 + \gamma_a^2)\tau_d^2 + \frac{\tau_d}{\tau_p}Q^2N_s \mp \gamma_a\tau_d\left(\frac{\tau_d}{\tau_J} + 2Q^2\right)\right)\lambda \\ &+ 4\tau_d^2(\gamma_{bf}^2 + \gamma_a^2)(\tau_d\tau_J^{-1} + 2Q^2) \pm 8\frac{\tau_d^2}{\tau_p}Q^2N_s(\alpha_H\gamma_{bf} - \gamma_a), \end{aligned} \tag{4.107}$$

where $Q_{s+} = Q_{s-} = Q$ is assumed. The upper and lower signs denote the stabilities of v and h modes, respectively, in steady state. Hence using (4.107), the influence of gain and phase anisotropies on the stability of h and v modes can be investigated.

Figure 4.12a shows the stability diagram of h and v modes for the case $\gamma_a\tau_d = 0.1$. It is shown that the h mode is stable below the solid line, while the v mode

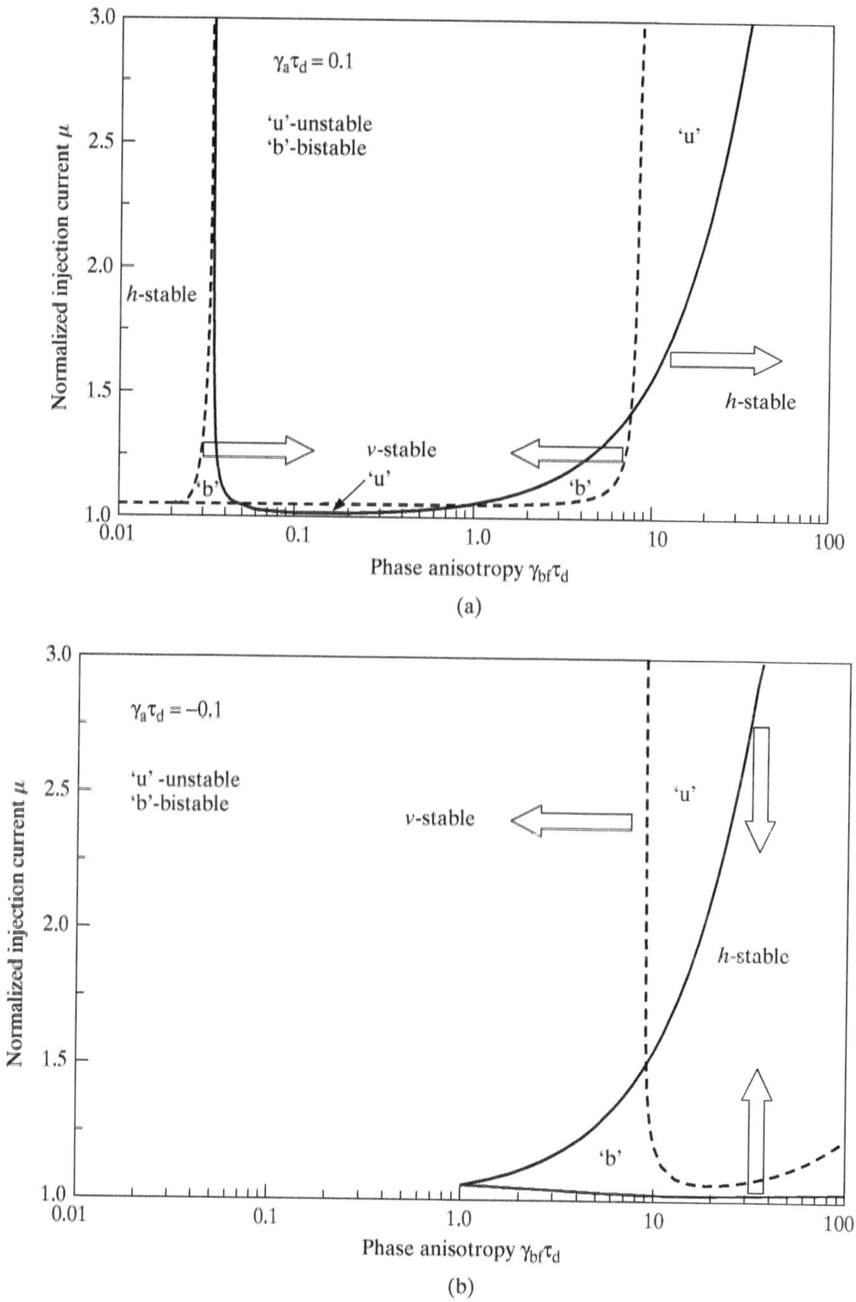

Figure 4.12 Stability diagram of the two orthogonally polarizations for (a) $\gamma_a \tau_d = 0.1$ and (b) $\gamma_a \tau_d = -0.1$; the parameters used in the calculation are similar to those given in Figure 4.11. (After Ref. 43.)

is stable inside the zone bounded by the dashed curves. As γ_{bf} goes to zero, only the h mode is stable as the h mode has a lower threshold current. On the other hand, for the zone with $\mu > 1.1$ and $0.03 < \gamma_{bf}\tau_d < 8.0$, only the v mode is stable. This indicates that despite the favoring by the gain anisotropy for the h mode, the emission will switch to v-mode emission as the current is increased near threshold at this zone. The zones of bistability and zones in which neither polarization is stable are also shown in the figure. Figure 4.12b shows the stability diagram of h and v modes for the case $\gamma_a\tau_d = -0.1$. It is shown that the h mode is stable in the region between the solid lines, while the v mode is stable to the left and below the dashed curve. Where γ_{bf} goes to zero, only the v mode is stable as it has the lowest threshold current. For $1.0 < \gamma_{bf}\tau_d < 10$, there is a switching of stability from the h mode favored near threshold to the v mode as the current μ increases from the threshold. The zones of bistable and zones in which neither polarization is stable are also shown in the figure. The other parameters utilized in the preceding calculations are the same as those used in Figure 4.11.

The preceding analysis of VCSELs using the four-level model indicates that the threshold and oscillation frequency of the two orthogonal polarizations are affected by gain anisotropy and birefringence. Hence, this four-level model is more flexible in explaining the polarization phenomena of VCSELs than is the two-level model. In fact, the four-level model may explain the overestimation of frequency splitting shown in Section 4.3, which is due to the ignorance of gain anisotropy. From Figure 4.12a, if $\gamma_{bf}\tau_d > 10$ and the laser is operating above threshold, the h mode of high frequency is stable and the corresponding split in frequency between the lasing and nonlasing modes, $\nu_v - \nu_h$, is $2(\gamma_{bf} - \gamma_a\alpha_H)$. The frequency splitting is now under the influence of gain anisotropy, which has been ignored in the assumption of Section 4.3 (i.e., frequency splitting is due only to birefringence). However, the value of $\gamma_a\alpha_H$ may be too small to explain the overestimation of frequency splitting. This is because if $\nu_v - \nu_h \sim +15$ GHz (i.e., $\gamma_{bf} \sim 47\tau_d^{-1}$), this implies $\gamma_a\alpha_H \sim 0.3\tau_d^{-1} \sim +0.04$ GHz, which is not large enough to reduce the value of frequency splitting to about +10 GHz as observed from the experiment. Hence, it is believed that the extra value of frequency splitting may also attribute to the influence of small cavity size, which is ignored in the derivation of the four-level model.

4.5 POLARIZATION NOISE IN VCSELs

Polarization noise is presented in all semiconductor lasers but is exceptionally strong in VCSELs. This is because spontaneous emission noise, which is the driving force of the polarization noise, is relatively strong in VCSELs because of the enhancement of the spontaneous emission factor by the small cavity size. In addition, the deterministic forces of the polarizations (i.e., intrinsic optical anisotropies) are relatively small (i.e., compared with facet emitting lasers), due to the cylindrical symmetry of the corresponding waveguide geometry. Hence, the combination of strong stochastic noise and weak restoring force creates relatively large polarization fluctuation. In the practical applications of VCSELs, it is necessary to understand the corresponding mechanisms of polarization fluctuation in order to avoid unnecessary generation of intensity noise. In this section, the influence of the initial conditions and Langevin noise sources on the time evolution of the two orthogonal polarizations is studied. Furthermore, it can be shown that the polarization characteristics of VCSELs such as optical spectra can be utilized to extract the nonlinear parameters (i.e., gain anisotropy and birefringence) of the modified four-level model.

4.5.1 Effects of Spontaneous Emission Noise

In Section 4.4, the polarization stability of VCSELs with isotropy medium is analyzed analytically. In the following paragraphs, the time evolution of both polarizations is calculated numerically with the influence of initial conditions and spontaneous emission noise included in the calculation. It can be shown that the excitation of polarizations is dependent on their initial conditions. In addition, the randomly generated spontaneous emission causes the hopping of the two orthogonal polarizations.

Spontaneous emission fluctuation can be introduced into the model by adding the complex noise term, $\xi(t)$ onto the RHS of (4.76) and (4.77). It is assumed that $\xi(t)$ has zero mean and a correlation time of the order of the dephasing time for atomic spontaneous emission or the inverse homogeneous linewidth. Since this correlation time is typically much shorter than noise spectrum, which can be treated as being effectively white and the following Langevin description for the noise is given by [25]

$$\langle \xi(t) \rangle = \langle \xi(t)\xi(t') \rangle = 0, \tag{4.108a}$$

$$\langle \xi^*(t)\xi(t') \rangle = \tau_p^{-2}\beta_s\Gamma_z\delta(t-t'), \tag{4.108b}$$

where $\beta_s(\sim 10^{-3} - 10^{-4})$ is the fraction of spontaneous emission that enters the lasing mode [44]. In (4.108), it is assumed that the electric fields are normalized in the rate equations and the stimulated emission rate into the lasing mode is equal to the total spontaneous emission rate into all modes.

The polarization properties of VCSELs under the influence of the initial selection of polarizations as well as normalized injection current μ are analyzed numerically. It is assumed the laser has parameters (i.e., $\gamma_{bf}\tau_d < 8$ and μ slightly greater than 1) favorable for bistable operation of h and v modes (i.e., region I of Fig. 4.11). $\xi(t)$ is initially set to zero for both orthogonal modes, and a small amount of spontaneous emission noise is allowed to couple with either h or v mode only at $t = 0$. Hence, the time evolution of both orthogonal polarizations is calculated numerically for different initial selection of polarizations and injection current. It can be shown that

- If the system begins with a v mode and the initial value ($t = 0$) of h polarization is set to zero, the v mode retained for μ either increase or decrease. This is because the v mode is stable for the whole range of injection currents (i.e., regions I and IV of Fig. 4.11) [43].
- If the system begins with the h-mode emission (i.e., v mode set to zero at $t = 0$), the optical emission switches to v mode at $\mu = \mu_h$. Further increase in μ (i.e., in region IV of Fig. 4.11) will not affect v-mode stability. It must be noted that once the laser reaches v-mode emission, it will remain stable for any further variation of injection current (i.e., larger or smaller than μ_h or inside the bistable region I of Fig. 4.11).

Hence, it is found that the excitation of modes in region I of Figure 4.11 is dependent on the corresponding initial value of the polarization emission so that h or v mode can be excited independently. However, in region IV of Figure 4.11, only the v mode is excited and is independent of the initial value of the polarization emission.

Now, if the laser has a large birefringence (i.e., $\gamma_{\mathrm{bf}}\tau_{\mathrm{d}} \sim 10$) comparable to the spin relaxation rate but the other parameters remain unchanged and start to operate at region III of Figure 4.11 for small μ (i.e., $\mu < \mu_h$), the emission will always be of h mode and is independent on the initial value of the emission polarization. At $\mu \geq \mu_h$ (i.e., region II of Fig. 4.11), the elliptical polarization state is observed, where the h mode is the predominant component. Further increase in μ will reduce the difference between the magnitude of the two polarizations, and again the time evolution of both orthogonal polarizations is independent of the initial value of the emission polarization. It is interesting to further explore the dynamic behavior of both polarizations at the unstable region II of Figure 4.11 under the influence of spontaneous emission.

Assuming that the lasers are in region II (i.e., $\gamma_{\mathrm{bf}}\tau_{\mathrm{d}} \sim 8$), the initial value of both polarized fields is randomly generated and the $\mu(\sim 1.51)$ is kept constant throughout the whole time interval. Figure 4.13a shows a sample of the time behavior of the mode intensities of the laser, and Figure 4.13b shows the corresponding statistical distribution of the duration of both orthogonal polarizations [45]. It can be shown that the two polarizations hop randomly with the increase of time. If the probability distribution of the two polarizations is plotted as a function of interval duration, it is found that the distributions are of exponential profile and have approximately the same decay rate (i.e., the larger the decay rate, the less probability to be found in the time evolution). In fact, it can be shown that if the values of μ and $\gamma_{\mathrm{bf}}\tau_{\mathrm{d}}$ are reduced, the magnitude of h mode as well as its probability of appearing in the time evolution will be reduced. Hence, mode hopping that appears in the unstable region II is due to the randomly generated spontaneous emission. However, their probability of appearance (i.e., in the time evolution at a constant injection current) is dependent on its birefringence as well as the injection current, which agree with the conclusions of Figure 4.11.

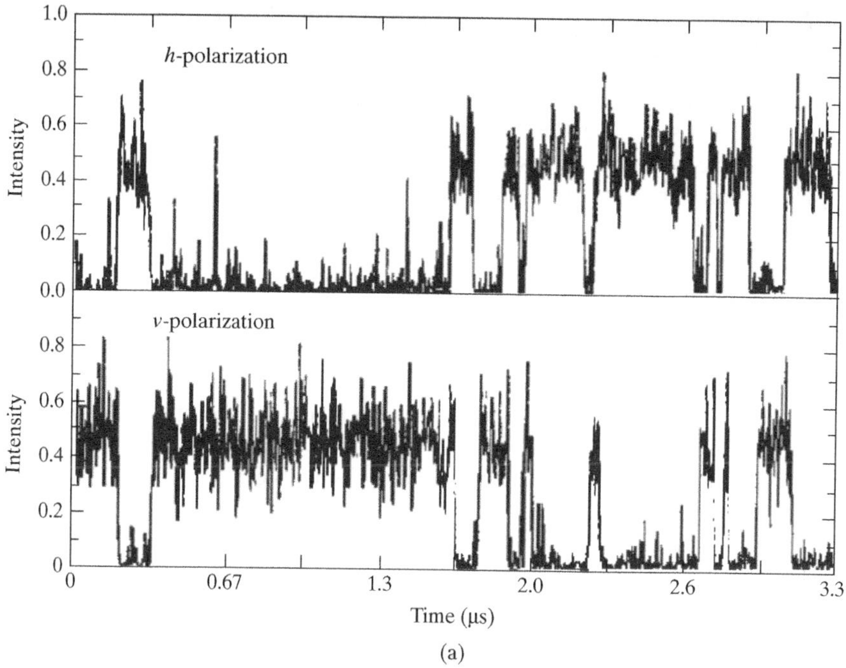

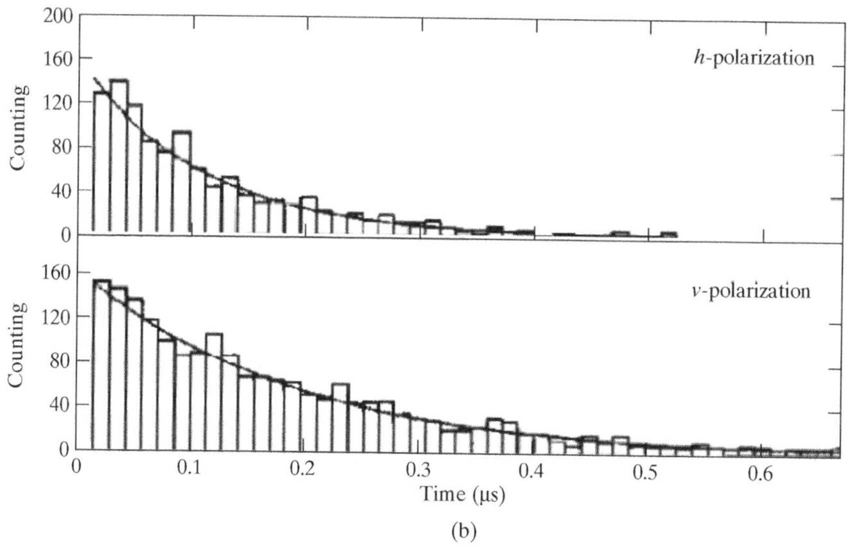

Figure 4.13 (a) Intensity of the h- and v-mode runs for a specific time interval; (b) corresponding statistical distributions of the durations of the h- and v-polarization intervals. (After Ref. 45).

4.5.2 Influence of Optical Anisotropies on the Polarization Fluctuation of VCSELs

In the preceding section, it has been shown that the fluctuation of polarization noise in VCSELs is dependent on the birefringence of laser cavity. In order to design a stable polarization laser, it is necessary to investigate the dependence of polarization dynamics on the intrinsic optical anisotropies of VCSELs. In this section, a simple one-dimensional model for the polarization dynamic of VCSELs is deduced from the four-level model by adiabatically eliminating spin inversion. Using this model, the optical anisotropies of VCSELs are evaluated from their polarization spectra obtained experimentally.

For practical VCSELs as discussed in the preceding sections, it is assumed that the lasing mode at steady state is polarized along the h axis. Therefore, the slowly varying transverse components of the optical field vector (4.57) due to the polarization noise can be approximated by suitable selection of intensity, phase, polarization direction, and polarization ellipticity as [46]:

$$\boldsymbol{E} \approx |F_h| \exp(-j\theta)\{\hat{\boldsymbol{h}} - (\phi + j\psi)\hat{\boldsymbol{v}}\}, \tag{4.109}$$

where $\phi, \psi \ll 1$ is assumed in the derivation of (4.109), $|F_h|^2$ is the optical intensity, and θ is the optical phase of F_h. The two Poincaré angles, ϕ and ψ, characterize the optical polarization of the v mode, where $\phi(0 \le \phi \le \psi)$ is the direction of the polarization ellipse and $\psi(-\pi/4 \le \psi \le \pi/4)$ is the ellipticity angle. Hence, the presence of ϕ and ψ is due to the fluctuation of polarization noise. It is shown in Section 4A.3 that under the conditions, $\phi, \psi \ll 1$, $\tau_d/\tau_J \gg 1$ and with constant optical intensity, the four-level model can be reduced to

$$\frac{\partial}{\partial t}\begin{bmatrix} \phi - \phi_s \\ \psi - \psi_s \end{bmatrix} = \begin{bmatrix} -\gamma_{\text{lin}} \cos 2\phi_r & -\omega_{\text{lin}} - 2\alpha_H \gamma_{\text{non}} \\ \omega_{\text{lin}} & -\gamma_{\text{lin}} \cos 2\phi_r - 2\gamma_{\text{non}} \end{bmatrix} \begin{bmatrix} \phi - \phi_s \\ \psi - \psi_s \end{bmatrix} + \begin{bmatrix} f_\phi \\ f_\psi \end{bmatrix}, \tag{4.110}$$

where the subscript "s" stands for steady state. The Langevin noise sources f_ϕ and f_ψ represent the spontaneous emission that perturbs the system away from the h mode's equilibrium state ($\phi = \psi = 0$) through $\phi(t)$ and $\psi(t)$. The complex noise terms, f_ϕ and f_ψ, comprise four independent real-valued numbers, which can be divided into phase noise, intensity noise, and two forms of polarization noise. The phase and amplitude noise has been given in (4.108), and the two polarization components are similar, uncorrelated, and real-valued Langevin noise sources of identical strength, which satisfy

$$\langle f_\psi(t_1) f_\psi(t_2)\rangle = \langle f_\phi(t_1) f_\phi(t_2)\rangle = \frac{n_{\text{sp}}\delta(t_1 - t_2)}{(p\tau_p)}, \tag{4.111a}$$

$$\langle |f_\psi(\omega)|^2\rangle = \langle |f_\phi(\omega)|^2\rangle = \frac{n_{\text{sp}}}{(p\tau_p)}, \tag{4.111b}$$

where p is the photon number and $n_{\text{sp}}(\ge 1)$ is the spontaneous emission number [47].

Equation (4.110) can be solved in the frequency domain by Fourier transform, and the corresponding solutions are given by

$$\phi(\omega) = \frac{(j\omega - \gamma_\beta - 2\gamma_{\mathrm{non}}) f_\phi(\omega) + (\omega_{\mathrm{lin}} + 2\alpha_{\mathrm{H}}\gamma_{\mathrm{non}}) f_\psi(\omega)}{(\omega - \omega_0 - j\gamma_0)(\omega + \omega_0 - j\gamma_0)}, \tag{4.112}$$

$$\psi(\omega) = \frac{-\omega_{\mathrm{lin}} f_\phi(\omega) + (j\omega - \gamma_\beta) f_\psi(\omega)}{(\omega - \omega_0 - j\gamma_0)(\omega + \omega_0 - j\gamma_0)}, \tag{4.113}$$

where $\lambda = -\gamma_0 \pm j\omega_0$ are the eigenvalues of (4.110), with

$$\gamma_0 = \gamma_\beta + \gamma_{\mathrm{non}}, \tag{4.114}$$

$$\omega_0 = \sqrt{(\omega_{\mathrm{lin}} + \alpha_{\mathrm{H}}\gamma_{\mathrm{non}})^2 - (\alpha_{\mathrm{H}}^2 + 1)\gamma_{\mathrm{non}}^2}, \tag{4.115}$$

where $\gamma_\beta = \gamma_{\mathrm{lin}} \cos 2\phi_{\mathrm{r}}$, γ_0, and ω_0 can be interpreted as effective loss anisotropy and effective birefringence, respectively. By combining these equations with the expressions for the polarization noise, it is relatively straightforward to calculate the experimentally accessible polarization-resolved optical spectra and intensity noise.

4.5.3 Polarization-Resolved Optical Spectra

It has been noted from (4.109) that the optical phase and intensity of the optical signal are dependent on the dominant $\hat{h}$ component of the optical field vector but the $\hat{v}$ component of the optical field vector gives the information on the polarization dynamics of the polarized noise. If the $\hat{h}$ component of the optical field vector is blocked, the corresponding optical field vector can be written as [46]

$$E_v(t) \approx -(\phi(t) + j\psi(t))F_h(t). \tag{4.116}$$

Furthermore, if the amplitude and phase of $F_h(t)$ are assumed constant, the v-mode spectrum, which dominates the polarization dynamic of VCSELs, is given by [48]

$$\langle |E_v(\omega)|^2\rangle \approx E_0^2 \langle |\phi(\omega) + j\psi(\omega)|^2\rangle$$
$$= \frac{n_{\mathrm{sp}}}{\tau_{\mathrm{p}}} \frac{E_0^2}{S} \frac{(\omega - \omega_{\mathrm{lin}})^2 + (\omega - \omega_{\mathrm{lin}} - 2\alpha_{\mathrm{H}}\gamma_{\mathrm{non}})^2 + \gamma_\beta^2 + (\gamma_\beta + 2\gamma_{\mathrm{non}})^2}{(\omega^2 - \omega_0^2 - \gamma_0^2)^2 + 4\gamma_0^2\,\omega^2}, \tag{4.117}$$

where E_0 is the h-mode amplitude. As is shown in (4.117), two peaks are observed from the optical spectrum. There is strong peak at $\omega \approx -\omega_0$, which corresponds to the "nonlasing v mode" and a much weaker peak at $\omega \approx \omega_0$, which is produced in a polarization type of four-wave mixing (FWM) between the v mode that peaks at $\omega \approx -\omega_0$ and the dominant h mode, which peaks at

$\omega = 0$. Furthermore, the intensity of the FWM peak, relative to that of the nonlasing peak, can be used to estimate the effective gain anisotropy and birefringence via [48]

$$\frac{\langle |E_v(\omega_0)|^2 \rangle}{\langle |E_v(-\omega_0)|^2 \rangle} \approx \frac{(\alpha_H^2 + 1)\gamma_{\text{non}}^2}{4\omega_0^2} + \frac{\gamma_0^2}{4\omega_0^2}. \tag{4.118}$$

The separation of the two degenerated orthogonal polarizations is due to anisotropies of laser cavity. Hence, the measurement of polarization spectra can be utilized to investigate the anisotropy parameters from the equations derived above.

The VCSEL predicted above can be verified easily through experiment. In the experiment, the VCSEL is enclosed in a temperature-stabilized box and driven by a stable current source in order to minimize external noise. The collimated laser light is first passed through a rotatable $\lambda/4$ plate and subsequently through a combination of a rotatable $\lambda/2$ plate and an optical isolator, which together effectively act as a rotatable polarizer. By setting the angles of the $\lambda/4$ and $\lambda/2$ plates, the polarization state on which the laser light is projected is selected. The spectrum of the polarized light can be measured using a planar Fabry–Perot interferometer that allows detailed measurement of the optical spectrum. Figure 4.14 shows the measured optical spectrum of the v-nonlasing mode (peak v_1) of 1.9 mW output power [49]. The h-lasing mode (peak h) is also shown as the dashed curve, which is largely suppressed by a factor of 10^5. A $100\times$ magnification clearly shows the presence of another nonlasing peak (peak v_2), which is a four-wave mixing (FWM) signal. The lasing peak is associated with the steady-state polarization of the laser, the nonlasing peak is a result of amplified spontaneous emission in the orthogonal polarization, and the FWM peak results from nonlinear mixing between these two.

From (4.117), it is noted that the optical spectra of Figure 4.14 contains information of some laser parameters. First, the frequency difference between the lasing and nonlasing peaks gives the effective birefringence ω_0, whereas the difference in their half width at half maximum (HWHM) spectral widths gives the effective loss anisotropy γ_0. It is found in Figure 4.14 that the effective birefringence is relatively small at $\omega_0/2\pi \approx +3.4$ GHz (i.e., plus sign because the high-frequency mode lases). In addition, the effective loss anisotropy has a more typical value of $\gamma_0/2\pi \approx 0.38$ GHz. The corresponding spectral width of the lasing mode is an instrument limited to 0.06 GHz (HWHM) by resolution of the Fabry–Perot interferometer. It must be noted that for most other VCSELs, $\omega_o/2\pi$ ranged between -3 and $+15$ GHz and γ_0 is always below 1 GHz [48]. The relative strength between the FWM peak and the nonlasing peak can be used to quantify the nonlinear anisotropy, $\gamma_{\text{non}}(\equiv P\tau_J/(\tau_d\tau_p))$, in VCSELs using (4.118). Furthermore, it is found that a combined nonlinear anisotropy of $(\alpha_H^2 + 1)\gamma_{\text{non}}^2 = 3.5\ \text{ns}^{-1}$. Hence, it is shown that the polarization fluctuation of VCSELs can be utilized to determine the corresponding intrinsic optical.

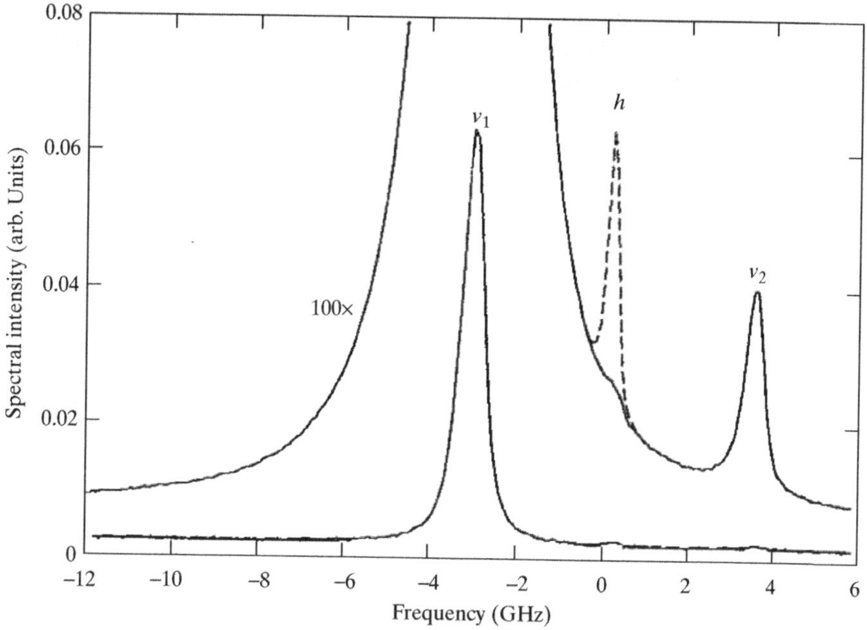

Figure 4.14 Polarization-resolved optical spectra of VCSEL at a constant current taking. For the solid curve the lasing h-mode is fully removed from the noisy spectrum and the lasing h-mode spectrum given in the dashed curve is suppressed by a factor of 10^5, which allows it to serve as a marker. The peaks v_1 and v_2 represent the nonlasing peak and four-wave mixing peak, respectively. (After Ref. 49).

4.6 SUMMARY

In summary, this chapter investigates the intrinsic optical anisotropy of VCSELs. First, the formation of in-plane gain anisotropies in VCSELs is studied. It is shown that the in-plane gain anisotropy in VCSELs can be obtained from

- The saturation of optical gain arising from the high intensity of the two orthogonal polarizations
- The growth of QW epitaxial layers on the non (001) substrate
- The influence of birefringence on the modal gain of the two orthogonal polarizations

Hence, a simple rate-equation model (i.e., two-level model) is derived with in-plane gain anisotropies factored in to analyze the polarization dynamics of VCSELs. In the model, the amplitudes of the two orthogonal polarizations are coupled together through self- and cross-saturation coefficients of the nonlinear gain so that the conditions of polarization switching and polarization bistability of VCSELs can be investigated.

Second, the electrooptically induced birefringence due to the inevitable internal electric field inside the laser cavity of VCSELs is studied theoretically. A theoretical model based on the assumption of electrooptic birefringence is derived to analyze the frequency splitting between the two orthogonal polarizations. It is shown that the electrooptic birefringence is dominated mainly inside the multilayered mirrors of VCSELs. Using this model, the frequency splitting between the two orthogonal polarizations is calculated and compared with the experimental results. It is found that the frequency response of the two orthogonal polarizations is not only determined by the electrooptic birefringence but may also be affected by the gain anisotropy as well as other nonlinear effects due to the small cavity size of VCSELs.

Finally, a four-level model is described to analyze the polarization dynamics of VCSELs. This four-level model takes into account the spin sublevels of the conduction and valence bands of the QW materials. Therefore, the lasing field of different polarizations associated with the transition between different spin sublevels can be included in the calculation. It is assumed that the transition between the two orthogonal polarizations is coupled through the spin–flip relaxation process so that their phase information can be evaluated simultaneously. The model is utilized to study the stabilities of the two orthogonal polarizations under the influence of optical anisotropies. Furthermore, the four-level model is transformed into a one-dimensional polarization rate equation via spin elimination so that the noise properties of VCSELs at steady state are analyzed. Using this model, the laser parameters of VCSELs such as effective loss anisotropy and birefringence are evaluated from the measured polarization noise spectra of VCSELs.

4A APPENDIX

4A.1 Calculation of the Susceptibilities

The third-order susceptibilities associated with self-saturation coefficients in the frame of $(\hat{\boldsymbol{x}}, \hat{\boldsymbol{y}}, \hat{\boldsymbol{z}})$ are given by [20]

$$\begin{aligned}
\varepsilon_0 \chi^{(3)}_{xxxx}(\omega_{\mathrm{v}}, \omega_{\mathrm{v}}, \omega_{\mathrm{v}}) = {} & \frac{1}{L_{\mathrm{w}}(2\hbar)^3} \int \frac{d\boldsymbol{k}^2}{(2\pi)^2} \\
& \times \left\{ \sum_{\substack{c=c_\pm,\, c'=c_\pm \\ v=v_{1\pm},\, v'=v_{1\pm}}} \frac{M^x_{\mathrm{v,c}} M^x_{\mathrm{c,v'}} M^x_{\mathrm{v',c'}} M^x_{\mathrm{c',v}}}{\omega_\nu - \mu_{\mathrm{c,v}}} \right. \\
& \times \left[\frac{1}{\mu_{\mathrm{v',v}}} \left(\frac{\rho^{(0)}_{\mathrm{c',c'}} - \rho^{(0)}_{\mathrm{v',v'}}}{\omega_{\mathrm{v}} - \mu^*_{\mathrm{c',v'}}} - \frac{\rho^{(0)}_{\mathrm{c',c'}} - \rho^{(0)}_{\mathrm{v,v}}}{\omega_{\mathrm{v}} - \mu_{\mathrm{c',v}}} \right) \right. \\
& \left. + \frac{1}{\mu_{\mathrm{c,c'}}} \left(\frac{\rho^{(0)}_{\mathrm{c',c'}} - \rho^{(0)}_{\mathrm{v',v'}}}{\omega_{\mathrm{v}} - \mu^*_{\mathrm{c',v'}}} - \frac{\rho^{(0)}_{\mathrm{c,c}} - \rho^{(0)}_{\mathrm{v',v'}}}{\omega_{\mathrm{v}} - \mu_{\mathrm{c,v'}}} \right) \right] \quad \text{(i)} \quad (4A.1)
\end{aligned}$$

$$+ \sum_{\substack{c=c_\pm, c'=c_\pm \\ v=v_{2\pm}, v'=v_{2\pm}}} \text{(Same as above)} \quad \text{(ii)}$$

$$\left. + \sum_{\substack{c=c_\pm, c'=c_\pm \\ v=v_{3\pm}, v'=v_{3\pm}}} \text{(Same as above)} \quad \text{(iii)} \right\}.$$

There are three blocks, (i)–(iii), in (4A.1), each associated with the transition between the conduction band and the lowest, the second lowest, and the third lowest energy subbands in the valence band, respectively. A similar expression holds for $\chi^{(3)}_{yyyy}$, $\chi^{(3)}_{xxxx}$, $\chi^{(3)}_{xxyy}$, $\chi^{(3)}_{xyxy}$, and $\chi^{(3)}_{xyyx}$, which can be obtained by replacing the directions of dipole transition matrix elements. Hence, the self-saturation coefficients in VCSELs can be obtained after appropriate coordinating transformations.

The third-order susceptibilities associated with cross-saturation coefficients are quite long and tedious. First consider $\chi^{(3)}_{vhhv}(\omega_h, \omega_h, \omega_v)$, which yields a cross-saturation coefficient ζ_{cvh} of the VCSELs for the two orthogonal optical fields, one with frequency ω_v in v mode and the other with frequency ω_h in h-mode. This component consists of $\chi^{(3)}_{xyyx}$ and $\chi^{(3)}_{yxxy}$. The former is given by

$$\begin{aligned}
\varepsilon_0 \chi^{(3)}_{xyyx}(\omega_h, \omega_h, \omega_v) = \frac{1}{L_{\mathrm{w}}(2\hbar)^3} \int \frac{d\boldsymbol{k}^?}{(2\pi)^2} \times \Bigg\{ &\text{(i)} \begin{bmatrix} \text{a term associated with} \\ c_\pm - v_{1\pm} \text{ transitions} \end{bmatrix} \\
&+ \text{(ii)} \begin{bmatrix} \text{a term associated with} \\ c_\pm - v_{2\pm} \text{ transitions} \end{bmatrix} \\
&+ \text{(iii)} \begin{bmatrix} \text{a term associated with} \\ c_\pm - v_{3\pm} \text{ transitions} \end{bmatrix} \\
&+ \text{(iv)} \begin{bmatrix} \text{a cross-term associated with} \\ c_\pm - v_{1\pm} \text{ and } c_\pm - v_{2\pm} \text{ transitions} \end{bmatrix} \\
&+ \text{(v)} \begin{bmatrix} \text{a cross-term associated with} \\ c_\pm - v_{1\pm} \text{ and } c_\pm - v_{3\pm} \text{ transitions} \end{bmatrix} \\
&+ \text{(vi)} \begin{bmatrix} \text{a cross-term associated with} \\ c_\pm - v_{2\pm} \text{ and } c_\pm - v_{3\pm} \text{ transitions} \end{bmatrix} \Bigg\}
\end{aligned} \quad (4A.2)$$

where the term (i) associated with the transition between the lowest conduction subbands and the lowest valence subbands is given by the following expression:

$$\text{(i)} = \sum_{\substack{c=c_\pm, c'=c_\pm \\ v=v_{1\pm}, v'=v_{1\pm}}} \frac{1}{\omega_v - \mu_{c,v}}$$

$$
\begin{aligned}
&\times \left[\frac{M^x_{\mathrm{v,c}} M^y_{\mathrm{c,v'}} M^y_{\mathrm{v',c'}} M^x_{\mathrm{c',v}}}{\omega_{\mathrm{v}} - \omega_h - \mu_{\mathrm{v',v}}} \left(\frac{\rho^{(0)}_{\mathrm{c',c'}} - \rho^{(0)}_{\mathrm{v',v'}}}{\omega_h - \mu^*_{\mathrm{c',v'}}} - \frac{\rho^{(0)}_{\mathrm{c',c'}} - \rho_{\mathrm{v}},\mathrm{v}^{(0)}}{\omega_{\mathrm{v}} - \mu_{\mathrm{c',v}}} \right) \right. \\
&+ \frac{M^x_{\mathrm{v,c}} M^x_{\mathrm{c,v'}} M^y_{\mathrm{v',c'}} M^y_{\mathrm{c'},\nu}}{\omega_{\mathrm{v}} - \omega_h - \mu_{\mathrm{c,c'}}} \left(\frac{\rho^{(0)}_{\mathrm{c',c'}} - \rho^{(0)}_{\mathrm{v',v'}}}{\omega_h - \mu^*_{\mathrm{c',v'}}} - \frac{\rho^{(0)}_{\mathrm{c,c}} - \rho^{(0)}_{\mathrm{v',v'}}}{\omega_{\mathrm{v}} - \mu_{\mathrm{c,v'}}} \right) \\
&+ \frac{M^x_{\mathrm{v,c}} M^x_{\mathrm{c,v'}} M^y_{\mathrm{v',c'}} M^y_{\mathrm{c',v}}}{\mu_{\mathrm{v',v}}} \left(\frac{\rho^{(0)}_{\mathrm{c',c'}} - \rho^{(0)}_{\mathrm{v,v}}}{\omega_h - \mu_{\mathrm{c',v}}} - \frac{\rho^{(0)}_{\mathrm{c',c'}} - \rho^{(0)}_{\mathrm{v',v'}}}{\omega_h - \mu^*_{\mathrm{c',v'}}} \right) \\
&\left. + \frac{M^x_{\mathrm{v,c}} M^y_{\mathrm{c,v'}} M^y_{\mathrm{v',c'}} M^x_{\mathrm{c',v}}}{\mu_{\mathrm{c,c'}}} \left(\frac{\rho^{(0)}_{\mathrm{c',c'}} - \rho^{(0)}_{\mathrm{v',v'}}}{\omega_h - \mu^*_{\mathrm{c,v'}}} - \frac{\rho^{(0)}_{\mathrm{c',c'}} - \rho^{(0)}_{\mathrm{v',v'}}}{\omega_h - \mu_{\mathrm{c',v'}}} \right) \right]
\end{aligned}
\tag{4A.3}
$$

This term is triply resonant when $\hbar\omega_{\mathrm{v}} = \hbar\omega_h = E_{\mathrm{c}}(\boldsymbol{k}) - E_{\mathrm{v}}(\boldsymbol{k})$. The terms (ii) and (iii) are given by expressions similar to (4A.3) by replacing the summation indices. These three terms are related to the main peaks at $\omega_{\mathrm{v}} = \omega_h$ in the cross-saturation coefficients.

The term (iv) becomes triply resonant when $\hbar\omega_{\mathrm{v}} = E_{\mathrm{c}}(\boldsymbol{k}) - E_{\mathrm{v}}(\boldsymbol{k})$ and $\hbar\omega_h = E_{\mathrm{c}}(\boldsymbol{k}) - E_{\mathrm{v}}(\boldsymbol{k})$, or vice versa. It is explicitly given by

$$
\begin{aligned}
(\mathrm{iv}) = \sum_{\substack{c=c\pm \\ v=v_{1\pm}, v'=v_{2\pm}}} &\left\{ \frac{1}{\omega_{\mathrm{v}} - \mu_{\mathrm{c,v}}} \left[\frac{M^x_{\mathrm{v,c}} M^y_{\mathrm{c,v'}} M^y_{\mathrm{v',c}} M^x_{\mathrm{c,v}}}{\omega_{\mathrm{v}} - \omega_h - \mu_{\mathrm{v',v}}} \left(\frac{\rho^{(0)}_{\mathrm{c,c}} - \rho^{(0)}_{\mathrm{v',v'}}}{\omega_h - \mu^*_{\mathrm{c,v'}}} - \frac{\rho^{(0)}_{\mathrm{c,c}} - \rho^{(0)}_{\mathrm{v,v}}}{\omega_{\mathrm{v}} - \mu_{\mathrm{c,v}}} \right) \right] \right. \\
&\left. + \frac{1}{\omega_{\mathrm{v}} - \mu_{\mathrm{c,v'}}} \left[\frac{M^x_{\mathrm{v,c}} M^y_{\mathrm{c,v'}} M^y_{\mathrm{v',c}} M^x_{\mathrm{c,v}}}{\omega_{\mathrm{v}} - \omega_h - \mu_{\mathrm{v,v'}}} \left(\frac{\rho^{(0)}_{\mathrm{c,c}} - \rho^{(0)}_{\mathrm{v',v'}}}{\omega_h - \mu^*_{\mathrm{c,v}}} - \frac{\rho^{(0)}_{\mathrm{c,c}} - \rho^{(0)}_{\mathrm{v',v'}}}{\omega_{\mathrm{v}} - \mu_{\mathrm{c,v'}}} \right) \right] \right\}
\end{aligned}
\tag{4A.4}
$$

Other cross-terms are given in a similar manner.

4A.2 Band Structure Model

In order to calculate the electron and hole wavefunctions in QWs, the multiband effective mass theory is used [22]. For most III–V semiconductor materials such as GaAs-based materials, it is a good approximation that the conduction and valence bands are decoupled. A parabolic band model and a Luttinger–Kohn Hamiltonian with strain components are used to calculate the conduction and valence bands, respectively [22]. The electron states near the conduction subband edge are assumed to be almost purely s-like and nondegenerate (excluding spin), while the hole states near the valence subband edge are almost purely p-like and fourfold degenerate (including spin). The envelope function scheme is adopted to describe the slowly varying part of the wavefunction. In the following paragraphs, the slowly varying part of the wavefunction for the electron and hole are calculated.

The influence of QW confinement potential on the energies and envelope functions of the electron subband edge at the zone center of the Brillouin zone can be calculated separately using the one-dimensional Schrödinger-type equation:

$$-\frac{\hbar^2}{2}\frac{d}{dz}\left(\frac{1}{m^*(z)}\frac{d\Psi_{cl}(z)}{dz}\right)+U_c(z)\Psi_{cl}(z)=\left(E_{cl}(k_t)-\frac{\hbar^2k_t^2}{2m^*(z)}\right)\Psi_{cl}(z), \tag{4A.5}$$

where $k_t^2 = k_x^2 + k_y^2$, Ψ_{cl} is the envelope function of the lth subband for electrons, m^* is the effective mass in the z direction, E_{cl} is the subband edge energy, and U_c is the QW confinement potential of electrons. This equation can be solved numerically using a finite difference method with the corresponding confinement profile for the approximated parabolic band.

The valence band structure in QWs is more complicated as there is a fourfold degeneracy (including spin degeneracy) at the top of the valence band. The periodic part of the Bloch function (not including the spin degeneracy) at the top of the valence band has the symmetry of a p-type wavefunction that is threefold degenerate. Combining with the spin, there are six valence bands (the heavy hole band, the light hole band, and the spin–orbit splitoff band) just below the conduction band. The spin–orbit splitoff band is split from the heavy hole and light hole bands by the spin–orbit interaction. In addition, the heavy and light hole valence bands are split as a result of the quantum confinement effect. If the energy separation of the spin–orbit splitoff band is far away from the heavy and light hole bands, the corresponding envelope function can be obtained by

$$(H_V^{LK}+U_v)\Psi_{vl}(z)=E_{vl}(k_t)\Psi_{vl}(z), \tag{4A.6}$$

where Ψ_{vl} is the envelope function of the lth subband for holes, H_V^{LK} is the Luttinger–Kohn Hamiltonian with strain introduced, and U_v is the confinement potential for hole, which are given by

$$H_V^{LK}=\begin{bmatrix} P+Q & -S & R & 0 \\ -S & P-Q & 0 & R \\ R^+ & 0 & P-Q & S \\ 0 & R^+ & S^+ & P+Q \end{bmatrix} \begin{matrix} |\frac{3}{2},\frac{3}{2}\rangle \\ |\frac{3}{2},\frac{1}{2}\rangle \\ |\frac{3}{2},-\frac{1}{2}\rangle \\ |\frac{3}{2},-\frac{3}{2}\rangle \end{matrix}, \tag{4A.7}$$

$$U_v=\begin{bmatrix} U_{HH} & 0 & 0 & 0 \\ 0 & U_{LH} & 0 & 0 \\ 0 & 0 & U_{LH} & 0 \\ 0 & 0 & 0 & U_{HH} \end{bmatrix}, \tag{4A.8}$$

and the parameters P, Q, R and S are given by

$$P=\frac{\hbar^2}{2m_0}\left(\gamma_1(k_x^2+k_y^2)-\frac{\partial}{\partial z}\left(\gamma_1\frac{\partial}{\partial z}\right)\right), \tag{4A.9a}$$

$$Q = \frac{\hbar^2}{2m_0}\left(\gamma_2(k_x^2 + k_y^2) + 2\frac{\partial}{\partial z}\left(\gamma_2 \frac{\partial}{\partial z}\right)\right), \tag{4A.9b}$$

$$R = \frac{\hbar^2}{2m_0}\left(\sqrt{3}\gamma_2(-k_x^2 + k_y^2) - 2j\sqrt{3}\gamma_2 \frac{\partial^2}{\partial z^2}\right), \tag{4A.9c}$$

$$S = \frac{\hbar^2}{2m_o}\left(\sqrt{3}\gamma_3(-jk_x - k_y)\frac{\partial^2}{\partial z^2}\right). \tag{4A.9d}$$

where γ_1, γ_2, γ_3 are the Luttinger–Kohn parameters, which are dependent on the position z and m_o is the relative mass of electron. The Luttinger–Kohn Hamiltonian is actually a set of coupled linear differential equations for the envelope functions, which can also be solved using the finite difference approximation.

The periodic part of the Bloch state (basis function) for electrons is given by

$$|u^{\rm c}_{1/2}\rangle = |s \uparrow\rangle \quad \text{and} \quad |u^{\rm c}_{-1/2}\rangle = |s \downarrow\rangle, \tag{4A.10}$$

where $\uparrow$ and $\downarrow$ denote the up and down electron spinors and $|s\rangle$ is the s-like conduction band Bloch states. For holes, the Bloch states are represented by the linear combinations of the products of the spinor and the p-like valence band Bloch states $|p_x\rangle$, $|p_y\rangle$, and $|p_z\rangle$ and are given by

$$|u^{\rm v}_{3/2}\rangle = -\frac{1}{\sqrt{2}}|p_x + jp_y \uparrow\rangle \equiv \left|\frac{3}{2}, \frac{3}{2}\right\rangle, \tag{4A.11a}$$

$$|u^{\rm v}_{1/2}\rangle = -\frac{1}{\sqrt{6}}(|p_x + jp_y \downarrow\rangle - 2|p_z \uparrow\rangle) \equiv \left|\frac{3}{2}, \frac{1}{2}\right\rangle, \tag{4A.11b}$$

$$|u^{\rm v}_{1/2}\rangle = \frac{1}{\sqrt{6}}(|p_x - jp_y \uparrow\rangle + 2|p_z \downarrow\rangle) \equiv \left|\frac{3}{2}, -\frac{1}{2}\right\rangle, \tag{4A.11c}$$

$$|u^{\rm v}_{-3/2}\rangle = \frac{1}{\sqrt{2}}|p_x - jp_y \downarrow\rangle \equiv \left|\frac{3}{2}, -\frac{3}{2}\right\rangle. \tag{4A.11d}$$

When the subband envelope functions are obtained, the optical matrix elements can be calculated by the following expression

$$M_{p,q} = \langle v_p|\boldsymbol{M}|c_q\rangle, \tag{4A.12}$$

where p and q denote the Bloch functions of electron or holes and $|v_p\rangle = \Psi_{vl}(z)|u^{\rm v}_q\rangle \exp(j\boldsymbol{k}\cdot\boldsymbol{r})$. $\Psi(z)$ is the envelope function along the z direction, and $|u\rangle$ is the periodic and rapidly varying part of the Bloch function. A similar expression of $|c_q\rangle$ can also be obtained in a similar manner. The squared of optical matrix elements for the in-plane optical field is obtained within the envelope function approximation

$$\begin{aligned}|M_{p,q}|^2 = |\langle u^{\rm v}_p|\boldsymbol{M}|u^{\rm c}_q\rangle|^2 \Big\{ &\tfrac{1}{2}\left(|\langle\Psi_{{\rm c}p}|\Psi_{{\rm v}3/2,q}\rangle|^2 + |\langle\Psi_{{\rm c}p}|\Psi_{{\rm v}-3/2,q}\rangle|^2\right) \\ &+ \tfrac{1}{6}\left(|\langle\Psi_{{\rm c}p}|\Psi_{{\rm v}-1/2,q}\rangle|^2 + |\langle\Psi_{{\rm c}p}|\Psi_{{\rm v}1/2,q}\rangle|^2\right)\Big\},\end{aligned} \tag{4A.13}$$

where $\langle \Psi_{cp} | \Psi_{vs,q} \rangle$ (for $s = -\frac{3}{2}, -\frac{1}{2}, \frac{1}{2}, \frac{3}{2}$) is the overlap integral of the envelope functions and the expression of $\Psi_{vs,q}$ is obtained from the one-dimensional Schrödinger type equation (4A.5).

4A.3 Adiabatic Elimination of Spin Dynamics

To identify the role of carrier dynamics in the polarization behavior of the laser, the variable n will be eliminated adiabatically from (4.74) with the approximated expression valid under the conditions [46]

$$\frac{\tau_d}{\tau_J} \gg |E_+|^2 + |E_-|^2, \tag{4A.14}$$

$$1 \gg N - 1 \qquad \text{or} \qquad N \sim 1. \tag{4A.15}$$

In (4.74), the adiabatically eliminated n becomes

$$n \approx -\frac{\tau_J}{\tau_d}(|E_+|^2 - |E_-|^2), \tag{4A.16}$$

where $\partial n/\partial t \to 0$ is assumed due to the slowly varying components on the RHS of (4.74).

It is convenient to separate the optical intensity and polarization by introducing the so-called Stokes vector $\vec{s} = (s_0, s_1, s_2, s_3)$ as

$$s_0 = |E_+|^2 + |E_-|^2 = S, \tag{4A.17a}$$

$$s_1 = 2\mathrm{Re}[E_+^* E_-] = S\cos 2\psi \cos 2\phi, \tag{4A.17b}$$

$$s_2 = 2\mathrm{Im}[E_+^* E_-] = S\cos 2\psi \sin 2\phi, \tag{4A.17c}$$

$$s_3 = |E_+|^2 - |E_-|^2 = S\sin 2\psi, \tag{4A.17d}$$

where S is the photon intensity and the angles ψ and ϕ represent the polarization state. The angle $0 \leq \phi \leq \pi$ characterizes the polarization; it is the angle between the long axis of the polarization ellipse and the x-axis. The angle $-\pi/4 \leq \psi \leq \pi/4$ characterizes the ellipticity of the light; $\psi = 0$ corresponds to the linearly polarized light, whereas $\psi = \pm\pi/4$ corresponds to circularly polarized light. Using the angles $(2\psi, 2\phi)$ as spherical coordinates, the polarization state can be conveniently depicted as a single point on the Poincaré sphere [43].

Now, substituting (4A.17) into (4.72) and (4.73) to replace the variable n gives the following version of rate equation with spin eliminated:

$$\frac{\partial \psi}{\partial t} = \gamma_{bf}\sin 2\phi + \gamma_a \sin 2\psi \cos 2\phi - \frac{\tau_J}{\tau_d \tau_p} S \sin 2\psi \cos 2\psi, \tag{4A.18a}$$

$$\cos 2\psi \frac{\partial \phi}{\partial t} = -\gamma_{bf}\sin 2\psi \cos 2\phi + \gamma_a \sin 2\phi - \alpha_H \frac{\tau_J}{\tau_d \tau_p} S \sin 2\psi \cos 2\psi, \tag{4A.18b}$$

$$\frac{\partial S}{\partial t} = 2\tau_{\mathrm{p}}^{-1}(N-1)S - 2\gamma_{\mathrm{a}} S \cos 2\psi \sin 2\phi - 2S\frac{\tau_{\mathrm{J}}}{\tau_{\mathrm{d}}\tau_{\mathrm{p}}}\sin^2 2\psi, \tag{4A.18c}$$

$$\frac{\partial N}{\partial t} = -\tau_{\mathrm{d}}^{-1}\left((N-\mu) + NS - S^2\frac{\tau_{\mathrm{J}}}{\tau_{\mathrm{d}}}\sin^2 2\psi\right). \tag{4A.18d}$$

In (4A.18a) and (4A.18b), N disappears as any variation of N from its equilibrium will lead to equal gain or loss for all Stokes parameters and thus cannot affect the field polarization. It is noted that $\tau_{\mathrm{d}}/\tau_{\mathrm{J}} > 100$ is commonly found in VCSELs, and S can be assumed to be less dependent on the angles $(2\psi, 2\phi)$, so that S in (4A.18a) and (4A.18b) can be considered as a constant. If the loss anisotropy and birefringence are separated from (4A.18a) and (4A.18b), a more general spin-eliminated version can be simplified to

$$2\frac{\partial \psi}{\partial t} = \omega_{\mathrm{lin}} \sin 2\phi + \gamma_{\mathrm{lin}} \sin 2\psi \cos 2(\phi - \phi_{\mathrm{r}}) - 2\gamma_{\mathrm{non}} \sin 2\psi \cos 2\psi, \tag{4A.19a}$$

$$2\cos 2\psi \frac{\partial \phi}{\partial t} = -\omega_{\mathrm{lin}} \sin 2\psi \cos 2\phi + \gamma_{\mathrm{lin}} \sin 2(\phi - \phi_{\mathrm{r}}) - 2\alpha_{\mathrm{H}}\gamma_{\mathrm{non}} \sin 2\psi \cos 2\psi, \tag{4A.19b}$$

where ϕ_{r} represents the orientation of the loss anisotropy and birefringence with respect to the crystal axis. The other parameters are defined as $\omega_{\mathrm{lin}} = 2\gamma_{\mathrm{bf}}$, $\gamma_{\mathrm{lin}} = 2\gamma_{\mathrm{a}}$, $\gamma_{\mathrm{non}} = S\tau_{\mathrm{J}}/(\tau_{\mathrm{d}}\tau_{\mathrm{p}})$. The parameters ω_{lin} and γ_{lin} are the linear birefringence and linear anisotropy, respectively. To remove the various sine and cosine functions in (4A.19), ϕ and ψ can be expanded to first order as $\phi, \psi \ll 1$. The steady-state angles thus found are

$$\psi_{\mathrm{s}} \approx \frac{\gamma_{\mathrm{lin}} \sin 2\phi_{\mathrm{r}}}{2(\omega_{\mathrm{lin}} + 2\alpha_{\mathrm{H}}\gamma_{\mathrm{non}})} \ll 1, \tag{4A.20a}$$

$$\phi_{\mathrm{s}} \approx \left(\frac{\gamma_{\mathrm{lin}} \cos 2\phi_{\mathrm{r}} + 2\gamma_{\mathrm{non}}}{\omega_{\mathrm{lin}}}\right)\psi_{\mathrm{s}} \ll 1. \tag{4A.20b}$$

Equation (4A.20a) is asymmetric in ω_{lin}; large ellipticity is most likely for negative ω_{lin}, specifically for the case of dominant linear birefringence ($\omega_{\mathrm{lin}} >> \gamma_{\mathrm{lin}}, \gamma_{\mathrm{non}}$). It is also found that $\phi_s \ll \psi_s$.

For $\psi_s, \phi_s \ll 1$ the linearized polarization rate equations, including noise, are

$$\frac{\partial}{\partial t}\begin{bmatrix}\phi - \phi_s \\ \psi - \psi_s\end{bmatrix} = \begin{bmatrix}-\gamma_{\mathrm{lin}} \cos 2\phi_{\mathrm{r}} & -\omega_{\mathrm{lin}} - 2\alpha_{\mathrm{H}}\gamma_{\mathrm{non}} \\ \omega_{\mathrm{lin}} & -\gamma_{\mathrm{lin}} \cos 2\phi_{\mathrm{r}} - 2\gamma_{\mathrm{non}}\end{bmatrix}\begin{bmatrix}\phi - \phi_{\mathrm{s}} \\ \psi - \psi_{\mathrm{s}}\end{bmatrix} + \begin{bmatrix}f_\phi \\ f_\psi\end{bmatrix}, \tag{4A.21}$$

where the Langevin noise sources f_ϕ and f_ψ have been introduced into linearized polarization rate equations. The simplicity of these results is due to the fact that, after spin elimination, the polarization dynamics (ϕ, ψ) are separated almost completely from other dynamics such as intensity S and its corresponding phase, average inversion N. The only coupling is via the intensity dependence of γ_{non}, and this coupling disappears when the intensity is reasonably constant (i.e., under the condition that the fluctuations are limited or at frequencies very different from those of the polarization dynamics).

REFERENCES

1. C. J. Chang-Hasnain, J. P. Harbison, G. Hasnain, A. C. Von Lehmen, L. T. Florez, and N. G. Stoffel, "Dynamic, polarization, and transverse mode characteristics of vertical cavity surface emitting lasers," *IEEE J. Quantum Electron.*, Vol. 27, no. 6, pp. 1402–1409, 1991.
2. T. Ohtoshi, T. Kuroda, A. Niwa, and S. Tsuji, "Dependence of optical gain on crystal orientation in surface emitting lasers with strained quantum wells," *Appl. Phys. Lett.*, Vol. 65, no. 15, pp. 1886–1888, 1994.
3. D. Sun, E. Towe, P. H. Ostdiek, J. W. Grantham, and G. J. Vansuch, "Polarization control of vertical cavity surface emitting lasers through use of an anisotropic gain distribution in [110]-oriented strained quantum well structures," *IEEE J. Select. Topics Quantum Electron.*, Vol. 1, no. 2, pp. 674–680, 1995.
4. M. Takahashi, N. Egami, T. Mukaihara, F. Koyama, and K. Iga, "Lasing characteristics of GaAs(311)A substrate InGaAs-GaAs vertical cavity surface emitting lasers," *IEEE J. Select. Topics Quantum Electron.*, Vol. 3, no. 2, pp. 372–378, 1997.
5. N. Nishiyama, A. Mizutani, N. Hatori, M. Arai, F. Koyama, and K. Iga, "Lasing characteristics of InGaAs-GaAs polarization controlled vertical cavity surface emitting laser grown on GaAs (311) B substrate," *IEEE J. Select. Topics Quantum Electron.*, Vol. 5, no. 3, pp. 530–535, 1999.
6. H. Kawaguchi, T. Irie, and M. Murakami, "Pitchfork bifurcation polarization bistability in laser diodes with external cavities," *IEEE J. Quantum Electron.* Vol. 31, no. 3, pp. 447–455, 1995.
7. M. S. Park, B. T. Ahn, B. S. Yoo, H. Y. Chu, H. H. Park, and C. J. Chang-Hasnain, "Polarization control of vertical cavity surface emitting lasers by electro-optic birefringence," *Appl. Phys. Lett.*, Vol. 76, no. 7, pp. 813–815, 2000.
8. K. D. Choquette, and R. E. Leibenguth, "Control of vertical cavity laser polarization anisotropic transverse cavity geometries," *IEEE Photon. Technol. Lett.*, Vol. 6, no. 1, pp. 40–42, 1994.
9. T. Yoshikawa, T. Kawakami, H. Saito, H. Kosaka, M. Kajita, K. Kurihara, Y. Sugimoto, and K. Kasahara, "Polarization controlled single mode VCSEL," *IEEE J. Quantum Electron.*, Vol. 34, no. 6, pp. 1009–1014, 1998.
10. H. Y. Chu, B. S. Yoo, M. S. Park, and H. H. Park, "Polarization characteristics of index-guided surface emitting lasers with tilted pillar structure," *IEEE Photon. Technol. Lett.*, Vol. 9, no. 8, pp. 1066–1068, 1997.
11. J. H. Ser, Y. G. Ju, J. H. Shin, and Y. H. Lee, "Polarization stabilization of vertical cavity surface emitting lasers by inscription of fine metal-interlaced gratings," *Appl. Phys. Lett.*, Vol. 66, no. 21, pp. 2769–2771, 1995.

12. P. Dowd, P. J. Heard, J. A. Nicholson, L. Raddatz, I. H. White, R. V. Penty, J. C.C. Day, G. C. Allen, S. W. Corzine, and M. R.T. Tan, "Complete polarization control of GaAs gain-guided top surface emitting vertical cavity lasers," *Electron. Lett.*, Vol. 33., no. 15, pp. 1315–1317, 1997.
13. K. H. Ha, Y. H. Lee, H. K. Shin, K. H. Lee, and S. M. Whang, "Polarization anisotropy in asymmetric oxide aperture VCSELs," *Electron. Lett.*, Vol. 34., no. 14, pp. 1401–1402, 1998.
14. T. Mukaihara, F. Koyama, and K. Iga, "Engineering polarization control of GaAs/AlGaAs surface emitting lasers by anisotropic stress from elliptical etched substrate hole," *IEEE Photon. Technol. Lett.*, Vol. 5, no. 2, pp. 133–135, 1993.
15. T. Numai, K. Kurhara, K. Kuhn, H. Kosaka, I. Ogura, M. Kajita, H. Saito, and K. Kasahara, "Control of light output polarization for surface emitting laser type device by strained active layer grown on misoriented substrate," *IEEE J. Quantum Electron.*, Vol. 31, no. 4, pp. 636–641, 1995.
16. Y. H. Chen, C. I. Wilkinson, J. Woodhead, J. P. R. David, C. C. Botton, and P. N. Robson, "Influence of ordering on the polarization characteristics of GaInP vertical cavity surface emitting lasers," *IEEE Photon. Technol. Lett.*, Vol. 9, no. 2, pp. 143–145, 1997.
17. Y. G. Ju, Y. H. Lee, H. K. Shin, and I. I Kim, "Strong polarization selectivity in 780 nm vertical cavity surface emitting lasers grown on misoriented substrate," *Appl. Phys. Lett.*, Vol. 71, no. 6, pp. 741–743, 1997.
18. G. P. Agrawal and N. K. Dutta, *Long Wavelength Semiconductor Lasers*, Van Nostrand Reinhold, New York, 1986.
19. Y. Takahashi, A. Neogi, and H. Kawaguchi, "Polarization dependent nonlinear gain in semiconductor lasers," *IEEE J. Quantum Electron.*, Vol. 34, no. 9, pp. 1660–1672, 1998.
20. Y. Takahashi and H. Kawaguchi, "Polarization dependent gain saturations in quantum well lasers," *IEEE J. Quantum Electron.*, Vol. 36, no. 7, pp. 864–871, 2000.
21. P. S. Zory, ed., *Quantum Well Lasers*, Academic Press, 1993.
22. S. L. Chuang, *Physics of Optoelectronic Devices*, Wiley Series in Pure and Applied Optics, Wiley, New York, 1995.
23. Y. R. Shen, *The Principles of Nonlinear Optics*, Wiley, New York, 1984.
24. A. Niwa, T. Ohtoshi, and T. Kuroda, "Orientation dependence of optical properties in long wavelength strained quantum well lasers," *IEEE J. Select. Topics Quantum Electron.*, Vol. 1, no. 2, pp. 211–217, 1995.
25. C. H. Chong and J. Sarma, "Lasing mode selection in vertical cavity surface emitting laser diodes," *IEEE Photon. Technol. Lett.*, Vol. 5, no. 7, pp. 761–763, 1993.
26. A. Valle, "Selection and modulation of high order transverse modes in vertical cavity surface emitting lasers," *IEEE J. Quantum Electron.*, Vol. 34, no. 10, pp. 1924–1932, 1998.
27. H. Uenohara, K. Tateno, T. Kagawa, Y. Ohiso, H. Tsuda, T. Kurokawa, and C. Amano, "Polarization controlled 850 nm wavelength vertical cavity surface emitting lasers grown on (311)B substrates by metal-organic chemical vapor deposition," *IEEE J. Select. Topics Quantum Electron.*, Vol. 5, no. 3, pp. 537–544, 1999.

28. C. J. Chang-Hasnain, J. P. Harbison, L. T. Florez, and N. G. Stoffel, "Polarization characteristics of quantum well vertical cavity surface emitting lasers," *Electron. Lett.*, Vol. 27, no. 2, pp. 163–165, 1991.

29. M. J. Adams, *An Introduction to Optical Waveguide*, Wiley, New York, 1981.

30. A. Valle, K. A. Shore, and L. Pesquera, "Polarization selection in birefringence vertical cavity surface emitting lasers," *IEEE J. Lightwave Technol.*, Vol. 14, no. 9, pp. 2062–2068, 1996.

31. J. Faist, and F. K. Reinhart, "Phase modulation in GaAs/AlGaAs double heterostructures. I. Theory," *J. Appl. Phys.*, Vol. 67, no. 11, pp. 6998–7005, 1990.

32. A. Yariv, *Quantum Electronics*, 3rd ed., Wiley, New York, 1989.

33. M. P. van Exter, A. K. Jansen van Doorn, and J. P. Woerdman, "Electro-optic effect and birefringence in semiconductor vertical cavity lasers," *Phys. Rev. A*, Vol. 56, no. 1, pp. 845–853, 1997.

34. M. San Miguel, Q. Feng, and J. V. Moloney, "Light-polarization dynamics in surface emitting semiconductor lasers," *Phys. Rev. A*, Vol. 52, no. 2, pp. 1728–1739, 1995.

35. C. Etrich, P. Mandel, N. B. Abraham, and H. Zeghlache, "Dynamics of a two-mode semiconductor laser," *IEEE J. Quantum Electron.*, Vol. 28, no. 4, pp. 811–821, 1992.

36. T. C. Damen, L. Viña, J. E. Cunningham, J. Shah, and L. J. Sham, "Subpicosecond spin relaxation dynamics of exciton and free carriers in GaAs quantum wells," *Phys. Rev. Lett.*, Vol. 67, no. 24, pp. 3432–3435, 1991.

37. M. Z. Maialle, E. A. de Andrada e Silva, and L. J. Sham, "Exciton spin dynamics in quantum wells," *Phys. Rev. B*, Vol. 47, no. 23, pp. 15776–15788, 1993.

38. F. Meier, and B. P. Zakharchenya, eds. *Optical Orientation*, North Holland, Amsterdam, 1984.

39. T. Uenoyama and L. J. Sham, "Carrier relaxation and luminescence polarization in quantum wells," *Phys. Rev. B*, Vol. 42, no. 11, pp. 7114–7123, 1990.

40. L. Muñoz, E. Pérez, V. Bellani, S. Zimmermann, L. Viña, K. Ploog, E. S. Koteles, and K. M. Lau, "Exciton dynamics and spin relaxation in unstrained and tensile-strained quantum wells," *J. Opt. Soc. Am. B*, Vol. 13, no. 5, pp. 994–999, 1996.

41. R. Ferreira and G. Bastard, "Spin-flip scattering of holes in semiconductor quantum wells," *Phys. Rev. B*, Vol. 43, no. 12, pp. 9687–9691, 1991.

42. V. Srinivas, Y. J. Chen, and C. E. C. Wood, "Intrinsic and extrinsic processes in photoluminiscence, reflectivity and spin dynamics of GaAs quantum wells," *J. Opt. Soc. Am. B*, Vol. 13, no. 5, pp. 989–993, 1996.

43. J. Martin-Regalado, F. Prati, M. San Miguel, and N. B. Abraham, "Polarization properties of vertical cavity surface emitting lasers," *IEEE J. Quantum Electron.*, Vol. 33, no. 5, pp. 765–783, 1997.

44. G. Björk and Y. Yamamoto, "Analysis of semiconductor microcavity lasers using rate equations," *IEEE J. Quantum Electron.*, Vol. 27, no. 11, pp. 2386–2396, 1991.

45. M. Travagnin, M. P. van Exter, and J. P. Woerdman, "Influence of carrier dynamics on the polarization stability and noise-induced polarization hopping in surface emitting semiconductor lasers," *Phys. Rev. A*, Vol. 56, no. 2, pp. 1497–1507, 1997.

46. M. P. van Exter, R. F. M. Hendriks, and J. P. Woerdman, "Physical insight into the polarization dynamics of semiconductor vertical cavity lasers," *Phys. Rev. A*, Vol. 57, no. 3, pp. 2080–2090, 1997.

47. K. Petermann, *Laser Diode Modulation and Noise*, Kluwer, Dordrecht, 1998.
48. M. P. van Exter, M. B. Willemsen, and J. P. Woerdman, "Polarization fluctuations in vertical cavity surface semiconductor lasers," *Phys. Rev. A*, Vol. 58, no. 5, pp. 4191–4205, 1998.
49. M. P. van Exter, A. Al-Remawi, and J. P. Woerdman, "Polarization fluctuations demonstrate nonlinear anisotropy of a vertical cavity semiconductor lasers," *Phys. Rev. Lett.*, Vol. 80, no. 22, pp. 4875–4878, 1998.

CHAPTER 5

Thermal Characteristics of Vertical Cavity Surface Emitting Lasers

The steady-state characteristics, including lasing frequency, threshold current, and output power, of vertical cavity surface emitting lasers (VCSELs) under the influence of temperature are analyzed. A brief review of the most recent development of thermal models of VCSELs is included. Both simple and comprehensive methods to calculate heat distribution inside the laser cavity are studied in this chapter.

5.1 INTRODUCTION

Vertical cavity surface emitting lasers (VCSELs) have attracted considerable interest since the mid-1990s because of their single longitudinal-mode operation, circular output beams with low divergence, the possibility of monolithic two-dimensional integration, and the compatibility with on-wafer probe testing. However, despite these advantages, VCSELs still exhibit a number of undesirable features such as the excitation of higher-order transverse modes and the undetermined polarization properties [1]. In fact, the most widely recognized limitation on the performance of VCSELs is the generation of heat inside the laser cavity [2]. Self-heating in VCSELs can be attributed to the excessive heatsource and the accumulation of heat inside the laser cavity. Excessive heatsource is due to the high series resistance (i.e., a value of few hundred ohms) of the doped semiconductor distributed Bragg reflectors (DBRs), which is much greater than that of facet emitting lasers [3]. On the other hand, heat dissipation from the laser cavity of VCSELs into the heatsink is relatively slow when compared to that of facet emitting lasers. This is due to the high current density, high thermal resistivity, as well as the junction up configuration of VCSELs. The temperature at the active layer T_{active} can be estimated by writing $T_{\text{active}} = T_{\text{HS}} + R_{\text{T}}IV$, where T_{HS} is the heatsink temperature, R_{T} [in Kelvin per watts (K/W)] is the total thermal resistance, and IV is the equivalent electrical power dissipated inside the VCSEL. Hence, serious heat accumulation inside the laser cavity is unavoidable

Analysis and Design of Vertical Cavity Surface Emitting Lasers By Siu Fung Yu
ISBN 0-471-39124-7

because of high thermal resistance and injection current density. This is why VCSELs usually operate at a higher temperature than do facet emitting lasers. The substantial increase in temperature causes the increase in threshold current density, the reduction of output optical power, and the shift of resonant frequency to the longer wavelength. As a result, the influence of thermal effect on device performance is more pronounced in VCSELs than in facet emitting lasers [3]. Furthermore, thermal problems are even more pressing in VCSEL arrays where long-range thermal crosstalk is of great concern [4]. Therefore, the challenge for designing VCSELs operating at CW condition is to improve the heat dissipation efficiency, reduce the threshold current density, and increase the output optical power. In addition, the operating lifetime of semiconductor lasers usually decreases exponentially with temperature. Therefore, it is essential to design lasers with consistently low self-heating for better reliability.

In this chapter, the optical and electrical characteristics of VCSELs under the influence of self-heating are reviewed and various approaches to model the thermal properties of VCSELs are discussed. This chapter is organized as follows. First, the dependence of lasing wavelength, threshold current and output power on the temperature of laser cavity is discussed. The design consideration for the optimum threshold current and output power is also studied. Second, the methods to evaluate the effective thermal resistance and effective heatsource of VCSELs as well as their limitation to calculate the average temperature change inside the laser cavity are investigated. A simple thermal rate equation model is also developed using the concept of effective thermal resistance. Using this model, the steady-state characteristics such as thermal rollover of the $L-I$ curves as well as the modulation response of VCSELs can be analyzed. Third, two numerical methods, the finite difference method and the finite element method, are applied to calculate the detailed heat distribution inside the laser cavity of VCSELs. The approach to implement these numerical methods as well as the corresponding boundary conditions is also described. Finally, a brief summary is given.

5.2 INFLUENCE OF TEMPERATURE ON STEADY-STATE PERFORMANCE OF VCSELs

In this section, the steady-state performance of VCSELs under the influence of self-heating is reviewed. The dependence of lasing wavelength, threshold current, as well as output optical power on the temperature of laser cavity is described. The design consideration of VCSELs at continuous-wave (CW) operation is also discussed.

5.2.1 Influence of Temperature on Lasing Wavelength of VCSELs

A typical VCSEL structure (i.e., see Fig. 4.8) consists of two DBRs separated by a spacer layer, which incorporates an active layer. If the lasing wavelength of the VCSEL is designed to be λ_R, the thickness of the spacer layer and all layers of the two DBRs should be selected as λ_R and $\lambda_R/4$, respectively. In this case, the peak of the standing wave intensity is located at the position of the active layer so that λ_R can be expressed as

$$\lambda_R = n_{spacer} h_{spacer}, \tag{5.1}$$

where n_{spacer} is assumed to be the effective refractive index of the spacer layer, including the active layer and h_{spacer} is the corresponding physical thickness.

It must be noted that the refractive index and thickness of all layers of the laser cavity are temperature-dependent. Therefore, the exact value of λ_R can be calculated only by the numerical technique as described in Section 2.4.1 with the change of refractive index Δn, and thickness Δh of all layers with temperature included in the calculation. The variation of Δn and Δh can be approximated by

$$\Delta n \cong \left.\frac{\partial n}{\partial T}\right|_{\lambda} \Delta T + \left.\frac{\partial n}{\partial \lambda}\right|_{T} \Delta\lambda_R, \tag{5.2a}$$

$$\Delta h \cong \lambda_R \left.\frac{\partial h}{\partial T}\right|_{\lambda} \Delta T, \tag{5.2b}$$

where ΔT and $\Delta\lambda_R$ are the change in temperature and resonant wavelength, respectively. The partial derivatives $\partial n/\partial T|_{\lambda}$ and $\partial n/\partial\lambda|_T$ are to be evaluated at constant wavelength and temperature, respectively, for the calculation of the change in refractive index. The remaining partial derivative, $\partial h/\partial T|_{\lambda}$, is the linear thermal expansion coefficient. The thermal dispersion of the DBRs and spacer layer are matched, and are given by [5]

$$\frac{1}{n_{spacer}} \left.\frac{\partial n_{spacer}}{\partial T}\right|_{\lambda} = \frac{1}{n_L h_L + n_H h_H} \left[h_L \left.\frac{\partial n_L}{\partial T}\right|_{\lambda} + h_H \left.\frac{\partial n_H}{\partial T}\right|_{\lambda} \right], \tag{5.3}$$

where n_L (n_H) is the refractive index and h_L (h_H) is the thickness of layers. The subscripts L and H represent the low and high refractive indices of the layers of DBRs, respectively. The rate of change of λ_R with temperature can be determined from the temperature sensitivity of the refractive index and thermal expansion of the spacer layer

$$\frac{\partial \lambda_R}{\partial T} = \frac{\lambda_R}{n_g} \left[\left.\frac{\partial n_{spacer}}{\partial T}\right|_{\lambda} + n_{spacer} \left.\frac{\partial h_{spacer}}{\partial T}\right|_{\lambda} \right], \tag{5.4}$$

where n_g is the group refractive index of the spacer layer and is expressed as

$$n_g = n_{spacer} - \lambda_R \left.\frac{\partial n}{\partial \lambda}\right|_{T}. \tag{5.5}$$

Figure 5.1 shows the measured lasing wavelength and threshold current of gain-guided VCSELs under the influence of temperature [6]. The VCSEL is

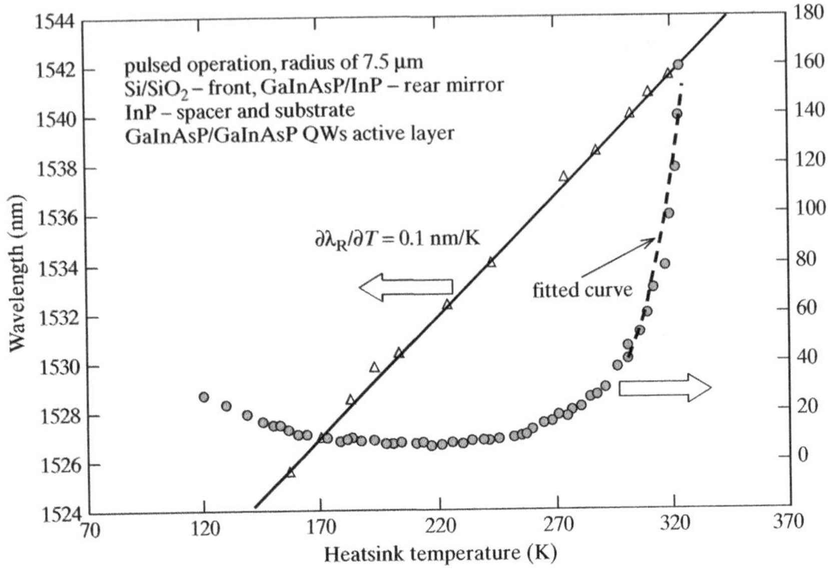

Figure 5.1 Lasing wavelength and threshold current (under pulsed operation) as a function of temperature for a 15-μm-diameter VCSEL (the triangles depict the temperature shift of the laser line) (after Ref. 6).

TABLE 5.1 VCSEL Temperature–Lasing Wavelength Profile

Material	$\partial h/\partial T$ (10^{-6}°C^{-1})	$\partial n/\partial T$ ($\times 10^{-4}$ K^{-1})	n (RT)	n_g (RT)	κ (W cm^{-1} K^{-1})
GaAs	6.4 – 6.9	~4 at 1.25 μm	~3.4 at 1.25 μm	4.4 at 1.25 μm	0.44
$Al_xGa_{1-x}As$ [42]	6.4 – 1.2x	~4 at 0.95 μm	~3.5 at 0.95 μm	4.3 at 0.95 μm	0.44/ $(1 + 12.7x - 13.22x^2)$
InP	4.59	~3 at 1.5 μm	~3.6 at 1.5 μm	4.6 at 1.5 μm	0.68
$In_{1-x}Ga_xAs_yP_{1-y}$	4.56 + 1.18y	~3 at 1.3 μm	~3.6 at 1.5 μm	4.6 at 1.55 μm	~0.382

constructed by Si/SiO_2–GaInAsP/InP DBRs, which are separated by an InP spacer layer. The corresponding active layer inside the spacer layer consists of GaInAsP/GaInAsP MQWs, and the laser is designed to operate at 1.54 μm wavelength at room temperature. During the measurement, the temperature is maintained approximately uniform over the entire laser cavity for pulsed operation so that the shift of wavelength can be linearly proportionated according to the temperature change. It is found that the VCSEL $\partial\lambda_R/\partial T$ is about 0.1 nm/K. Substitute the parameters given in Table 5.1 into (5.4) and (5.5), and it can be shown that the calculated $\partial\lambda_R/\partial T$ is close to that given from the measurement. This indicates that the thermal dispersion matching condition given in (5.3) is satisfied and the detuning of λ_R can be estimated by the change of refractive index inside the spacer layer. In fact, (5.4) is a good approximation of, the temperature dependence of lasing wavelength of VCSELs under pulsed operation provided the compositions of spacer layer and the DBRs materials are of the same type. In fact, it is shown experimentally that the value of $\partial\lambda_R/\partial T$ for VCSELs (1) with AlGaAs/GaAs mirrors plus $Al_{0.4}Ga_{0.6}As$ spacer layer operating at 0.95 μm wavelength [7] and (2) with AlGaAs/GaAs mirrors plus GaAs spacer layer operating at 1.25 μm wavelength [8] are found to be 0.084 and 0.088 nm/K, respectively, which are matched with the calculated values obtained from (5.4) using the parameters given in Table 5.1.

5.2.2 Influence of Temperature on Threshold Current and Output Power of VCSELs

It is observed in Figure 5.1 that the threshold current I_{th} convexes downward and reaches minimum at a temperature around 220 K. This profile of I_{th} is in fact a typical characteristic of VCSELs, which is different from those of facet emitting lasers. For facet emitting lasers, I_{th} decreases monotonically with the reduction of temperature. This is because the facet emitting lasers have a small cavity mode spacing compared to the linewidth of the gain profile so that the resonant mode λ_R is always close to the gain peak wavelength, λ_p (i.e., $\lambda_p \cong \lambda_R$). Hence, it is expected that I_{th} decreases but the value of optical gain peak increases with the reduction of temperature. However, in VCSELs, the variation of temperature leads to an offset between λ_p and λ_R as $\partial\lambda_R/\partial T < \partial\lambda_p/\partial T$.

Figure 5.2 compares the variation of λ_p and λ_R with temperature, and the corresponding temperature dependence of optical gain spectra is also inserted in the Figure [9]. It is observed that near 270 K, $\lambda_p = \lambda_R$. For temperatures higher than 270 K, optical gain reduces monotonically at λ_R so that the corresponding threshold current is increased. On the other hand, for temperatures between 170 and 270 K, the optical gain at λ_R as well as the threshold current remains unchanged. However, further reduction of temperature from 170 K reduces optical gain at λ_R so that the threshold current is increased. Hence, the profile of I_{th} exhibits a downward convex over this range of temperature. It is observed from Figure 5.1 that the minimum value of I_{th} occurs near 220 K but not at 270 K, which implies that other temperature-dependent loss mechanisms influence the overall cavity loss. Nevertheless, the condition of $\lambda_p = \lambda_R$ gives a close estimation on the minimum value of I_{th}. It is noted that VCSELs under the condition of $\lambda_p = \lambda_R$ can produce a reasonably low threshold current, but the required operating temperature may not be appropriate for normal application. In order to optimize I_{th} at a desired operating temperature, suitable selection of gain offset wavelength (i.e., $\lambda_p - \lambda_R$) is required, which can be obtained by varying the design of DBRs for different λ_R.

Analysis of AlGaInP-based 670–690-nm VCSELs [13] shows that suitable selection of gain offset wavelength can provide a minimum threshold current as well as a maximum output power over a wide temperature range. The VCSELs under investigation have planar gain-guided structure, which are fabricated in

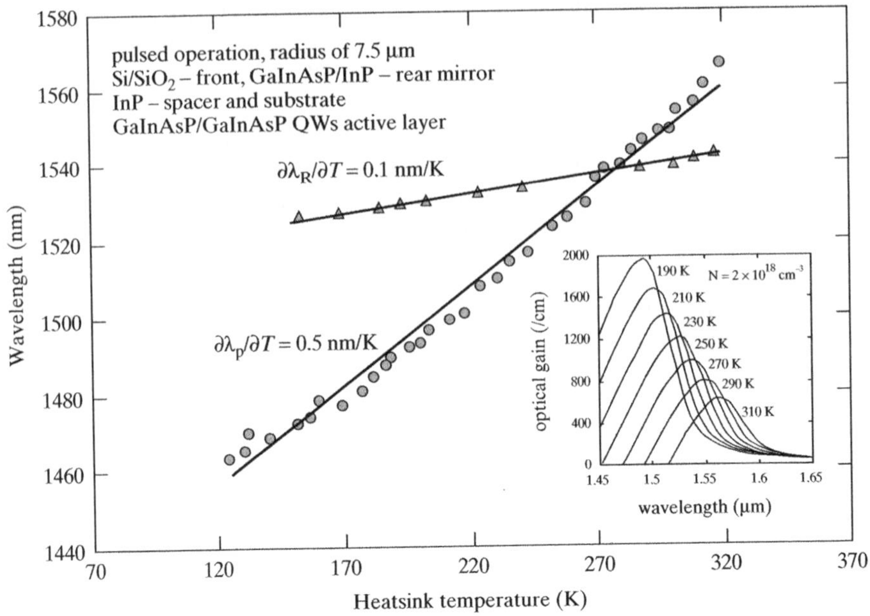

Figure 5.2 Variation of cavity mode λ_R and optical gain peak λ_p with temperature. At room temperature, λ_p is 8 nm above the λ_R. The inset diagram gives the QW's gain spectrum at different ambient temperatures showing the thermal shift of the peak gain wavelength λ_p as well as of the magnitude of peak gain. (After Ref. 9).

a front emitting geometry with proton implantation to define device diameters and to channel current into the active region (see Figure 1.10). The front and rear DBRs are composed of $Al_{0.5}Ga_{0.5}As$ and AlAs quarter-wave layers sandwiched between the $In_{0.56}Ga_{0.44}P$ QWs with barriers and cladding layers of $(Al_{0.5}Ga_{0.5})_{0.5}In_{0.5}P$ on each side of the wells. Figure 5.3 shows the threshold current for the four 20-μm-diameter devices as a function of heatsink temperature [13]. The devices have four resonant wavelengths λ_R, which varies from 675 to 690 nm, and the peak gain wavelength λ_p of the QW active layer is 688 nm, where both λ_R and λ_p are measured at room temperature. It is observed that the threshold current of VCSELs with λ_R equal to 675, 680, and 685 nm increases monotonically with temperature. In contrast, the threshold current of VCSELs with $\lambda_R = 690$ nm initially decreases with temperature but increases with increase of temperature beyond 30°C, which is the heatsink temperature of an optimal gain peak/cavity mode overlap. Furthermore, it is noted that the threshold current of the 690-nm lasers over the temperature range of 20–50°C is relatively constant when compared with the other lasers. Hence, the required gain wavelength offset for VCSELs with uniform threshold current over an operating temperature range is $\lambda_p - \lambda_R < 0$ at room temperature.

Another benefit for such an optimal gain peak/cavity mode overlap (i.e., uniform threshold current over a range of operation temperature) is the improvement

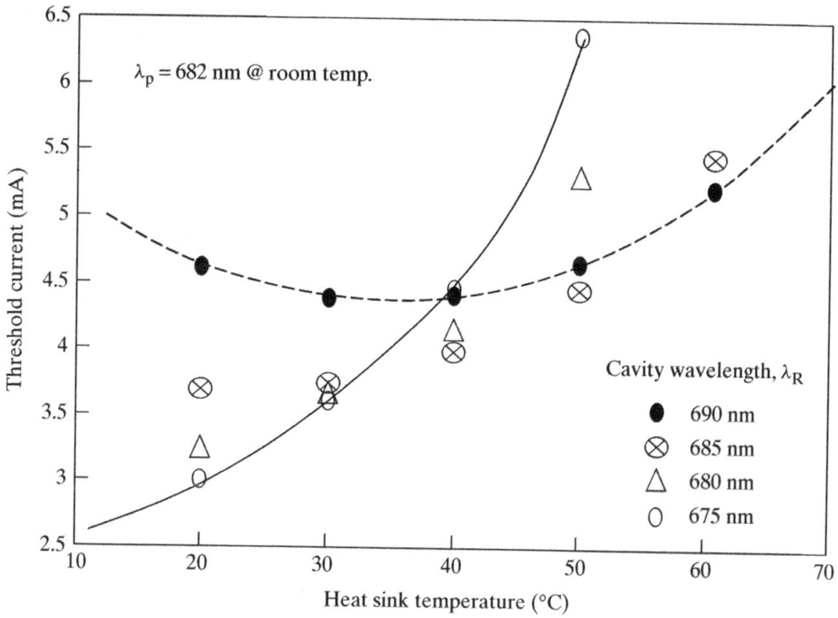

Figure 5.3 Threshold current as a function of heatsink temperature for various 20-μm-diameter VCSELs; the wavelength designation refers to the VCSEL emission wavelength at subthreshold current (after Ref. 13).

of output power. In Figure 5.4, the light–current ($L-I$) characteristics of 15-μm AlGaInP-based VCSELs are at two different emission wavelengths for various heatsink temperatures [13]. Figure 5.4a shows data for the laser with $\lambda_R =$ 675 nm at room temperature. The laser has a relatively low threshold current of 3.4 mA at 20°C, but the threshold current increases significantly with increasing temperature, and the output power drops approximately linearly with a slope of −0.88 mW/K. In contrast, the laser with $\lambda_R = 690$ nm (at room temperature) has a much smaller variation in threshold current and a slower power drop of −0.075 mW/K, as shown in Figure 5.4b. From the 690-nm device, it is observed that a peak output power of 1.5 and 0.9 mW is achieved at 50°C and 60°C, respectively, representing the highest powers achieved to date at these operating temperatures. This indicates that devices with lesser threshold current may not maximize the output power. It is more preferable to design VCSELs that have constant threshold current over a wider range of operating temperature with which the output power can be maximized.

The $L-I$ curves shown in Figure 5.4 indicate some other distinct characteristics of VCSELs under the influence of thermal effects. It is observed that (1) the threshold current, which is also dependent on the selection of gain offset wavelength, increases with the increase of temperature; (2) the differential quantum efficiency decreases with the increase in temperature; and (3) the peak output power reduces with the increase in heatsink temperature but reduces to zero with

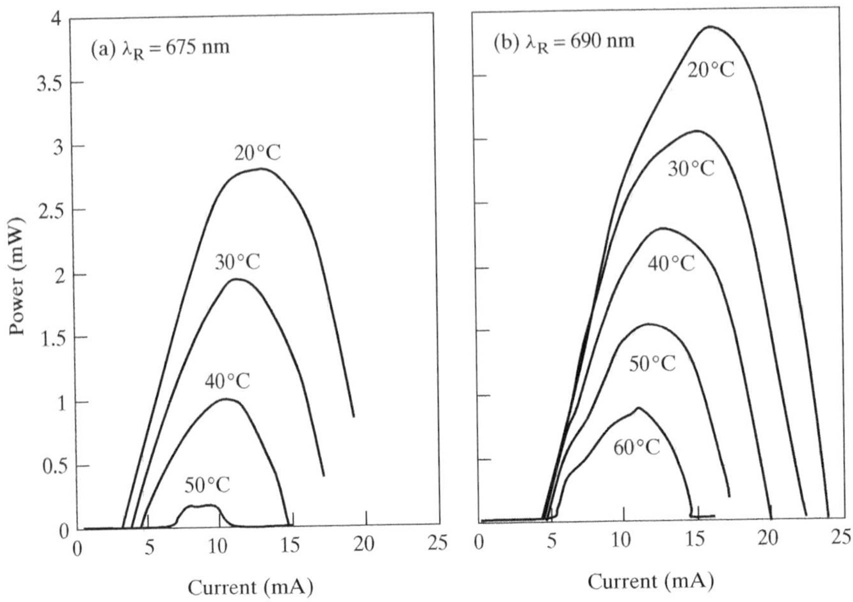

Figure 5.4 Light/current curves as a function of heatsink temperature for 15-μm-diameter VCSELs with resonant wavelength λ_R of (a) 675 nm and (b) 690 nm at room temperature.

the increase in injection current (i.e., thermal rollover of $L-I$ curves). In fact, these phenomena are observed in devices with implementation of gain-guided transverse confinement [14] as well as the index-guided oxide confined structure [15]. In addition, these $L-I$ curve characteristics of the are also observed in VCSELs with different gain offset wavelengths.

In order to model the dependence of threshold current on temperature as shown in Figure 5.1, the total injected current into the QW active layer is assumed to consist of three components. Figure 5.5 shows the corresponding schematic of the recombination mechanism of injection current inside the QW active region. The injection current is assumed to consist of (1) the recombination of carrier concentration inside the QW active layer, (2) the recombination of carrier concentration inside the barriers of the active layer, and (3) the leakage of injection current over the barriers between the quantum wells [10]. Hence, the threshold current of VCSELs can be expressed as

$$I_{\mathrm{th}}(N,T) = \eta_{\mathrm{inj}}(I_{\mathrm{NR,a}}(N,T) + I_{\mathrm{NR,b}}(N,T) + I_{\mathrm{leak}}(N,T)), \tag{5.6}$$

where η_{inj} is the injection efficiency into the QW active layer and can be assumed to be a constant; I_{leak} is the carrier leakage over the heterostructure; $I_{\mathrm{NR,a}}$ and $I_{\mathrm{NR,b}}$ are the nonradiative recombination, including Auger recombination inside the QWs and barriers, respectively, in the active layer. The recombination process

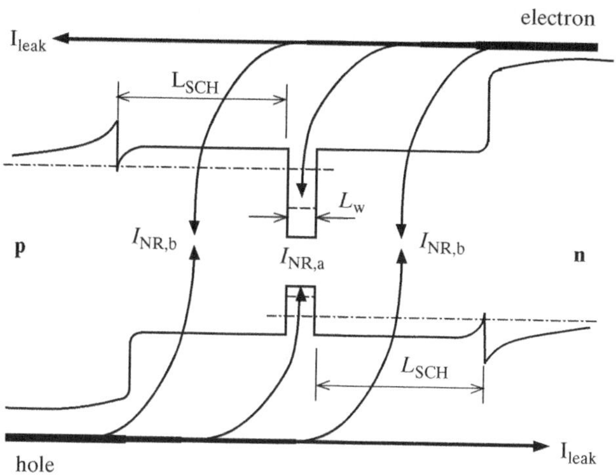

Figure 5.5 Energy diagram for a QW laser diode with both SCH and cladding regions. The recombination mechanism of injected electrons and holes is also indicated in the diagram.

due to spontaneous and stimulated emission inside the active region is ignored at threshold because the dominant factor is the nonradiative recombination.

The corresponding nonradiative recombination inside the QWs and barriers are Auger recombination and intervalence band absorption as well as Shockley–Read–Hall. It is given that the nonradiative current inside the active layer is [3]

$$I_{\mathrm{NR,a}} = qL_{\mathrm{w}}A_{\mathrm{eff}}\left(\frac{N_{\mathrm{th}}}{\tau_n} + C_{\mathrm{Aug}}N_{\mathrm{th}}^3\right), \tag{5.7}$$

where N_{th} is the carrier concentration of the QW at threshold, τ_n is the carrier lifetime, C_{Aug} is the Auger recombination coefficient, L_{w} is the thickness of the QW, and A_{eff} is the area of the injection region. In this case, the threshold gain at λ_{R} produced by N_{th} is required to satisfy the round-trip conditions at the corresponding operation temperature. However, N_{th} can be approximated by using a "no k" selection rule and is given by [11]

$$N_{\mathrm{th}} = \frac{D_n k_{\mathrm{B}} T}{L_{\mathrm{w}}} \log\left[\left(1 + \exp\left(\frac{E_{\mathrm{F}n}}{k_{\mathrm{B}}T}\right)\right)\cdot\left(1 + \exp\left(\frac{E_{\mathrm{F}n} - \Delta E_n}{k_{\mathrm{B}}T}\right)\right)\right], \tag{5.8}$$

where ΔE_n $(= E_{n1} - E_{n2})$ is the separation between the first and second electron energy subbands of the quantum well and $E_{\mathrm{F}n}$ is the expression of electron quasi-Fermi level of the main mode. The derivation of $E_{\mathrm{F}n}$ under the assumption of no-k selection rule is given in the Appendix.

The nonradiative recombination at the barriers can also be estimated using (5.7) but with N_{th} replaced by N_{b}, which is the carrier concentration at barriers.

In addition, the leakage current I_{leak} over the barriers can be calculated with the assumption that the carrier concentration at the cladding layer heterostructure has sufficient energy to overcome the barrier. The electric field is negligible across the cladding region I_{leak} and is given by [10]

$$I_{\text{leak}} = qL_{\text{SCH}}A_{\text{eff}}\frac{N_{\text{b}}}{\tau_{\text{n}}}, \tag{5.9a}$$

where L_{SCH} is the width of the barrier region (i.e., separate confinement layer). The carrier concentration at barriers is approximated by the homojunction diode, as given by [12]

$$N_{\text{b}} = \sqrt{N_{\text{c}}N_{\text{v}}}\exp\left(\frac{-(E_{\text{g,b}} - \Delta E_{\text{F}np})}{k_{\text{B}}T}\right), \tag{5.9b}$$

where N_{c} and N_{v} are the effective conduction and valence edge density of states, respectively; $E_{\text{g,b}}$ is the bandgap of the confinement layers surrounding the active region; and $\Delta E_{\text{F}np}$ $(= E_{\text{g,w}} + E_{n1} + E_{p1} + E_{\text{FT}})$ is the quasi-Fermi-level separation in the active region. The value of $\Delta E_{\text{F}np}$ can be obtained by performing the no-k selection calculation for E_{FT} as discussed in the Appendix. From the equations above, it is noted that N_{b} is more sensitive to temperature than is N_{th}. Furthermore, if $L_{\text{w}} \sim 100$ Å and $L_{\text{SCH}} \sim 1.4$ μm for GaAs/AlGaAs QWs, the magnitude of I_{leak} is much larger than that of $I_{\text{NR,b}}$. Hence, it is reasonable to assume that the variation of threshold current is dominated by the leakage current under the influence of temperature.

The preceding calculation of threshold current can be simplified if the Arrhenius-type relation is employed, which is given by [2]

$$I_{\text{th}}(T) = I_{\text{th}}(T_{\text{ref}})\exp\left(\frac{T - T_{\text{ref}}}{T_0}\right) \tag{5.10}$$

where T_{ref} and T_0 are the reference and characteristic temperatures, respectively, in degrees Kelvin. The Arrhenius-type relation can be interpreted as an approximation to leakage current. In general, T_{ref} is usually selected at 300 K as this is the normal laser operating temperature, and T_0 can be a function of temperature varying between −200 and 400 K. In fact, T_0 can be a constant at some temperature range in VCSELs. A fitted curve using (5.10) is also shown in Figure 5.1 for comparison with the assumptions of I_{th} (300 K) $\sim$ 40 mA, $T_{\text{ref}} = 300$ K, and $T_0 = 18$ K in the calculation. It is shown that the variation of threshold current over a range of temperature can be well fitted by (5.10) with suitable selection of T_{ref} and T_0. The main advantage of such a simple approximation is to avoid studying the complicated transport process inside the QW active layer.

The thermal rollover of L–I curves may be attributed to the influence of temperature on the optical gain and leakage current of VCSELs:

1. The gain spectrum broadens, the location of λ_{p} shifts to longer wavelength, and the corresponding optical gain reduces with the increase in temperature.

However, the temperature dependence of gain spectrum can be minimized by suitable selection of gain offset at a desired operating temperature.

2. The thermal leakage of carriers outside the active region can lead to a reduction of injection efficiency. In fact, leakage current contributes significantly to the thermal rollover of the $L-I$ curves. This is because when the temperature increases, the position of the active layer's Fermi levels rises relative to the bandgap. Consequently, the active layer becomes decreasingly capable of confining carriers. In fact, the leakage current given in (5.10) has been modeled as a function of carrier density and temperature implicitly.

In order to model the output power of VCSELs, a simple formula predicting the injection current dependence of optical output power is taken from (2.11), which is given below,

$$P = A_{\mathrm{eff}}(J - J_{\mathrm{th}}(T)) \cdot h\nu \cdot \frac{\eta_{\mathrm{d}}}{q}. \tag{5.11}$$

where the differential quantum efficiency η_{d}, which is the ratio between the output loss and the total cavity loss of VCSELs, determines the slope of the $L-I$ curves. For the increase in heatsink temperature, η_{d} should decrease because the total cavity loss increases with temperature. However, the dependence of η_{d} on temperature is less sensitive than that of J_{th}. Hence, the nonlinear characteristics of the $L-I$ curves can be assumed to be determined mostly by the temperature dependence of J_{th} in (5.11), and η_{d} can be considered as temperature-independent. As the variation of J_{th} is due mainly to the leakage current from the heterostructure [16], the expression of J_{th} given in (5.6) can be substituted into (5.11) to calculate the thermal rollover of the $L-I$ curves.

The threshold current and output power are also affected by the profile of the transverse modes, which are dependent on the temperature of the laser cavity. For gain-guided VCSELs under CW operation, active region heating results in a nonuniform, bell-shaped temperature distribution, and the temperature distribution becomes more pronounced with the increasing pump current. Hence, the refractive index at the radiation regions increases due to the increases in temperature (i.e., $\partial n/\partial T > 0$), causing the focus of the radiation fields. This self-focusing of the transverse modes into the core region of the active layer is well known as *thermal lensing* [17]. The consequence of self-focusing due to thermal lensing is the reduction of threshold gain of the higher-order transverse modes, which can be excited at lower injection levels. In fact, it has been shown that gain-guided VCSELs with proton implantation excited higher-order transverse modes due to the influence of the thermal lensing effect [13,18]. Furthermore, a "kink" occurred in the $L-I$ curves, which indicates the onset of higher-order transverse modes. This phenomenon is usually observed in gain-guided devices because the active layer has no built-in refractive index profile to confine the transverse modes so that any variation of refractive index profile will change the transverse mode profile or excite higher-order transverse modes. However, for devices with index-guided structure such as oxide-confined structures [18], a single transverse mode can be maintained in VCSELs even at high power. This is because the thermal lensing effect has less significant influence on the transverse optical confinement structure of VCSELs because of the built-in refractive index profile. It must be noted that the spatial hole burning effect can amplify the thermal lensing effects in VCSELs as $\partial n/\partial N < 0$. For accurate calculation of self-focusing on the transverse mode behavior of VCSELs, a self-consistent analysis technique is required to simultaneously account for the optical field, carrier concentration, and temperature profile. This will be discussed further in another chapter of this book.

5.2.3 Other Optimization Designs for Stable Threshold Current of VCSELs

Has discussed in Section 5.2.2, that the gain offset wavelength can be applied to optimize the threshold current and output power of VCSELs. The optimum operation condition of VCSELs is achieved by thermal-heating-induced wavelength blueshifted of gain peak wavelength toward the cavity mode so that the VCSELs are aligned at the operation temperature. However, the threshold current still varies with temperature because of the intrinsic properties of optical gain. The wavelength dependence on optical gain can in fact be minimized if the optical gain spectrum can be flattened. This is because the broad gain bandwidth provides a constant optical gain for the cavity mode over a temperature range so that the threshold current can be maintained at a constant low level. Therefore, it is proposed that the gain bandwidth be extended by using active layers consisting of plural types of multiple quantum wells [19,20]. The InGaAs the MQW active layer consists of four QWs with different energy levels. The corresponding gain peak wavelengths λ_p of the four quantum wells are set to 960, 960, 640, and 920 nm [19,20] so that a broadened gain bandwidth can be achieved and the VCSEL threshold current can also be maintained over a range of temperature. However, this arrangement of the active layer has to be optimized with temperature to produce an overall flat optical gain spectrum over a range of wavelength.

It is noted that the heat generation inside the laser cavity can be reduced significantly, provided the series resistance of the DBRs is minimized. Increasing the doping level inside the semiconductor layers of DBRs can reduce the corresponding series resistance but at the cost of the low optical absorption (i.e., high doping increases free-carrier absorption losses) [21]. Alternatively, this problem can be avoided if a current shortcut is built to bypass the injection carrier into the active region through the DBRs so that the high series resistance of the mirrors does not contribute to the generation of heat. This can be realized using intracavity contacts for current injection into the active region [22,23]. As shown in Figure 5.6 [23], the region between the mirrors consists of an active region bounded by the p-type and n-type contact layers, which promote current spreading across the aperture. Hence, the carrier is directly injected into the active region without passing through the DBRs. In this case, no current is injected into the active layer via the n- and p-type DBRs, which means that the heat generation inside the DBRs can be removed.

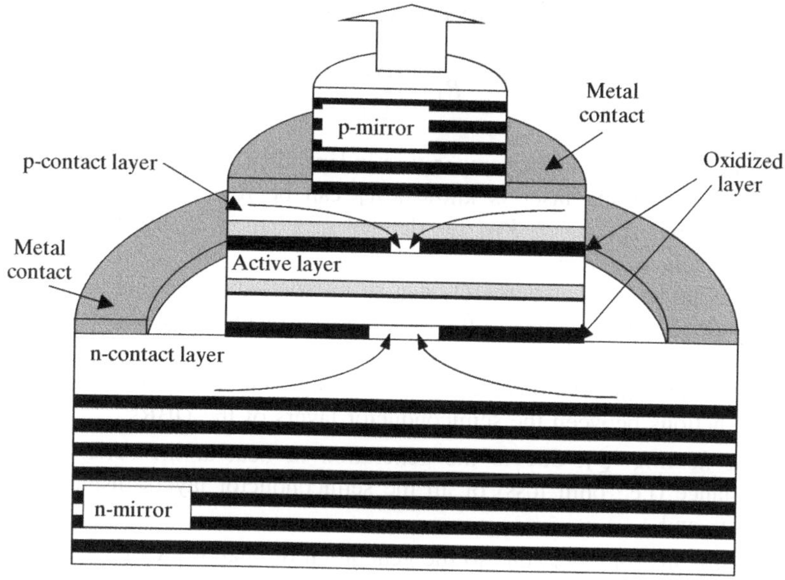

Figure 5.6 Schematic diagram of a VCSEL with intracavity contacts for current injection into the active region.

Furthermore, increasing the heatsink capability of VCSELs can reduce the corresponding thermal resistance. This can be realized by using a huge block of copper heatsink [24] or an Au-plated heat spreading layers [25] attached on the p side of the DBRs where the light is emitted through the n-substrate layer. Hence, the effective thermal resistance of the VCSELs decreases so that the variation of temperature due to self-heating can be minimized. The improvement in heatsinking increases the maximum output power of lasers, and the dependence of threshold current on temperature can be minimized.

5.3 SIMPLE THERMAL MODELS OF VCSELs

In this section, the concept of effective thermal resistance and heatsource is applied to analyze the self-heating effects in VCSELs. Effective thermal resistance of VCSELs with nonuniform heat flow due to the complicated configuration of the dielectric/semiconductor layers is derived. The main advantage of the effective thermal resistance is the simplified calculation of the average temperature generation inside the laser cavity. Furthermore, a simple thermal rate equation model can also be deduced using the concept of effective thermal resistance so that steady state as well as the modulation response of VCSELs can be calculated easily.

5.3.1 Effective Thermal Resistance and Heatsource

In this chapter, the effective thermal resistance R_T (K/W) is defined as the ratio between the average temperature change ΔT (K) and the total effective heatsource Q_T (W) inside the laser cavity, which is given by

$$R_T = \frac{\Delta T}{Q_T}. \tag{5.12}$$

Hence, R_T is assumed to be temperature-independent. If the effective thermal conductivity κ_T of the laser is known, R_T can be related to the normalized effective thermal resistance R_{TH} (as defined in Chapter 2), by multiplying κ_T by R_T (i.e., $R_{TH} = \kappa_T R_T$).

The total effective heatsource Q_T can be interpreted as the total nonradiative power inside the laser cavity. In fact, Q_T consists of three heatsources:

- *Mirror heterojunction heating* Q_m due to the generation of heat at the heterojunctions between the semiconductor layers of the DBRs,
- *Joule heating* Q_J, due to the generation of heat at the series electrical resistance (i.e., ohm loss) of all the semiconductor layers inside the laser cavity, and
- *Junction heating* Q_{ac}, due to the generation of heat at the active layer.

By applying the principle of superposition, the effective heatsource Q_T can be written by

$$Q_T = Q_m + Q_J + Q_{ac}. \tag{5.13}$$

The mirror heterojunction heating can be expressed as

$$Q_m = I V_{J,het} \times \text{number of heterojunction}, \tag{5.14a}$$

where I (A) is the injection current and $V_{J,het}$ (V) is the voltage drop across the heterointerfaces within the DBRs. The Joule heating is given by

$$Q_J = I^2 R_s, \tag{5.14b}$$

where R_s (Ω) is the total ohmic series resistance of the laser cavity. Finally, the active region junction heating can be written as

$$Q_{ac} = I_{th} V_J, \tag{5.14c}$$

where I_{th} is the threshold current and V_J is the p-n junction voltage drop. Hence, Q_T can be determined from (5.14), provided $V_{J,het}$, I_{th}, and V_J are known. In order to calculate the temperature rise in VCSELs, R_T should be determined under the assumption that Q_T is a single heatsource located at a particular position inside the cavity of VCSELs.

Figure 5.7a shows a VCSEL with a cylindrical airposted configuration. It is assumed that the device has a uniform heatsource, uniform heat flow from the heatsource down through a medium to the heatsink, and isotropic thermal

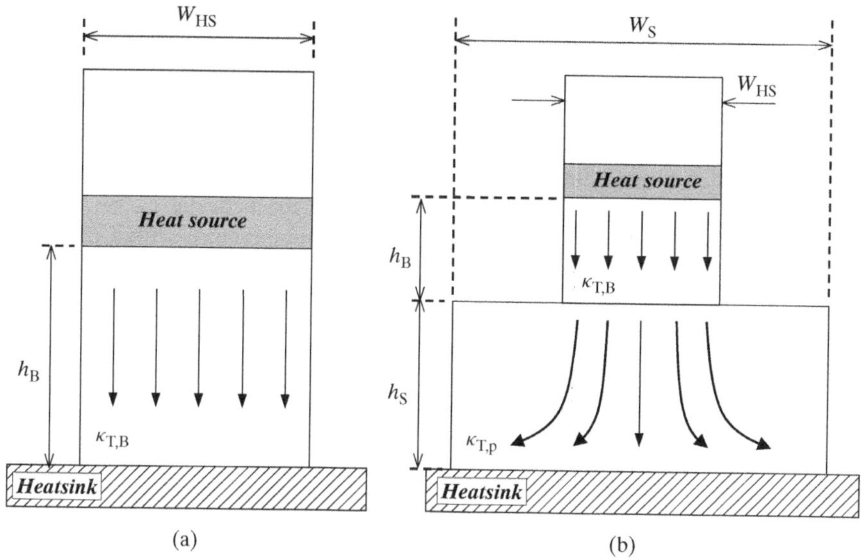

Figure 5.7 Schematic representation of heat flux spreading in VCSELs (a) with a cylindrical airposted structure and (b) with a pillar on a large-diameter pedestal.

conductivity. In this case, all the separated heatsources as indicated in (5.13) are grouped together to form a flat uniform heatsource. As it is assumed that the heat is flowing uniformly inside the laser cavity, the thermal resistance $R_{T,unif}$ (K/W) can be expressed as [26]

$$R_{T,unif} = \frac{4}{\pi W_{HS}^2} \frac{h_B}{\kappa_{T,B}}, \tag{5.15a}$$

where W_{HS} (cm) is the diameter of the heatsource and h_B (cm) and $\kappa_{T,B}$ (W cm^{-1} K^{-1}) stand for the thickness and thermal conductivity, respectively, of the bottom layer between the heatsource and the heatsink. Hence using (5.13) and (5.15a), the average temperature rise inside the laser cavity can be calculated. Furthermore, if the bottom layer consists of layers, $R_{T,unif}$ can be modified to

$$R_{T,unif} = \frac{4}{\pi W_{HS}^2} \sum_i \frac{h_i}{\kappa_{T,i}}, \tag{5.15b}$$

where h_i and $\kappa_{T,i}$ stand for the thickness and thermal conductivity, respectively, of the ith layer between the heatsource and heatsink.

Figure 5.7b shows another possible configuration of VCSELs, which comprises a pillar on a large-diameter pedestal. It is expected that the heat flow inside the pillar will be uniformly from the heatsource to the surface of the pedestal but the heat flow inside the pedestal will be a spreading of heat flux. Hence, the total thermal resistance R_T (K/W) of VCSELs should be considered as the summation of the thermal resistance of uniform heat flow $R_{\mathrm{T,unif}}$ and that of nonuniform heat flow $R_{\mathrm{T,non}}$, which is given by

$$R_{\mathrm{T}} = R_{\mathrm{T,unif}} + R_{\mathrm{T,non}}, \tag{5.16}$$

where $R_{\mathrm{T,unif}}$ can be calculated using (5.15). The expression for $R_{\mathrm{T,non}}$ can be obtained with the assumptions that the thickness of the pedestal $h_{\mathrm{S}} \to \infty$, and with an imposed boundary condition of a constant temperature within the disk. If the thermal conductivity of the pedestal is $\kappa_{\mathrm{T,p}}$, the corresponding thermal resistance $R_{\mathrm{T,non}}$ is given by [27]

$$R_{\mathrm{T,non}} = \frac{1}{2\kappa_{\mathrm{T,p}} W_{\mathrm{HS}}}. \tag{5.17}$$

On the other hand, for $h_{\mathrm{B}} \to 0$ and h_{S} finite (i.e., a flat disk uniform heat-source located on top of a uniform cylinder with finite dimension), $R_{\mathrm{T,non}}$ can be approximated by fitting to exact numerical solutions, which describes the two-dimensional heat flux spreading inside the pedestal, and is given by [28]

$$R_{\mathrm{T,non}} \cong \frac{4h_{\mathrm{S}}}{\pi \kappa_{\mathrm{T,p}} W_{\mathrm{S}}^2} + \frac{1}{2\kappa_{\mathrm{T,p}} W_{\mathrm{HS}}} \left(1 - \frac{W_{\mathrm{HS}}}{W_{\mathrm{S}}}\right)^{3/2}. \tag{5.18}$$

The first term, analogous to (5.15), describes the thermal resistance of uniform heat flow in the vertical direction through the cylinder of diameter W_{S}. The second term approximates the spreading resistance of thermal contact. Within the limit of very large W_{S}, the uniform heat flow vanishes and (5.18) describes the thermal contact resistance of a semiinfinite medium (5.17). For individual VCSELs located on large-diameter chips (i.e., $W_{\mathrm{S}} \gg W_{\mathrm{HS}}$), (5.18) would therefore predict the same thermal resistance as that given by (5.17). The uniform heat flow term in (5.18) describes the heat flow through the cylinder of diameter W_{S}. Hence, (5.18) implies that the substrate medium is thick enough for complete transverse spreading of the heat flux to take place. However, (5.18) can be applied only to VCSELs mounted junction-up. It also neglects the upward heat flow, which may lead to a slight overestimate of R_{T} for devices with small active region diameters.

Furthermore, if the pedestal consists of layers, (5.18) can be used to calculate the $R_{\mathrm{T,non}}$ by replacing h_{p} and $\kappa_{\mathrm{T,p}}$ with the corresponding effective values for the layers, which are given by

$$\kappa_{\mathrm{T,eff}} \cong \frac{\sum_i h_i / \kappa_{\mathrm{T},i}}{\sum_i h_i}, \tag{5.19a}$$

$$h_{\mathrm{eff}} \cong \kappa_{\mathrm{T,eff}} \sum_i \left(\frac{h_i}{\kappa_{\mathrm{T},i}}\right), \tag{5.19b}$$

where $\kappa_{T,eff}$ and h_{eff} represent the effective thermal conduction and thickness of the multilayer media. Hence, the effective thermal resistance $R_{T,non}$ is given by

$$R_{T,non} \approx \frac{4h_{eff}}{\pi \kappa_{T,eff} W_S^2} + \frac{1}{2\kappa_{T,eff} W_{HS}} \left(1 - \frac{W_{HS}}{W_S}\right)^{3/2}. \tag{5.20}$$

It must be noted that (5.20) will no longer apply when, for a fixed W_{HS}, either the cylinder height h_{eff} is too small or the cylinder diameter W_S is too large for the complete radial spreading of the heat flow to occur. In such a case, (5.20) overestimates the actual thermal resistance. For any fixed W_{HS} and W_S, there is a critical thickness h_c, below which (5.20) is no longer applicable. The value of h_c can be estimated by examining (5.20) when $R_{T,non}$ reaches a minimum as a function of W_S. This leads to the following condition for h_{eff}:

$$h_{eff} \geq h_c = \frac{3\pi W_S}{32} \left(1 - \frac{W_{HS}}{W_S}\right)^{1/2}. \tag{5.21}$$

In the preceding calculation of thermal resistance of VCSELs, a uniform heat-source is used to represent all the thermal heatsources inside the cavity. For a more appropriate analysis of $R_{T,non}$, the distribution of uniform heatsource should be assumed. In the device shown in Figure 5.8, the uniform heatsources are located in three sections: the active layer and the p- and n-doped DBRs of a VCSEL. Heat spreading occurs at these sections, which consist of layers with different thermal conductivity. In order to calculate the effective thermal resistance

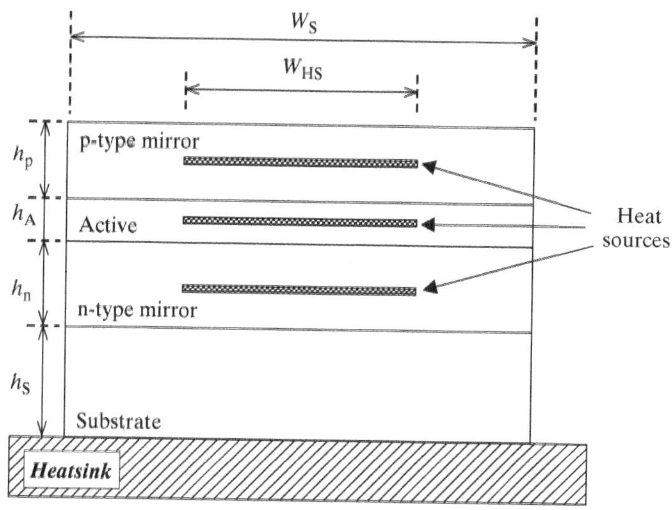

Figure 5.8 Schematic diagram of a VCSEL mounted substrate-down. Three heatsources are assumed to be located at the center of the n mirror, p mirror, and active layer, respectively.

of the device, the effective thermal conductivity and thickness of each section should be evaluated using (5.19) so that the vertical heat flow can be described using (5.20). Hence, the thermal resistance of each section $R_{T,i}$ can be found by substituting the corresponding $\kappa_{T,eff}$ and h_{eff} of the ith section into (5.20). Note that the summation is extended from the heatsources all the way down to the substrate. Using this approach, contributions to thermal resistance associated with each heatsource can be calculated separately and then, on the basis of the superposition principle, added together with appropriate weights determined by their relative power and position. Hence, the effective thermal resistance $R_{T,eff}$ can be written as

$$R_{T,eff} \cong \frac{\sum R_T Q_T + h_{eff,ac} \sum R_T Q_T / h_{eff}}{\sum Q_T}, \tag{5.22}$$

where the first summation of the denominator of (5.22) extends over the heat-sources located no higher than the active region and the second summation extends over the heatsources located above the active region. Hence, Q_T is the power dissipated by the heatsource below and above the active region and $h_{eff,}$ and $h_{eff,ac}$ are the position of the corresponding heatsource and active region, respectively, in the transformed coordinates. The potential problem with the uniform cylinder model is that the transverse spreading is implied to occur in the medium of thermal conductivity $\kappa_{T,eff}$, determined by considering all layers between the heatsource and the heatsink. As long as thermal conductivities of these layers along the transverse direction do not differ too much, (5.20) and (5.22) can be expected to give a reasonably good estimate of thermal resistance.

5.3.2 Simple Rate Equation–Based Thermal Model

From the preceding section, the L–I curves of VCSELs can be evaluated using the approach of effective thermal resistance. However, this method of calculation would not permit the evaluation of the corresponding modulation response under the influence of selfheating. In fact, simple rate equations can be utilized to calculate the steady state as well as the dynamic response simultaneously.

The thermal characteristics of VCSELs can be described by the thermal rate equation, which comes from the thermal conduction Equation [10]

$$\rho_m C_p \frac{\partial T}{\partial t} = \nabla \cdot (\kappa_T \nabla T) + \rho_T, \tag{5.23}$$

where ρ_m (g/cm^3) is the mass density, C_p (Jg^{-1} K^{-1}) is the heat capacity, κ_T (W cm^{-1} K^{-1}) is the thermal conductivity of the laser cavity, $\rho_T (= Q_T/V_S)$ (W/cm^3) is the power density, Q_T (W) is the total heatsource generated inside the laser cavity, and V_S (cm^3) is the volume between the heatsource and heatsink. In order to introduce the concept of thermal resistance as described in Section 5.3.1,

(5.23) is integrated over the cavity volume in between the heatsource and heatsink, which gives

$$C_T \frac{\partial T}{\partial t} = (P_{IV} - P) - \frac{T - T_0}{R_T}, \tag{5.24}$$

where T_0 is the background temperature, $C_T(= \rho_m C_p V_S)$ is the thermal capacitance of the laser cavity. It is assumed that an equivalent thermal resistor of resistance, R_T (K/W), represents the divergence of heat flux within the volume V_S. In (5.24), P_{IV} and P represent the total input electrical power and output optical power of the device, respectively. P_{IV} can be expressed as the product of injection current and total biased voltage, which is given by

$$P_{IV} = I \cdot \sum V_{J,het} + IV_J + I^2 R_S, \tag{5.25}$$

where the summation is extended over all the heterojunctions. In (5.25), I is the total injection current, V_J is the junction voltage of QW active layer, $V_{J,het}$ is the voltage drop at the heterointerface of the DBRs, and R_S is the total series resistance of the laser. In the derivation of (5.24), it is assumed that Q_T is directly proportional to $(P_{IV} - P)$, which implies that the input power not contributing to light will be converted into heat. The L–I curves of VCSELs under the influence of temperature can be evaluated by substituting (5.24) into (5.11), giving

$$P = A_{eff}(J - J_{th}(T_0 + (P_{IV} - P) \cdot R_T)) \cdot h\nu \cdot \frac{\eta_d}{q}, \tag{5.26}$$

which implies that the rollover of the L–I curves of VCSELs is affected mainly by the temperature dependence of J_{th} (i.e., via leakage current).

For the analysis of steady state as well as the modulation response of VCSELs influenced by the thermal effect, (5.24) has to be solved simultaneously with the carrier and photon rate equations. In this case, the corresponding rate equations of the simplest form are modified to [29]

$$\frac{\partial N}{\partial t} = \eta_d \frac{J - J_{th}(T)}{qd} - G(N, T)S, \tag{5.27a}$$

$$\frac{\partial S}{\partial t} = \Gamma_z G(N, T)S - \frac{S}{\tau_p} + \beta_s \Gamma_z B_{sp}(T)N^2, \tag{5.27b}$$

where N (cm^{-3}) is the carrier concentration, S (cm^{-3}) is the photon density, G (ns^{-1}) is the normalized optical gain, and the other parameters have their usual meanings. This simple rate equation model has been applied to analyze the steady-state as well as modulation response of VCSELs [12,16,29,30].

The photon power P, in watts, of the laser given in (5.24) can be expressed in terms of S

$$P = \tfrac{1}{2} v_g (1 - |r_{eff}|^2) A_{eff} \Gamma_z S, \tag{5.28}$$

where $|r_{\text{eff}}|$ is the effective reflectivity of the DBRs and Γ_z is the longitudinal confinement factor of the laser. In the rate equations, the optical gain G is also a function of temperature. If the cavity mode and gain peak are assumed to be matched at the range of operation, G, which is a function of carrier concentration and temperature, can be expressed as

$$G = v_g a_N(T) \log\left(\frac{N - N_0}{b_N(T)}\right), \tag{5.29}$$

where a_N and b_N are functions of temperature and can be obtained by curve fitting to the bandstructure calculations of QW gain peak [12]. A more general expression for G is given by mena et al. [16] using a linear relation between G and N, and the temperature dependence on G can be written as

$$G = G_N(T) \cdot (N - N_t(T)), \tag{5.30}$$

where the temperature dependences of $G_N(T)$ and $N_t(T)$ are expressed as

$$G_N(T) = G_0 \cdot \frac{a_{g0} + a_{g1}T + a_{g2}T^2}{b_{g0} + b_{g1}T + b_{g2}T^2}, \tag{5.31a}$$

$$N_t(T) = N_{t0} \cdot (c_{n0} + c_{n1}T + c_{n2}T^2), \tag{5.31b}$$

and the coefficients of T in (5.31) can be found in Table 5.2.

TABLE 5.2 Parameters of Thermal Model Based on Simple Rate Equation

Parameters	(a)	(b)	(c)	(d)
$G_o(s^{-1})$	2.112×10^4	6.5092×10^4	2.199×10^4	1.901×10^5
a_{g0}	-1.598×10^4	-6.941×10^3	-9.490×10^3	-1.972×10^4
a_{g1} (K^{-1})	7.344	8.282	44.34	73.91
a_{g2} (K^{-2})	0.2092	0.08846	0.01355	0.02669
b_{g0}	1.717×10^5	1.719×10^3	3.337×10^4	1.597×10^4
b_{g1} (K^{-1})	-1.093×10^3	-49.41	-204.5	-119.4
b_{g2} (K^{-2})	1.743	0.182	0.3405	0.2322
N_{t0}	2.206×10^7	6.208×10^6	8.133×10^6	2.726×10^6
c_{N0}	1.417	6.521	0	1
c_{N1} (K^{-1})	-0.006279	-0.03651	0.001723	-0.007715
c_{N2} (K^{-2})	1.539×10^{-5}	6.012×10^{-5}	0	2.068×10^{-5}
I_0 (A)	3.935	22.44	2.073	1.923×10^4
a_0 (K)	3700	6773	3016	2422
a_1 (K)	1.259×10^{-5}	1.980×10^{-4}	1.799×10^{-6}	8.465×10^{-6}
a_2	6.522×10^{-9}	9.377×10^{-9}	1.854×10^{-8}	5.570×10^{-8}
a_3 (K)	2.471×10^9	6.634×10^8	7.662×10^8	7.472×10^9
R_T (°C/mW)	1.647	5.5	2.4	0.9

The voltage across the p-n junction of the QW layer can be expressed in terms of carrier concentration [30]

$$V_J = \frac{1}{q}\left(E_g(T) + k_B T \cdot \log\left\{\left(\exp\left(\frac{N}{N_c}\right) - 1\right) \cdot \left(\exp\left(\frac{N}{N_v}\right) - 1\right)\right\}\right), \tag{5.32}$$

where E_g is the energy bandgap of the QW active layer. V_J is now a function of temperature as well as carrier concentration, and the self-consistent calculation of heat generation inside the laser cavity can be obtained using (5.32) and the rate equation model. Alternatively, V_J can be represented by an empirical expression with the injection current and temperature as the variables [12,16,29,30]. If the series resistance and voltage drop across the DBRs are also included in the calculation, the most appropriate approximation of V_J can be written as [16]

$$V_J = IR_s + V_T(T) \cdot \log\left(1 + \frac{I}{I_s}\right), \tag{5.33}$$

where V_T is the diode's thermal voltage and I_S is the diode's saturation current. In general, V_J can be expressed as a polynomial function of current and temperature

$$V_J = (e_0 + e_1 T + e_2 T^2 + \cdots) \cdot (f_0 + f_1 I + f_2 I^2 + \cdots), \tag{5.34}$$

where a_i and f_i are the coefficients to be determined experimentally. This simplified approach not only allows the voltage's current and temperature dependence to be accurately modeled but also permits the characteristics of the optical and electrical devices to be largely decoupled from one another, thereby simplifying the extraction of model parameter values from experimental data.

The threshold current density J_{th} in (5.27) can be calculated from (5.6). In order to determine the threshold current, the corresponding leakage current should be carefully deduced. It is suggested that the leakage current be calculated by (5.9) using the no-k selection approach. Alternatively, instead of using a self-consistent calculation of (5.9), ΔE_{Fnp} can be expressed in terms of a linear combination of carrier concentration and temperature [12]. The corresponding curve fit to ΔE_{Fnp} is given by [16]

$$\Delta E_{Fnp} - E_{g,b} = -a_0 + a_1 N + a_2 NT - \frac{a_3}{N}, \tag{5.35}$$

where a_0 to a_3 are constants, which can be found in Table 5.2. Hence, the thermal leakage current as a function of carrier concentration and temperature under the assumption of homojunction diode equation is given by [16]

$$I_{leak} = I_0 \exp \frac{-a_0 + a_1 N + a_2 NT - a_3/N}{k_B T}. \tag{5.36}$$

This simple rate-equation-based thermal model has been applied to analyze the thermal rollover of the L–I characteristics of VCSELs with different QW material gain and configuration structures. The following are the four types of VCSELs found in the literature and the models using the rate equations above and the corresponding parameters deduced from the measurement are given in Table 5.2. The four types of VCSELs are

- Index-guided InGaAs VCSELs with vertical contact of 100 μm^2 area operating at 0.98 μm and composed of GaAs-AlAs DBRs, three $In_{0.2}Ga_{0.8}As$ QWs, and $Al_{0.2}Ga_{0.8}As$ confinement layers. Transverse carrier confinement is provided through an etched mesa design [12]. The corresponding current–voltage (I–V) curve is approximated by the modified expression of (5.33):

$$V = \frac{I R_S}{T - 198\ \text{K}} + V_T \cdot \log\left[1 + \frac{I}{I_S(T - 198\ \text{K})}\right], \tag{5.37a}$$

  where $V_T = 1.468$ V, $I_s = 3.907 \times 10^{-6}$A and $R_s = 12.928$ kΩ.

- An AlGaInP-based 0.68-μm selectively oxidized VCSEL with a 9-μm^2-area device [31]. This device consists of compressively strained InGaP QWs, an AlGaInP barrier and cladding layers, and AlGaAs-graded DBRs. The corresponding I–V curve is approximated by (5.34) using the following expression:

$$\begin{aligned} V &= (1.285 - 0.00442T + 1.13 \times 10^{-5}T - 1.03 \times 10^{-8}T^3) \\ &\quad \times (2.718 + 308.7I - 3.141 \times 10^4 I^2 + 2.765 \times 10^6 I^3). \end{aligned} \tag{5.37b}$$

- A 0.86-μm bottom emitting AlGaAs VCSEL of a 16-μm-diameter device was grown on an $Al_{0.1}Ga_{0.9}As$ substrate and consists of an Si-doped $Al_{0.15}Ga_{0.85}As$-AlAs, GaAs–$Al_{0.2}Ga_{0.8}As$ n-type DBR, six QWs, and a C-doped $Al_{0.15}Ga_{0.85}As$-$Al_{0.5}Ga_{0.5}As$–AlAs p-type DBR [14]. The corresponding I–V curve is approximated by (5.34) using the following expression:

$$\begin{aligned} V &= 1.721 + 275I - 2.439 \times 10^4 I^2 + 1.338 \times 10^6 I^3 \\ &\quad - 4.154 \times 10^7 I^4 + 6.683 \times 10^8 I^5 - 4.296 \times 10^9 I^6. \end{aligned} \tag{5.37c}$$

- Thin oxide apertured VCSEL of 3.1 μm diameter composed of an $Al_{0.9}Ga_{0.1}As$–GaAs p-type DBR, three $In_{0.17}Ga_{0.83}As$–GaAs QWs, an $Al_{0.3}Ga_{0.7}As$ cavity, and an AlAs–GaAs n-type DBR [32]. The corresponding I–V curve is approximated by (5.33) with $V_T = 0.9366$ V, $R_s = 149.8$ Ω, and $I_s = 7.918 \times 10^{-5}$ A.

There is one important issue observed from modeling of the L–I characteristics of VCSELs using the rate equation model. Once the empirical fitting of the I–V curve (i.e., including the leakage current) is determined, the steady state and modulation response calculated from the model can be matched well with the experimental data. This implies that this simple rate equation model is appropriate to further development of the circuit-level VCSEL models, which allows simple and accurate simulations for design optimization and verification of a large amount. In addition, the rate equation model allows us to implicitly model a VCSELs spatial mechanism without resorting to computationally intensive finite element or finite difference calculations. This approach should be a useful tool for the simulation of VCSELs in the context of optoelectronic system design and simulation.

5.4 QUASI-THREE-DIMENSIONAL THERMAL MODELS OF VCSELs

Detailed heat distribution inside the laser cavity of VCSELs can be calculated by solving the heat equation with the finite difference method or finite element method. However, the numerical calculations require the knowledge of the distribution of heatsources, which have to be evaluated with the current flow and electric potential inside the laser cavity. In this section, the use of the finite difference method and finite element method is described to solve the heat equation with the appropriate definition of heatsources and boundary conditions.

5.4.1 Nonuniform Distribution of Heatsources

For a typical VCSEL with of n- and p-doped DBRs, it is expected that heat is generated at (1) the heterojunction between the layers of the DBRs, (2) the overall ohmic loss inside the laser cavity, and (3) the p-n junction loss of the QW active layer. The reabsorption of light inside the laser cavity is assumed to be negligible. In the following derivation of the distribution of heatsources, the profile of injection current density is assumed to be $J(r, z)$ (A/cm^2), where r and z are the radial and longitudinal directions of the laser cavity, respectively the detailed derivation of $J(r, z)$ will be discussed in Chapter 6 of this book.

The heatsource density $\rho_{T,\text{het}}$ (W/cm^3) at the heterojunction between the layers of the DBRs is given by

$$\rho_{T,\text{het}}(r, z) = V_{J,\text{het}}(r, z)J^2(r, z)\Delta z, \tag{5.38}$$

where Δz is the thickness of the cylinder segment. However, the expression of $\rho_{T,\text{het}}$ can be approximated by an effective resistivity of the heterojunction, R_{het} ($\Omega \cdot \text{cm}^2$) and is given by [33]

$$\rho_{T,\text{het}}(r, z) = \frac{R_{\text{het}}(r, z)J^2(r, z)}{\Delta z}. \tag{5.39}$$

In general, R_{het} is a function of biased voltage but is assumed to be constant in this model. The effective resistance of the heterostructure in the n-doped DBR is approximately 10 times lower than that of the p-doped DBR because of the difference in effective masses and band offsets. The spatial variation of ohmic power loss $\rho_{T,\mathrm{ohm}}$ (W/cm^3) is given by [33]

$$\rho_{T,\mathrm{ohm}}(r, z) = \frac{J^2(r, z)}{\sigma(r, z)}, \tag{5.40}$$

where σ (Ω^{-1} cm^{-1}) is the electrical conductivity of the semiconductor materials. The power dissipated inside the QW active layer $\rho_{T,\mathrm{act}}$ (W/cm^3) is described by

$$\rho_{T,\mathrm{act}}(r, z_a) = V_J J^2(r, z_a) L_z - P_d(r, z_a), \tag{5.41}$$

where z_a is the location of the active layer, V_J is the voltage drop across the QW active layer, and P_d (W/cm^3) is the power density inside the active layer and is defined as

$$P_d(r, z_a) = \frac{\frac{1}{2} v_g h \nu_R S(z_a) |\Psi(r, z_a)|^2}{L_w}, \tag{5.42}$$

where ν_R is the operation frequency, S (cm^{-3}) is the photon density inside the active layer, L_w is the thickness of the QW, and Ψ is the transverse optical field profile of the lasing guided mode. In this equation, it is assumed that the electrical power not contributes to the generation of light will be dissipated in form of heat.

5.4.2 Solving the Heat Equation: Finite Difference Method

The heat equation, which governs the steady-state heat flow inside the laser cavity of VCSELs, is given by

$$\nabla \cdot (\kappa_T(r, z) \nabla T(r, z)) = -\rho_T(r, z), \tag{5.43}$$

where T is the spatial variation of the temperature distribution, κ_T (W cm^{-1} K^{-1}) is the material thermal conductivity, and ρ_T (W/cm^3) is the distribution of internal heatsource density, which has been defined in Section 5.4.1. The implementation of the correct boundary conditions is important to obtain a unique solution of the heat equation. The simplest approximation boundary conditions are given below [34]:

- $T(r, 0) = T_{HS}$ at the junction of the heatsink, where T_{HS} is the temperature of heatsink
- $\partial T/\partial r = \partial T/\partial z = 0$, at the surface of the device

The first boundary condition is the Dirichlet type, based on the assumption of an infinite capacity for the heatsink. It has been verified numerically that a value of 100 μm for the thickness of the copper layer is good enough to approximate a very large infinite heatsink. The second boundary condition is the Neumann type, which specifies that the rate of heat flow across the boundary is zero (adiabatic type) and no thermal radiation will escape from the surface of the lasers.

For an azimuthally symmetric anisotropic medium, the heat equation given above can be simplified to cylindrical variables of two dimensions and is written as [18,35]

$$\frac{1}{r}\frac{\partial}{\partial r}\left(\kappa_{\mathrm{T,r}} r \frac{\partial T(r,z)}{\partial r}\right) + \frac{\partial}{\partial z}\left(\kappa_{\mathrm{T,z}} \frac{\partial T(r,z)}{\partial z}\right) = -\rho_{\mathrm{T}}(r,z), \tag{5.44}$$

where $\kappa_{\mathrm{T,r}}$ and $\kappa_{\mathrm{T},z}$ denote the thermal conductivity of bulk material in the radial and longitudinal directions, respectively, and ρ_{T} is assumed to be rotationally symmetric. For the laser cavity of VCSELs similar to those in Figure 5.7, the thermal conductivity is isotropic (i.e., $\kappa_{\mathrm{T},z} = \kappa_{\mathrm{T,r}}$) and uniformly distributed inside each section of the layers. Hence, the heat distribution can be solved by using (5.44) with $\kappa_{\mathrm{T},z} = \kappa_{\mathrm{T,r}}$ in the calculation. However, if the p and n mirrors of VCSELs are constructed by 20–40 pairs of thin layers (i.e., of thicknesses h_{H} and h_{L}, respectively) with different bulk thermal conductivities (i.e., $\kappa_{\mathrm{T,H}}$ and $\kappa_{\mathrm{T,L}}$), the calculation of heat distribution inside the entire mirror can also be simplified by considering the anisotropic behavior (i.e., $\kappa_{\mathrm{T},z} \neq \kappa_{\mathrm{T,r}}$) of the mirrors. $\kappa_{\mathrm{T,r}}$ and $\kappa_{\mathrm{T},z}$ can be expressed in terms of serial and parallel thermal resistance, respectively, and the two components of the DBR thermal conductivities can be calculated as below [36]:

$$\kappa_{\mathrm{T},z} = \frac{h_{\mathrm{H}}\kappa_{\mathrm{T,H}} + h_{\mathrm{L}}\kappa_{\mathrm{T,L}}}{h_{\mathrm{H}} + h_{\mathrm{L}}} \quad \text{and} \quad \kappa_{\mathrm{T,r}} = \frac{h_{\mathrm{H}} + h_{\mathrm{L}}}{h_{\mathrm{H}}/\kappa_{\mathrm{T,H}} + h_{\mathrm{L}}/\kappa_{\mathrm{T,L}}}. \tag{5.45}$$

The heat conduction within multilayered mirrors is stronger in the radial direction than in the vertical direction.

A simple way to solve the heat equation is to use the finite difference method with the corresponding boundary conditions. The derivatives given in the heat equation can be expressed in the second-order differencing approach [37]

$$\frac{\partial T}{\partial x} \equiv \frac{T(x+\Delta x) - T(x-\Delta x)}{\Delta x}, \tag{5.46a}$$

$$\kappa_{\mathrm{T},x}\frac{\partial^2 T}{\partial x^2} \equiv \kappa_{\mathrm{T},x}\frac{T(x+\Delta x) - 2T(x) + T(x-\Delta x)}{\Delta x^2}, \tag{5.46b}$$

where $x = r$ or z, Δx is the grid separation, and $\kappa_{\mathrm{T},x}$ is assumed to be uniform over the boundaries. If $\kappa_{\mathrm{T},x}$ changes at the boundary x, we obtain

$$\frac{\partial}{\partial}\left(\kappa_{\mathrm{T},x}\frac{\partial T}{\partial x}\right) \equiv \frac{1}{\Delta x^2}\Big(\kappa_{\mathrm{T},x}(x+\tfrac{1}{2}\Delta x)\cdot[T(x+\Delta x) - T(x)] - \kappa_{\mathrm{T},x}(x-\tfrac{1}{2}\Delta x)\cdot[T(x) - T(x-\Delta x)]\Big). \tag{5.46c}$$

Hence, the heat equation can be solved using the relaxation method for boundary value problems and the convergence efficiency of the calculation is monitored by successive overrelaxation as described in *Numerical Recipes in Fortran* [37].

In the preceding calculation, the thermal conductivity is assumed to be heat-independent, but in general it is dependent on temperature. For example, the temperature dependence of thermal conductivity for GaAs/AlGaAs can be written as [38]

$$\kappa_T(x, T) = \kappa_T(x, 300\text{ K})\left(\frac{300}{T}\right)^{5/4}, \tag{5.47}$$

where $\kappa_T(x)$ is valid for any composition of ternary $Al_xGa_{1-x}As$ and its expression can be found in Table 5.1. In order to solve $T(r, z)$ and $\kappa_T(r, z, T)$ self-consistently, an iterative approach should be used. First, the temperature distribution inside the laser cavity is solved by the finite difference method with κ_T assumed to be independent of temperature. Then, the new value of κ_T is updated after evaluation of the temperature profile. These calculation procedures for temperature profile and κ_T are repeated iteratively until T and κ_T converge to a stable solution.

Figure 5.9 shows typical heat distribution inside an ion-implanted GaAs QW VCSEL biased above threshold [35]. It is assumed that the ion implantation is penetrated through the Bragg reflector into the QW active layer to form the

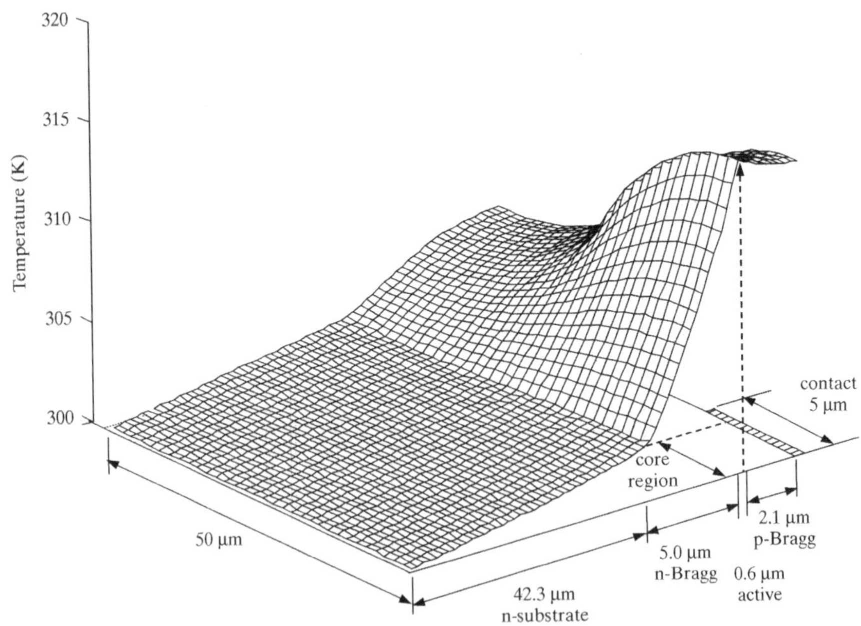

Figure 5.9 Typical above-threshold quasi-3D distribution of temperature inside the cavity of an ion-implanted QW VCSEL.

current confinement region inside the cladding region. The diameter of the metal contact on the p-DBR reflector has a diameter of 10 μm, but the core region (i.e., without ion implantation) has a diameter of only 7 μm. In the diagram, only half of the profile is shown, due to the symmetric distribution of temperature along the transverse direction. It is observed that the magnitude is maximized inside the core region of the QW active layer. This is because of the high-joule heating and nonradiative recombination, which occurs inside the QW active layer. The discontinuity of temperature distribution between the n-DBR reflector and the substrate is due to the large difference of thermal conductivity between these layers. In order to give a detailed description of the heat distribution inside the laser cavity as shown in the diagram, the number of grid points should be at least 70 × 100 in dimension. It is also noted that the corresponding memory and computational time required for the calculation are approximately 10 Mbytes and an hour of central processing unit (CPU) time, respectively, on an IBM-compatible PC Pentium II (200-MHz) computer. The computer program is written by FORTRAN 77, and the corresponding codes generated by the compiler are optimized for the processor of the computer.

5.4.3 Solving the Heat Equation: Finite Element Method

With the finite element method, the structure concerned is first suitably divided into a patchwork of smaller subregions, called *elements*. In a two-dimensional problem, the elements can be triangles or rectangles. Using many elements of different shapes and sizes, any complicated structure can be accurately represented. Each element can have different material properties, such as thermal conductivity including the tensor representation. Any heatsource density profile can also be considered, since each element can have any given value representing the heat generated inside the VCSEL. Hence, a wide range of practical lasing structures, with any given cross section, heat profile, or temperature-dependent material properties, can be modeled without restrictions. The equation governing axisymmetric heat conduction is given in (5.44) and can be written in the following format [39]:

$$\frac{\partial}{\partial r}\left(\kappa_{\mathrm{T,r}} r \frac{\partial T(r,z)}{\partial r}\right) + \frac{\partial}{\partial z}\left(\kappa_{\mathrm{T,z}} r \frac{\partial T(r,z)}{\partial z}\right) + r\rho_{\mathrm{T}}(r,z) = 0, \tag{5.48a}$$

with the general expression of the boundary conditions

$$T = T(r,z) \qquad \text{on} \qquad S_1, \tag{5.48b}$$

$$\kappa_{\mathrm{T,r}} r \frac{\partial T(r,z)}{\partial r}\hat{n}_r + \kappa_{\mathrm{T,z}} r \frac{\partial T(r,z)}{\partial z}\hat{n}_z + r q_{\mathrm{T}}(r,z) = 0 \quad \text{on} \quad S_2, \tag{5.48c}$$

$$\kappa_{\mathrm{T,r}} r \frac{\partial T(r,z)}{\partial r}\hat{n}_r + \kappa_{\mathrm{T,z}} r \frac{\partial T(r,z)}{\partial z}\hat{n}_z + r h'(T - T_0(r,z)) = 0 \quad \text{on} \quad S_3, \tag{5.48d}$$

where q_T (W/cm^2) is a specified surface heat flow due to conduction and $h'(T - T_0)$ is a surface heat flux due to convection; $\hat{n}_r$ and $\hat{n}_z$ are unit vectors normal to the r and z directions, respectively; and S_1, S_2, and S_3 are the boundaries enclosing the two-dimensional heat conduction problem.

In this example, the temperature distribution is defined by a set of complete polynomials over each triangle, which is continuous over the problem domain. The spatial variation of the temperature $T_e(r, z)$ within each first-order triangle [40] can be expressed in terms of three nodal temperature values given by

$$T_e(r, z) = \sum_{i=1}^{3} N_i T_i, \tag{5.49}$$

where N_i represents the ith interpolation matrix (i.e., its elements are functions of r and z) and T_i is the corresponding nodal temperature at the ith triangle vertex. N_i can be written as

$$N_i = \frac{1}{2\Delta}(a_i + b_i r + c_i z), \tag{5.50}$$

where Δ is the area of the triangle element and a_i, b_i, and c_i are some constants to be determined. The Galerkin formulation [39] can be used to derive the finite element equation for the axisymmetric case, and finally the matrix equation can be written as

$$[\mathbf{M}][\mathbf{T}] = [\rho_T], \tag{5.51}$$

where $[\mathbf{T}]$ is an unknown column matrix representing all the nodal temperature values. The ith row and kth column of the square matrix $[\mathbf{M}]$ may be evaluated by summing contributions from all the contributing elements representing the laser structure and can be obtained on the basis of (5.44) by

$$[\mathbf{M}]_{i,k} = \sum_{\text{element}} \kappa_T \int_\Delta \frac{\partial N_i}{\partial r}\frac{\partial N_k}{\partial r} + \frac{\partial N_i}{\partial z}\frac{\partial N_k}{\partial z} r\, dr\, dz, \tag{5.52}$$

where the integration is carried out over each triangular element, and the summation over all such triangles used to represent the full structure. Here, κ_T is the thermal conductivity of each element, which can be a tensor quantity, if necessary, for a QW region. The element $\rho_{T,i}$ of the heatsource column matrix $[\rho_T]$ can be obtained from

$$[\rho_T]_i = \sum \int_\Delta \rho_T(r, z) N_i r\, dr\, dz. \tag{5.53}$$

In this derivation, it is assumed that no heat escapes from the surface of the device and the junction between heatsink and substrate is maintained at heatsink temperature so that q_T and h' in (5.48) are set to zero. Once $[\mathbf{M}]$ and $[\rho_T]$

are obtained, the temperature distribution can be calculated. In order to achieve sufficient accuracy, more than 3000 triangular elements have to be used to represent the laser r, z domain, and a fast sparse matrix algorithm is used to achieve high computational efficiency [40,41]. Again, if the dependence of heat on the value of κ_T is taken into calculation, the heat equation has to be solved in an iterative manner similar to the finite difference method.

In comparison with the finite difference method, the finite element method is a powerful engineering tool, valued for its flexibility and versatility in modeling VCSELs with complicated thermal structure. However, the finite element method is more difficult to realize than the finite difference method. It is believed that if uniform grid points are used, the corresponding computational time and memory occupied by finite element and finite difference methods should be roughly the same. Hence, for analysis of VCSELs with uniform thermal structure, the finite difference method should be selected; otherwise, the finite element should be used.

5.5 SUMMARY

It is shown that the redshift of the resonant wavelength of VCSELs with the increase in laser cavity temperature can be predicted from the temperature dependence of the spacer region. However, this is true only under the condition that the multilayered mirrors consist of material composition similar to the spacer layer. For example, if a VCSEL has GaInAsP/InP mirrors and an InP spacer layer, the temperature dependence of the resonant wavelength can be approximated by the properties of InP material. This relationship can simplify the design of resonant wavelength VCSELs under pulsed operation.

For the design of VCSELs with optimum threshold current and output power at a range of operation temperature, a negative gain offset wavelength (i.e., $\lambda_p - \lambda_R < 0$) at room temperature is required. Hence, λ_R can be aligned with λ_p at a desired range of operation temperature. It is also shown that if the threshold current can maintain uniformity over a range of operation temperature (i.e., with suitable selection of offset wavelength), the output power can be maximized. To further improve the temperature insensitivity of VCSELs, it is also possible to design optical gain with flattened spectrum so that λ_R can maintain a uniform optical gain over a range of operation temperature. Alternatively, the thermal resistance of VCSELs can be reduced by introducing a huge block of heatsink onto the p side of VCSELs. Hence, the generation of temperature inside the laser cavity can be minimized.

The simplest way to analyze the heat generation inside the laser cavity is to define the effective thermal resistance, which is defined as the ratio between the average temperature change and the total heatsource. Hence, the heat generation can be easily deduced in VCSELs with complicated configuration of the DBRs/dielectric layers. However, the effective thermal resistance is only limited to the evaluation of the L–I curves of VCSELs. In order to calculate the steady state as well as the modulation response of VCSELs, the rate equation model should be applied with the concept of effective thermal resistance. Two parameters need to be carefully derived: (1) the leakage current, which dominates the threshold current under the influence of temperature as well as the rollover of the $L-I$ curvers, and (2) the $I-V$ curve, which determines the amount of heat generated inside the laser cavity.

To obtain a detailed heat profile inside the laser cavity of VCSELs, the finite difference method or finite element method should be used for calculation. The main difference between the two numerical methods is the allocation of grid points. It must be noted that the finite element method is more sophisticated for the definition of grid points and is therefore suitable for analyzing the heat distribution inside VCSELs with complicated structure. However, the elements used in the finite element method are difficult to generate. Hence, for analysis of heat distribution inside the laser cavity, the finite difference method is usually used. Under normal circumstances, the computational time and the required computing memory of the two methods are approximately the same as long as they have the same amount of grid points as well as the same profiles of the grid points.

5A APPENDIX

The transition $I(\nu)$ from the first electron subband E_{n1} to the first hole subband E_{p1} is shown in Figure 5.10. It is assumed that the modified valance band edge is offset from the 3D band edge by E_{p1} so that the corresponding modified energy gap can be approximated by $E_{\mathrm{g,w}} + E_{\mathrm{n1}} + E_{\mathrm{p1}}$, where $E_{\mathrm{g,w}}$ is the bandgap of the QW active layer. The vertical axes, E and E', as indicated in Figure 5.10, have the relationship $E' = h\nu - E_{\mathrm{g,w}} - E_{\mathrm{p1}} - E_{\mathrm{n1}} \equiv h\overline{\nu} - E$, where ν is the lasing frequency. If the k selection is ignored, the transitions between subbands can be described by a transition function $I(\nu)$, which can be written as [11]

$$I(\nu) = \frac{1}{e^{-(h\overline{\nu}-E_{\mathrm{F}p})/k_{\mathrm{B}}T} - e^{-E_{\mathrm{F}n}/k_{\mathrm{B}}T}} \times \log\left(\frac{(1+e^{E_{\mathrm{F}p}/k_{\mathrm{B}}T})(1+e^{-E_{\mathrm{F}n}/k_{\mathrm{B}}T})}{(1+e^{(h\overline{\nu}-E_{\mathrm{F}n})/k_{\mathrm{B}}T})(1+e^{-(h\overline{\nu}-E_{\mathrm{F}p})/k_{\mathrm{B}}T})}\right), \tag{5A.1}$$

where $E_{\mathrm{F}n}$ and $\mathrm{E}_{\mathrm{F}p}$ are the quasi-Fermi levels for electrons and holes measured positively into the band from the position of the appropriate first subband energy. From (5A.1), it can be shown that (1) $I(\nu) = 0$ if $h\overline{\nu} = 0$ or E_{FT}, where $E_{\mathrm{FT}} = E_{\mathrm{F}n} + E_{\mathrm{F}p}$, implying no absorption or emission of photon; and (2) $I(\nu) < 0$ if $h\overline{\nu} > E_{FT}$, $I(\nu) < 0$, indicating that stimulated absorption is exceeding emission. Hence, the lasing regime can be verified to be $0 < h\overline{\nu} < E_{\mathrm{FT}}$.

Using (5A.1), the threshold condition of a Fabry–Perot-type laser can be described by the following Equation [11]

$$\log\left(\frac{(1+\exp(E_{\mathrm{F}p}/k_{\mathrm{B}}T))(1+\exp(-E_{\mathrm{F}n}/k_{\mathrm{B}}T))}{(1+\exp(h\overline{\nu} - E_{\mathrm{F}n}/k_{\mathrm{B}}T))(1+\exp(-(h\overline{\nu} - E_{\mathrm{F}p})/k_{\mathrm{B}}T))}\right) = K_{\mathrm{QW}}, \tag{5A.2}$$

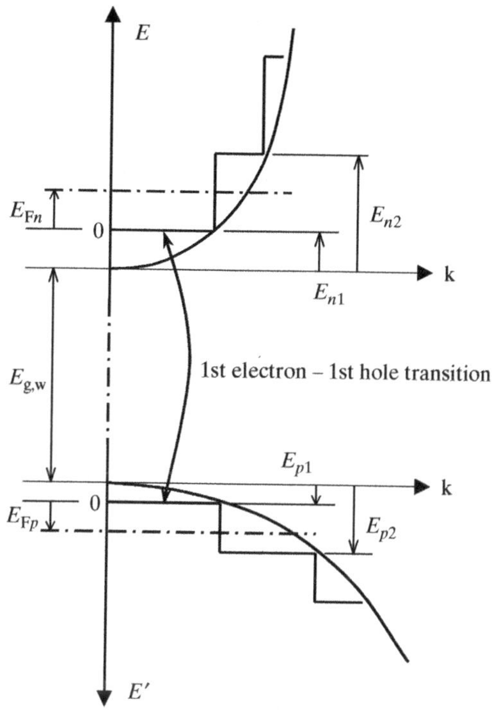

Figure 5.10 Energy momentum diagram of a quantum-well laser: E_{n1} and E_{p1} (E_{n2} and E_{p2}) are the first (second) subband energies for electrons and heavy holes, respectively.

where K_{QW} is a unique design parameter for a QW laser and can be written as

$$K_{\mathrm{QW}} = \frac{L_{\mathrm{w}}^2}{2\Gamma_z \tau_{\mathrm{p}} h D_{np} k_{\mathrm{B}} T \overline{\mathrm{B}}_i}, \tag{5A.3}$$

where D_{np} is the joint density of states for a QW, $\overline{\mathrm{B}}_i$ is the modified Einstein coefficient for stimulated emission, and the other constants have their usual meanings. It is assumed that $\Gamma_z \sim d/L_{\mathrm{eff}}$, where d is the thickness of the active layer including, the QWs, and L_{eff} is the effective cavity length of VCSELs. τ_{p} is the photon lifetime of VCSELs and is expressed as

$$\tau_{\mathrm{p}} = \nu_{\mathrm{g}} \left(\frac{1}{L_{\mathrm{eff}}} \log \left(\frac{1}{R} \right) + \alpha \right), \tag{5A.4}$$

where α is the total cavity loss and R is the average reflectivity of front and rear mirrors. Equation (5A.2) gives the relationship between the lasing frequency and Fermi levels for the first electron and first hole subband that also satisfy the particular threshold condition of the QW laser as given in (5A.4).

The dependence of $h\bar{\nu}$ on the Fermi energies, E_{Fn} and E_{Fp}, in (5A.2) can be found by solving

$$\exp\frac{h\bar{\nu}}{k_B T} = X_r \pm \sqrt{X_r^2 - \exp\left(\frac{E_{FT}}{k_B T}\right)}, \tag{5A.5}$$

where

$$X_r = \frac{1}{2}\exp(-K_{QW})\left((1-\exp(K_{QW}))\left(\exp\left(\frac{E_{Fn}}{k_B T}\right)\right.\right.$$
$$\left.\left.+\exp\left(\frac{E_{Fp}}{k_B T}\right)\right) + 1 + \exp\left(\frac{E_{FT}}{k_B T}\right)\right). \tag{5A.6}$$

The Fermi energies E_{Fn} and E_{Fp} as well as their sum, E_{FT}, are related together through the charge neutrality by [11]

$$\left(1+\exp\frac{E_{Fn}}{k_B T}\right)\left(1+\exp\frac{E_{Fn}-\Delta E_n}{k_B T}\right)$$
$$= \left(1+\exp\frac{E_{Fp}}{k_B T}\right)^Z\left(1+\exp\frac{E_{Fp}-\Delta E_p}{k_B T}\right)^Z, \tag{5A.7}$$

where $\Delta E_n = E_{n2} - E_{n1}$ and ($\Delta E_p = E_{p2} - E_{p1}$) is the separation between the electron (hole) first and second subbands, and Z ($= m_h/m_e$) is the ratio of the effective masses. Hence, the possible frequency $h\bar{\nu}$ with a Fermi energy E_{Fn} (or E_{Fp}) at a particular threshold condition (i.e., K_{QW}) of VCSELs can be determined. However, E_{Fn} derived from (5A.2) and (5A.6) may not be substituted into (5.8) to calculate N_{th} because the possible lasing frequency of the cavity mode inside the laser cavity of VCSELs has not been specified.

Because of the short cavity length in VCSELs, it is assumed that there is only one longitudinal mode, which satisfies the round-trip phase relation, supported inside the cavity. The corresponding roundtrip phase relation is given by

$$\phi_f + \phi_r + \frac{4\pi h_{spacer} n_{spacer}}{\lambda_R} = 2\pi, \tag{5A.8}$$

where λ_R is the wavelength of the lasing mode and ϕ_f and ϕ_r are the phase shifts associated with the two DBRs. It must be noted that n_{spacer}, ϕ_f, and ϕ_r are all functions of temperature. Hence, the mode frequency

$$\frac{1}{\lambda_R} = \frac{1-(\phi_f+\phi_r)/2\pi}{2h_{spacer} n_{spacer}}. \tag{5A.9}$$

As all the transitions are assumed to be from the first electron subband to the first heavy hole subband, the frequency $\nu_R = c/\lambda_R$ of a mode is contained in the parameter $h\overline{\nu}$, which is rewritten as

$$h\overline{\nu} = hc\frac{(1-(\phi_f+\phi_r)/2\pi)}{2h_{\text{spacer}}n_{\text{spacer}}} - E_{g,w}(T) - E_{n1} - E_{p1}. \tag{5A.10}$$

Now, $h\overline{\nu}$ of VCSELs can be specified according to (5A.10), and so it becomes necessary to solve (5A.2) for E_{Fn} (or E_{Fp}) in terms of $h\overline{\nu}$ for a particular K_{QW}. The preceding description of equations does not enforce strict k selection, so this means that $h\overline{\nu}$ for the response maximum varies with temperature and τ_p. It is noted that this differs from the situation of strict k selection, where $h\overline{\nu}$ is always zero for lasing from the first subband.

REFERENCES

1. C. J. Chang-Hasnain, J. P. Harbison, G. Hasnain, A. C. Von Lehmen, L. T. Florez, and N. G. Stoffel, "Dynamic, polarization and transverse mode characteristics of VCSELs," *IEEE J. Quantum Electron.*, Vol. 27, no. 6, pp. 1402–1408, 1991.
2. M. Osinski and W. Nakwaski, "Thermal effects in vertical cavity surface emitting lasers," in *Current Trends in Vertical Cavity Surface Emitting Lasers*, T. P. Lee, ed., Selected Topics in Electronics and Systems, Vol. 3, World Scientific, Singapore, 1995.
3. G. P. Agrawal and N. K. Dutta, *Long Wavelength Semiconductor Lasers*, Van Nostrand Reinhold, New York, 1986.
4. T. Wipiejewski, D. B. Young, B. J. Thibeault, and L. A. Coldren, "Thermal crosstalk in 4×4 vertical cavity surface emitting laser arrays," *IEEE Photon. Technol. Lett.*, Vol. 8, no. 8, pp. 980–982, 1996.
5. J. J. Dudley, D. L. Crawford, and J. E. Bowers, "Temperature dependence of the properties of DBR mirrors used in surface emitting normal optoelectronic devices," *IEEE Photon. Technol. Lett.*, Vol. 4, no. 4, pp. 311–313, 1992.
6. K. Streubel, S. Rapp, J. Andre, and J. Wallin, "Room temperature pulsed operation of 1.5 μm vertical cavity surface emitting lasers with an InP-based bragg reflector," *IEEE Photon. Technol. Lett.*, Vol. 8, no. 9, pp. 1121–1123, 1996.
7. W. Schmid, D. Wiedenmann, M. Grabberr, R. Jager, R. Michalzik, and K. J. Ebeling, "CW operation of a diode cascade InGaAs quantum well VCSEL," *Electron. Lett.*, Vol. 34, no. 6, pp. 553–554, 1998.
8. Y. Qian, Z. H. Zhu, Y. H. Lo, D. L. Huffaker, D. G. Deppe, H. Q. Hou, B. F. Hammons, W. Lin, and Y. K. Tu, "Submilliamp 1.3 μm vertical cavity surface emitting lasers with threshold current density of <5004A/cm^2," *Electron. Lett.*, Vol. 33, no. 12, pp. 1052–1053, 1997.
9. S. Rapp, J. Piprek, K. Streubel, J. Andre, and J. Wallin, "Temperature sensitivity of 1.54 μm vertical cavity surface emitting lasers with an InP-based Bragg reflector," *IEEE J. Quantum Electron.*, Vol. 33, no. 10, pp. 1839–1845, 1997.
10. N. Bewtra, D. A. Suda, G. L. Tan, F. Chatenoud, and J. M. Xu, "Modeling of quantum well lasers with electro-opto-thermal interaction," *IEEE J. Select. Topics Quantum Electron.*, Vol. 1, no. 2, pp. 331–340, 1995.

11. G. W. Taylor and P. A. Evaldsson, "Temperature dependent operation of the vertical cavity surface emitting lasers," *IEEE J. Quantum Electron.*, Vol. 30, no. 10, pp. 2262–2270, 1994.

12. J. W. Scott, R. S. Geels, S. W. Corine, and L. A. Coldren, "Modeling temperature effects and spatial hole burning to optimize vertical cavity surface emitting laser performance," *IEEE J. Quantum Electron.*, Vol. 29, no. 5, pp. 1295–1308, 1993.

13. M. H. Crawford, R. P. Schneider, Jr., K. D. Choquette, and K. L. Lear, "Temperature dependent characteristics and single mode performance of AlGaInP-based 670-690 nm vertical cavity surface emitting lasers," *IEEE Photon. Technol. Lett.*, Vol. 7, no. 7, pp, 724–726, 1995.

14. Y. Ohiso, K. Tateno, Y. Kohama, A. Wakatsuki, H. Tsunetsugu, and T. Kurokawa, "Flip-chip bonded 0.85 μm bottom-emitting vertical cavity surface emitting laser array on an AlGaAs substrate," *IEEE Photon. Technol. Lett.*, Vol. 8, no. 9, pp. 1115–1117, 1996.

15. Z. Zou, D. L. Huffaker, and D. G. Deppe, "Ultralow threshold cryogenic vertical cavity surface emitting laser," *IEEE Photon. Technol. Lett.*, Vol. 12, no. 1, pp. 1–3, 1999.

16. P. V. Mena, J. J. Morikuni, S. M. Kang, A. V. Harton, and K. W. Wyatt, "A comprehensive circuit level model of vertical cavity surface emitting lasers," *J. Lightwave Technol.*, Vol. 17, no. 12, pp. 2612–2631, 1999.

17. M. Brunner, K. Gulden, R. Hovel, M. Moser, and M. Iiegems, "Thermal lensing effects in small oxide confined vertical cavity surface emitting lasers," *Appl. Phys. Lett.*, Vol. 76, no.1, pp. 7–9, 2000.

18. G. R. Hadley, K. L. Lear, M. E. Warren, K. D. Choquette, J. W. Scott, and S. W. Corzine, "Comprehensive numerical modeling of vertical cavity surface emitting lasers," *IEEE J. Quantum Electron.*, Vol. 32, no. 4, pp. 607–616, 1996.

19. M. Kaitia, K. Kurihara, M. Nido, A. Kimura, T. Yoshikawa, K. Kurihara, Y. Sugimoto, and K. Kasahara, "Temperature characteristics of a vertical-cavity surface-emitting laser array with a broad-gain bandwidth," *IEEE Select. Topics Quantum Electron.*, Vol. 1, no. 2, pp. 654–660, 1995.

20. M. Kajita, K. Kurihara, H. Saito, T. Yoshikawa, Y. Sugimoto, and K. Kasahara, "Temperature insensitive vertical cavity surface emitting laser array with a broad gain bandwidth," *Electron. Lett.*, Vol. 31, no. 22, pp. 1925–1926, 1995.

21. D. B. Young, J. W. Scott, F. H. Peters, M. L. Majewski, B. J. Thibeault, S. W. Corinze, and L. A. Colden "Enhanced performance of offset-gain high-barrier vertical-cavity surface-emitting lasers," *IEEE J. Quantum Electron.*, Vol 29, no. 6, pp. 2013–2021, 1993.

22. M. H. MacDougal, P. D. Dapkus, V. Pudikov, H. Zhao, and G. M. Yang, "Ultralow threshold current vertical cavity surface emitting lasers with AlAs oxide-GaAs distributed bragg reflectors," *IEEE Photon. Technol. Lett.*, Vol. 7, no. 3, pp. 229–231, 1995.

23. M. H. MacDougal, J. Geske, C. K. Lin, A. E. Bond, and P. D. Dapkus, "Low resistance intracavity-contacted oxide-aperture VCSEL's," *IEEE Photon. Technol. Lett.*, Vol. 10, no. 1, pp. 9–11, 1998.

24. M. Grabherr and M. Miller, *Bottom Emitting VCSELs for High cw Optical Output Power*, Annual report 1997, Dept. of Optoelectronics, Univ. Ulm, Germany, pp. 52–56.

25. T. Wipiejewski, M. G. Peters, B. J. Thibeault, D. B. Young, and L. A. Coldren, "Size dependent output power saturation of vertical cavity surface emitting laser diodes," *IEEE Photon. Technol. Lett.*, Vol. 8, no. 1, pp. 10–12, 1996.

26. F. P. Incropera and D. P. DeWitt, *Fundamentals of Heat Transfer*, Wiley, New York, 1981.

27. H. S. Carslaw and J. C. Jaeger, *Conduction of Heat in Solids*, 2nd ed., Clarendon Press, Oxford, 1959.

28. W. Nakwaski and M. Osinski, "Thermal resistance of top-surface-emitting vertical cavity semiconductor lasers and monolithic two-dimensional arrays," *Electron. Lett.*, Vol. 28, no. 6, 572–574, 1992.

29. P. V. Mena, J. J. Morikuni, S. K. Kang, A. V. Harton, and K. W. Wyatt, "A simple rate equation based thermal VCSEL model," *J. Lightwave Technol.*, Vol. 17, no. 5, pp. 865–872, 1999.

30. S. F. Yu, W. N. Wong, P. Shum, and E. L. Li, "Theoretical analysis of modulation response and second-order harmonic distortion in vertical cavity surface emitting lasers," *IEEE J. Quantum Electron.*, Vol. 32, no. 12, pp. 2139–2147, 1996.

31. M. H. Crawford, K. D. Choquette, H. Q. Hou, R. J. Hickman, K. M. Geib, and B. E. Hammons, "Visible VCSELs: Recent advances and applications," in *Proc. 1997 Digest LEOS Summer Topical Meetings Vertical Cavity Lasers*, pp. 17–18, 1997.

32. B. J. Thibeault, K. Bertilsson, E. R. Hegblom, E. Strzelecka, P. D. Floyd, R. Naone, and L. A. Coldren, "High speed characteristics of low optical loss oxide apertured vertical cavity surface emitting lasers," *IEEE Photon. Technol. Lett.*, Vol. 9, no. 1, pp. 11–13, 1997.

33. R. Michalzik and K. J. Kbeling, "Modeling and design of proton-implanted ultralow-threshold vertical cavity laser diodes," *IEEE J. Quantum Electron.*, Vol. 29, no. 6, pp. 1963–1974.

34. W. B. Joyce and R. W. Dixon, "Thermal resistance of heterostructure lasers," *J. Appl. Phys.*, Vol. 46, no. 2, pp. 855–862, 1975.

35. W. M. Man and S. F. Yu, "Comprehensive modeling of diffused quantum wells vertical cavity surface emitting lasers," *IEEE J. Select. Topics Quantum Electron.*, Vol. 4, no. 4, pp. 715–722, 1998.

36. J. Piprek and S. J. B. Yoo, "Thermal comparison of long wavelength vertical cavity surface emitting laser diodes," *Electron. Lett.*, Vol. 30, no. 11, pp. 866–868, 1994.

37. W. H. Press, S. A. Teukolsky, W. T. Vetterling, and B. P. Flannery, *Numerical Recipes in Fortran*, 2nd ed., Cambridge Univ. Press, 1992.

38. J. S. Blakemore, "Semiconducting and other major properties of gallium arsenide," *J. Appl. Phys.*, Vol. 53, pp. R123–R181, 1982.

39. K. H. Huebner, *The Finite Element Method for Engineers*, Wiley, New York, 1975.

40. B. M. A. Rahman, S. P. Lepkowski, and K. T.V. Grattan, "Thermal modeling of vertical cavity surface emitting lasers using the finite-element method," *IEE Proc. Optoelectron.*, Vol. 142, no. 2, pp. 82–86, 1995.

41. R. P. Sarazala, W. Nakwaski, and M. Osinski, 'Comprehensive thermal-electrical self-consistent model of proton implanted top surface emitting lasers," *Int. J. Optoelectron.*, Vol. 10, no. 5, pp. 357–371, 1995.

42. S. Adachi, "Lattice thermal resistance of III-V compound alloys," *J. Appl. Phys.*, Vol. 54, pp. 1844–1848, April 1983.

CHAPTER 6

Electrical Characteristics of Vertical Cavity Surface Emitting Lasers

Methods of modeling spatial hole burning using rate equations of carrier concentration are discussed. The influence of transverse modes and injection current on the formation of spatial hole burning is investigated. Detailed calculation of current transport mechanism at the heterointerfaces of the multilayered mirrors is also explained. The approach to model the quasi-3D distribution of electrical potential and current density using Poisson's equation and resistance network, respectively, is described. A comprehensive self-consistent calculation of VCSELs including the quasi-3D distribution of electric potential, current flow, and temperature, as well as the transverse distribution of transverse modes and carrier concentration inside the active layer, is also presented.

6.1 INTRODUCTION

Single mode operation, low threshold current and high wallplug efficiency can be achieved in vertical cavity surface emitting lasers (VCSELs) if the distribution of carrier concentration inside the active region is controlled carefully. It is noted that the performance of VCSELs with gain-guided [1,2] and index-guided waveguides [3–5] deteriorates significantly in the presence of spatial hole burning (SHB). SHB is excited by stimulated recombination of carrier concentration inside the active layer, and the shape of SHB follows the profile of transverse fields. For VCSELs operating at high power, SHB reduces the modal gain of the fundamental mode so that higher-order transverse mode can be excited. This is due to the reduction of overlap between the transverse distribution of optical gain and the fundamental mode. As a result, the wallplug efficiency of VCSELs is reduced and the corresponding threshold current is increased. The self-focusing effect is also observed in VCSELs due to SHB (i.e., carrier-induced index change) inside the active layer. The self-focusing effect narrows the beamwidth of the transverse modes so that the stability operation of the fundamental mode is further reduced.

Analysis and Design of Vertical Cavity Surface Emitting Lasers By Siu Fung Yu
ISBN 0-471-39124-7

Current spreading is the diffusion of carrier concentration toward the cladding region of the active layer. Current spreading reduces the wallplug efficiency and increases the threshold current so that VCSELs with a strong confinement of injection current inside the core region of the active layer are desired [6]. It is expected that VCSELs with buried heterostructure have better wallplug efficiency than that with ion-implanted [7] or oxide-confined structures [8,9]. Current spreading is critically dependent on the detailed confinement structure of injection current such as the injection current path and the resistance of heterointerfaces. In VCSELs with buried heterostructure [3,4], the injection current is strongly confined inside the core region and the diffusion of carriers into the cladding region of the active layer is minimized. In VCSELs with double-oxide-confinement structure of injection current, the diffusion of carriers into the cladding region of the active layer can be minimized, provided the separation between the two oxide layers is optimized [8,9].

Current crowding, which has arisen from the accumulation of carriers at a particular location of the active layer, has significant influence on the uniformity of the injection current profile. The performance of VCSELs can deteriorate by current crowding through the excitation of SHB as well as thermal lensing effect (i.e., due to the excessive heat sources) [1,7,8]. It must be noted that thermal lensing increases the strength of self-focusing. Current crowding usually occurs in VCSELs with a strong current confinement structure of carrier concentration such as those with buried heterostructure or oxide-confined structure. Most of the time, the crowded current supplies higher carrier density at the periphery of the core region of active layer.

As mentioned above, SHB, current spreading, and current crowding significantly influence the performance of VCSELs. In addition, the excessive injection of carriers generates heat inside the laser cavity. In order to obtain an accurate analysis of the performance of VCSELs, the electrical, thermal, and optical mechanisms have to be calculated simultaneously through their linkage of optical gain and refractive index inside the active layer. In this chapter, the models using a rate equation of carrier concentration, with the consideration of injection current, transverse modes, and carrier transport effect, are developed to analyze SHB. In addition, the influence of SHB on threshold current and transverse-mode profiles of VCSELs is investigated. The approach to calculate the resistivity of heterointerfaces is explained, and the methods employed to minimize the resistance of heterointerfaces using graded interface and modulation doped are discussed. The dependence of current spreading and current crowding on the geometry of VCSELs is also investigated. This can be done by solving the quasi-3D distribution of electrical potential using Poisson's equation and the appropriated boundary conditions. In addition, a simple resistance network is developed to solve the quasi-3D current flow. It is noted that the SHB, current flow, and temperature distribution are mutually linked together through the optical gain and the refractive index distribution inside the active layer. A self-consistent calculation is suggested to simultaneously solve the electrical, thermal, and optical behaviors of VCSELs, including calculation of SHB, current spreading, and crowding, as well as self-focusing.

6.2 CHARACTERISTICS OF SPATIAL HOLE BURNING (SHB) IN VCSELs

In this section, the rate equation model of carrier concentration is used to analyze the SHB inside the active layer of VCSELs. The dependence of SHB on the injection current profile and carrier transport effects is also studied theoretically. It can be shown that SHB can be minimized provided that the injection current is confined inside the core region of the active layer. Furthermore, the influence of SHB on the operation stability of the transverse mode is investigated. Self-focusing of the fundamental mode is observed to be due to the SHB as well as the thermal lensing effect. As a result, the optical gain of the fundamental mode reduces and higher-order transverse modes excite above threshold.

6.2.1 Modeling of SHB Using Rate Equations of Carrier Concentration

The nonuniform distribution of carrier concentration N inside the active layer of VCSELs with cylindrical coordinates can be calculated self-consistently with the transverse modes accounted for by solving the following rate equations

$$\frac{\partial N}{\partial t} = \frac{J}{qd} - R_{\mathrm{t}}(N) - R_{\mathrm{st}}(N) + D_n \nabla^2 N, \tag{6.1a}$$

$$\frac{\partial S_{lm}}{\partial t} = \left[v_{\mathrm{g}} \Gamma_{z,lm} \langle g_{l,m} \rangle - \frac{1}{\tau_{p,lm}} \right] S_{lm} - \beta_{\mathrm{s}} \Gamma_z B_{sp} \langle N \rangle^2, \tag{6.1b}$$

where $\Gamma_{z,lm}$ is the longitudinal confinement factor, $\tau_{p,lm}$ is the photon lifetime for the $\mathrm{LP}_{l,m}$ mode, and R_{t} is the carrier recombination term including nonradiative recombination and radiative recombination due to spontaneous emission. R_{t} can be written as

$$R_{\mathrm{t}}(N) = \frac{N}{\tau_n} + B_{\mathrm{sp}} N^2 + C_{\mathrm{Aug}} N^3. \tag{6.2}$$

The stimulated recombination term R_{st} has a simplified expression:

$$R_{\mathrm{st}}(r, \theta) = v_{\mathrm{g}} \sum_{l,m} g(N) |\Psi_{lm}(r, \theta)|^2 S_{lm}. \tag{6.3}$$

The terms $\langle g_{l,m} \rangle$ and $\langle N \rangle$ are the normalized optical gain and carrier concentration given by

$$\langle g_{l,m} \rangle = \frac{\int_0^{2\pi} \int_0^{\infty} \Psi_{lm}(r, \theta) g(N) \Psi_{lm}^*(r, \theta) r \, dr \, d\theta}{\int_0^{2\pi} \int_0^{\infty} \Psi_{lm}(r, \theta) \Psi_{lm}^*(r, \theta) r \, dr \, d\theta}, \tag{6.4a}$$

$$\langle N \rangle = \frac{\displaystyle\int_0^{2\pi}\int_0^{\infty} \Psi_{lm}(r,\theta)N(r,\theta)\Psi_{lm}^*(r,\theta)r\,dr\,d\theta}{\displaystyle\int_0^{2\pi}\int_0^{\infty} \Psi_{lm}(r,\theta)\Psi_{lm}^*(r,\theta)r\,dr\,d\theta}, \tag{6.4b}$$

where $g(N)$ can be expressed as (2.6) or (2.18), $\Psi_{lm}(r,\theta)$ is the profile of the normalized transverse field, and S_{lm} is the photon density of the LP_{lm} mode. Ψ_{lm} can be calculated using the methods mentioned in Chapter 3 and is dependent on the waveguide geometry of the laser cavity. The last term on the RHS of (6.1a) describes the diffusion of carrier concentration along the transverse direction of the active layer. If the distribution of N along the radial and azimuthal directions is considered, the diffusion term of (6.1a) can be expressed as

$$\nabla^2 N = \frac{1}{r}\frac{\partial}{\partial r}\left(r\frac{\partial N}{\partial r}\right) + \frac{1}{r^2}\frac{\partial^2 N}{\partial \theta^2}. \tag{6.5}$$

Hence, the profile of carrier concentration is determined mainly by the profile of injection current density, stimulated recombination inside the active layer, and the diffusion of the carrier concentration inside the active layer. However, the influence of nonradiative recombination on the profile of carrier concentration is less significant.

If Ψ is approximated by LP modes, the corresponding radial direction of Ψ can be written in terms of Bessel functions and the azimuthal direction can be expressed as sinusoidal functions (see Section 3.2.2). For a first-order approximation, a series of Bessel and sinusoidal functions can be used to describe the radial and azimuthal variation of carrier concentration, respectively. The azimuthal variation of carrier concentration can be linearized using the perturbation method to obtain the nonuniform distribution of carrier concentration analytically, which can be written as [10]

$$\overline{N}(r,\theta,t) = \overline{N}(r,t) + \overline{n}(r,\theta,t), \tag{6.6}$$

where $\overline{N} \gg |\overline{n}|$ is assumed. Substituting (6.6) into (6.1a) and collecting the terms, which are dependent on θ, we obtain the following equations [10]

$$\frac{\partial \overline{N}}{\partial t} = D_n\frac{1}{r}\frac{\partial}{\partial r}\left(r\frac{\partial \overline{N}}{\partial r}\right) - R_{\mathrm{t}}(\overline{N}) + \frac{J}{qd}, \tag{6.7a}$$

$$\frac{\partial \overline{n}}{\partial t} = \frac{D_n}{r}\frac{\partial}{\partial r}\left(r\frac{\partial \overline{n}}{\partial r}\right) + \frac{D_n}{r^2}\frac{\partial^2 \overline{n}}{\partial \theta^2} - R_{\mathrm{t}}' - R_{\mathrm{st}}, \tag{6.7b}$$

where

$$R_{\mathrm{t}}' \cong \overline{n}(\tau_n^{-1} + 2\overline{N}B_{\mathrm{sp}} + C_{\mathrm{Aug}}(2\overline{n}\,\overline{N} + \overline{N}^2)). \tag{6.8}$$

From (6.7), $\overline{N}(r)$ can be interpreted as the carrier distribution created by the specified injection current density distribution, and $\overline{n}(r,\theta)$ is the perturbation of the carrier density due to stimulated recombination. Now, the nonuniform distribution of radial and azimuthal directions of $\overline{n}(r,\theta)$ can be decomposed because $\overline{N} \gg |\overline{n}|$.

If $\Psi_{lm}(r,\theta)$ is written as $\Psi_{lm}(r,\theta) = \Psi_l(\mathrm{r})\cos(l\theta)$ and $\overline{N} \gg |\overline{n}|$, then R_{st} can be expressed as

$$R_{\mathrm{st}}(\mathrm{r},\theta) \cong v_{\mathrm{g}} \sum_{l,m} g(\overline{N})|\Psi_l(r)|^2 S_{lm}\cos^2(l\theta). \tag{6.9}$$

Using expression (6.9), $\overline{n}$ can be decomposed into equations that are independent of radial and azimuthal direction, as shown below

$$\begin{aligned}\frac{\partial \tilde{n}_{\mathrm{c},0}}{\partial t} &= D_n \frac{1}{r}\frac{\partial}{\partial r}\left(r\frac{\partial \tilde{n}_{\mathrm{c},0}}{\partial r}\right) - R_{\mathrm{t}}''(\overline{N})\tilde{n}_{\mathrm{c},0} \\ &\quad - \frac{1}{2}v_{\mathrm{g}}\sum_{l,m} a_{l,0}\, g(\overline{N})|\Psi_l(r)|^2 S_{lm},\end{aligned} \tag{6.10}$$

$$\begin{aligned}\frac{\partial \tilde{n}_{\mathrm{c},p}}{\partial t} &= D_n \frac{1}{r}\frac{\partial}{\partial r}\left(r\frac{\partial \tilde{n}_{\mathrm{c},p}}{\partial r}\right) - D_n\frac{p^2}{r^2}\tilde{n}_{\mathrm{c},p} \\ &\quad - R_{\mathrm{t}}''(\overline{N})\tilde{n}_{\mathrm{c},p} - \frac{1}{2}v_{\mathrm{g}}\sum_{l,m} b_{l,p}\, g(\overline{N})|\Psi_l(r)|^2 S_{lm},\end{aligned} \tag{6.11}$$

$$\begin{aligned}\frac{\partial \tilde{n}_{\mathrm{s},p}}{\partial t} &= D_n \frac{1}{r}\frac{\partial}{\partial r}\left(r\frac{\partial \tilde{n}_{\mathrm{s},p}}{\partial r}\right) - D_n\frac{p^2}{r^2}\tilde{n}_{\mathrm{s},p} \\ &\quad - R_{\mathrm{t}}''(\overline{N})\tilde{n}_{\mathrm{s},p} - \frac{1}{2}v_{\mathrm{g}}\sum_{l,m} c_{l,p}\, g(\overline{N})|\Psi_l(r)|^2 S_{lm},\end{aligned} \tag{6.12}$$

where p is an integer and $\tilde{n}_{\mathrm{c},p}$ and $\tilde{n}_{\mathrm{s},p}$ are the coefficients of the series expansion of $\overline{n}$. The detailed formulation of (6.10)–(6.12) is given in Section 6A.1. The coefficients $a_{l,p}$, $b_{l,p}$, and $c_{l,p}$ have the following expressions:

$$\begin{aligned} a_{l,p} &= a_{l,0} = 1 + \frac{\sin(4l\pi)}{4l\pi}, \\ b_{l,p} &= \frac{\sin(2\pi(2l-p))}{4\pi(2l-p)} + \frac{\sin(2\pi p)}{2\pi p} + \frac{\sin(2\pi(2l+p))}{4\pi(2l+p)} \\ c_{l,p} &= \frac{1}{4\pi(2l-p)}(\cos(2\pi(2l-p)) - 1) + \frac{1}{2\pi p}(1-\cos(2\pi p)) \\ &\quad + \frac{1}{4\pi(2l+p)}(1-\cos(2\pi(2l+p))). \end{aligned} \tag{6.13}$$

If two dominant modes, LP_{01} and LP_{11}, are considered only in the calculation of mode competition in VCSELs, substitute $l = 0$ and 1 into $a_{l,p}$, $b_{l,p}$, and $c_{l,p}$. It can be shown that $a_{0,p} = a_{1,p} = 1$, $b_{0,p} = b_{1,p} = 0$ for $p \neq 2$, but $b_{0,p} = 0$ and $b_{1,p} = 1$ for $p = 2$, and $c_{0,p} = c_{1,p} = 0$ for all p. In this case, if the initial conditions are set for the integration in the form

$$\tilde{n}_{c,p}(r, t = 0) = 0 \quad \text{and} \quad \tilde{n}_{s,p}(r, t = 0) = 0, \tag{6.14}$$

then $\tilde{n}_{c,p}$ and $\tilde{n}_{s,p}$, which satisfy a homogeneous equation, will remain zero all the time because of $b_{0,p} = c_{0,p} = c_{1,p} = 0$. It is seen that the only nonzero terms are $\tilde{n}_{c,0}$ and $\tilde{n}_{s,2}$. Hence, the nonuniform distribution of $\bar{n}$ along the azimuthal direction can be simplified and described by two rate equations with variables $\tilde{n}_{c,0}$ and $\tilde{n}_{s,2}$.

In the preceding equations, the analysis of azimuthal variation of the carrier concentration has been simplified using the perturbation method. However, the radial variation of carrier concentration has to be solved numerically such as using the finite difference method with the appropriate boundary conditions for VCSELs with different types of structures. Therefore, it is necessary to obtain a simplified expression for the radial variation of the carrier concentration. In order to do so, the lasing modes are assumed to be azimuthally independent, and the expansion of carrier concentration dependent only on the radial coordinate can be approximated by a Bessel series, which is given by [11]

$$N(r, t) = N_0(t) - \sum_{k=1}^{\infty} N_k(t) J_0\left(\frac{a_k r}{W_{\text{eff}}}\right), \tag{6.15}$$

where N_0 and N_k are the expansion coefficients, W_{eff} is the effective active layer radius, and a_k is the kth root of the first-order Bessel function J_1 [i.e., $J_1(a_k) = 0$]. In (6.15), $\partial N/\partial r = 0$ satisfies at $r = 0$ because of the symmetric nature of the laser waveguide. The definition of W_{eff} is to impose the most appropriated boundary conditions on the carrier concentration: (1) if W_{eff} is set to W (i.e., radius of active region), (6.15) simply forces the carrier profile to remain flat at the edge of the active layer—this is equivalent to the carrier distribution inside VCSELs with an etched post structure; and (2) if $W < W_{\text{eff}}$, then W corresponds to an effective active layer boundary defined by the current injection profile, where the majority of the current is near the center of the devices. This is similar to those with a strong confinement of injection current. However, for devices with a weak confinement of injection current, (6.15) may not be an appropriated approximation.

Substitute (6.15) into (6.1a), with which the azimuthal variation of the carrier concentration is ignored in the calculation. Integrate the equation over the radius W of the active region, thereby producing the rate equation for N_0, with the approximation that $\int_0^W J_0(a_k r/W) r\, dr \approx 0$ for all kth mode, gives [11]

$$\frac{\partial N_0}{\partial t} = \frac{J}{qd} - R_t(N_0) - \frac{2}{W^2} \int_0^W R_{st}(N) r\, dr. \tag{6.16}$$

Again, multiplying $J_0(a_k r/W)$ to (6.1a) and integrating the equation from 0 to W yields the rate equations for each term $N_k (k > 0)$ in the Bessel series expansion, which is given by [11]

$$\frac{\partial N_k}{\partial t} = -N_k\left(R'_{t,k}(N_0) + D_n\left(\frac{a_k}{W}\right)^2\right) + \frac{2}{W^2 J_0^2(a_k)}\int_0^W R_{st}(N) J_0\left(\frac{a_k r}{W}\right) r\,dr, \tag{6.17}$$

and

$$\begin{aligned} R'_{t,k}(N_0) &\cong (\tau_n^{-1} + 2N_0 B_{sp} + C_{Aug}(2N_0 N_k + N_0^2)) \\ &\cong (\tau_n^{-1} + 2N_0 B_{sp} + C_{Aug} N_0^2), \end{aligned} \tag{6.18}$$

where $N_0 \gg N_k$ is assumed in the derivation of (6.18). Furthermore, the following relations have been used in the derivation of (6.16) and (6.17):

$$\int_0^W J_0\left(\frac{a_m r}{W}\right) J_0\left(\frac{a_n r}{W}\right)\cdot r\,dr = \begin{cases} \dfrac{W^2}{2} J_0^2(a_m) & m = n \\ 0 & m \neq n \end{cases}, \tag{6.19}$$

$$\nabla^2 N = \sum_{k=1}^{\infty}\left(\frac{a_k}{W}\right)^2 N_k(t) J_0\left(\frac{a_k r}{W}\right), \tag{6.20}$$

where it is assumed that $W_{eff} \sim W$.

Equations (6.16) and (6.17) can be further simplified if the single transverse mode LP_{01} is assumed to be excited only inside the laser cavity. In this case, all carrier concentration components except N_0 and N_1, can be ignored in (6.16) and (6.17). Hence, the carrier rate equation can be written as [12]

$$\frac{\partial N_0}{\partial t} = \frac{J}{qd} - R_t(N_0) - v_g(\Gamma_1 g(N_0) - \gamma_N \xi_{1,1} N_1)\cdot S_{01}, \tag{6.21}$$

$$\frac{\partial N_1}{\partial t} = -\frac{N_1}{\tau_{eff}} + v_g(\Gamma_{1,1} g(N_0) - \gamma_N \xi_{1,1,1} N_1)\cdot S_{01}, \tag{6.22}$$

$$\tau_{eff} = \left(R'_{t,1}(N_0) + D_n\left(\frac{a_1}{W}\right)^2\right)^{-1}, \tag{6.23}$$

where τ_{eff} is the effective carrier lifetime that includes the diffusion effect arising from the SHB. The parameters $\gamma_N, \Gamma_1, \Gamma_{1,1}, \xi_{1,1}$, and $\xi_{1,1,1}$ given in (6.21) and (6.22) can be found in Section 6A.2. For the single-mode approximation given above, N_0 can be interpreted as the average carrier concentration along the radial direction and N_1 is the perturbation of the carrier concentration arising from Ψ_{01}. These equations can be used to explain the influence of SHB on the electrical properties of VCSELs.

In the preceding modeling of the carrier concentration, the distribution of the carrier concentration profile has not been treated properly. This is because the boundary conditions have not been seriously considered in the derivation of the simplified rate equations. For etched post VCSELs, especially those with small cavity size, at the sidewall of the active layer, the carrier concentration is of a large magnitude and the surface recombination (i.e., nonradiative recombination at the sidewall) has significant influence on the threshold current and quantum efficiency of the lasers. However for VCSELs (i.e., oxide confined or ion-implanted) in which the carrier concentration is confined inside the core region of the active layer and only a small amount of carrier concentration reaches the edge of the active layer, the influence of the surface recombination effect can be ignored. In order to apply appropriate boundary conditions to VCSELs with different current confinement structures, the finite difference method may be applied to solve (6.1). The partial derivative of N with respect to r or θ can be written as

$$\frac{\partial N}{\partial x} \equiv \frac{N(x+\Delta x) - N(x-\Delta x)}{\Delta x}, \tag{6.24}$$

$$\frac{\partial^2 N}{\partial x^2} \equiv \frac{N(x+\Delta x) - 2N(x) + N(x-\Delta x)}{\Delta x^2}, \tag{6.25}$$

where $x = r$ or θ, and Δx is the grid separation. If N is assumed to have cylindrical symmetry along the active layer and $r = 0$ is the center of the active layer, the boundary condition at the center is given by

$$\left.\frac{\partial N}{\partial r}\right|_{r=0} = 0. \tag{6.26}$$

This boundary condition is applied to all types of VCSELs with cylindrical symmetry. For devices such as those etched with post structure, the boundary condition at the sidewall of the active region $r = W_t$ can be written as [14,15]

$$\left.\frac{\partial N}{\partial r}\right|_{r=W_t} + \left.\frac{v_{surf}}{D_n} N\right|_{r=W_t} = 0, \tag{6.27}$$

where v_{surf} is the surface recombination velocity and (6.27) is applied to $r = W_t$ for all θ. On the other hand, for devices with a strong confinement of injection current, only a small amount of carrier concentration is diffused toward to the edge of the active layer. The boundary condition (6.27) should be replaced using $N \to 0$ for $r \to \infty$. In fact, this boundary condition can be implemented by approximating the carrier concentration at $r = W_t$ using a Bessel function of zeroth order K_0

$$N = N|_{r=W_t} \cdot K_0 \cdot k_e \cdot (r - W_t) \text{ for } r > W_t, \tag{6.28}$$

where the decay constant of N, namely, k_e, is deduced from the continuity requirement of carrier concentration; in other words, N and its derivative should continue at $r = W_t$. It must be noted that (6.28) is also applied to $r = W_t$ for all θ. It should be understood that the profile of the carrier concentration could be significantly different if a different boundary condition [i.e., (6.27) or (6.28)] is used.

6.2.2 Influence of Injection Current Profile and Carrier Transport Effects on SHB

The profile of the carrier concentration is also affected by the injection current profile, which is dependent on the current injection contact as well as the current confinement structure inside the laser cavity. For VCSELs with ion-implanted [16] or etched mesa [17,18] structures with circular metal contact on the p-side of the distributed Bragg reflector (DBR) for current injection (see Fig. 1.10), the schematic shown in Figure 6.1 can be used for the investigation. The shadowed region represents a high-resistivity region created by oxygen ion or proton implantation or etched mesa, which reduces current spreading. In fact, the corresponding distribution of the injection current is similar to that in facet emitting lasers with stripe geometry [19]. This is because the injection current inside the VCSELs is concentrated inside the core region of the active layer, but the carriers are diffused toward the edge of the active layer because of the weak confinement of current inside that layer. Hence, the profile of injection current density will be dependent on the applied voltage at the circular contact, the total resistance of the multilayered mirrors, as well as the current–voltage ($I-V$) characteristic of the active layer (i.e., diode layer).

The injection current distribution along the radial direction of the active layer can be approximated by assuming that the total current I is the sum of uniform injection current I_1 under the circular-disk contact of radius W and the spreading

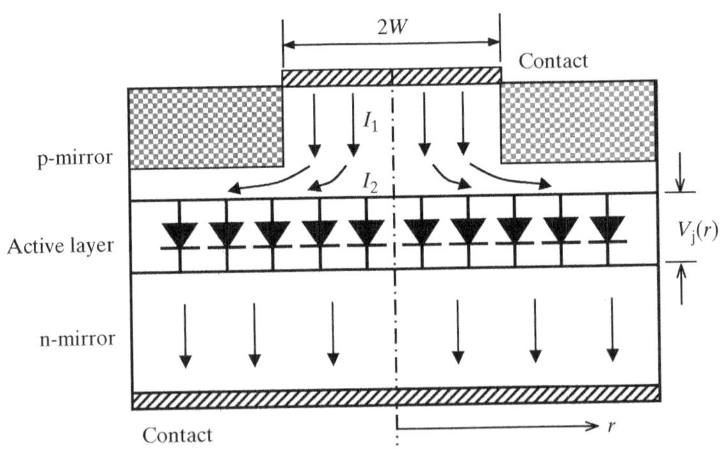

Figure 6.1 Electrical confinement structure of a VCSEL with ion implantation or etched mesa electrical confinement structure.

current I_2. The current density J across the junction is given by

$$J(r) = J_s \exp\left(\frac{qV_J}{2k_B T}\right), \tag{6.29}$$

where J_s is a constant and V_J is the voltage across the junction. As the current moves radially outward through a small annulus (of width dr), it encounters a voltage drop $-dV_J$, due to the resistive layer [16]

$$-dV_J = \frac{R_{sh}}{2\pi r} I_2\, dr. \tag{6.30}$$

The decrease of radial current, caused by the current sinking vertically by the diode, is

$$-dI_2 = 2\pi r J(r)\, dr, \tag{6.31}$$

where R_{sh} is the sheet resistance of the entire p layers, including the mirror. R_{sh} is given by

$$\frac{1}{R_{sh}} = \sum_i h_i \sigma_i, \tag{6.32}$$

where h_i is the thickness of the ith layer and σ is its conductivity. The summation of (6.32) is over all layers in the p region. Hence, from (6.29)—(6.31), a general equation that describes the nonuniform distribution of V_J can be deduced

$$\frac{1}{r}\frac{\partial}{\partial r}\left(r\frac{\partial V_J}{\partial r}\right) = J_s \exp\left(\frac{qV_J}{2k_B T}\right). \tag{6.33}$$

Now, solving (6.33) gives V_J, and using (6.29), the radial variation of the injection current along the active layer can be obtained. To solve (6.33) also requires two boundary conditions, $J(r) \to 0$ as $r \to \infty$ and $J = I_1/\pi W^2$ as $r \to W$. The solution of (6.33) describes the rapid decrease of J with the increase in r for various values of sheet resistance R_{sh}. In fact, the general analytic solution of (6.33) can be simplified by using the following format of approximation [16]:

$$J(r) = \begin{cases} J_0 & r \le W \\ J_0 \exp\left[-\dfrac{r-W}{r_0}\right] & r > W \end{cases}, \tag{6.34}$$

where $r_0^{-1} = \sqrt{R_{sh} J_s q / 2k_B T}$ and $J_0 = I_1/\pi W^2$. This equation is applied only to estimate the injection current profile of ion-implanted or etched mesa VCSELs in which the corresponding cladding region of the active layer has not been ion-implanted or etched.

However, for devices with strong injection current inside the core region of the active layer such as etched post VCSELs (see Fig. 1.11), the injection current is uniformly confined within the core region (i.e., aperture) but zero elsewhere [20].

This is because in the etched post devices, the injection current is strongly confined inside the laser cavity. For some oxide confined VCSELs with no diffusion of carriers into the cladding region of the active layer through the oxide aperture, this approximation is also valid [20].

In QW active layers, it is known that the injection carrier is required to penetrate through the undoped separate confinement heterostructure (SCH) before recombining inside the QWs. Therefore, it is expected that the distribution of the carrier concentration inside the QWs will be affected by the transportation mechanism between the SCH and QWs. Hence, the rate equation analysis should include the following two carrier transport constants in the calculation: (1) the carrier diffusion and capture time across the SCH τ_s and (2) the thermionic emission time from the QW τ_{te}. After modification of (6.1a), the rate equations for the carrier concentration in the undoped SCH region M and the QW active region N can be expressed as [21]

$$\frac{\partial M}{\partial t} = \frac{J}{qn_w L_{SCH}} - R_{t,SCH}(M) + \frac{L_w}{L_{SCH}}\frac{N}{\tau_{te}} + D_{n,SCH}\frac{1}{r}\frac{\partial}{\partial r}\left(r\frac{\partial M}{\partial r}\right), \quad (6.35a)$$

$$\frac{\partial N}{\partial t} = \frac{L_{SCH}}{L_w}\frac{M}{\tau_s} - R_{t,w}(N) - R_{st}(N) + D_{n,w}\frac{1}{r}\frac{\partial}{\partial r}\left(r\frac{\partial N}{\partial r}\right), \quad (6.35b)$$

where the azimuthal variation of the carrier concentration has been ignored in the derivation and the nonradiative recombination of the carrier concentration inside the SCH and QW regions are given by

$$R_{t,SCH}(M) = \frac{M}{\tau_{n1}} + \frac{M}{\tau_s}, \quad (6.36a)$$

$$R_{t,w}(N) = \left(\frac{N}{\tau_{n2}} + B_{sp}N^2 + C_{Aug}N^3\right) + \frac{N}{\tau_{te}}, \quad (6.36b)$$

where τ_{n1} and τ_{n2} are the carrier lifetimes inside the SCH and QW regions, respectively. Now consider VCSELs with injection current confinement such as those with ion-implanted or mesa etched structures [16,17,22] and the devices with a wide radius of active layer so that the influence of surface recombination can be ignored in the analysis. In fact, it can be shown that with suitable selection of τ_s and τ_{te}, SHB can be minimized [21]. The reduction mechanism of SHB can be explained as the SCH layer forms a reservoir of carrier concentration, which allows the transportation of M to refill the depletion region of the carrier concentration inside the active layer. Therefore, with a suitable design on the dimension of SCH and QWs layers, SHB can be minimized using carrier transport effects.

6.2.3 Influence of Spatial Hole Burning on the Stability of Transverse Modes

In the sections above, the methods used to model the dependence of SHB on the nonradiative recombination at the sidewall of the active layer and the profile and confinement structure of the injection current have been discussed. In this section, the characteristics of transverse modes in VCSELs (e.g., LP modes) are analyzed with respect to SHB. In addition, the stability of transverse modes in VCSELs with different waveguide designs is investigated.

For VCSELs with weakly-guided waveguide (i.e., gain guiding or weak index guiding), the characteristics of transverse modes are affected mainly by its modal gain $\langle g_{l,m}\rangle$, which is the overlap between the radial distribution of LP modes and the active region [i.e., see (6.3)]. Therefore, the shape of p-metal contact as well as the profile of injection current can have a significant influence on the stability of the LP_{01} mode operating above threshold [10,11,23]. It is shown that stable LP_{01} mode operation at high output power can be maintained for devices with injection current concentrated inside the core region of the active layer (i.e., using a circular-disk contact of small radius). This is because SHB can be minimized by the injection current through the small circular-disk contact so that modal gain of LP_{01} is maximized. However if a circular-ring contact is used, higher-order modes can be excited at high power if the VCSEL has no electrical confinement along the transverse direction. This is because the circular-ring contact allows the diffusion of injected carrier concentration toward the edge of the active layer (i.e., influence of SHB is increased) and the corresponding modal gain of LP_{01} mode is reduced [3]. Figure 6.2 shows the calculated $L-I$ curves for the competition of the LP_{01} and LP_{11} modes at various biases [10]. It is noted that for weakly-guided VCSELs with circular-disk contact, LP_{01} is the dominant mode but for the device with circular-ring contact, LP_{01} can be completely suppressed. Hence, it is shown that for VCSELs with weakly-guided structure, stable fundamental mode operation can be maintained at high power if the injected carrier concentration is concentrated at the center of the active layer (i.e., using a circular disc contact).

In the analysis above, the influence of the self-focusing effect has been ignored. The self-focusing effect arises from the carrier induced index change through the serious SHB. As a result, the profile of the transverse modes will be collimated to the center of the active region and the corresponding modal gain will be varied. For VCSELs operating at high power, the self-focusing effect increases the modal gain of high-order modes and suppresses the domination of the LP_{01} mode (fundamental mode). This phenomenon is most likely to occur in weakly-guided VCSELs than that with strong guiding because SHB is a dominant factor of the former case. Hence, the self-focusing effect can have a significant influence on the stability of VCSELs with weakly-guided waveguides. Furthermore, the self-focusing effect is affected by self-heating inside the laser cavity. This is because the refractive index increases with the increase of temperature. The self-focusing effect will be amplified by self-heating in VCSELs because the heat distribution profile is maximum at the center of the active layer [24].

Figure 6.3 shows a self-consistent calculation of the transverse modes, carrier concentration, and temperature as well as refractive index change of ion-implanted VCSELs (i.e., gain-guided waveguide laser; see Fig. 1.10) [24]. The VCSEL has a circular-ring contact, but the current confinement structure allows the assumption of an equivalent circular-disk contact with radius 5 μm.

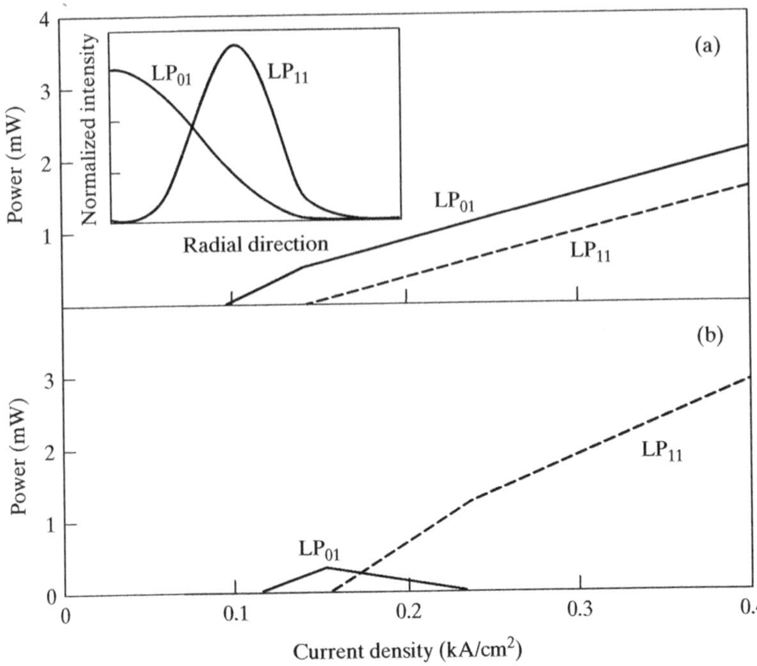

Figure 6.2 Light/current curves of VCSELs with weakly-guided waveguide: VCSELs with (a) circular disk contact, and (b) circular ring contact on the p-type mirror for the external injection of carrier concentration; the diagram inset inside the L/I curves indicates the profile of LP_{01} and LP_{11} modes. (After Ref. 10.)

Figure 6.3a shows a perfect LP_{01} mode operating at an injection current of 1.1 times the threshold. It is noted that if the injection current increases further, the LP_{11} mode starts to dominate. At an injection level of 1.5 times the threshold, the LP_{01} mode is completely suppressed and only the LP_{11} is observed. Figure 6.3b shows the variation of temperature profile with different injection currents. The temperature profile exhibits a peak at the center of the active layer. This is because the injection current is concentrated at the center region. Figure 6.3c shows the profile SHB inside the active region. It is noted that the higher the injected current density, the greater the SHB. When the injection current LP_{11} mode equals 1.5 times the threshold, SHB moves away from the center. This implies that the LP_{11} mode contributes to the stimulated recombination. Figure 6.3d shows the refractive index changes Δn_{T}, Δn_N, and Δn_{r} induced by heating, SHB, and both heating and SHB, respectively. For injection current greater than 1.3 times the threshold, the laser becomes more index-guided. Therefore, it is proved that the characteristics of transverse modes are dependent on both the temperature and SHB inside the active layer. Further, it is noted that large current spreading in the active layer will increase the corresponding threshold current density of VCSELs. This is because the diffused current due to current spreading will not

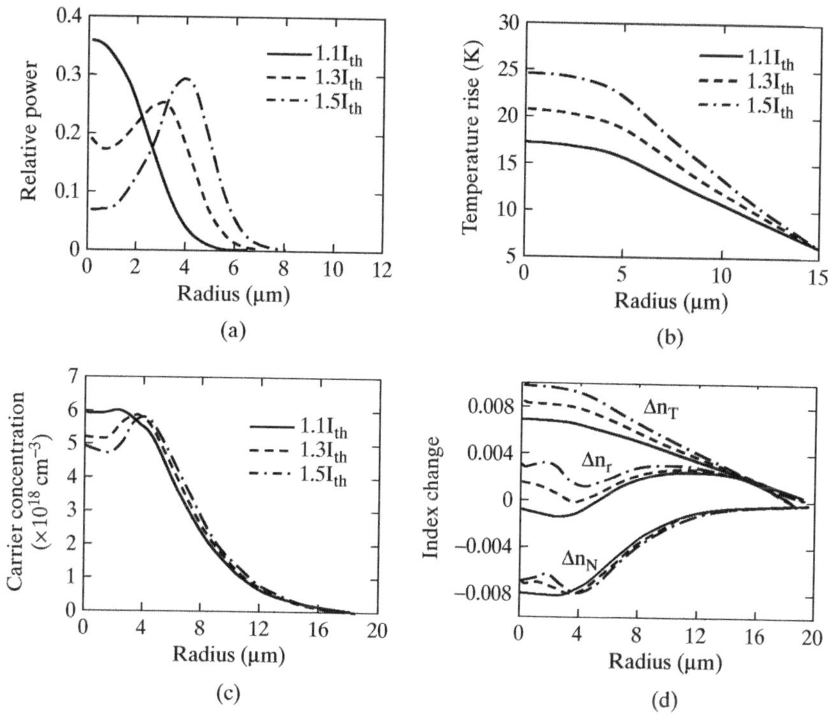

Figure 6.3 (a) Intensity distribution of the total transverse modes; (b) temperature profile; (c) carrier concentration distribution; (d) refractive index changes induced by carrier concentration and heat inside the active layer, for different injection current density, where Δn_T and Δn_N are the changes of refractive indexes induced by heating and carrier concentration, respectively and Δn_r is their sum. (After Ref. 24.)

contribute to the pumping of optical gain so that current spreading increases the threshold current density and temperature. Therefore, in the design of VCSELs with gain-guided waveguides, adequate injection current confinement should be ensured for stable single-mode operation (i.e., minimizing SHB) as well as the reduction of threshold current density (i.e., maximizing wallplug efficiency).

For VCSELs with strongly index-guided waveguides (i.e., etched post lasers [4] or some buried heterostructure lasers [5]), the design requirement of single-mode operation may be different to that of weak index-guided devices. This is because two phenomena, diffraction loss and surface recombination, are observed in strongly index-guided lasers. The influence of diffraction loss on the threshold current, quantum efficiency, and reflectivity of DBRs has been discussed in Chapters 2 and 3. The following paragraphs concentrate on the influence of surface recombination (i.e., nonradiative recombination at the sidewall) on the excitation of transverse modes and threshold.

It is assumed that a buried heterostructure VCSEL with low diffraction loss configuration is considered in the calculation [24]. The corresponding refractive

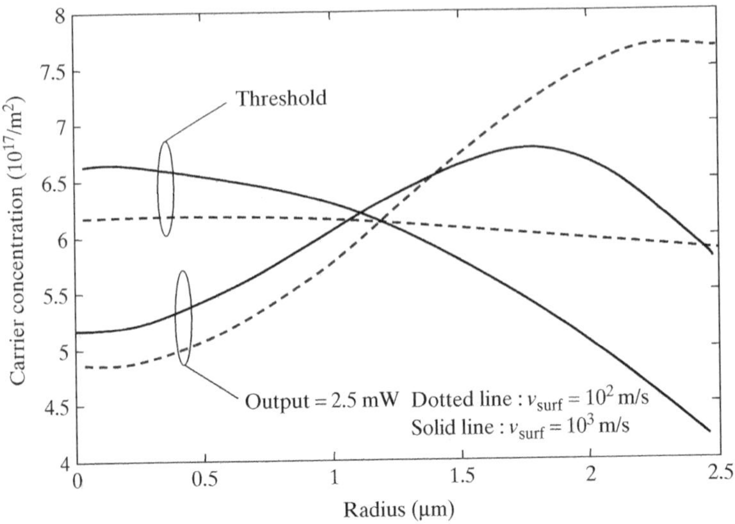

Figure 6.4 Carrier concentration profile of VCSELs with index-guided waveguide and the injection current strongly confined inside the core region of the laser cavity: solid line—$v_{surf} = 10^3$ m/s; dashed line—$v_{surf} = 10^2$ m/s (after Ref. 14).

index step between the core and cladding layers of the laser is set at 0.007, where diffraction loss is minimized. Figure 6.4 shows the SHB characteristics of the VCSELs with a core radius of 2.5 μm [16]. At threshold, the profile of carrier concentration for the case $v_{surf} = 10^3$ m/s has a gradient steeper than that for the case of $v_{surf} = 10^2$ m/s. Therefore, the LP_{01} mode is more favorable than the TE_{01} (i.e., equivalent to LP_{11}) mode at low operating currents. The preference is stronger for the case of higher surface recombination velocity 10^3 m/s, where the carrier concentration has a strong peak at the center of the optical cavity. At output power of 2.5 mW, it is shown that the SHB due to the LP_{01} mode is more pronounced at a low value of surface recombination velocity. In this case, the TE_{01} mode becomes favorable at high operating currents because the SHB due to the TE_{01} is less pronounced than that due to the LP_{01} mode. Hence, it is believed that the large value of v_{surf} may help stabilize a single fundamental mode operation in VCSELs with index-guided structure and small cavity size. In the calculation above, the TE_{01} mode is used instead of LP_{11} as the side mode of the laser. This is because for devices with relatively large refractive index step between the core and cladding regions, the LP modes approximation is not appropriated (i.e., see also Chapter 3).

It is observed experimentally that a strongly index-guided structure with a small cavity radius supports a single fundamental mode operation [4]. This can be explained by the influence of diffraction loss and surface recombination inside the laser cavity. Furthermore, it is expected that the threshold current density increases with reduction of the core diameter because the surface recombination does not follow the scaling rule. As the core diameter is reduced, the high threshold current density as well as an increase in thermal resistance can become a serious problem. Hence, in order to reduce the threshold current, the surface recombination should be reduced at the same time as the cavity diameter is being reduced.

6.3 ELECTRICAL CHARACTERISTICS OF MULTILAYERED MIRRORS

High reflectivities in semiconductor quarter-wavelength DBRs are demonstrated in GaAs/AlAs [25,26], InP/InGaAsP [27], AlAsSb/GaAlAsSb [28], and InAlAs/InAlGaAs [29] material systems. It is noted that the large difference in refractive index between the two constituents' layers is responsible for high optical reflectivity. However, the accompanying energy bandgap difference, which scales roughly linearly with the index difference, results in potential barriers in the heterointerfaces. These potential barriers impede the carrier flowing in the mirror and result in large series resistance especially in the p-type doping case due to its low carrier mobility. Hence, methods have been proposed to engineer flat valence band profiles for low-resistance p-doped multilayered mirror, such as GaAs (high index)/$Al_{0.7}Ga_{0.3}As$ (low index) DBRs with an intermediate $Al_{0.35}Ga_{0.65}As$ layer or a 200-Å superlattice of GaAs(10 Å)/$Al_{0.7}Ga_{0.3}As$ (10 Å) inserted at the heterointerfaces [30]. This gradual change in the energy gap by varying the Ga/As ratio flattens the valence band of the heterointerfaces and increases the series resistance problem of the mirror. Alternatively, modulation doping on the DBRs with interface grading can also reduce the corresponding series resistance such as using periodically doped mirrors [31–33]. This can be done by periodically applying a dopant of high level ($\sim 10^{19}$ cm^{-3}) at the heterointerfaces of the mirror. The doping profile can also be a delta-type doping peak at the heterointerface superimposed on the modulated doping profiles [34]. The total series resistance of the periodically doped mirrors can be effectively reduced as the increase in doping concentration reduces the thickness of the depletion layers of the heterointerfaces. Using heavily doped regions at the location of the heterointerfaces where the bandgap increases toward the active layer also improves the optical characteristics of the DBRs. These heterointerfaces correspond to the nodes of the optical standing waves in the DBRs, so the free-carrier absorption is not large, due to the low optical intensity at those heavily doped regions. Hence, periodically doped mirrors lead to low resistance and high reflectivity. Si and Be can be used as the corresponding dopants in the n- and p-type GaAs/AlAs mirrors, respectively [31–34]. For more detailed information on the fabrication of DBRs with different types of dopants and modulation profiles, see the text by Wilmsen et al. [35].

In the analysis of a p mirror with steps and superlattice heterointerfaces, it is found that the series resistance of the p abrupt mirror strongly depends on the temperature but that of a p mirror with graded heterointerface is independent of its temperature [32]. Figure 6.5a shows the schematic energy band diagram of

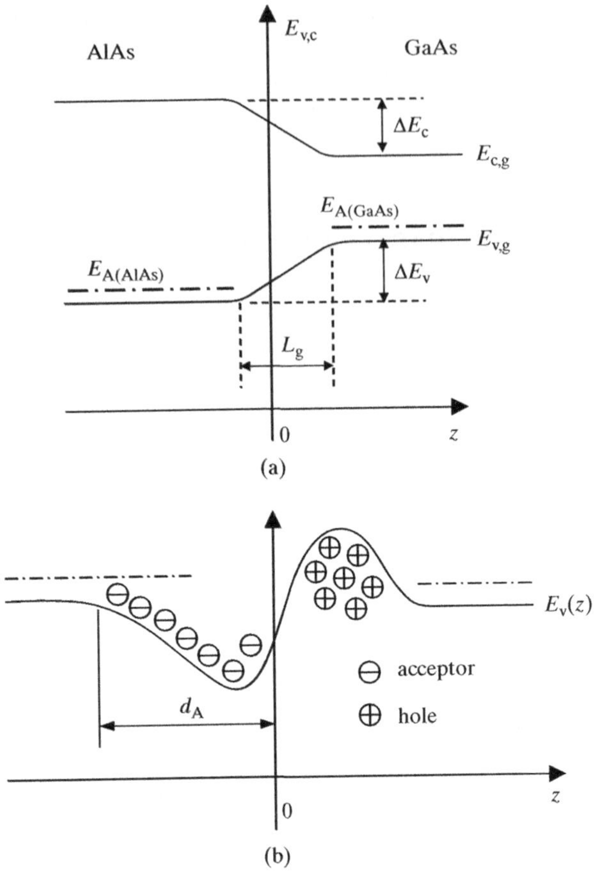

Figure 6.5 Schematic energy band diagram of the AlAs/GaAs heterointerface: (a) grading potentials $E_{v,g}$ and $E_{c,g}$; (b) total valence band potential $E_v(z) = E_{v,g} + qV_C$.

the p-doped graded AlAs/GaAs heterointerface in which the band bending at the interface is determined by the grading potentials $E_{v,g}$ and $E_{c,g}$, respectively. It is noted that the thermal current density $J_{thermal}$ flowing over a potential barrier (e.g., say p-doped GaAs/AlAs heterointerface) can be expressed as [32]

$$J_{thermal} = A_p^* T^2 \exp\left(-\frac{q\phi_{b,v}}{k_B T}\right)\left(\exp\left(\frac{qV_b}{k_B T}\right) - 1\right), \tag{6.37}$$

where ($A_p^* = qm_h^* k_B^2 / 2\pi^2\hbar^3$) is the effective Richardson coefficient for holes, $m_h^* = (1/m_{h,GaAs}^* + 1/m_{h,AlAs}^*)^{-1}$ is the effective mass of holes, $\phi_{b,v}$ is the potential barrier height related to hole density, and V_b is the applied biased voltage. Equation (6.37) can be further simplified if $\exp(qV_b/k_B T) \gg 1$ is assumed. The contribution of electrons in $J_{thermal}$ is ignored in the derivation of (6.37) as the hole is the dominant mechanism of current transportation. On the other hand, at thermal equilibrium the tunneling current from the valence band to the empty state of the conduction band and from the conduction band to the empty state of the valence band should be in detailed balance. Hence, the total tunneling current density J_{tunnel} in the same potential barrier of the p-doped GaAs/AlAs heterointerface can be expressed as [36]

$$\begin{aligned} J_{\text{tunnel}} &= J_{\text{tunnel}}^{\text{GaAs}\rightarrow\text{AlAs}} - J_{\text{tunnel}}^{\text{AlAs}\rightarrow\text{GaAs}} \\ &\cong A_{\text{p}}^{*}\frac{T}{k_{\text{B}}}\left(\int_{-\infty}^{-\phi_{\text{b,v}}} \mathcal{T}_{p}(E_z)\log\left(1+\exp\left(\frac{E_{\text{Fp(GaAs)}}-E_z}{k_{\text{B}}T}\right)\right)\cdot dE_z \right. \\ &\quad \left. - \int_{-\infty}^{-\phi_{\text{b,v}}} \mathcal{T}_{\text{p}}(E_z)\log\left(1+\exp\left(\frac{E_{\text{Fp(AlAs)}}-E_z-qV_{\text{b}}}{k_{\text{B}}T}\right)\right)\cdot dE_z \right), \end{aligned} \tag{6.38}$$

where E_z is the kinetic energy of the carriers and $E_{\text{Fp(GaAs)}}$ ($E_{\text{Fp(AlAs)}}$) is the Fermi level of GaAs (AlAs). The derivation of the general expression of the transmission probability for hole, T_{p}, is given in Section 6A.3 [37]. From (6.37) and (6.38), it is noted that J_{thermal} is more dependent on temperature than that of J_{tunnel} as J_{thermal} is directly proportional to T^2 but that of J_{tunnel} is only directly proportional to T. Therefore, it is expected that the tunneling current accounts for the majority of current flowing in the DBRs with p-graded superlattice heterointerface.

6.3.1 Calculation of Tunneling Current Flowing across p-Doped Graded Heterointerfaces

Equation (6.38) can be used to calculate the tunneling current through the p-doped graded heterointerface, provided the potential energy across the heterointerface is known. Figure 6.5b illustrates the valence band edge when p-type doping for $z < 0$, is included [36]. Acceptors near the interface are ionized, the free holes drift to the GaAs side, and on the AlAs side a negative space-charge layer (i.e., depletion layer) builds up. The separation of the carriers results in a Coulomb potential V_{C}. The total valence band energy E_{v} at the heterointerface can be approximated by the linear superposition of the grading and the Coulomb potential:

$$E_{\text{v}}(z) = E_{\text{v,g}}(z) + qV_{\text{C}}(z). \tag{6.39}$$

In order to describe the energy diagram of Figure 6.5, $E_{\text{v,g}}(z)$ can be written as [36]

$$E_{\text{v,g}}(z) = \Delta E_{\text{v}} \begin{cases} 0, & z \le \dfrac{-L_g}{2} \\ 1+\sin\left(\dfrac{\pi z}{L_g}\right) & -\dfrac{L_g}{2} < z < \dfrac{L_g}{2}, \\ 1, & z \ge \dfrac{L_g}{2} \end{cases} \tag{6.40}$$

where ΔE_v is the energy barrier and L_g is the grading length or the width of the transition region between GaAs and AlAs materials. The charge distribution $\rho(z)$ is assumed to have the following profile

$$\rho(z) = \begin{cases} -qP(z) & z \le -d_\mathrm{A} \\ -q(P(z) - N_A) & -d_\mathrm{A} < z \le 0\,, \\ -qP(z) & z > 0 \end{cases} \tag{6.41}$$

where d_A is the width of depletion region on the AlAs side of the heterointerface in which all acceptors are ionized and $P(z)$ is the distribution of hole concentration along the z direction. It is assumed that the doping concentration of N_A is homogeneous for $z < 0$ but $p(z)$ contributes to the charge distribution outside this region. Hence, $P(z)$ is a function of z and is dependent on the shape of the potential energy.

From the distribution of charge, the Coulomb potential can be obtained from Poisson's equation as follows:

$$\frac{\partial^2 V_\mathrm{C}}{\partial z^2} = -\frac{\rho(z)}{\varepsilon}. \tag{6.42}$$

In order to solve (6.42), it is assumed at zero bias voltage, the Fermi level is aligned across the heterointerface, and the valence band energy relative to the Fermi level at both ends of the junction is determined by the local charge neutrality condition in the GaAs and AlAs layers. Deep inside the AlAs (e.g., for $z \to -\infty$) and the GaAs (e.g., for $z \to +\infty$) for low temperatures and heavily doped semiconductors, the Fermi level is pinned at the acceptor energy levels $E_{\mathrm{A(AlAs)}}$ and $E_{\mathrm{A(GaAs)}}$, respectively. Hence, the boundary conditions are

$$E_\mathrm{v}(-\infty) = E_{\mathrm{A(AlAs)}} \quad \text{and} \quad E_\mathrm{v}(+\infty) = E_{\mathrm{A(GaAs)}} + qV_\mathrm{b}, \tag{6.43}$$

where V_b is the biased voltage across the junction and the quasi-Fermi level in the GaAs is shifted by qV_b as a result of the applied voltage. Integrating (6.42) and assuming a constant dielectric constant ε gives the Coulomb potential [36]

$$qV_\mathrm{C}(z) = \begin{cases} h_1(z) + E_{\mathrm{A(AlAs)}} & z \le -d_\mathrm{A} \\ h_1(z) - \dfrac{1}{2\varepsilon} q^2 N_\mathrm{A}(z + d_\mathrm{A})^2 + E_{\mathrm{A(AlAs)}}, & -d_\mathrm{A} < z \le 0\,, \\ h_2(z) + E_{\mathrm{A(AlAs)}} - \Delta E_\mathrm{v} + qV_\mathrm{b} & z > 0 \end{cases} \tag{6.44}$$

where

$$h_1(z) = \int_{-\infty}^{z} \int_{-\infty}^{z'} \frac{q^2}{\varepsilon} P(z'')\,dz''\,dz' \quad \text{and} \quad h_2(z) = \int_{z}^{\infty} \int_{z'}^{\infty} \frac{q^2}{\varepsilon} P(z'')dz''\,dz'. \tag{6.45}$$

Continuity of V at $z = 0$ leads to the neutrality condition

$$\int_{-\infty}^{\infty} P(z)\,dz = N_{\mathrm{A}} d_{\mathrm{A}}. \tag{6.46}$$

Similarly, continuity of $\partial V/\partial z$ at $z = 0$ yields the depletion width:

$$d_{\mathrm{A}} = \left(\frac{2\varepsilon}{q^2 N_{\mathrm{A}}}(h_1(0) - h_2(0) + E_{\mathrm{A(AlAs)}} - E_{\mathrm{A(GaAs)}} + \Delta E_{\mathrm{v}})\right)^{1/2}. \tag{6.47}$$

Equations (6.46) and (6.47) are the necessary conditions but do not suffice to describe the system completely because additional information is needed to evaluate the spatial distribution of hole concentration $P(z)$.

The distribution of hole concentration along the z direction, $P(z)$, can be approximated by a linear superposition of the squared absolute values of the eigenfunctions $\Psi_i(z)$

$$P(z) = \sum_{i=0}^{\infty} c_i \Psi_i^*(z)\Psi_i(z), \tag{6.48}$$

where c_i are expansion coefficients. The eigenfunctions $\Psi_i(z)$ are solutions of the one-dimensional Schrödinger equation

$$H\Psi_i(z) = E_i\Psi_i(z), \tag{6.49}$$

where the eigenenergy E_i and Hamiltonian H are given by

$$H = -\frac{\hbar^2}{2m_{\mathrm{h}}^*}\frac{\partial^2}{\partial z^2} + E_{\mathrm{v,g}}(z) + qV_{\mathrm{C}}(z). \tag{6.50}$$

It is noted that the profile of the eigenfunctions $\Psi_i(z)$ is determined by $V_{\mathrm{C}}(z)$ but is also a function of $\Psi_i(z)$ because $\Psi_i(z)$ describes $p(z)$ inside the heterointerface. Therefore, the Schrödinger and Poisson equations have to be solved self-consistently for $\Psi_i(z)$ and $V_{\mathrm{C}}(z)$. In addition, the finite difference method or finite element method can be employed to evaluate the general solution for $\Psi_i(z)$ and $V_C(z)$.

If the semiconductor material is not heavily doped and all holes can be assumed to be in the ground state, the corresponding ground-state wavefunction can be described by $\Psi_0(z)$ and the localized charge distribution density near the interface can be written as $P(z) = c_0|\Psi_0(z)|^2$. In this case, the ground-state energy and wavefunction can be approximated by the Rayleigh–Ritz variation principle (i.e., an approximated analytic solution of Ψ_0 can be estimated). This can be done by expressing a set of variational test functions Ψ_{T} to approximate Ψ_0. Ψ_{T} can be written in terms of a simple asymmetric Gaussian function, with maxima at $z = b$ and different decay constants a_1 and a_2 for both directions [36]:

$$\Psi_T(z, b, a_1, a_2) = \frac{2}{(a_1 + a_2)\sqrt{\pi}} \begin{cases} \exp\left(-\frac{1}{2}\left(\frac{z-b}{a_1}\right)^2\right), & z < b \\ \exp\left(-\frac{1}{2}\left(\frac{z-b}{a_2}\right)^2\right), & z \geq b \end{cases}. \tag{6.51}$$

The maximum expectation value of the Hamiltonian H using Ψ_T gives the best approximation for the ground-state energy for some a_1, a_2, b:

$$E_0 \sim \max_{a_1, a_2, b} \langle \Psi_T | H | \Psi_T \rangle. \tag{6.52}$$

Hence, the corresponding ground-state hole wavefunction can be obtained from the same set of a_1, a_2, and b. The value of c_0 can be simply determined from the charge neutrality condition (6.45). Now, the self-consistent calculation of the Schrödinger and Poisson equations gives the appropriate values of a_1, a_2, b, and c_0. This potential profile can also be used to calculate the hole transmission probability and the tunneling current through the barrier.

6.3.2 Total Current Flowing across *n*- and *p*-Doped Multilayered Mirrors

In the above analysis of current through the barrier between a heterointerface, current tunneling is considered to be the dominant mechanism. In fact, an accurate modeling of current–voltage characteristics of the entire structures with n and p heterointerfaces requires a self-consistent solution of Poisson's equation, continuity equations for carriers, and the drift diffusion Equation [38,39]. Poisson's equation can be generally expressed as

$$\frac{\partial}{\partial z}\left(\varepsilon(z)\frac{\partial V}{\partial z}\right) = -\rho(z), \tag{6.53}$$

where $\rho(z) = q(P(z) - N(z) + N_D^+(z) - N_A^-(z))$, $n(z)$ is the electron concentration, and N_D^- is the concentration of ionized donors. The continuity equation, which assumed no accumulation of charge in any position at any time, governs the flow of charge particles. The continuity equations for electrons and holes can be written as

$$\frac{1}{q}\frac{\partial J_n}{\partial z} - R_t(N, P) = 0, \tag{6.54a}$$

$$\frac{1}{q}\frac{\partial J_p}{\partial z} + R_t(N, P) = 0, \tag{6.54b}$$

where J_n and J_p are the current density of electrons and holes, respectively. In this analysis, R_t consists of the recombination of carriers and bimolecular radiative recombination, which can be reexpressed as

$$R_t = \frac{NP - N_i^2}{(N + N_i)\tau_p + (P + N_i)\tau_n} + B_{sp}(NP - N_i^2), \tag{6.55}$$

where N_i is the intrinsic carrier concentration. If InP-based materials are used as the DBRs, the Auger recombination should also be considered.

Three types of transportation mechanisms (i.e., drift diffusion, thermionic emission, and tunneling) should be included in the calculation. Drift diffusion current determines the transportation mechanism inside the regions where either the valence or the conduction bands change continuously. These currents consist of the gradients in both material parameters in addition to the usual terms for the electric field and the carrier concentration gradients, which are given by

$$J_n = qD_n\frac{\partial N}{\partial z} - q\mu_n N(z)\frac{\partial V}{\partial z} - \frac{3}{2}qD_n N(z)\frac{\partial \log(m_e^*)}{\partial z}, \tag{6.56a}$$

$$J_p = -qD_p\frac{\partial P(z)}{\partial z} - q\mu_p P(z)\frac{\partial V}{\partial z} + \frac{3}{2}qD_p P(z)\frac{\partial \log(m_h^*)}{\partial z}, \tag{6.56b}$$

where μ_n and μ_p are the mobilities of electrons and holes, respectively. In addition, at the heterointerface where a discontinuity in the valence or conduction bands appears, the thermionic emission and tunneling currents are applied as the extra boundary conditions at the heterointerface. Hence, the corresponding injection current J_n and J_p can be written as

$$J_n = J_{\text{thermal, n}}^{\text{GaAs}\to\text{AlAs}} \quad J_{\text{thermal, n}}^{\text{AlAs}\to\text{GaAs}} + J_{\text{tunnel, n}}^{\text{GaAs}\to\text{AlAs}} - J_{\text{tunnel, n}}^{\text{AlAs}\to\text{GaAs}}, \tag{6.57a}$$

$$J_p = J_{\text{thermal, p}}^{\text{GaAs}\to\text{AlAs}} - J_{\text{thermal, p}}^{\text{AlAs}\to\text{GaAs}} + J_{\text{tunnel, p}}^{\text{GaAs}\to\text{AlAs}} - J_{\text{tunnel, p}}^{\text{AlAs}\to\text{GaAs}}. \tag{6.57b}$$

The expressions for the thermionic emission and tunneling currents from GaAs to AlAs depend strictly on the parameters on the GaAs side of the band discontinuity. Using (6.37) and (6.38), the equivalent expressions for the current from GaAs to AlAs are given by [39]

$$J_{\text{thermal, n}}^{\text{GaAs}\to\text{AlAs}} = A_n^* T^2 \exp\left(-\frac{q\phi_{b,c}}{k_B T}\right)\exp\left(\frac{E_{Fn} - E_c}{k_B T}\right), \tag{6.58a}$$

$$J_{\text{thermal,p}}^{\text{GaAs}\to\text{AlAs}} = A_p^* T^2 \exp\left(-\frac{q\phi_{b,v}}{k_B T}\right)\exp\left(\frac{E_v - E_{Fp}}{k_B T}\right), \tag{6.58b}$$

$$J_{\text{tunnel,n}}^{\text{GaAs}\to\text{AlAs}} = A_n^* \frac{T}{k_B}\int_0^{\phi_{b,c}} \mathcal{T}_n(E_z)\log\left(1 + \exp\left[\frac{E_{Fn} - E_c - E_z}{k_B T}\right]\right) dE_z, \tag{6.59a}$$

$$J_{\text{tunnel, p}}^{\text{GaAs}\to\text{AlAs}} = A_p^* \frac{T}{k_B}\int_{-\phi_{b,v}}^{0} \mathcal{T}_p(E_z)\log\left(1 + \exp\left[\frac{E_v - E_{Fp} - E_z}{k_B T}\right]\right) dE_z, \tag{6.59b}$$

where the subscripts n and p stand for electrons and holes, respectively. The transmission probability for electrons and holes (i.e., $\mathcal{T}_n$ and $\mathcal{T}_p$) to transport across the conduction band and valence band spike, respectively, can be approximated by using the 1D Wentzel–Kramers–Brillouin (WKB) approximation [39]

$$\mathcal{T}_n(E_z) = \exp\left(-\frac{2\pi}{\hbar}\int_0^{L_b(E_z)} \sqrt{2m^*(E_c - E_z)}\,dz\right), \tag{6.60a}$$

$$\mathcal{T}_p(E_z) = \exp\left(-\frac{2\pi}{\hbar}\int_0^{L_b(E_z)} \sqrt{2m^*(E_z - E_v)}\,dz\right), \tag{6.60b}$$

where $L_b(E_z)$ is the displacement of energy band from the interface at energy E_z.

To solve potential and carrier distribution of the entire DBRs, certain boundary conditions for the potential and carrier concentrations should be assessed. If ohmic contacts are applied on the left and right sides of the mirror, there is no voltage drop across the contact and an infinite surface recombination rate at the contact is assumed. In addition, it is assumed that the condition of charge neutrality is applied at the contacts and no space charge exists at the contacts. The preceding equations can be solved simultaneously using the finite difference method. The corresponding detailed implantation of the model can be found in the book by Kurata [40].

There is difference between the model with continuity equations and that with the Schrödinger equation used in the calculation of carriers at the heterointerface. In the model with continuity equations, the gradient of the current density as well as the rate of recombination determine the profile of carriers. However, for the model with the Schrödinger equation, the capture of carriers inside the potential well at the heterointerface determines the distribution of carriers. For more accurate analysis of the distribution of carrier concentration at the heterointerface, the continuity and Schrödinger equations should be solved simultaneously [41].

6.4 QUASI-3D ANALYSIS OF ELECTRIC POTENTIAL DISTRIBUTION INSIDE VCSELs

The 3D distribution of potential and current density can be evaluated using an approach similar to the model with continuity equations given in the preceding Section [36]. However, the detailed calculation of energy bands and Fermi levels of multilayered structures for the determination of the carrier concentration require extensive computational effort. In fact, 3D calculation of the distribution of electric potential and current flow can be solved using the above mentioned approach but ignoring the continuity equations [42]. Hence, the influence of the current crowding effect [7,8] due to different geometry of metal contacts can be easily determined. In addition, the proposed approach has assumed that the local current density across the junction depends on the carrier concentration as well as the recombination rate inside the active layer. This model includes calculation of the current self-distribution effect [8]. In the following paragraphs, the simplified model used to solve the quasi-3D distribution of electric potential as well as current flow is described.

6.4.1 Modeling of Quasi-3D Distribution of Electric Potential

The nonuniform distribution of electric potential over the entire laser cavity can be deduced using the approach given in Section 6.2.2. This is done by solving Poisson's equation and the rate equation of carrier concentration inside the active region simultaneously. However, the information of conduction band, valence band and the corresponding Fermi energy levels is ignored. Poisson's equation and carrier rate equation are coupled with the appropriate boundary conditions between the interfaces of spacer and active layer [2]. Poisson's equation used in this model is given by

$$\frac{1}{r}\frac{\partial}{\partial r}\left(\varepsilon(r)r\frac{\partial V}{\partial r}\right)+\frac{\partial}{\partial z}\left(\varepsilon(z)\frac{\partial V}{\partial z}\right)=-\rho(r,z), \tag{6.61}$$

which can be solved numerically using the finite difference method. In order to simplify the calculation, it is assumed that $\rho(r, z) = 0$ inside the laser cavity. This is roughly valid as the DBRs are usually doped to reduce electrical resistance so that the external current injection has less influence on the distribution of electric field. The corresponding boundary conditions for V are

- $V|_{\text{contacts}} = V_b$, where V_b is the biased voltage at the metal contacts,
- $\nabla V = 0$, which implies no current leakage in the direction perpendicular to the surface of the laser cavity

Figure 6.6 shows the schematic of a VCSEL assumed for the numerical computation with the appropriated boundary conditions. It must be noted that the boundary conditions are varied for different designs of metal contacts in VCSELs. The remaining boundary condition, which coupled Poisson's equation together with the rate equation of the carrier concentration, will be determined in the following paragraph.

The carrier concentration $N(r)$ along the transverse direction of the active layer, can be calculated by solving (6.1a)

$$\frac{\partial N}{\partial t}=\frac{J(r)-J_{\text{leak}}(r)}{qd}-R_t(N)-R_{st}(N)+D_n\frac{1}{r}\frac{\partial}{\partial r}\left(r\frac{\partial N}{\partial r}\right), \tag{6.62}$$

where J_{leak} is the leakage current density. For VCSELs with a different current confinement structure, appropriated boundary conditions should be selected to solve the nonuniform distribution of the carrier concentration using (6.62). The selection of boundary conditions for VCSELs has been discussed in Section 6.2.1.

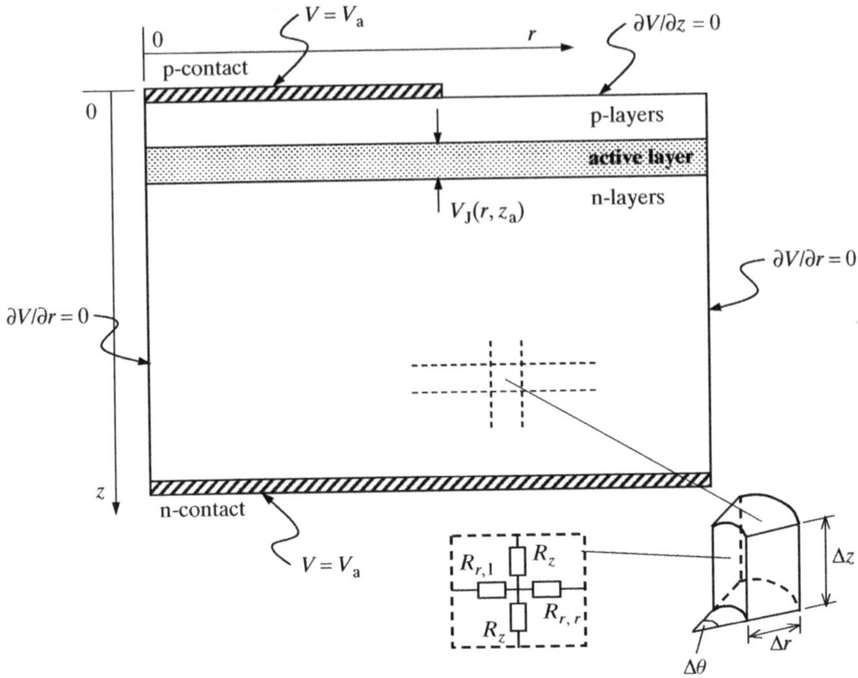

Figure 6.6 (a) The 2D VCSEL model with boundary conditions assumed for the numerical computation; (b) discretization of the laser structure of cylindrical symmetry and corresponding resistance network.

In order to couple (6.61) and (6.62), it is required to calculate the voltage drop across the active layer with the corresponding carrier concentration. For VCSELs with a bulk active layer, the corresponding junction voltage V_J across the active layer can be written in the form of a Joyes–Dixon expression, which is given by [42]

$$qV_J = E_{F,c} - E_{F,v}$$
$$= E_g + k_B T \cdot \left[\log\left(\frac{N^2}{N_c N_v}\right) + \sum_{i=1} K_i \left\{ \left(\frac{N}{N_c}\right)^i + \left(\frac{N}{N_v}\right)^i \right\} \right], \quad (6.63)$$

where $K_1 = 1/\sqrt{8}$, $K_2 = -4.95009 \times 10^{-3}$, $K_3 = 1.48386 \times 10^{-3}$, and $K_4 = -4.42563 \times 10^{-6}$. The corresponding conduction and valence edge density of states is given by

$$N_{c/v} = 2\left(\frac{2\pi m^*_{e/h} k_B T}{h^2}\right). \quad (6.64)$$

On the other hand, if the QW active layer is used, the approximate expression of junction voltage is given by [43]

$$
\begin{aligned}
qV_J &= E_{F,c} - E_{F,v} \\
&= E_g + k_B T \cdot \log\left[\left(\exp\left(\frac{N(r)}{N_c}\right) - 1\right)\left(\exp\left(\frac{N(r)}{N_v}\right) - 1\right)\right], \quad (6.65)
\end{aligned}
$$

where N_c and N_v are the effective conduction and valence edge density of states, respectively, and can be expressed as

$$
N_{c/v} = \frac{m^*_{e/h} k_B T}{\pi \hbar^2 L_w}. \quad (6.66)
$$

In the derivation of (6.65), only the first quantized energy level of the conduction and valence bands of QWs is considered. In addition, the charge neutrality is assumed inside the intrinsic active region. Now the remaining boundary condition can be written as [i.e., using 6.27]

$$
J(r, z = z_a) = J_s \exp\left(\frac{qV_J}{2kT}\right), \quad (6.67)
$$

where z_a is the longitudinal displacement of the active layer. As is shown, the injection current density J at the active layer is coupled to the voltage distribution, which is in fact also a function of the carrier concentration inside the active layer. Hence, using Poisson's equation and the rate equation of the carrier concentration, the quasi-3D distribution of electric potential inside the laser cavity as well as the carrier concentration inside the active layer can be evaluated self-consistently.

6.4.2 Modeling of Quasi-3D Current Flow

The knowledge of thermal emission and tunneling of carriers across the heterointerfaces of the multilayer stacks is required to determine the current flow inside the laser cavity of VCSELs. However, the exact calculation of the 3D current flow requires complicated implementation and extensive computational effort, which might not even be of interest for practical analysis. In order to improve the computational efficiency of the 3D analysis of current flow, a resistance network approach is proposed to evaluate the quasi-3D current flow inside the laser cavity [1]. The resistance network is also schematically sketched in Figure 6.6. The entire laser structure, including the substrate, is discretized in a nonequidistant rectangular grid. As the device has a circular symmetry, the modeling can be restricted to the r–z plane of a cylinder coordinate system. Each rectangle in this grid is represented by a cylinder segment of an arbitrary azimuthal angle $0 < \Delta\theta \leq 2\pi$, which is itself electrically modeled by four ohmic resistances with a common branch point. The values of the left and right radial ohmic resistance $R_{\mathrm{r,l}}^{(n,m)}$ and $R_{\mathrm{r,r}}^{(n,m)}$, and the longitudinal resistance $R_z^{(n,m)}$ can be expressed as [1]

$$R_z^{(n,m)} = \frac{\Delta z_m}{2\Delta\theta r_n \Delta r_n \sigma(r_n, z_m)}, \tag{6.68}$$

$$R_{\mathrm{r,l}}^{(\mathrm{n},m)} = \frac{-\log(1 - \Delta r_n/2r_n)}{\Delta\theta\,\Delta z_m \sigma(r_n, z_m)} \text{ and } R_{r,r}^{(\mathrm{n},m)} = \frac{\log(1 + \Delta r_n/2r_n)}{\Delta\theta\,\Delta z_m \sigma(r_n, z_m)}, \tag{6.69}$$

where r_n and z_m are the center coordinates of the cylinder segment in the nth column and mth row of the grid. The local electric conductivity $\sigma(r_n, z_m)$ is constant in each segment of sufficiently small dimension Δr_n and Δz_m. The electric potential obtained from Poisson's equation is applied at the corresponding branch point of the resistance network for the calculation of current flow. If no ion implantation is applied to confine the injection current, σ can be interpreted as the conductivity in the homogenous regions σ_0 of the laser, governed by the local doping concentration, and thus independent of the radius [i.e., $\sigma(r_n, z_n) \equiv \sigma_0(r_n)$]. For p- and n-GaAs/GaAlAs DBRs, GaAs spacer layer and GaAs substrate, the corresponding background electrical conductivity is estimated to be 1.5, 7, 3, and 500 $\mathrm{cm}^{-1}\,\Omega^{-1}$, respectively.

If a current is injected from the metal contact through the DBRs, the heterointerfaces in the DBRs will play an important role in the determination of the current flow along the z direction. Figure 6.7 shows the corresponding heterointerface resistances and ohmic resistance of p- and n-type DBRs. As has been discussed, the resistance of the current flow is affected mainly by the corresponding potential barrier so that the corresponding resistance is voltage-dependent. However, to a first-order approximation, the heterointerface resistance R_{het} is assumed to be a constant (as well as independent to temperature) and is added with the ohmic resistances in the longitudinal z direction to the resistance network at each of these low–high interfaces. This procedure is equivalent to introducing an anisotropic conductivity in the multilayer stacks. Because of the difference in effective masses and band offset, the resistance of an n-type heterointerface can be estimated to be a factor of ≥ 10 lower than that of a p-type heterointerface in the same material system and a doping regime considered. For p-doped GaAs/AlAs layers, $R_{\mathrm{het,p}}$ is approximately equal to $2.5 \times 10^{-4}\ \Omega.\mathrm{cm}^2$ [1].

If the conductivity of each layer as well as the heterointerface resistance are known, the quasi-3D distribution of current flow can be calculated, provided the electric potential distribution has been solved using the method given in the above section. Hence, the current flow inside devices such as those with airpost, etched mesa, or oxide-confined structures can be estimated. However, for the calculation of the current flow inside the devices with ion-implanted structure, the profile of the implanted ions should also be included in the calculation of conductivity of the layers. It must be noted that the profile of the implanted ions is dependent on the kinetic energy of the implanted ions as well as the shape of the mask. It is assumed that the energetic ions penetrating into the semiconductor material

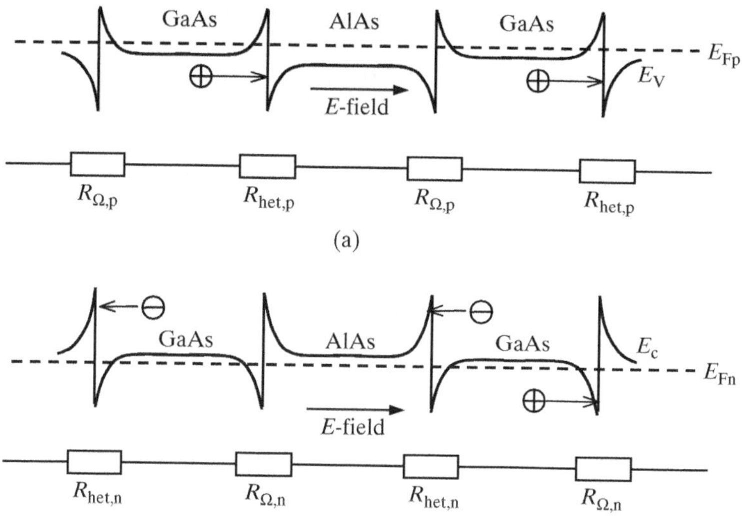

Figure 6.7 Schematic of the influence of heterointerfaces in a GaAs/AlAs multilayered structure on the current flow of (a) a hole and (b) an electron. The corresponding linear network of heteroresistance is also given in the figure. E -field is the electric field strength; E_c, E_v, and $E_{Fn,p}$ denote the conduction band edge, valence band edge, and Fermi level, respectively.

(i.e., unshielded region) will lose energy through collisions with the target lattice and electrons and finally come to rest. Therefore, the depths of penetration will depend on the ion used, the kinetic energy of the ion, as well as the structure of the target material. Using the approach given by Furukawa et al. [44], the distribution of ions $F(r, z)$, implanted through a circular mask of radius W, is given by

$$F(r, z) = \frac{N_{ID}}{\sqrt{2\pi}\langle\Delta R_{pr}\rangle\langle\Delta R_{pz}\rangle} \exp\left\{-\frac{(z - \langle R_{pz}\rangle)^2}{2\langle\Delta R_{pz}\rangle^2}\right\}$$
$$\times \left\{\frac{\langle\Delta R_{pr}\rangle}{2} \exp\left\{-\frac{(W - r)^2}{\langle\Delta R_{pr}\rangle^2}\right\} + \frac{\sqrt{\pi}r}{2} \operatorname{erfc}\left(\frac{W - r}{\langle\Delta R_{pr}\rangle}\right)\right\}, \tag{6.70}$$

where $\langle R_{pz}\rangle$ is the projection range and $\langle\Delta R_{pz}\rangle$ and $\langle\Delta R_{pr}\rangle$ are the vertical and transverse projected standard deviation of the implanted ion, respectively. The parameters $\langle R_{pz}\rangle$, $\langle\Delta R_{pz}\rangle$, and $\langle\Delta R_{pr}\rangle$ are the functions of implanting ions, implantation energy, and substrate material, respectively; N_{ID} is the implantation dose, and erfc( ) is the complementary error function. After ion implantation, diffusion of impurities takes place via rapid thermal annealing. The ions diffuse thermally into the active layer as a result of the concentration gradient. The annealing process can be described by the following diffusion equation

$$\frac{\partial N_{\mathrm{I}}(r,z,t)}{\partial t} = D_{\mathrm{imp}}\left\{\frac{1}{r}\frac{\partial}{\partial r}\left(r\frac{\partial N_I(r,z,t)}{\partial r}\right) + \frac{\partial^2 N_I(r,z,t)}{\partial z^2}\right\}, \qquad (6.71)$$

where N_{I} is the impurity concentration and $D_{\mathrm{imp}}(= 10^{-11}\ \mathrm{cm^2\ s^{-1}})$ is the diffusion coefficient of the impurity at a specific annealing temperature. Equation (6.71) can be solved numerically using the finite difference method using (6.70) as the initial distribution of ions. In addition, it is assumed that no impurity has escaped from the device to the surrounding area, so that the corresponding boundary conditions are given by [2]

- $\dfrac{\partial N_{\mathrm{I}}}{\partial r} = \dfrac{\partial N_{\mathrm{I}}}{\partial z} = 0$ at the edge of the device
- $\dfrac{\partial N_{\mathrm{I}}}{\partial r} = \dfrac{\partial N_{\mathrm{I}}}{\partial z} = 0$ at the top and bottom surfaces of the device

The nonuniform distribution of N_{I} has a significant influence on the p-side DBR region as the destruction of lattice alters the electrical conductivity and optical absorption loss inside the p-side DBR.

In the ion-implanted region of the p-side DBR, it is assumed that the electrical conductivity is dependent on the distribution of impurity concentration $N_{\mathrm{I}}(r,z)$. Therefore, the electrical conductivity of the ion-implanted area $[\sigma_{\mathrm{imp}}(\Omega^{-1}\ \mathrm{cm}^{-1})]$ can be approximated by [1]

$$\sigma_{\mathrm{imp}}(r,z) \approx \sigma_0(z)\left(1 - \left(1 - \frac{1}{K_{\mathrm{I}}}\right)\frac{\Delta R_{\mathrm{p}z} N_{\mathrm{I}}(r,z)}{N_{\mathrm{ID}}}\right), \qquad (6.72)$$

where $K_I(\sim 10^6)$ is the reduction factor [1]. From (6.72), σ_{imp} is a reducing function of N_{I} due to the strength of lattice damage. Furthermore, the absorption loss, $\alpha(\mathrm{cm}^{-1})$, due to the implanted ion inside the p-side DBR can also be approximated by

$$\alpha(r,z) \approx \alpha_{\mathrm{o}}\left(1 - \gamma\left(1 - \frac{1}{K_{\mathrm{I}}}\right)\frac{\Delta R_{\mathrm{p}z} N_{\mathrm{I}}(r,z)}{N_{\mathrm{ID}}}\right)^{-1}, \qquad (6.73)$$

where $\alpha_{\mathrm{o}}(= 10\ \mathrm{cm}^{-1})$ is the absorption loss without the implanted ion and $\gamma(= 0.05)$ is a scaling factor. Hence, the extra reduction in conductivity due to the implanted ion can be calculated using the above mentioned method for VCSELs such as those with proton implantation for transverse carrier confinement.

6.5 SELF-CONSISTENT CALCULATION OF VCSELs

For VCSELs operating above threshold, the carrier concentration and photon density inside the active layer perform complicated interactions, which cause the temperature variation. The carrier concentration, photon density, and heat are coupled together through the transverse distribution of optical gain and refractive index, which are dependent on the carrier concentration, photon density, and temperature of the active layer. Hence, SHB due to stimulated recombination and thermal lensing, a result of the self-heating effect have to be calculated through the self-consistent calculation of the carrier concentration, photon density, and heat. The temperature dependence of resistance and thermal conductivity may also have to be included the calculation.

The model given in Section 5.3.2 is the simplest form of equations capable of performing the self-consistent calculation of the carrier concentration, photon density, and temperature, which can be rewritten as

$$\frac{\partial S_{lm}}{\partial t} = [\nu_g \Gamma_z \langle g_{l,m} \rangle - 1/\tau_{p,lm}] S_{lm} + \beta_s \Gamma_z B_{sp} \langle N^2 \rangle, \tag{6.74a}$$

$$\frac{\partial N}{\partial t} = \frac{J - J_{leak}(T)}{qd} - R_t(N) - \underbrace{\nu_g \sum_{l,m} \langle g_{l,m} \rangle S_{lm}}_{\text{stimulated recombination}} + D_n \nabla^2 N, \tag{6.74b}$$

$$C_p \frac{dT}{dt} = (P_{IV} - P) - \frac{T - T_0}{R_T}. \tag{6.74c}$$

Although this model of rate equations is simple, it is still adequate to take into account the linkage between the carrier concentration, photon density, and heat through the variation of optical gain and refractive index inside the active layer.

The rate equation of photon density [i.e., (6.74a)] is derived from the traveling-wave equation. S_{lm} represents the average photon density inside the active layer. The longitudinal confinement factor Γ_z and the modal gain $\langle g_{l,m} \rangle$ in the longitudinal and transverse distributions of the optical field, respectively. Γ_z and $\langle g_{l,m} \rangle$ are affected by the variation of the carrier concentration (i.e., SHB) and temperature (i.e., thermal lensing) inside the active layer through the self-focusing effect. Furthermore, the photon lifetime τ_p is obtained from the reflectivity of the multilayered mirrors; τ_p can be deduced using the mode-matching technique as given in Section 3.3, with which the influence of diffraction loss can be included in the calculation. Hence, the 3D effect of the optical field has been simply introduced into (6.47a) as the photon density is coupled to the carrier concentration and temperature through Γ_z and $\langle g_{l,m} \rangle$.

The approach described above to implement the quasi-3D optical field profile inside the laser cavity into the rate equation model has been discussed in Chapter 3. Alternatively, if the 3D distribution of the optical field needs to be calculated, the rate equation of photon density should be replaced by the scalar wave equation [i.e., such as the beam propagation method (BPM) given Section 3.4] or the vectorial wave equation (i.e., such as that given in Section 3.5). Again, the nonuniform distribution of the carrier concentration and temperature and their influence on the optical gain and refractive index inside the active layer have to be considered.

The rate equation of the carrier concentration [i.e., (6.74b)] describes the transverse distribution of the carrier concentration inside the active layer. The profile of the carrier concentration is affected by the injection current density, stimulated recombination, and diffusion of the carrier concentration. It is noted that SHB is dependent on the profile of transverse fields as a result of stimulated recombination. In addition, the corresponding threshold current is dominated by the leakage current, which is dependent on the temperature and cavity size of VCSELs. Hence, (6.74b) describes the linkage between the carrier concentration and photon density as well as temperature. Section 6.2 has discussed the simplest way to linearize the nonuniform distribution of the carrier concentration inside the active layer and the method to incorporate the transverse distribution of optical field and injection current. However, for calculation of the quasi-3D distribution of electric potential and current flow inside the laser cavity, Poisson's equation and the resistance network (i.e., as discussed in the sections above), have to be solved in conjunction with the rate equation of the carrier concentration. The necessity of solving the quasi-3D distribution of electric potential and current flow is to determine the current spreading and current crowding as well as the quasi-3D distribution of heatsources inside the laser cavity.

In (6.74c), it is assumed that heat is generated by the difference between the total input electrical power and the output optical power. It is noted that P_{IV} is a function of the injection current and junction voltage V_J of the active layer. In addition, V_J is a function of the carrier concentration. Hence, the average temperature rise inside VCSELs is coupled with the carrier concentration through the electrical power and the optical power. On the other hand, the quasi-3D distribution of heat can also be calculated from the method described in Section 5.4.2. The 3D distribution of heatsources can be evaluated from the quasi-3D distribution of electric potential and current flow as discussed above.

The transverse distribution of optical gain and refractive index along the active layer is also dependent on the temperature profile (i.e., thermal lensing). Therefore, the temperature profile, optical fields, and the electrical potential (i.e., including the carrier concentration inside the active layer, current flow, and potential distribution inside the laser cavity) are strongly coupled together. It must be noted that change in any one of the quantities will affect the others, and these quantities have to be solved self-consistently. Therefore, the approach given in (6.74) can be utilized to develop a self-consistent quasi-3D model of VCSELs [7–9,45–47]. Figure 6.8 explains the operation mechanism of a quasi-3D model of VCSELs with temperature, optical field, and electric potential (i.e., including carrier concentration) under self-consistent calculation. It is assumed that these varying quantities are coupled together via the variation of optical gain and refractive index change inside the active layer. In addition, the dependence of thermal and electrical resistivities of the laser cavity on the temperature can be considered. In the implementation of the model, the variation of optical gain with the carrier concentration and temperature can be approximated by the empirical fitted equation as given in Chapter 5. The variation of refractive index

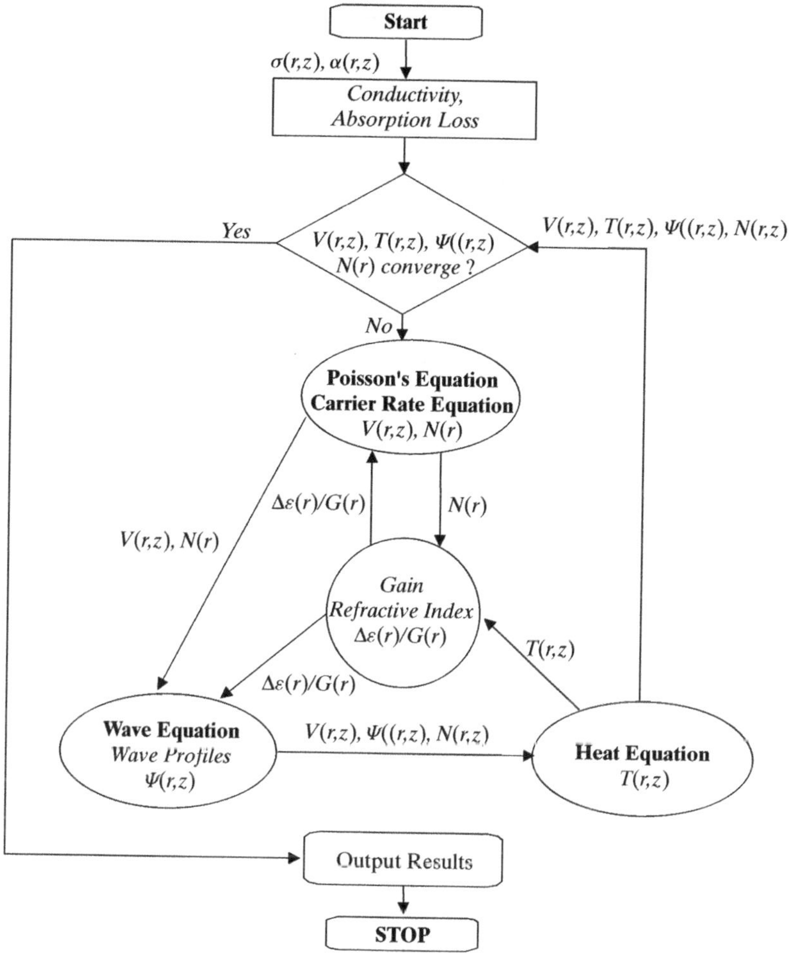

Figure 6.8 Flowchart for the self-consistent calculation of the quasi-3D distribution of current density, potential energy, and temperature inside the laser cavity as well as the quasi-2D distribution of carrier concentration inside the active layer.

inside the active layer with the carrier concentration and temperature can also be approximated by (3.44).

This self-consistent method has been applied to analyze VCSELs with proton-implanted structures [1,2,7], oxide confined structures [8,9], undercut ridge structures [46], as well as double-fused lasers [47]. In order to demonstrate the current spreading and current crowding in VCSELs, devices with two different types of current confinement structures are considered in the following investigation. It is assumed that the VCSELs have proton implantation [1,2,7] and double oxide-confined [8,9] structures for the transverse confinement of injection carriers. The devices, however, have similar contact geometry, which allows transverse

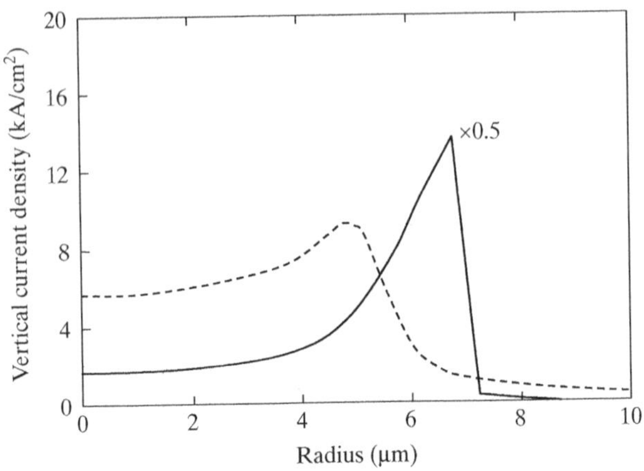

Figure 6.9 Radial variation of the longitudinal component of current density inside the active layer: VCSEL with proton implantation structure (solid line), and VCSEL with double-oxide-confined structure (dashed line) for the electrical confinement of injection current (after Refs.1 and 8).

injection of current. See Figures 1.10 and 1.14 for detailed configuration of the VCSELs used in this investigation. For both VCSELs, the p-contact metal is evaporated on the top of the p-doped contact layer to inject carriers into the core region of the active layer. Figure 6.9 shows the typical current profile inside the active layer of the VCSELs. It is observed that for both devices, current crowding occurs at the periphery between the core and cladding regions of the active layer. No current spreading is observed for the case with proton implantation structure. In order to maintain the uniformity of the injection current in devices with proton implantation structure, the equivalent differential resistance of the laser should increase. On the other hand, the uniformity of the injection current can be optimized in oxide-confined VCSELs by varying the separation between the oxide layers. It must be noted that the increase in separation between the oxide layers will also increase the degree of current spreading. Hence, devices with ion implantation structure have better immunity to current crowding and current spreading than do those of oxide confined structure.

Experiments have shown that the maximum wallplug (electrical-to-optical) efficiency of VCSELs with ion implantation structure and those of oxide-confined structure is approximately 21% and 57%, respectively [35]. This is because wallplug efficiency is high in VCSELs with oxide-confined structure (i.e., strongly index-guided devices) as the overlap between the transverse field and the optical gain is confined inside the core region of the active layer. However for devices with ion implantation structure, the transverse field is weakly guided inside the core region of the active layer. As a result, the transverse field, which extends into the cladding region of the active layer, is strongly absorbed so that the electrical conversion efficiency is significantly reduced. Hence, VCSELs with oxide-confined structure have a better wallplug efficiency even with relatively weak current confinement than that of VCSELs with ion implantation structure.

If both devices have similar composition of p- and n-doped DBRs as well as material at the active layer, most likely the total power will dissipate at the active layer. Only less than 10% of the total power will dissipate in terms of heat inside the n-doped multilayered mirror and substrate. Now, the heat distribution of devices with ion implantation structure inside the active layer will depend on the thermal conductivity of the active layer (i.e., if the proton implantation has no influence on the thermal conductivity of the active layer). However, for devices with oxide-confined structure, the heat distribution inside the active layer is dependent on the thermal conductivity of both the active and oxide layers. If both devices support a fundamental transverse mode with regular beamwidth, the heat distribution inside the active layer will be of Gaussian shape in VCSELs with ion implantation structure. However, in VCSELs with oxide confinement structure, the heat distribution will be relatively uniform inside the core region but decay exponentially toward the cladding region. Hence, thermal lensing is more serious in VCSELs with ion implantation structure. This may be another reason why VCSELs with oxide-confined structure have better wallplug efficiency than do those ion implantation structure.

6.6 SUMMARY

This chapter describes the derivation of electrical models as well as the numerical analysis of VCSELs with different waveguide, p-contact configuration, and current confinement structures.

First, the models using rate equations of the carrier concentration are developed to analyze SHB. The influence of the injection current profile and the carrier transport effect on the profile of SHB is also taken into calculation. It is noted that SHB is due to the stimulation recombination of the carrier concentration inside the core region of active layer. However, SHB can be minimized if the appropriated dimension of circular-disk contact on the p-type mirror is used to confine the injected carrier concentration inside the core region of the active layer. It is also shown that SHB has significant influence on the stability of a single LP_{01} mode operation at high power due to the presence of the self-focusing effect. The self-focusing effect is caused by the carrier-induced index change inside the core region of the active layer so that the beamwidth of the transverse modes is reduced. It is believed that the influence of self-focusing on the transverse modes is more significant in gain-guided or anti-index-guided devices than in strongly index-guided waveguides. Self-focusing can be more serious if thermal lensing is included in the calculation.

Second, the transport mechanisms of carriers across the heterointerfaces of the DBRs are analyzed. It is shown that the tunneling current across the barrier of heterointerfaces is the dominating transport mechanism of carriers in p-type mirrors. This is because the tunneling current of the p-doped multilayered mirror is less temperature-dependent. The energy barrier at the heterointerfaces can be calculated by solving Poisson's equation so that the tunneling current can be obtained. In addition, the carrier captured inside the potential well can be calculated using the Schrödinger equation, which has to be solved self-consistently with Poisson's equation. Alternatively, continuity equations can be used to calculate the distribution of carriers over the heterointerfaces of the entire laser cavity. In this case, both thermal and tunneling currents should be applied as the transport mechanism of carriers across the heterointerfaces. In addition, for more accurate calculation of carrier distribution over the heterointerfaces of the entire laser cavity, the capture of carriers inside the potential well of the heterointerfaces as well as the continuity equation of carriers has to be solved self-consistently.

Finally, the quasi-3D distribution of potential energy and current flow inside the laser cavity is modeled. Again, Poisson's equation is used to analyze the quasi-3D distribution of electric potential. Suitable boundary conditions should be selected at the metal contacts and the surface of the laser cavity. In addition, the ideal diode equation is applied as the boundary condition at the junction of the active layer. Hence, the relation between the electric potential and the carrier distribution inside the active layer can be coupled together. Furthermore, the quasi-3D profile of the current flow is calculated using the resistance network and the electric potential obtained from Poisson's equation. The quasi-3D distribution of electric potential, current flow, heat, as well as the carrier concentration inside the active layer has to be solved self-consistently through the linkage of optical gain and refractive index inside the active layer. Therefore, the influence of the self-focusing effect, thermal lensing, as well as current spreading and current crowding on the electricooptical performance of VCSELs with different waveguide geometry and current confinement structures can be studied. It is found that for VCSELs with weak current confinement structure, current spreading is caused by the diffusion of the carrier into the cladding region of the active layer. As a result, wallplug efficiency is reduced significantly, and the corresponding threshold current is increased. However, for VCSELs with strong current confinement, current crowding may occur near the periphery of the core region of the active layer. This nonuniform injection of the carrier concentration leads to increased SHB as well as excess thermal heating, which causes the thermal lensing effect in VCSELs. Hence, the stability operation of the fundamental transverse-mode operation can deteriorate via the self-focusing effect. In VCSELs with ion-implanted structure for current confinement, current crowding can be minimized with suitable selection of differential resistance of the DBRs. On the other hand, for VCSELs with double-oxide structure, the separation between the oxide aperture determines the degree of current spreading and current crowding inside the active layer.

In conclusion, in order to design VCSELs with maximum immunity to SHB (i.e., self-focusing effect) and maximum wallplug efficiency, VCSELs must have a strong current confinement structure such as ion implantation (i.e., ion implanted into the active layer) or an oxide-confined structure, but with a uniform current profile inside the core region of the active layer. It must be noted that the role of circular-disk contact is important only in VCSELs without electrical confinement structure. The index-guided waveguide is preferred to maintain a stable single fundamental mode operation so that the influence of self-focusing effect and thermal lensing can be minimized. Buried heterostructure or oxide-confined structure is the suitable choice as the waveguide of VCSELs. However, the influence of current crowding may affect the uniformity of the injection profile of the devices.

6A APPENDIX

6A.1 Perturbation Solution for Carrier Concentration Profile in Azimuthal Direction

The azimuthal direction dependence of $\overline{n}(r, \theta)$ can be separated from its radial direction by expressing $\overline{n}(r, \theta)$ as a perturbation solution in the form [10]

$$\overline{n}(r, \theta, t) = \sum_{k=0}^{\infty} \tilde{n}_{c,k}(r, t)\cos(k\theta) + \tilde{n}_{s,k}(r, t)\sin(k\theta). \tag{6A.1}$$

Substituting this expression into (6.7b) gives

$$\begin{aligned}
&\frac{\partial \tilde{n}_{c,0}}{\partial t} + \sum_{k=1}^{\infty}\left[\frac{\partial \tilde{n}_{c,k}}{\partial t}\cos(k\theta) + \frac{\partial \tilde{n}_{s,k}}{\partial t}\sin(k\theta)\right] \\
&\quad = D_n \frac{1}{r}\frac{\partial}{\partial r}\left(r\frac{\partial \tilde{n}_{c,0}}{\partial r}\right) + D_n \frac{1}{r}\frac{\partial}{\partial r}\sum_{k=1}^{\infty}\left[\left(r\frac{\partial \tilde{n}_{c,k}}{\partial r}\right)\cos(k\theta) + \left(r\frac{\partial \tilde{n}_{s,k}}{\partial r}\right)\sin(k\theta)\right], \\
&\quad - \frac{D_n}{r^2}\sum_{k=1}^{\infty}[k^2\tilde{n}_{c,k}\cos(k\theta) + k^2\tilde{n}_{s,k}\sin(k\theta)] \\
&\quad - R_t''(\overline{N})\sum_{k=1}^{\infty}[\tilde{n}_{c,k}\cos(k\theta) + \tilde{n}_{s,k}\sin(k\theta)] - R_{st}
\end{aligned} \tag{6A.2}$$

where $R_t'' \cong 2\overline{N}B_{sp} + C_{Aug}(2\overline{N} + \overline{N}^2)$. Following integration of (6A.2) over θ, it is found that

$$\frac{\partial \tilde{n}_{c,0}}{\partial t} = D_n \frac{1}{r}\frac{\partial}{\partial r}\left(r\frac{\partial \tilde{n}_{c,0}}{\partial r}\right) - R_t''(\overline{N})\tilde{n}_{c,0} - \frac{1}{2\pi}\int_0^{2\pi} R_{st}\, d\theta. \tag{6A.3}$$

Multiplying (6A.2) by $\cos(p\theta)$ and integrating over θ gives

$$\begin{aligned}
\frac{\partial \tilde{n}_{c,p}}{\partial t} &= D_n \frac{1}{r}\frac{\partial}{\partial r}\left(r\frac{\partial \tilde{n}_{c,p}}{\partial r}\right) - D_n \frac{p^2}{r^2}\tilde{n}_{c,p} - R_t''(\overline{N})\tilde{n}_{c,p} \\
&\quad - \frac{1}{2\pi}\int_0^{2\pi} R_{st}\cos(p\theta)\, d\theta,
\end{aligned} \tag{6A.4}$$

where p is an integer. Again, multiplying (6A.2) by $\sin(p\theta)$ and integrating over θ gives

$$\frac{\partial \tilde{n}_{s,p}}{\partial t} = D_n \frac{1}{r}\frac{\partial}{\partial r}\left(r\frac{\partial \tilde{n}_{s,p}}{\partial r}\right) - D_n \frac{p^2}{r^2}\tilde{n}_{s,p} - R_t''(\overline{N})\tilde{n}_{s,p} - \frac{1}{2\pi}\int_0^{2\pi} R_{st}\sin(p\theta)\,d\theta. \tag{6A.5}$$

6A.2 Perturbation Solution for Carrier Concentration Profile in Radial Direction

If the azimuthal variation of the carrier concentration is ignored in the calculation, the stimulation recombination can be expressed as follows [11]:

$$\int_0^W R_{st}(N) r\,dr = v_g \sum_m^{\infty} \int_0^W \Psi_{0m}^*(r) g(N) \Psi_{0m}(r) r\,dr \cdot S_{0m}, \tag{6A.6a}$$

$$\int_0^W R_{st}(N) J_0\left(\frac{a_k r}{W}\right) r\,dr = v_g \sum_m^{\infty} \int_0^W \Psi_{0m}^*(r) g(N) J_0\left(\frac{a_k r}{W}\right) \Psi_{0,m}(r) r\,dr \cdot S_{0m}. \tag{6A.6b}$$

Again, if $g(N)$ is approximated by a linear relation as given in (2.6) or by a logarithmic relation as given in (2.18), then (6A.6) can be simplified to

$$\frac{2}{W^2}\int_0^W R_{st}(N) r\,dr = v_g \sum_m^{\infty}\left(\Gamma_m g(N_0) - \gamma_N \sum_{i=1}^{\infty} N_i \xi_{m,i}\right) \cdot S_{0m}, \tag{6A.7a}$$

$$\frac{2}{W^2 J_0^2(a_k)}\int_0^W R_{st}(N) J_0\left(\frac{a_k r}{W}\right) r\,dr = v_g \sum_m^{\infty}\left(\Gamma_{m,k} g(N_0) - \gamma_N \sum_{i=1}^{\infty} N_i \xi_{m,k,i}\right) \cdot S_{0m}, \tag{6A.7b}$$

where γ_N represents the approximation of the gain function used in the derivation and is given by

$$\gamma_N = \begin{cases} g_N & \text{linear} \\ \dfrac{a_N}{N_0} & \text{logarithmic} \end{cases}. \tag{6A.8}$$

The other parameters given in (6A.7) have the following expression:

$$\Gamma_m = \frac{2}{W^2}\int_0^W |\Psi_{0m}(r)|^2 r\,dr,\ \xi_{m,k} = \frac{2}{W^2}\int_0^W J_0\left(\frac{a_k r}{W}\right)|\Psi_{0m}(r)|^2 r\,dr, \tag{6A.9a}$$

$$\Gamma_{m,k} = \frac{2}{W^2 J_0^2(a_k)}\int_0^W |\Psi_{0m}(r)|^2 J_0\left(\frac{a_k r}{W}\right) r\,dr \text{ and } \xi_{m,k,i} = \frac{2}{W^2 J_0^2(a_k)}\int_0^W J_0\left(\frac{a_i r}{W}\right)|\Psi_{0m}(r)|^2 J_0\left(\frac{a_k r}{W}\right) r\,dr. \tag{6A.9b}$$

Substituting (6A.7) into (6.16) and (6.17) gives

$$\frac{\partial N_0}{\partial t} = \frac{J}{qd} - R_t(N_0) - v_g \sum_m^{\infty} \left(\Gamma_m g(N_0) - \gamma_N \sum_{k=1}^{\infty} N_k \xi_{m,k} \right) \cdot S_{0m}, \quad (6A.10a)$$

$$\frac{\partial N_k}{\partial t} = -N_k \left[R'_{t,k}(N_0) + D_n \left(\frac{a_k}{W}\right)^2 \right] + v_g \sum_m^{\infty} \left(\Gamma_{m,k} g(N_0) - \gamma_N \sum_{i=1}^{\infty} N_i \xi_{m,k,i} \right) \cdot S_{0m}. \quad (6A.10b)$$

6A.3 Transmission Probability Across Arbitrary Potential Barriers

Figure 6.10 shows an arbitrary potential barrier, which is approximated by a sequence of m small segments. The corresponding potential barrier, $U(z)$, and the effective mass m∗(z) are approximated by the multistep functions [37]:

$$U(z) = U_i = U\left[\frac{z_{i-1} + z_i}{2}\right], \quad (6A.11a)$$

$$m^*(z) = m_i^* = m^*\left[\frac{z_{i-1} + z_i}{2}\right], \quad (6A.11b)$$

for $z_{i-1} < z < z_i (i = 0, 1, 2, \ldots, m, m+1)$. The wavefunction Ψ_i in the ith region, associated with an electron with energy E moving normally on the barrier, is given by

$$\Psi_i(z) = A_i \exp(jk_i z) + B_i \exp(-jk_i z), \quad (6A.12)$$

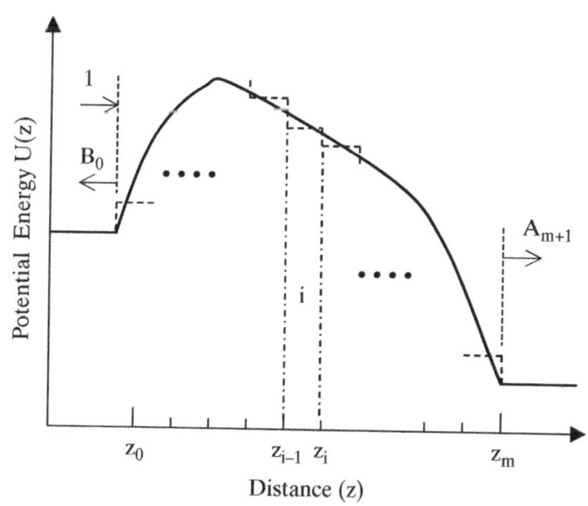

Figure 6.10 Energy band diagram (solid line) and approximated potential function (broken line) for the potential barrier.

where

$$k_i = \sqrt{\frac{[2m_i^*(E - U_i)]}{\hbar}}. \tag{6A.13}$$

From the continuity of $\Psi_i(z)$ and $(1/m_i^*)(d\Psi_i/dz)$ at each boundary, the determining A_i and B_i in (6A.12) can be reduced to the multiplication of the following $m + 1(2 \times 2)$ matrices:

$$\begin{pmatrix} A_n \\ B_n \end{pmatrix} = \prod_{i=0}^{n-1} M_i \begin{pmatrix} A_0 \\ B_0 \end{pmatrix}, \tag{6A.14}$$

where

$$M_i = \frac{1}{2}\begin{bmatrix} (1+s_i)e^{[-j(k_{i+1}-k_i)z_i]} & (1-s_i)e^{[-j(k_{i+1}+k_i)z_i]} \\ (1-s_i)e^{[j(k_{i+1}+k_i)z_i]} & (1+s_i)e^{[j(k_{i+1}-k_i)z_i]} \end{bmatrix} \text{ and } s_i = \frac{m_{i+1}^*}{m_i^*}\frac{k_i}{k_{i+1}}. \tag{6A.15}$$

By setting $A_0 = 1$ and $B_{m+1} = 0$ (i.e., no injection of carrier from z_m) in (6A.14) for $n = mn + 1$, the transmission amplitude A_{m+1} of the transmitted wavefunction can be written as

$$A_{m+1} = \frac{m_{m+1}^*}{m_0^*}\frac{k_0}{k_{m+1}}\frac{1}{M_{22}} \tag{6A.16}$$

and the transmission probability T(E) can be given as

$$T(E) = \frac{m_0^*}{m_{m+1}^*}\frac{k_{m+1}}{k_0}|A_{m+1}|^2, \tag{6A.17}$$

where

$$\begin{pmatrix} M_{11} & M_{12} \\ M_{21} & M_{22} \end{pmatrix} = \prod_{i=0}^{m} M_i. \tag{6A.18}$$

REFERENCES

1. R. Michalzik and K. J. Ebeling, "Modeling and design of proton-implanted ultralow threshold vertical cavity laser diode," *IEEE J. Quantum Electron.*, Vol. 29, no. 6, pp. 1963–1974, 1993.
2. W. M. Man and S. F. Yu, "Comprehensive modeling of diffused quantum wells vertical cavity surface emitting lasers," *IEEE J. Select. Topics Quantum Electron.*, Vol. 4, no. 4, pp. 715–722, 1998.
3. G. C. Wilson, D. M. Kuchta, J. D. Walker, and J. S. Smith, "Spatial hole burning and self-focusing in vertical cavity surface emitting laser diode," *Appl. Phys. Lett.*, Vol. 64, no. 5, pp. 542–544, 1994.

4. K. D. Choquette, G. Hasnain, Y. H. Wang, J. D. Wynn, R. S. Freund, A. Y. Cho, and R. E. Leibenguth, "GaAs vertical cavity surface emitting lasers fabricated by reactive ion etching," *IEEE Photon. Technol. Lett.*, Vol. 3, no. 10, pp. 859–862, 1991.
5. K. Mori, T. Asaks, H. Iwano, M. Ogura, S. Fujii, T. Okada, and S. Mukai, "Effect of cavity size on lasting characteristics of a distributed bragg reflector surface emitting laser with buried heterostructure," *Appl. Phys. Lett.*, Vol. 60, no. 1, pp. 21–23, 1992.
6. R. Papannareddy, W. Ferguson, and J. K. Butler, "Current spreading and carrier diffusion in zinc-diffused multiple-stripe-geometry lasers," *Appl. Phys. Lett.*, Vol. 50, no. 19, pp. 1315–1317, 1987.
7. W. Nakwaski and M. Osinski, "Current spreading in proton implanted vertical cavity top-surface emitting lasers," *Int. J. Optoelectron.*, Vol. 10, pp. 119–127, 1995.
8. M. Osinski, V. A. Smagley, G. A. Smolyakov, T. Svimonishvili, P. G. Eliseev, and G. Simonis, "Three dimensional simulation of oxide-confined vertical cavity surface emitting semiconductor laser," *Proc. SPIE*, Vol. 3419, pp. 196–207, 1998.
9. H. K. Bissessur, F. Koyama, and K. Iga, "Modeling of oxide-confined vertical cavity surface emitting lasers," *IEEE J. Select. Topics Quantum Electron.*, Vol. 3, no. 2, pp. 344–351, 1997.
10. A. Valle, J. Sarma, and K. A. Shore, "Spatial hole burning effects on the dynamics of vertical cavity surface emitting laser diodes," *IEEE J. Quantum Electron.*, Vol. 31, no. 8, pp. 1423–1431, 1995.
11. P. V. Mena, J. J. Morikuni, S. M. Kang, A. V. Harton, and K. W. Wyatt, "A comprehensive circuit level model of vertical cavity surface emitting lasers," *J. Lightwave Technol.*, Vol. 17, no. 12, pp. 2612–2631, 1999.
12. S. F. Yu, W. N. Wong, P. Shum, and E. H. Li, "Theoretical analysis of modulation response and second order harmonic distortion in vertical cavity surface emitting lasers," *IEEE J. Quantum Electron.*, Vol. 32, no. 12, pp. 2139–2147, 1996.
13. R. C. Tucker and D. J. Pope, "Circuit modeling of the effect of diffusion on damping in a narrow-stripe semiconductor lasers," *IEEE J. Quantum Electron.*, Vol. 19, pp. 1179–1183, 1983.
14. M. Ogura and H. C. Hsieh, "Effect of surface recombination velocity on the threshold current and differential quantum efficiency of the surface emitting laser diode," *IEEE J. Quantum Electron.*, Vol. 32, no. 4, pp. 597–606, 1996.
15. B. J. Thibeault, T. A. Strand, T. Wipiejewski, M. G. Peters, D. B. Young, S. W. Corzine, L. A. Coldren, and J. W. Scott, "Evaluating the effects of optical carrier losses in etched post vertical cavity lasers," *J. Appl. Phys.*, Vol. 78, no. 10, pp. 5871–5875, 1995.
16. N. K. Dutta, "Analysis of current spreading, carrier diffusion, and transverse mode guiding in surface emitting lasers," *J. Appl. Phys.*, Vol. 69, no. 5, pp. 1961–1963, 1990.
17. E. R. Hegblom, N. M. Margalit, B. J. Thibeault, L. A. Coldren, and J. E. Bowers, "Current spreading in apertured vertical cavity lasers," *Proc. SPIE*, Vol. 3003, pp. 176–180, 1997.
18. Y. Xiong and Y. H. Lo, "Current spreading and carrier diffusion in long wavelength vertical cavity surface emitting lasers," *IEEE Photon. Technol. Lett.*, Vol. 10, no. 9, pp. 1202–1205, 1998.

19. W. B. Joyce and S. H. Wemple, "Steady state junction current distributions in thin resistive films on semiconductor junctions," *J. Appl. Phys.*, Vol. 41, pp. 3818–3830, 1970.
20. J. H. Shin, H. E. Shin, and Y. H. Lee, "Effect of carrier diffusion in oxidized vertical cavity surface emitting lasers determined from lateral spontaneous emission," *Appl. Phys. Lett.*, Vol. 70, no. 20, pp. 2652–2654, 1997.
21. S. F. Yu, "Dynamic behavior of vertical cavity surface emitting lasers," *IEEE J. Quantum Electron.*, Vol. 32, no. 7, pp. 1168–1179, 1996.
22. D. Vakhshoori, J. D. Wynn, G. J. Zydzik, R. E. Leibenguth, M. T. Ason, K. Kojima, and R. A. Morgan, "Top-surface emitting lasers with 1.9 V threshold voltage and the effect of spatial hole burning on their transverse mode operation and efficiencies," *Appl. Phys. Lett.*, Vol. 62, no. 13, pp. 1448–1450, 1993.
23. C. H. Chong and J. Sarma, "Lasing mode selection in vertical cavity surface emitting laser diodes," *IEEE Photon. Technol. Lett.*, Vol. 5, no. 7, pp. 761–763, 1993.
24. Y. G. Zhao and J. G. McInerney, "Transverse-mode control of vertical cavity surface emitting lasers," *IEEE J. Quantum Electron.*, Vol. 32, no. 11, pp. 1950–1958, 1996.
25. K. Kojima, R. A. Morgan, T. Mullaly, G. D. Guth, M. W. Focht, R. E. Leibenguth, and M. T. Ason, "Reduction of p-doped mirror electrical resistance of GaAs/AlGaAs vertical cavity surface emitting lasers by delta doping," *Electron. Lett.*, Vol. 29, no. 20, pp. 1771–1772, 1993.
26. A. Mizutain, N. Hatoria, N. Nishiyama, F. Koyama, and K. Iga, "MOCVD grown InGaAs/GaAs vertical cavity surface emitting laser on GaAs(311)B substrate," *Electron. Lett.*, Vol. 33, no. 22, pp. 1877–1878, 1997.
27. H. Gebretsadik, P. K. Bhattacharaya, K. K. Kamath, O. R. Qasaimeh, D. J. Klotzkin, C. Caneau, and R. Bhat, "InP-based 1.5 μm vertical cavity surface emitting laser with epitaxially grown defect-free GaAs based distributed bragg reflectors," *Electron. Lett.*, Vol. 34, no. 13, pp. 1316–1318, 1998.
28. E. Hall, H. Kroemer, and L. A. Coldren, "AlAsSb-based distributed bragg reflectors using InAlGaAs as high-index layer," *Electron. Lett.*, Vol. 35, no. 5, pp. 425–427, 1999.
29. J. H. Baek, W. S. Han, H. K. Cho, B. Lee, and I. H. Choi, "Optically pumped continuous wave operation of InAlGaAs/InAlAs/InP based 1.55 μm vertical cavity surface emitting laser with SiO_2/TiO_2 dielectric mirror," *Electron. Lett.*, Vol. 35, no. 10, pp. 814–815, 1999.
30. K. Tai, L. Yang, Y. H. Wang, J. D. Wynn, and A. Y. Cho, "Drastic reduction of series resistance in doped semiconductor distributed bragg reflectors for surface emitting lasers," *Appl. Phys. Lett.*, Vol. 56, no. 25, pp. 2496–2498, 1990.
31. M. Sugimoto, H. Kosaka, K. Kurihara, I. Ogura, T. Numai, and K. Kasahara, "Very low threshold current density in vertical cavity surface emitting laser diodes with periodically doped distributed bragg reflectors," *Electron. Lett.*, Vol. 28, no. 4, pp. 385–387, 1992.
32. K. Kurihara, T. Numai, I. Ogura, A. Yasuda, M. Sugimoto, and K. Kasahara, "Reduction in series resistance of the distributed bragg reflector in vertical cavities by using quasi-graded supperlattices at the heterointerfaces," *J. Appl. Phys.*, Vol. 73, no. 1, pp. 21–27, 1993.

33. E. F. Schubert, L. W. Tu, G. J. Zydzik, R. F. Kopf, A. Benvenuti, and M. R. Pinto, "Elimination of heterojunction band discontinuities by modulation doping," *Appl. Phys. Lett.*, Vol. 60, no. 4, pp. 466–468, 1992.
34. G. Reinner, E. Zeeb, B. Moller, M. Ries, and K. J. Ebeling, "Optimization of planar Be-doped InGaAs VCSEL's with two-sided output," *IEEE Photon. Lett.*, Vol 7, no. 7, pp. 730–732, 1995.
35. C. Wilmsen, H. Temkin, and L. A. Coldren, *Vertical Cavity Surface Emitting Lasers—Design, Fabrication, Characterisation and Application*, Cambridge Univ. Press, Cambridge, U.K., 1999, Chapter 4.
36. E. Zeeb and K. J. Ebeling, "Potential barriers and current-voltage characteristics of p-doped graded AlAs-GaAs heterojunctions," *J. Appl. Phys.*, Vol. 72, no. 3, pp. 993–999, 1992.
37. Y. Anodo and T. Itoh, "Calculation of transmission tunneling current across arbitrary potential barriers," *J. Appl. Phys.*, Vol. 61, no. 4, pp. 1497–1502, 1987.
38. R. F. Nabiev and C. J. Chang-Hasnain, "Voltage drop in n- and p- type reflectors for vertical cavity surface emitting lasers," *IEEE Photon. Technol. Lett.*, Vol. 7, no. 7, pp. 733–735, 1995.
39. D. W. Winston and R. E. Hayes, "Optoelectronic device simulation bragg reflectors and their influence on surface emitting laser characteristics," *IEEE J. Quantum Electron.*, Vol. 34, no. 4, pp. 707–715, 1998.
40. M. Kurata, *Numerical Analysis for Semiconductor Devices*, Lexington Books, Lexington, MA, 1982.
41. G. Debaisieux, G. Herve-Gruyer, M. Filoche, S. Bouchoule, and J. F. Palmier, "Self-consistent 1-D solution of multiquantum well laser equations," *Opt. Quantum Electron.*, Vol. 29, pp. 651–660, 1997.
42. W. B. Joyce, "Analytic approximation for the Fermi energy of an ideal Fermi gas," *Appl. Phys. Lett.*, Vol. 53, no. 5, pp. 354–356, 1982.
43. P. S. Zory, *Quantum Well Lasers*, Academic Press, New York, 1993.
44. S. Furukawa, H. Matsumura, and H. Ishiwara, "Theoretical considerations on lateral spread of implanted ions," *Jpn. J. Appl. Phys.*, Vol. 11, pp. 134–142, 1972.
45. G. R. Hadley, K. L. Lear, M. E. Warren, K. D. Choquette, J. W. Scott, and S. W. Corzine, "Comprehensive numerical modeling of vertical cavity surface emitting lasers," *IEEE J. Quantum Electron.*, Vol. 32, no. 4, pp. 607–616, 1996.
46. A. Tsigopoulos, V. Paschos, C. Caroubalos, P. Salet, and J. Jacquet, "Numerical modeling of undercut ridge VCSEL's designed for CW operation at 1.3 μm: Design Optimization," *IEEE J. Quantum Electron.*, Vol. 33, no. 12, pp. 2221–2230, 1997.
47. J. Piprek, D. I. Babic, and J. E. Bowers, "Simulation and analysis of 1.55 mm double fused vertical cavity lasers," *J. Appl. Phys.*, Vol. 81, no. 8, pp. 3382–3390, 1997.

CHAPTER 7

Direct Modulation of Vertical Cavity Surface Emitting Lasers

Amplitude modulation response and second-harmonic distortion of VCSELs are derived analytically from a simple rate equation model. The transient response of multimode VCSELs is discussed and analyzed. In addition, the large-signal transient response of VCSELs such as gain switching is investigated. Comprehensive time-domain models of VCSELs including quasi-3D distribution of optical fields are also described. Finally, the effect of the hot carrier on the dynamic response of VCSELs is studied.

7.1 INTRODUCTION

Vertical cavity surface emitting lasers (VCSELs) are considered the key components in future high-speed optical fiber communication systems because of their potential capability for stable single-longitudinal-mode operation, extremely high intrinsic relaxation oscillation frequency, ultra-low-threshold current, and efficient coupling into optical fibers. The extremely high intrinsic relaxation oscillation frequency (ROF) of VCSELs is the result of high photon density inside the small laser cavity as the ROF is proportional to the square root of the photon density. It is demonstrated that the ROF of VCSELs can be as high as 71 GHz [1], but the maximum modulation bandwidth is limited to 16 GHz [2]. Because of the limitation of the modulation bandwidth, the transmission experiments (i.e., using 0.98-μm InGaAs/GaAs QWs proton-implanted/oxide-confined VCSELs) show that the maximum data rate has been restricted to 10 Gbits/s [3] and 12.5 Gbits/s [4] in a graded-index multimode fiber link 0.5 km long and in a standard single-mode fiber link 1.6 km long, respectively, with a minimum bit error rate of 10^{-11}. Therefore, in order to increase the transmission rate of optical fiber communication systems, obstacles that can occur in the modulation bandwidth of VCSELs have to be minimized.

Analysis and Design of Vertical Cavity Surface Emitting Lasers By Siu Fung Yu
ISBN 0-471-39124-7

The limitation of the maximum modulation bandwidth in VCSELs is attributed to the inherent nonlinearities inside the QW active region such as spatial hole burning (SHB) [5], carrier transport [20], carrier heating [6,7] and self-heating of the laser cavity (i.e., active region and distributed Bragg reflectors) [8]. This is because the differential optical gain of QW material, which is directly proportional to the modulation bandwidth of VCSELs, can be reduced significantly by these inherent nonlinearities. In addition, the modulation response of VCSELs is affected by the diffraction of light within the small aperture. This is because the reduction of reflectivity of distributed Bragg reflectors (DBRs) reduces the modulation bandwidth of VCSELs as a result of the increase in total cavity loss [9]. On the other hand, the increase in optical confinement due to self-focusing deteriorates the stability of transverse modes [10,11] as well as the modulation response [12,13] of VCSELs.

In the previous chapters, SHB, carrier transport, and self-heating as well as diffraction of light and self-focusing have been considered in the development of laser models. Using the models, the steady-state behavior of VCSELs including threshold current, quantum efficiency, and light/current curves have been investigated. In this chapter, analysis of the dynamic response of VCSELs is based on the theoretical models developed in the previous chapters. First, amplitude modulation (AM) response, second-harmonic distortion (SHD), and ROF of single-mode VCSELs are studied with the influence of SHB, carrier transport, and self-heating included in the calculation. Furthermore, the modulation response of multimode VCSELs is investigated. The large-signal transient response of VCSELs such as gain switching is also analyzed then a quasi 3D traveling-wave model is proposed to study the influence of light diffraction and the self-focusing effect in VCSELs. The traveling-wave model is solved numerically using the time-domain first-order difference approximation, which is based on the beam propagation method. Finally, the influence of carrier heating on the dynamic characteristics of VCSELs is investigated.

7.2 SMALL-SIGNAL MODULATION RESPONSE OF VCSELs

In the following paragraphs, a simple rate equation model is developed to analyze the modulation response of VCSELs. An airposted 0.98-μm InGaAs/GaAs QW VCSEL [14] is considered in the analysis, and the corresponding schematic of the lasers is shown in Figure 1.11b. In the model, it is assumed that (1) the p- and n-doped DBRs have equal core diameter in order to simplify our calculation, (2) the waveguide (core) region is surrounded by air and its effective refractive index equals 3.30, (3) the injection current is well confined and uniformly injected into the active region, and (4) only the fundamental mode is considered in the analysis, due to the influence of surface recombination loss. Using these assumptions, detailed derivation of AM and SHD responses of VCSELs is given. In addition, the proposed model is verified by comparison with the experimental results.

7.2.1 Derivation of AM Response and SHD of VCSELs

A simple rate equation model of VCSELs including the carrier transport effect can be written as [8]

$$\frac{\partial S}{\partial t} = \left(v_g \Gamma_z \langle g \rangle - \frac{1}{\tau_p}\right) S + \beta_s \Gamma_z B_{sp} N^2, \tag{7.1}$$

$$\frac{\partial N}{\partial t} = \frac{M}{\tau_s} - \frac{N}{\tau_n} - \frac{N}{\tau_{te}} - v_g g(N) |\Psi|^2 S + D_n \frac{1}{r}\frac{\partial}{\partial r}\left(r \frac{\partial N}{\partial r}\right), \tag{7.2}$$

$$\frac{\partial M}{\partial t} = \frac{J}{q L_w n_w} - \frac{M}{\tau_n} - \frac{M}{\tau_s} + \frac{N}{\tau_{te}}, \tag{7.3}$$

where S is the photon density; N and M are the carrier concentrations inside the active and separate confinement heterostructure (SCH) region, respectively; and Ψ represents the transverse distribution of the LP_{01} mode. In (7.2), the optical gain g of the QW active layer is expressed as a function of the carrier concentration

$$g(N) = a_N \log\left(\frac{N}{N_t}\right), \tag{7.4}$$

where a_N is the gain coefficient and N_t is the carrier concentration at transparency. $\langle g \rangle$ is the average optical gain and is given by

$$\langle g \rangle = \int_0^W \Psi(r) g(N) \Psi^*(r) r\, dr \bigg/ \int_0^\infty \Psi(r)\Psi^*(r) r\, dr, \tag{7.5}$$

where W is the core radius.

In (7.1)–(7.3), M and N are nonuniformly distributed along the transverse direction r. However, it is noted that the transverse distribution of the carrier concentration along the SCH layers is quite uniform. Furthermore, if the depletion of the carrier concentration (i.e., SHB) inside the active layer is less than 20% of its peak magnitude, it is reasonable to assume M is uniform along the transverse direction and the nonuniform distribution of N is approximated by a perturbation approach written as [15]

$$N = N_0 - N_k J_0(ar), \tag{7.6}$$

where a is a constant, N_0 is the average carrier concentration in the transverse direction, and N_k is the perturbation of the carrier concentration arising from the transverse optical profile. $J_0(ar)$ is the Bessel function of the first kind of zeroth order and satisfies $J_1(aW) = 0$. If N_k is smaller than N_0, the optical gain can be simplified to

$$g(N) = a_N \log\left(\frac{N}{N_t}\right) \approx a_N \left(\log\left(\frac{N_0}{N_t}\right) - \frac{N_k}{N_0} J_0(ar)\right). \tag{7.7}$$

Substituting (7.6) and (7.7) into (7.1)–(7.3) and integrating them along the transverse direction, the rate equations can be rewritten as

$$\frac{\partial S}{\partial t} = \left(v_g a_N \Gamma_z \left(\Gamma_r \log\left(\frac{N_0}{N_t}\right) - \frac{N_k}{N_0}\xi_r\right) - \frac{1}{\tau_p}\right) S + \beta_s \Gamma_z B_{sp} N_0^2, \quad (7.8)$$

$$\frac{\partial N_0}{\partial t} = \frac{M}{\tau_s} - \frac{N_0}{\tau'_{te}} - D_n \chi_1 N_k - v_g a_N \left(\Gamma_1 \log\left(\frac{N_0}{N_t}\right) - \frac{N_k}{N_0}\xi_1\right) S, \quad (7.9)$$

$$\frac{\partial N_k}{\partial t} = v_g a_N \left(\Gamma_2 \log\left(\frac{N_0}{N_t}\right) - \frac{N_k}{N_0}\xi_2\right) S + \left(D_n \chi_2 - \frac{1}{\tau'_{te}}\right) N_k, \quad (7.10)$$

$$\frac{\partial M}{\partial t} = \frac{J}{q L_w n_w} - \frac{M}{\tau'_s} + \frac{N_0}{\tau_{te}}, \quad (7.11)$$

where $1/\tau'_{te} = 1/\tau_n + 1/\tau_{te}$, $1/\tau'_s = 1/\tau_n + 1/\tau_s$ and the expression of other parameters Γ_r, ξ_r, Γ_1, ξ_1, Γ_2, ξ_2, χ_1, and χ_2 can be found in Section 7A.1. The self-heating effect can also be included using the auxiliary thermal equation $T = T_0 + R_{TH}(P_{IV} - P_{h\nu})/\kappa_T$. This is because the transient response of temperature is in order of microseconds, which is much slower than that of the modulation period so that the temperature can be assumed constant for VCSELs under small-signal modulation. See also Chapter 5 for more information on the implementation of the thermal rate Equation

The rate equations can be solved under small-signal assumptions. The injection current J, photon density S, carrier concentrations N_0 and N_k, as well as the carrier concentration inside the SCH layers M are expressed as the summation of DC and small-signal terms

$$J = J_s + J_1 e^{j\Omega t}, \quad (7.12a)$$

$$S = S_s + S_1 e^{j\Omega t} + S_2 e^{j2\Omega t}, \quad (7.12b)$$

$$M = M_s + M_1 e^{j\Omega t} + M_2 e^{j2\Omega t}, \quad (7.12c)$$

$$N_0 = N_{0s} + N_{01} e^{j\Omega t} + N_{02} e^{j2\Omega t}, \quad (7.12d)$$

$$N_k = N_{ks} + N_{k1} e^{j\Omega t} + N_{k2} e^{j2\Omega t}, \quad (7.12e)$$

where Ω is the angular modulation frequency and the small-signal conditions of $J_1 \ll J_s$, $(S_1, S_2) \ll S_s$, $(N_{01}, N_{02}) \ll N_{0s}$, $(N_{k1}, N_{k2}) \ll N_{ks}$ and $(M_1, M_2) \ll M_s$ are satisfied. By substituting (7.12) into (7.8)–(7.11), and neglecting the third-order small terms, the equations for the AM response and SHD of VCSELs can be deduced.

7.2.2 Simplified Expression of AM Response, SHD, and Resonant Oscillation Frequency

The analytic expressions of AM response and SHD deduced from (7.8)–(7.11) using (7.12) can be very complicated. In fact, the dependence of SHB on the

ROF can be observed from the simplified analytic expression of AM response and SHD, which is obtained by applying the quasistatic steady approximation on N_s to the rate equation model. Setting the time derivative of (7.10) to zero, for the reason that $(1 - D_n \chi_2 \tau'_{te})/\tau'_{te}$ in (7.10) has value in order of 100 ps and is short compared to the period of practical small-signal modulation waveforms [15], gives

$$N_k = \frac{\varepsilon \Gamma_2 \log(N_0/N_t) S}{1 + \varepsilon \xi_2 (S/N_0)}, \tag{7.13}$$

where $\varepsilon = v_g a_N \tau'_{te}/(1 - D_n \chi_2 \tau'_{te})$. The dependence of SHB on D_n is shown clearly in (7.13). For a negative value of χ_2, N_k (i.e., the depth of SHB) increases with the decrease of D_n but N_k is an increasing function of τ'_{te}. The influence of τ_s has not been included in (7.13) because M is uniformly distributed along the transverse direction. Now, the rate equations can be simplified by substituting (7.13) into (7.8) and (7.9), which for small β_s yields

$$\frac{\partial S}{\partial t} = \left(v_g a_N \Gamma_z \Gamma_r \log\left(\frac{N_0}{N_t}\right)\left(\frac{1 - \varepsilon_s S}{N_0}\right) - \frac{1}{\tau_p}\right) S + \beta_s B_{sp} N_0^2, \tag{7.14}$$

$$\frac{\partial N_0}{\partial t} = \frac{M}{\tau_s} - \frac{N_0}{\tau'_{te}} - v_g a_N \Gamma_p \log\left(\frac{N_0}{N_t}\right)\left(\frac{1 - \varepsilon_p S}{N_0}\right) S, \tag{7.15}$$

where $\varepsilon_s = \varepsilon(\Gamma_2 \xi_r/\Gamma_r + \xi_2)$, $\Gamma_P = \Gamma_1 + D_n \chi_1 \tau_{te} \Gamma_2$, and $\varepsilon_p = \varepsilon((\Gamma_2 \xi_r - \Gamma_1 \xi_2)/\Gamma_p + \xi_2)$. Hence, the AM response of VCSELs can be solved using (7.11), (7.14), and (7.15) with the small-signal assumptions, which is given by

$$\frac{S_1}{J_1} = \frac{1}{T_{12} \tau_{te} q L_z n_w} \times \frac{1}{\mathrm{Den}(\Omega)}, \tag{7.16}$$

$$\begin{aligned} \mathrm{Den}(\Omega) = {} & \frac{1}{\tau_s \tau_{te}}\left(\frac{1}{T_{11}} + j\Omega\right) - \frac{1}{T_{12} T_{21}}\left(\frac{1}{\tau'_s} + j\Omega\right) \\ & - \left(\frac{1}{T_{11}} + j\Omega\right)\left(\frac{1}{T_{22}} + j\Omega\right)\left(\frac{1}{\tau'_s} + j\Omega\right), \end{aligned} \tag{7.17}$$

and the SHD of VCSELs is equal to

$$\begin{aligned} \frac{S_2}{S_1} = {} & \{B_s(2\Omega)(1/T_{22}\tau'_s - 1/\tau_s \tau_{te} - 4\Omega^2 - 2j\Omega(1/T_{22} + 1/\tau'_s)) \\ & - B_p(2\Omega)(1/T_{12}\tau'_s + j\Omega/T_{12})\} \frac{S_1}{\mathrm{Den}(2\Omega)}. \end{aligned} \tag{7.18}$$

The expression of T_{11}, T_{12}, T_{21}, and T_{22} as well as B_s and B_p can be found in Section 7A.1, and the steady-state parameters of VCSELs can be calculated using the Newton–Raphson method [16].

The ROF ω_n of VCSELs can be obtained from (7.17) and is given by

$$\omega_n^2 \approx \frac{1}{T_{12}T_{21}} + \frac{1}{T_{11}T_{22}} + \frac{1}{\tau_s}\left(\frac{1}{T_{11}} + \frac{1}{T_{22}} - \frac{1}{\tau_{te}}\right), \tag{7.19}$$

where it is assumed that $\tau_s \approx \tau_s'$ in the derivation of (7.19). It is noted that ω_n is inversely proportional to τ_s if $1/T_{11} + 1/T_{22} > 1/\tau_{te}$. Hence, the ROF of VCSELs can be maximized by carefully designing the dimension of QW active and SCH layers such as $\tau_s \ll \tau_{te}$ and $1/T_{11} + 1/T_{22} \gg 1/\tau_{te}$ (i.e., $\tau_s < 10$ ps and $\tau_{te} > 400$ ps). If the terms representing the carrier transport effects (i.e., τ_{te} and τ_s) and the spontaneous emission rate are omitted, the ROF can be simplified to

$$\omega_n^2 \approx (v_g a_N)^2 \Gamma_r \Gamma_z \Gamma_p \log\left(\frac{N_{0s}}{N_t}\right)\left(\frac{S_s}{N_{0s}}\right)\left(1 - \varepsilon_s \frac{S_s}{N_{0s}}\left(3 - \log\left(\frac{N_{0s}}{N_t}\right)\right) + \frac{S_s}{N_{0s}}\right), \tag{7.20}$$

where $\varepsilon = v_g a_N \tau_n/(1 - D_n \chi_2 \tau_n)$, $\Gamma_p = \Gamma_1 + D_n \chi_1 \tau_n \Gamma_2 \approx \Gamma_1$ (i.e., as χ_1 is ~2.12), and $S_s \ll N_{0s}$ are used in the derivation of ω_n. In (7.20), the influence of SHB on ROF is introduced into (7.20) through ε_s as $3 > \log(N_{0s}/N_t)$ for the normal operating range of N_{0s}. Hence, SHB has less influence on the ROF of VCSELs as it is determined by the parameter ε_s, which is significant only at high S_s. It must be noted that the derivation of ω_n from (7.20) is valid only for $N_k < 0.2 \times N_0$. For the cases with severe SHB, exact calculation of the carrier concentration distribution is required.

7.2.3 Numerical Analysis of VCSELs

Figure 7.1a shows the measured light/current curves (dotted lines) of VCSELs [14]. It is shown that the output power is saturated at high injection current and the peak power is reduced with W. In addition, small-cavity VCSELs have faster rolloff light/current curves than large-cavity ones. Several effects are responsible for the reduction of output power as W is reduced: (1) the thermal resistance of VCSELs, which is inversely proportional to W (i.e., small devices suffer high thermal dissipation); (2) the size dependence of cavity loss due to the surface scattering and diffraction losses; and (3) the increase in threshold current density with the junction temperature. In fact, these phenomena can be introduced into the rate equation model through the normalized effective thermal resistance R_{TH}, the equivalent modal loss (i.e., τ_p), and the dependence of threshold current density on temperature, which can be approximated by the Arrhenius-type relation. See Chapter 5 for implementation of threshold current density using the Arrhenius-type approximation. Hence, the light/current curves of VCSELs can be calculated using the rate equation model developed in Section 7.2.2 with the size dependent parameters included in the calculation. The calculated light/current curves of VCSELs are also shown in Figure 7.1a (solid line). In the calculation, it is assumed that $D_n = 5\ \text{s}^{-1}\,\text{cm}^2$, $\tau_n = 3$ ps, $\tau_s = 10$ ps, $\tau_{te} = 200$ ps, and $\kappa_T = 0.45\ \text{W}\,\text{cm}^{-1}\,\text{K}^{-1}$. In addition, the expressions of optical gain and threshold

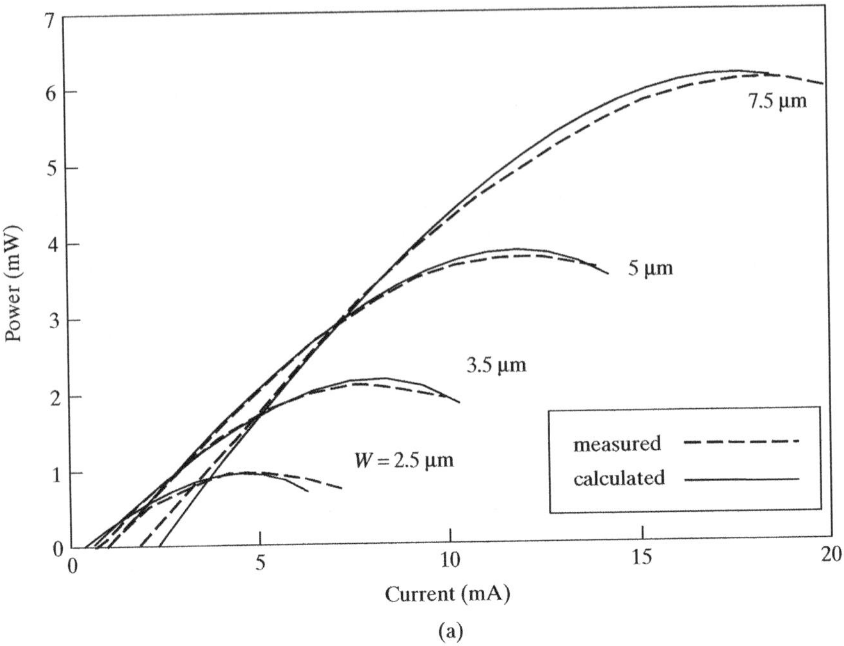

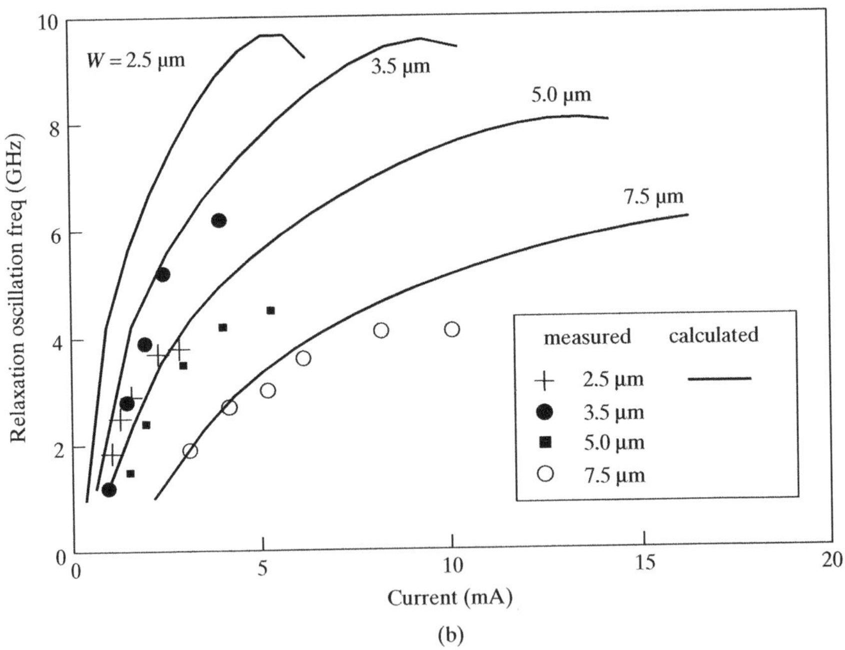

Figure 7.1 Comparison of measured and calculated (a) light/current curves and (b) relaxation oscillation frequency of VCSELs where W is 2.5, 3.5, 5.0, and 7.5 μm; it is assumed that $D_n = 5\ \mathrm{s^{-1}\ cm^2}$, $\tau_s = 10$ ps, and $\kappa_T = 0.45\ \mathrm{W\ cm^{-1}\ K^{-1}}$ in the calculation. (After Ref. 8.)

current density given in Chapter 5 are used in the model. However, the remaining size-dependent parameters R_{TH} and τ_p have to be deduced from the measured light/current curves. It can be shown that where W equals 2.5, 3.5, 5, and 7.5 μm, the corresponding value of R_{TH} (τ_p) is found to be 4500, 2600, 1660, and 990 cm^{-1} (1.71, 2.09, 2.68, and 3.25 ps), respectively. It is noted that both R_{TH} and $1/\tau_p$ exhibit an inverse relation with W, which indicated that the model of rate equations gives a satisfactory explanation on the physical phenomena of VCSELs. For example, the variation of the threshold current I_{th} with W can be explained by the following rate equation model

$$J_{th} \approx \frac{qL_w n_w N_{th}}{\tau_s'}, \tag{7.21}$$

$$\frac{1}{\tau_p} = v_g \Gamma_z a_N \log\left(\frac{N_{th}}{N_t}\right), \tag{7.22}$$

where J_{th} is the threshold current density and N_{th} is the carrier concentration at threshold. Equations (7.21) and (7.22), which are deduced from (7.1) and (7.3), respectively, represent the threshold condition of VCSELs. Obviously, the increase in W implies the decrease in N_{th} through τ_p so that the magnitude of J_{th} is also reduced. For airpost VCSELs, I_{th} is equal to AJ_{th}, where $A(=\pi W^2)$ is the junction area. Therefore, the decrease of J_{th} doesn't imply the decrease of I_{th}, as the reduction of J_{th} is compensated by the increase of A so that I_{th} can remain nearly constant for some range of W (i.e., 2.5–5 μm). However, the I_{th} increase for $W > 5$ μm due to the rate of change of $1/\tau_p$ is less than that of A. Hence, the discussion above explains the dependence of threshold current on the size of the laser cavity.

Figure 7.1b compares the calculated and measured ROF [14] of VCSELs. It must be noted that the ROF is calculated without optimizing the fitted parameters R_{TH} and τ_p (i.e., R_{TH} and τ_p are obtained from the empirical fitting of the light/current curves as given in Fig. 7.1a). In general, the measured results show that small-cavity VCSELs have higher ROF than do large devices because of the high photon density inside the small laser cavity. However, the ROF of the smallest laser ($W = 2.5$ μm) is not the highest (the $W = 3.5$ device has the highest ROF). This is attributed to the high optical loss of small VCSELs, which drives the threshold gain toward the QW saturation levels reducing the differential gain and hence the modulation efficiency [14]. This phenomenon has not been taken into account in the model. However, for devices with $W \geq 3.5$ μm, the agreement between the theoretical prediction and measured results (as shown in Fig. 7.1b) is satisfied considering the simplicity of the model. Therefore, the gain saturation effect due to the high optical loss has less influence on large-area devices ($W \geq 3.5$ μm) and is ignored in the calculation.

The influence of the carrier transport and SHB on the light/current curves and ROF of VCSELs is shown in Figures 7.2a and 7.2b, respectively. The nonlinearities are examined by changing D_n from 5 to 10 $cm^2\,s^{-1}$ and τ_s from 10 to 20 ps for devices with which W equals 3.5 and 7.5 μm. It is noted that the carrier

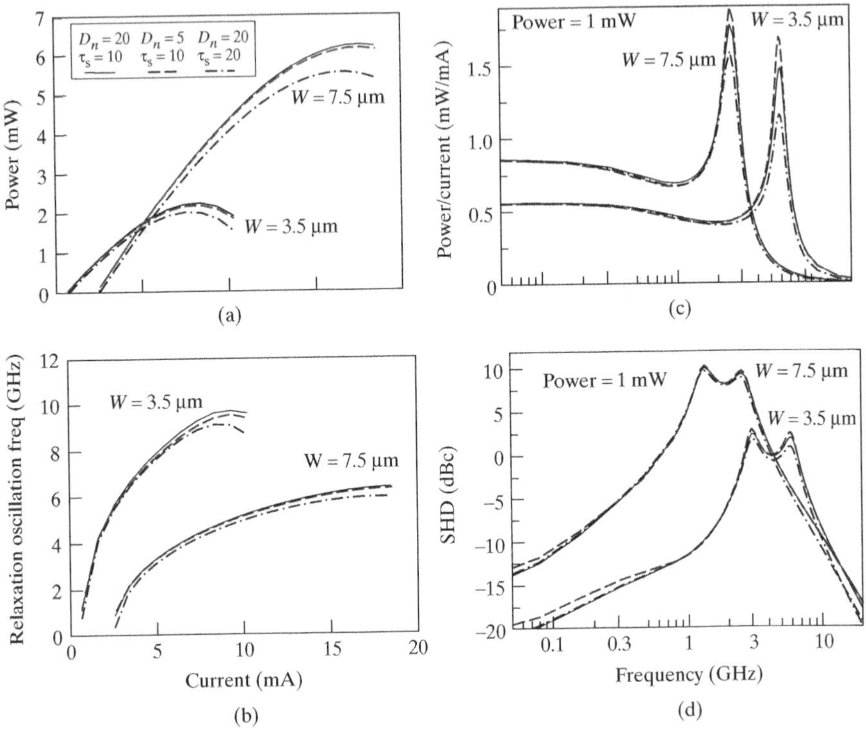

Figure 7.2 (a) Light/current curves and (b) relaxation oscillation frequency of VCSELs where W is 3.5 and 7.5 μm under the influence of D_n and τ_s; the corresponding amplitude modulation response (c) and second-harmonic distortion (d) are shown. The steady-state output power of the VCSELs is set to 1 mW. Solid lines—$D_n = 20\ s^{-1}\ cm^2$, $\tau_s = 10$ ps; dotted lines—$D_n = 5\ s^{-1}\ cm^2$, $\tau_s = 10$ ps; dashed lines—$D_n = 20\ s^{-1}\ cm^2$, $\tau_s = 20$ ps. (After Ref. 8).

transport and SHB has less influence on the static and dynamic behaviors of VCSELs. The corresponding AM response and SHD are shown in Figures 7.2c and 7.2d, respectively. It is assumed that the lasers are biased so that the output power is maintained at 1 mW. It is observed that a small device has a better modulation bandwidth than do large-area lasers and a VCSEL with small W exhibits less SHD. Two resonance peaks appear in SHD; the second peak is caused by resonance in the AM response, which will not occur if a constant optical modulation depth is considered [17].

As shown in the preceding paragraph, the static and dynamic behaviors of VCSELs are less dependent on the carrier transport and SHB but are determined mostly by the size of the laser cavity. This implies that the influence of self-heating is dominant over the carrier transport and SHB in single-mode VCSELs. In order to verify the preceding statement, the influence of thermal conductivity κ_T on the static and dynamic behaviors of VCSELs is analyzed. Figure 7.3a shows the light/current curves of devices with W equal to 3.5 and 7.5 μm and

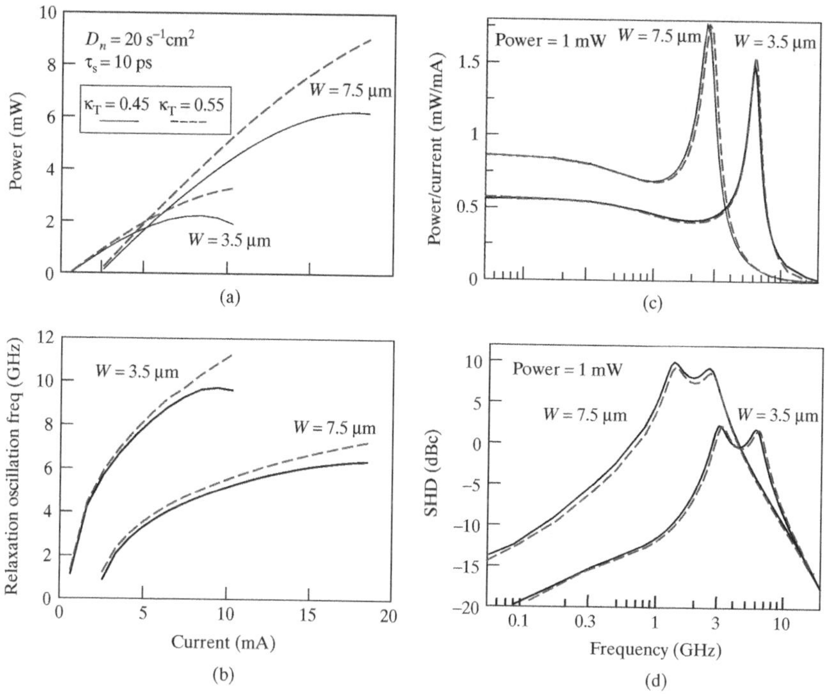

Figure 7.3 (a) Light/current curves and (b) relaxation oscillation with frequency of VCSELs where W is 3.5 and 7.5 μm under the influence of κ_T with the corresponding amplitude modulation response (c) and second-harmonic distortion (d). The steady-state output power of the VCSELs is set to 1 mW. Solid lines—$\kappa_T = 0.45$ W cm^{-1} K^{-1}; dotted lines—$\kappa_T = 0.55$ W cm^{-1} K^{-1}. (After Ref. 8.)

thermal conductivity varying between 0.45 and 0.55 W cm^{-1} K^{-1}. The other laser parameters, D_n and τ_s, are set to 20 cm^2/s and 10 ps, respectively. It is observed that the maximum output power is enhanced by several milliwatts. The corresponding ROF is also shown in Figure 7.3b. It is shown that the ROF is enhanced for both lasers but more so in a small cavity device. Slight increase of κ_T improves the saturated ROF by more than several gigahertz. The corresponding AM response and SHD under the influence of κ_T are also shown in Figures 7.3c and 7.3d, respectively. Again, the corresponding output power of the VCSELs is maintained at 1 mW. It is observed that κ_T has no negative effect on the modulation response of VCSELs. Therefore, the ROF and modulation bandwidth of single-mode VCSELs are strongly dependent on the self-heating effect.

7.3 SMALL-SIGNAL MODULATION RESPONSE OF MULTIPLE TRANSVERSE MODES IN VCSELs

Excitation of side modes is in fact an undesirable behavior of VCSELs. However, it is shown experimentally that multimode fiber interconnection systems using multimode VCSELs exhibit better immunity to intensity and modal noise than do those using single-mode VCSELs [18,19]. This is because multimode VCSELs act as a low coherent light source, which minimizes the coherence of the reflected signal so that the relative intensity noise and bit error rate can be reduced. In order to utilize multimode VCSELs in multimode fiber link, suitable control of transverse modes in VCSELs is required. Hence, it is necessary to carry out the study on the modulation response of multimode VCSELs. In the following paragraphs, the small-signal modulation response of broad-area VCSELs is analyzed. A simple multimode rate equation model explains the multimode modulation response of VCSELs observed from the experiment. It can be shown that SHB and the profile of the transverse modes have significant influence on the modulation response of VCSELs.

7.3.1 Modeling of Multiple Transverse Modes in VCSELs

The modulation response of VCSELs under direct modulation can be calculated by

$$\frac{\partial N}{\partial t} = \frac{J}{qd} - R_{\mathrm{t}}(N) - \nu_{\mathrm{g}} \sum_{l,m} g(N) S_{lm}(r, \theta) + D_n \nabla^2 N, \tag{7.23}$$

$$\frac{\partial S_{lm}}{\partial t} = \left(\nu_{\mathrm{g}} \Gamma_{z,lm} \langle g_{l,m} \rangle - \frac{1}{\tau_{\mathrm{p},lm}} \right) S_{lm} - \beta_{\mathrm{s}} \Gamma_{z,lm} B_{\mathrm{sp}} \langle N \rangle^2, \tag{7.24}$$

where the carrier transport and self-heating are omitted from the calculation for a simple reason. It must be noted that the carrier transport and self-heating vary only the magnitude of ROF [20] but the corresponding profiles of AM response and SHD will not be affected as indicated in Section 7.2. The rate equation model can be solved using the Runge–Kutta method [16] and the radial and azimuthal variation of the carrier concentration can be simplified using the approximation as given in Chapter 6. For example, if the current is injected through a disk and the modes considered are azimuthally independent, the azimuthal variation of N can be ignored in the calculation. In this case, the radial variation can be approximated by a Bessel series so that the rate equation of carrier concentration is transformed to a series of differential equations, which are radial independent. Alternatively, more accurate analysis on the radial distribution of the carrier concentration can be calculated numerically using the finite difference method and appropriated boundary conditions. In the following paragraph, the small-signal modulation response of a 20-μm-diameter proton implanted bottom emitting AlGaAs/GaAs QW VCSEL [21,22] is analyzed numerically. The schematic of the device structure can be found in Figure 1.10.

7.3.2 Numerical Analysis of Multiple Transverse Modes under Small-Signal Modulation

In broad-area gain-guided VCSELs, it can be shown that the excitation of higher order transverse modes with the increase of the injection current is caused by

(1) the increase in modal gain of higher-order modes and (2) the influence of self-focusing effects, which have resulted from the SHB of the carrier concentration and thermal lensing. These phenomena of transverse modes have been explained in Chapter 6 for the weakly-guided VCSELs.

In fact, it is observed experimentally that the fundamental mode (i.e., LP_{01} mode) evolves from the threshold but slightly increases above threshold and promotes a higher-order mode (i.e., most likely to be LP_{21} mode due to the large diameter of the active region and high bias). This is because the SHB increases the optical gain near the periphery of the active layer so that a higher-order mode is supported. In this case, the fundamental and higher order modes coexist above the threshold and a low overlapping transverse-mode pair is formed. The corresponding AM response exhibits two resonance peaks and a notch in between, which is believed to be caused by the low overlapping of the two transverse modes [22]. Figure 7.4 shows the calculated AM response of VCSELs with low overlapping between the LP_{01} and LP_{21} modes. In the calculation, the excitation of the two transverse modes is achieved by selecting a suitable injection current and D_n in the rate equation model. The AM response of VCSELs is calculated by

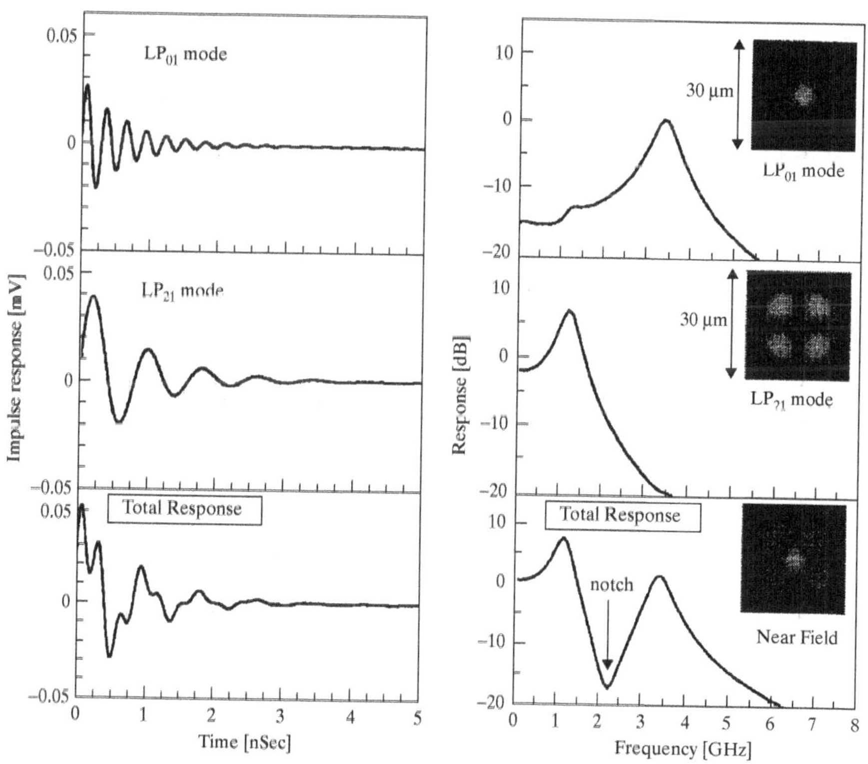

Figure 7.4 Calculated small-signal impulse and frequency response of LP_{01} and LP_{21} modes and of the overall lasing for a 20-μm-diameter VCSEL; insets are near-field profiles of the transverse modes. (After Ref. 22.)

considering the impulse response of the lasers so that the corresponding frequency response of each mode as well as for the overall output can be evaluated using fast Fourier transform (FFT). It is observed in the calculation that there is a mode-specific ROF, 3.38 GHz and 1.17 GHz for LP_{01} and LP_{21} modes, respectively, while the total response, the incoherent (intensity) superposition of them, results in two resonance peaks and a notch in between. It is shown that each mode evolves essentially as a single-mode laser, due to the low spatial overlap that has enabled each mode to interact independently with its own carrier pool. Hence, the AM response of VCSELs observed experimentally can be explained by the low overlapping between the two transverse modes (i.e., difference in modal gain).

In the same experiment, it is observed that the further increase in the injection current excites higher-order transverse modes, which have substantial overlapping with the fundamental mode. Furthermore, the corresponding AM response exhibits two resonance peaks but with no deep notch in between. This can be explained by the reduction of modal gain at the center of the core region due to SHB, which is accompanied by an increase of real refractive index [22]. The net result is a nonuniform cavity, which can support complex modes, some of which have a nonvanishing intensity in the center of the device. In order to verify this explanation, it is assumed that the simultaneous lasing of two modes: LP_{01} at the center of the device and a complex composed of LP_{21} and LP_{02} modes are partially overlapped, in the numerical investigation. Figure 7.5 shows the calculated AM response of the VCSELs with suitable selection of the injection current level and laser parameters as before. In contrast to the case of the low overlapping modes, here the response of each mode contains two resonance frequencies, which also appear in the total device response. The two resonance peaks are also less distinct relative to the case of the low overlapping modes, and no deep notch in between is observed at the total frequency response. Hence, it can be explained that the distinct AM response of VCSELs at high injection current is due to the partial overlapping of the transverse modes (i.e., slightly different in modal gain).

The transverse modes in gain-guided VCSELs (i.e., ion-implanted structure) are not likely to be highly overlapped because of the weak guiding mechanism of the transverse modes. In order to analyze the modulation response of VCSELs with highly overlapping modes, a device with index-guided structure (i.e., VCSEL grown by MBE over openings in a SiN_4 mask, deposited on a GaAs wafer with a rectangular geometry [23]) is considered in the investigation. It is observed experimentally that those two lasing modes, the TEM_{00} and TEM_{01} modes, are highly overlapping inside the laser cavity. Furthermore, the AM responses of the two modes exhibit the same resonance frequency, and the total frequency response is characterized by a single resonance frequency. Again, the observation above is analyzed using the multimode rate equation model with a suitable selection of the device's parameters and injection current level [22]. The AM response of VCSELs is calculated, and the corresponding results are shown in Figure 7.6. In the figure, a single resonant peak is observed in the AM response. Hence, the

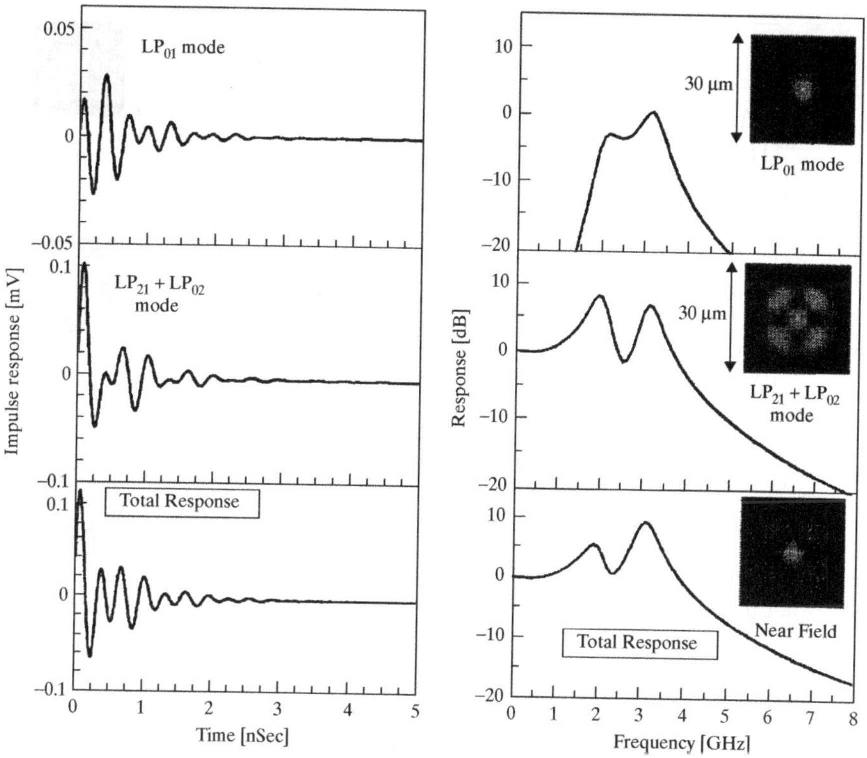

Figure 7.5 Calculated small-signal impulse and frequency response of LP_{01} and $LP_{21}+LP_{02}$ modes and of the overall lasing for a 20-μm-diameter VCSEL; insets are near-field profiles of the transverse modes. (After Ref. 22.)

high overlapping of transverse modes implies that the transverse modes interact dependently with one carrier pool (i.e., same modal gain) inside the laser cavity.

From the investigations described above, the influence of transverse modes on the AM response of VCSELs can be summarized as follows:

- For the case of low overlapping, the two transverse modes react as if they are two independent lasers within the same laser cavity and are fed by the same pumping source. This is because each mode interacts with a spatially distinct faction of the carrier distribution inside the active layer. Hence, two resonant frequencies and a deep notch in between are observed.
- For the case of a fundamental mode partially overlapping with higher-order modes, each mode exhibits two resonant peaks in the AM response. This coupled behavior is a result of the interaction between the optical modes and a share of the carrier concentration.
- For the device with mutually high overlap of transverse modes, the very strong coupling between the modes results in uniform carrier and photon

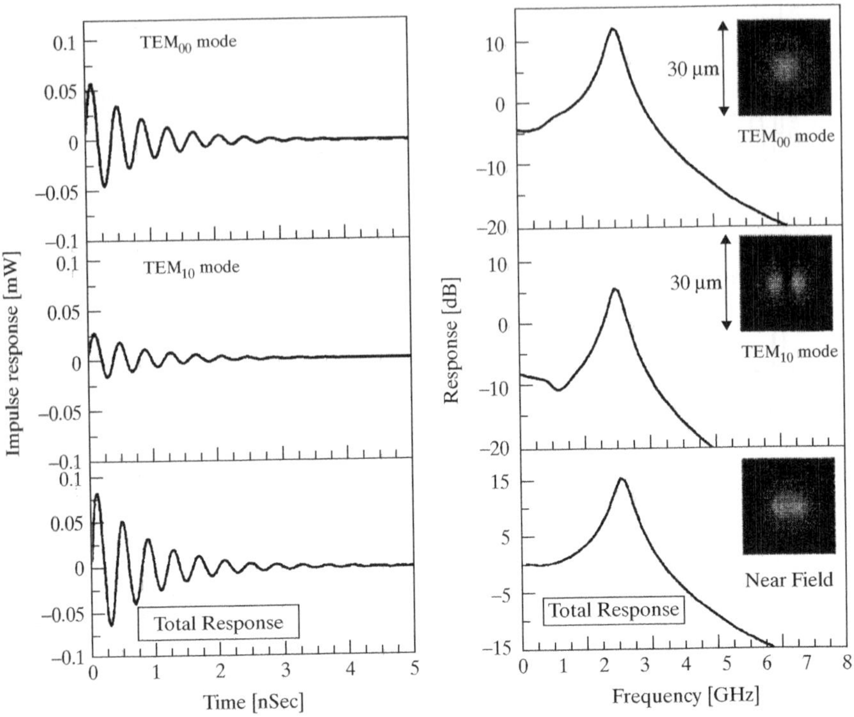

Figure 7.6 Calculated small-signal impulse and frequency response of TEM_{00} and TEM_{01} modes and of the overall lasing for a 15×15-μm^2 VCSEL; insets are near-field profiles of the transverse modes. (After Ref. 22.).

densities. The AM response exhibits a single resonant peak, which is very similar to that of a single mode laser.

Therefore, it is verified that the SHB as well as the profile of transverse modes have significant influence on the profile of AM response in multimode VCSELs. However, in single-mode VCSELs as discussed in the preceding section, the SHB varies only the modulation bandwidth, so the corresponding profile of AM response is not affected. Hence, this is the main difference between single-mode VCSELs and multimode VCSELs under small signal modulation response.

7.4 LARGE-SIGNAL MODULATION RESPONSE OF VCSELs

The small-signal modulation behavior of VCSELs with single- and multimode operations has been discussed above. In the following sections, the large-signal transient response of VCSELs under the excitation of electrical pulses modulation, sinusoidal modulation, and external injection of optical pulses are discussed and analyzed. Hence, a complete picture of the modulation response of VCSELs can be obtained.

7.4.1 Fast Electrical Pulse Modulation of VCSELs

In this section, the experimental results of the turnon transient response of a AlGaAs/GaAs QW VCSEL are presented. The gain-guided VCSEL under investigation has a current confinement structure achieved by proton implantation [24,25]. The device has top circular-ring contact for the injection of carriers and light emission from the top mirror. A similar schematic of the VCSELs used is shown in Figure 1.10. During the experiment, the laser is biased below threshold and modulated by electrically pumped square pulses of duration around 0.8 ns with 100 ps rise- and falltime, respectively.

Figure 7.7 shows the typical transient response of VCSELs with an 8 μm diameter of active layer biased at $0.4I_{th}$ and modulated by a square electrical pulse with amplitude of $20I_{th}$ [24]. Figure 7.7a shows the measured transient response of the Laguerre–Gaussian modes L_{00} and L_{10} [26], which are similar to the LP_{01} and LP_{11} modes, respectively, over the transverse direction. The fundamental L_{00} mode starts lasing, showing highly damped relaxation oscillation. After the onset of the fundamental mode for ~250 ps and the corresponding relaxation oscillation is already damped, the L_{01} mode (i.e., first-order mode) is excited because of the increase in modal gain that compensates for the cavity losses.

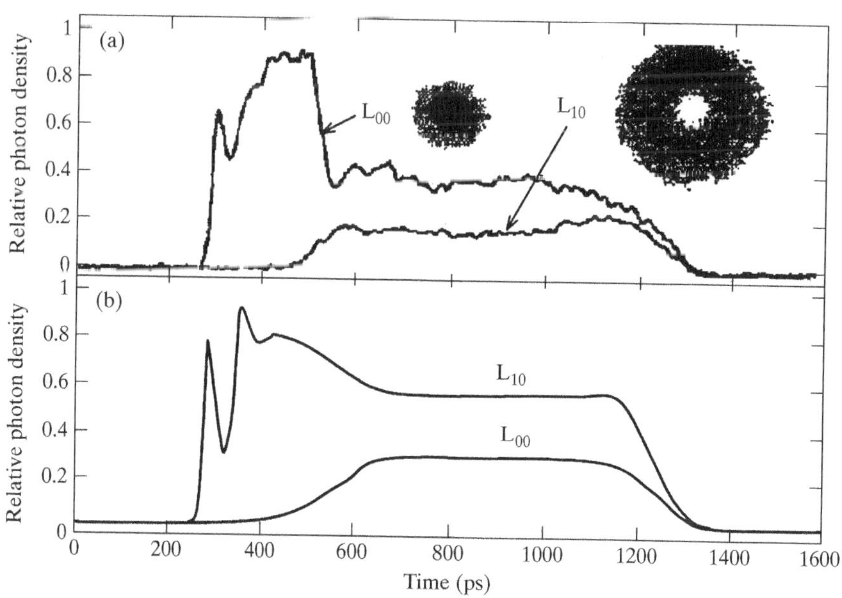

Figure 7.7 Transient response of L_{00} and L_{10} modes under electrical pulse modulation: (a) Measured and (b) calculated results; insets are near-field profiles of the transverse modes. (After Ref. 24.)

The presence of the first-order mode in the cavity produces a significant drop in the intensity of the fundamental mode, which suggests that both modes are sharing the available gain. In fact, this competition between the two modes is a result of SHB (i.e., overlap between the transverse distribution of optical gain and transverse modes). It can be shown that before the onset of the lasing mode, the carrier concentration is uniformly distributed along the active region. During this interval, the fundamental mode is excited because of its modal gain, which is higher than that of the higher-order modes, which is determined by the overlap between the transverse distribution of optical gain and transverse mode profile. For the time between 200 and 600 ps, the excitation of the fundamental mode causes the generation of SHB because of the fundamental mode's strong intensity at the center of the active region. Hence, the modal gain of the fundamental mode as well as its intensity drop so that the first-order mode is onset when the time equals 600 ps. This explains the dependence of the transient response of transverse modes on SHB and the reason for the turnon delay of the higher transverse mode. The transient behavior of VCSELs with multiple transverse modes can be modeled using (7.23) and (7.24), and the corresponding results are shown in Figure 7.7b.

The investigation described above is repeated but using VCSELs of different dimensions (i.e., diameter of top contact of 15, 18, 22, and 24 μm with top contact window diameter of 12, 15, 18, and 20 μm, respectively). The devices are biased at $0.85 I_{th}$ and excited with a 800-ps square electrical pulse with amplitude $4 I_{th}$. The transient response of VCSELs can be summarized as follows [25]:

- Multimode operation is observed for all VCSELs at this pumping condition, and the number of lasing modes increases with the diameter of the active region. This is because the modal gain of higher-order modes increases in larger-diameter VCSELs. In addition, the wavelength separation between the modes is observed to be inversely proportional to the diameter of the active region.
- For 15-μm VCSELs, three transverse modes are excited under electrical pulse modulation. The delay time for the onset of the first-order mode is 200 ps and then becomes dominant. The onset of the second-order mode is ~600 ps. The reason for such a mode competition was explained in the paragraph above. Similar dynamics are observed for 18-μm VCSELs, although up to three higher-order modes are excited with onset delays of 40, 136, and 630 ps, respectively.
- For 22- and 24-μm VCSELs, the turnon delay of higher-order modes is even shorter, giving rise to the simultaneous emission of several modes since the onset of the laser action. However, for the amplitude of electrical pulses higher than $4 I_{th}$, the sequence of the onset of transverse modes varies. The first-order mode will be excited first before the onset of the fundamental mode. This is because the circular-ring contact of VCSELs favors the excitation of higher-order modes at high injection level.

The preceding paragraphs explain the importance of SHB for the onset of transverse modes in VCSELs. It is noted that the delayed onset of higher-order modes in VCSELs depends on their excitation conditions and this behavior is qualitatively different from that of facet emitting lasers. This is because the delayed onset of higher-order modes in VCSELs is caused by the partial overlap of different transverse modes in the active region. The delay time τ_{delay} between the fundamental and higher-order modes can be roughly approximated by different transverse modes acting as independent lasers, which can be utilized to estimate the turnon delay of higher-order modes in large-area VCSELs (diameter >18 μm). Hence, $\tau_{\text{delay},i}$ of the ith mode can be written as [25]

$$\tau_{\text{delay},i} = \tau_{\text{eff}} \log\left(\frac{I/I_{\text{th},0} - 1}{I/I_{\text{th},0} - \Gamma_{\text{t},0}/\Gamma_{\text{t},i}}\right), \tag{7.25}$$

where τ_{eff} is the effective carrier lifetime, I is the amplitude of input current, $I_{\text{th},0}$ is the threshold current of the fundamental mode, and $\Gamma_{\text{t},0}$ and $\Gamma_{\text{t},i}$ are the transverse confinement factor of the fundamental and ith- order mode, respectively.

7.4.2 Electrical Gain Switching of VCSELs

One important application of high-speed modulation is the generation of short optical pulses of picosecond duration through gain switching. In this section, the gain switching properties of multimode VCSELs under sinusoidal modulation, including the influence of SHB, are studied numerically [27]. It is assumed that the index-guided VCSEL has a cylindrical geometry (see Fig. 1.13). The difference in refractive index between the core and cladding regions is set to 0.1, the radius of the core region is 4 μm, and the injection current is uniformly injected into the active layer.

The model given in (7.23) and (7.24) is used to analyze the large-signal modulation response of VCSELs. The sinusoidal modulation can be implemented into (7.23) by writing $J = J_{\text{b}} + J_{\text{m}} \sin(2\pi f_{\text{m}} t)$ for the injection current within the core region. $J_{\text{b}}(=0.95 \times J_{\text{th}})$ is the biased current, J_{m} $(-8 - 15 \times J_{\text{th}})$ and f_{m} $(=2.5$ GHz) are the modulation depth and frequency, respectively. To simplify the investigation, only LP modes are considered. The studies show that the SHB, geometry of disk contact and the profile of transverse modes have a significant influence on the pulsewidth of the generated pulses, which is summarized as follows [27]:

- The deep spatial hole (i.e., due to the small value of D_n) reduces the overlapping between the fundamental LP_{01} mode and the profile of the carrier concentration so that the mode sees effectively less gain compared with the case of less spatial hole. Hence, the corresponding pulsewidth is wider as a result of SHB. This infers that the pulsewidth differs between transverse modes because of their variation in modal gain (i.e., the overlapping between the transverse mode and the transverse distribution of the carrier concentration).

- It has been shown in Chapter 6 that the radius of the circular-disk contact also affects the profile of the carrier concentration. For the radius of a circular disk less than 4 μm (i.e., less than the radius of core region), the injected carriers will concentrate on the spatial hole of the carrier profile so that SHB can be reduced. Hence, the reduction of circular-disk contact from the dimension of the core region will minimize the pulsewidth of the generated pulse under gain switching with the increase in modal gain. However, further reduction of the radius below 1 μm will increase the pulsewidth because of a reduction in modal gain.
- For devices with a radius of disk contact 4 μm, two transverse modes, LP_{01} and LP_{11}, can be excited inside the laser cavity. If the devices are modulated with $J_m = 10 - 15 \times J_{th}$, gain-switched pulsetrains are observed. The calculation shows that the two modes exhibit mode competition at all values of modulation currents. In most situations, the onset of the LP_{11} mode is followed by the excitation of LP_{01}, which generates SHB. This is because significant SHB inside the active layer where the two transverse modes overlap leads to strong mode competition.

Therefore, SHB and the profile of transverse modes determine the pulsewidth as well as the excitation of the transverse modes generated by sinusoidal modulation.

7.4.3 Optical Gain Switching of VCSELs

VCSELs can generate short optical pulses optically pumped by a pulsed dye laser emitting a wavelength shorter than that of the generated short optical pulses. In the experiment, an AlGaAs/GaAs MQW VCSEL of lasing wavelength ~870 nm is excited by a pumped laser with wavelength ~790 nm and pulsewidth ≤6 ps. The pumped light is injected into VCSELs through the DBR, and the corresponding results are summarized in Figure 7.8, which shows the autocorrelation full width at half-maximum (FWHM) and peak-to-peak delay as a function of pump power [28]. The output pulsewidth and delay time clearly decrease with the increase of pump power but are saturated to 4 and 17 ps, respectively, at large pump power. In the experiment, the dependence of minimum FWHM and delay time on the pumped pulse and laser parameters has not been understood. Hence, it is necessary to develop a model that can accurately predict the output of VCSELs from a set of input conditions, so that a device design can be optimized for minimum FWHM and delay time.

In fact, the model of rate equations given in the preceding section can be used to analyze the dynamic behavior of VCSELs under optically pumped gain switching [28]:

$$\frac{\partial N}{\partial t} = \nu_g \frac{\alpha_{ab}(N)}{d} P_{pump}(t) - R_t(N) - \nu_g \frac{g(N)}{1 + \varepsilon S} S, \tag{7.26}$$

$$\frac{\partial S}{\partial t} = \left(\nu_g \Gamma_z \frac{g(N)}{1 + \varepsilon S} - \frac{1}{\tau_p} \right) S + \beta_s \Gamma_z B_{sp} N^2, \tag{7.27}$$

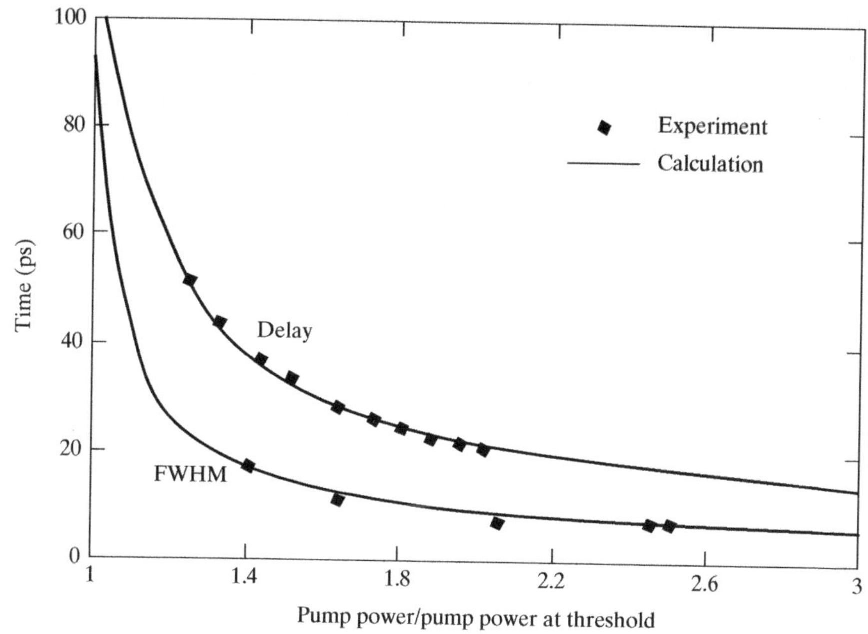

Figure 7.8 Experimental measurements of autocorrelation with full width at half-maximum (FWHM) and peak-to-peak delay versus normalized average input power; solid lines show the numerical simulation (after Ref. 28).

where α_{ab} ($\sim 1 - \exp(g(N) \cdot d)$ is the carrier-dependent absorption or bleaching, $P_{pump}(t)$ is the time-dependent pumped power, and ε is the gain suppression parameter required to be introduced into the optical gain because of the high output peak power of gain switching pulses. In this model, only the fundamental transverse mode is considered and the profile of the carrier concentration is assumed to be uniformly distributed along the transverse direction of the active layer for simple reasons. These two coupled differential equations can be easily solved by Runge–Kutta numerical integration [16]. Figure 7.9a shows the calculated time evolution of input pulse, carrier concentration, and output pulse at the region of constant FWHM and delay time as shown in Figure 7.8. It is assumed that the pumped pulse has a Gaussian temporal profile with pulsewidth less than 2 ps and high peak power so that the pulsewidth and delay time of the output pulse are minimized. From the diagram, it is observed that the leading edge of the output pulse is emitted coincidentally with the falling edge of the carrier concentration. This is because the carrier concentration rises as the integral of the pump pulse and then undergoes very slow spontaneous emission decay while the photon density of the output pulse is building up from a very low initial level. The value of the carrier concentration is at threshold at the time when the input and output pulses are at peak value. In addition, the output pulse is asymmetrical with a relatively fast rise and a slower decay tail.

In order to give a clear explanation of the dependence of output pulsewidth and delay time on the pumped pulses and parameters of VCSELs, the corresponding closed-form solutions need to be determined. This can be achieved by solving (7.26) and (7.27), including calculation of the particular transient characteristics of input pulse, carrier concentration, and output power. The corresponding approximations are shown in Figure 7.9b and are summarized as follows:

- At the beginning of the input pulse, there is no significant photon density (i.e., $S = 0$) and carrier concentration (i.e., $N = 0$ and $\alpha_{ab} \sim 1$ as $g \ll 0$). Hence, the rate equation of the carrier concentration can be simplified to

$$\frac{\partial N}{\partial t} = \nu_g \frac{\alpha_{ab}(N)}{d} P_{pump}(t) - R_t(N) - \nu_g \frac{g(N)}{1 + \varepsilon S} S \approx \frac{\nu_g}{d} P_{pump}(t). \quad (7.28)$$

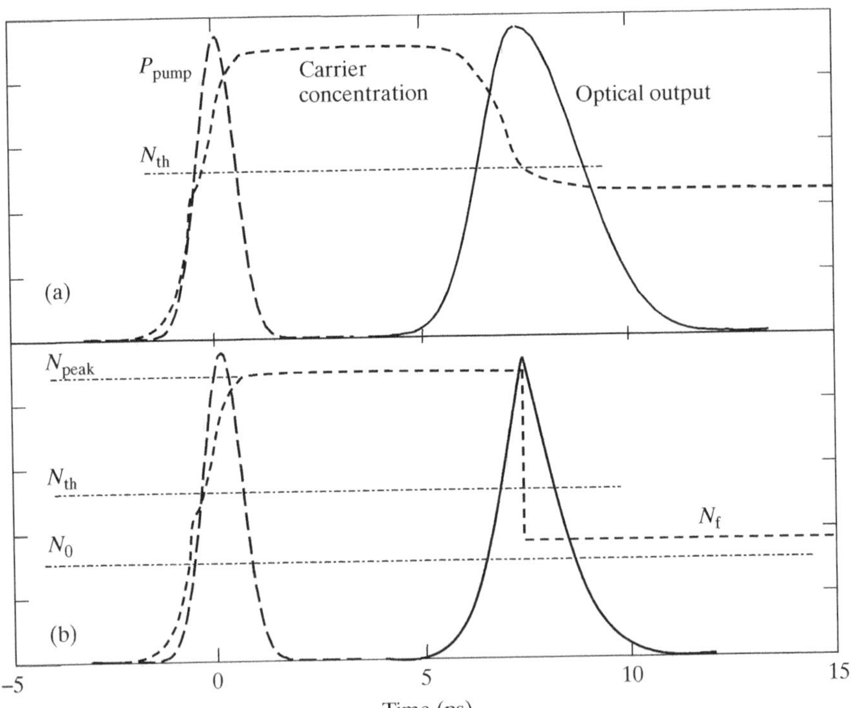

Figure 7.9 Graphical representation of (a) calculated gain switching dynamics and (b) dynamics showing approximation made to solve the rate equations in a closed form. (After Ref. 28.)

If a $(\text{sech})^2$ input pulse of FWHM T_{pump} is assumed, the peak carrier concentration generated is given by

$$N_{\text{peak}} = \frac{P_{\text{peak}} T_{\text{pump}} \nu_{\text{g}}}{0.882d}, \tag{7.29}$$

where P_{peak} is the peak power of the pumped pulse.

- The rising edge of the output pulse is assumed to occur after the falling edge of the input pulse so that a constant carrier concentration N_{peak}, remains until the peak of the output pulse occurs. In addition, the spontaneous emission term can be ignored in the excitation of output pulse so that the rate equation of photon density can be simplified to

$$\frac{\partial S}{\partial t} = \left(\nu_{\text{g}} \Gamma_z \frac{g(N)}{1+\varepsilon S} - \frac{1}{\tau_{\text{p}}}\right) S + \beta_{\text{s}} \Gamma_z B_{\text{sp}} N^2 \approx \left(\nu_{\text{g}} \Gamma_z g(N_{\text{peak}}) - \frac{1}{\tau_{\text{p}}}\right) S. \tag{7.30}$$

Hence, the corresponding solution to this first-order differential equation is given by [29]:

$$S(t) = S_0 \exp\left(\frac{t}{\tau_{\text{rise}}}\right), \tag{7.31}$$

where the rise time τ_{rise} is expressed as

$$\frac{1}{\tau_{\text{rise}}} = \Gamma_z \nu_{\text{g}} g(N_{\text{peak}}) - \frac{1}{\tau_{\text{p}}}. \tag{7.32}$$

The peak-to-peak delay occurs when S is maximized at time $t = \tau_{\text{p-p}}$. $S(t = \tau_{\text{p-p}})$ can be approximated by the difference in peak and threshold carrier concentration, provided the corresponding internal quantum efficiency is 100% (i.e., all the excess carriers convert to photon), this yields [28]

$$\tau_{\text{p-p}} = \frac{\log(S(\tau_{\text{p-p}})/S_0)}{\Gamma_z \nu_{\text{g}} g(N_{\text{peak}}) - 1/\tau_{\text{p}}} \approx \tau_{\text{p}} \frac{\log((N_{\text{peak}} - N_{\text{th}})/S_0)}{\tau_{\text{p}} \Gamma_z \nu_{\text{g}} g(N_{\text{peak}}) - 1}, \tag{7.33}$$

where S_0 is interpreted as one photon from spontaneous emission per entire active volume.

- If the output pulse is represented by a purely exponential rise and fall functions, the exponential risetime at half-height of the output pulse can be written as

$$\tau_{\text{rise}} \log(2) = (\Gamma_z \nu_{\text{g}} g(N_{\text{peak}}) - \tau_{\text{p}}^{-1})^{-1} \log(2). \tag{7.34}$$

Again, if the carrier concentration is approximated as a step function at the peak of the pump pulse, a constant value can be inserted into the photon rate equation to calculate the exponential photon decay. This implies that the photon density falltime is mostly a function of the final value of the carrier concentration N_{f}, rather than the decay rate of the carrier concentration. Hence, the exponential decay time at the half-height of output pulse can be derived from the falltime τ_{fall}, which is given by

$$\frac{1}{\tau_{\text{fall}}} = -\left(\Gamma_z v_g g(N_f) - \frac{1}{\tau_p}\right) \approx \frac{1}{\tau_p}. \tag{7.35}$$

Thus, the risetime is limited by input power, and the decay time is limited by photon lifetime. The entire expression for the FMHM becomes [28]

$$\begin{aligned}\tau_{\text{FWHM}} &= (\tau_{\text{rise}} + \tau_{\text{fall}})\log(2) \\ &= \tau_p \log(2)\left(1 - \frac{1}{\Gamma_z v_g g(N_{\text{peak}})\tau_p}\right)^{-1}.\end{aligned} \tag{7.36}$$

From (7.33) and (7.36), the dependence of delay time and pulsewidth on the pumped power can be explained quantitatively. It is noted that when the pump power is closed to the threshold, the denominators of (7.33) and (7.36) approach zero. This implies that the output pulse has a long delay time and wide pulsewidth. However at very high pump powers, the corresponding delay time is minimized and the pulsewidth of the output pulse approaches the photon lifetime of VCSELs. Therefore, the phenomena of VCSELs under optically gain-switched (i.e., Fig. 7.8) have been explained. Furthermore, it is noted that the critical design parameters are the device length and reflectivity of DBRs, since these will directly affect the internal photon lifetime and photon density of the laser cavity. Hence, for a given input pump power and pulsewidth, the constraints of the design are the cavity length and the reflectivity of the DBRs. The cavity length should, of course, be as small as possible without increasing the cavity losses significantly. When choosing the mirror reflectivities, however, their effect on the output power must be considered with respect to the design requirement. The mirrors must be designed carefully to allow minimum penetration depth and threshold without increasing the cavity lifetime or decreasing the power output to an unacceptable level.

7.5 COMPREHENSIVE MODELING OF DYNAMIC BEHAVIOR OF VCSELs

In the preceding sections, the transient response of VCSELs is analyzed using simple rate equation models. The transverse distribution of the carrier concentration and photon density has been taken into consideration so that the influence of mode competition can be analyzed under large- or small-signal modulation. However, the longitudinal distribution of optical fields under the influence of DBRs has been ignored in the analysis. It must be noted that the DBRs determine the threshold condition of the transverse modes under the influence of diffraction loss, with which the transient behavior of VCSELs can be modified significantly. In the following section, the approach to analyze the dynamic response of VCSELs is discussed with the longitudinal distribution of DBRs included in the calculation.

7.5.1 Time-Domain 1D and Quasi-3D Traveling-Wave Models of VCSELs

The time-dependent quasi-3D scalar traveling-wave equations along the laser cavity can be written as follows [12]:

$$\frac{1}{v_g}\frac{\partial \Psi^{\pm}}{\partial t} \pm \frac{\partial \Psi^{\pm}}{\partial z} = \frac{j}{2\beta}\left(\frac{1}{r}\frac{\partial}{\partial r}\left(r\frac{\partial}{\partial r}\right) + \frac{\partial^2}{\partial z^2} + \Delta\varepsilon(r, z, t)k_0^2\right)\Psi^{\pm}. \quad (7.37)$$

The small change in dielectric constant $\Delta\varepsilon$ as shown in (7.37) is given by

$$\Delta\varepsilon = n_2 \times \begin{cases} 2(\Delta n_b(r, z) + \Delta n(N)) + \dfrac{j(g(N) - \alpha_m)}{k_0} & \text{active layer} \\ 2\Delta n_b(r, z) - \dfrac{j\alpha_m}{k_0} & \text{elsewhere} \end{cases}, \quad (7.38)$$

where $\Delta n_b(r, z)$ is the difference of built-in refractive index distribution of the device, n_2 is the refractive index of the cladding region, and α_m is the effective material losses (including absorption and scattering losses) of the dielectric layers; $g(N)$ is the optical gain, Δn is the change of refractive index, and N is the carrier concentration inside the active layer.

Equation (7.37) can be transformed to 1D traveling-wave equations by separating the variables r and (z, t) so that the corresponding optical field can be expressed as [13]

$$\Psi^{\pm}(r, z, t) = \Psi(r; z)A^{\pm}(z, t), \quad (7.39)$$

where $A^{+}(z, t)$ and $A^{-}(z, t)$ are the slowly varying envelope of the forward and reverse traveling waves, respectively, along the longitudinal direction z and $\Psi(r; z)$ is the optical field distribution along the transverse direction. If (7.39) is used and the second derivatives of $A^{\pm}(z, t)$ with respect to z are neglected, (7.37) can be written as follows [13]:

$$\left(\frac{1}{v_g}\frac{\partial A^{\pm}}{\partial t} \pm \frac{\partial A^{\pm}}{\partial z}\right)\Psi = \frac{j}{2\beta}\left(\frac{1}{r}\frac{\partial}{\partial r}\left(r\frac{\partial \Psi}{\partial r}\right) + (\varepsilon(r, t; z)k_0^2 - \beta^2)\Psi\right)A^{\pm}. \quad (7.40)$$

This equation can be reduced to 1D traveling-wave equations in the z direction by the effective index method [30]. This can be done by assuming that the transverse field Ψ satisfies

$$\frac{1}{r}\frac{\partial}{\partial r}\left(r\frac{\partial \Psi(r; z)}{\partial r}\right) + (\varepsilon(r, t; z)k_0^2 - \beta_{\text{eff}}^2)\Psi(r; z) = 0, \quad (7.41)$$

where β_{eff} is the effective propagation coefficient in the transverse direction. Substituting (7.41) into (7.40) and integrating (7.40) over the transverse direction gives

$$\frac{1}{v_g}\frac{\partial A^{\pm}}{\partial t} \pm \frac{\partial A^{\pm}}{\partial z} = \frac{1}{2\beta}jk_0^2\Delta\varepsilon_{\text{eff}}A^{\pm}. \quad (7.42)$$

In (7.42), $\Delta\varepsilon_{\text{eff}}$ is defined as

$$\Delta\varepsilon_{\text{eff}} = \int_0^{2\pi}\int_0^{\infty} \Delta\varepsilon|\Psi(r;z)|^2 r\,dr\,d\theta \bigg/ \int_0^{2\pi}\int_0^{\infty} |\Psi(r;z)|^2 r\,dr\,d\theta\,. \quad (7.43)$$

7.5.2 First-Order Finite Difference Approximation of Traveling-Wave Equations

In general, the traveling-wave equations describing the propagation of the optical field inside the laser cavity can be written as

$$\frac{1}{v_g}\frac{\partial F^{\pm}(r,z,t)}{\partial t} \pm \frac{\partial F^{\pm}(r,z,t)}{\partial z} = M(r,z,t)F^{\pm}(r,z,t), \quad (7.44)$$

where $F^{\pm} = A^{\pm}$ or $\Psi^{\pm}$. Hence, the 1D (longitudinal direction) [13] and quasi-3D (longitudinal and transverse directions) [12] distribution of propagation fields can be calculated through $M(r,z,t)$. The corresponding expression of $M(r,z,t)$ is given by

$$M(r,z,t) = \frac{j}{2\beta} \times \begin{cases} \Delta\varepsilon_{\text{eff}}(z,t)k_0^2 & \text{for 1D} \\ \dfrac{1}{r}\dfrac{\partial}{\partial r}\left(r\dfrac{\partial}{\partial r}\right) + \Delta\varepsilon(r,z,t)k_0^2 & \text{for quasi-3D} \end{cases}. \quad (7.45)$$

Therefore, (7.45) can be used to describe 1D or quasi-3D distribution of the optical field inside the laser cavity depending on the choice of M.

Equation (7.44) can be solved by the time-domain traveling-wave method [31], which is shown in the following procedures:

- The first-order finite difference approximation is applied to the partial differential on the left-hand side (LHS) of (7.44), which is modified to

$$\left.\frac{\partial F^+}{\partial t}\right|_z \Delta t + \left.\frac{\partial F^+}{\partial z}\right|_t \Delta z \approx F^+(r, z+\Delta z, t+\Delta t) - F^+(r,z,t), \quad (7.46)$$

$$\left.\frac{\partial F^-}{\partial t}\right|_z \Delta t - \left.\frac{\partial F^-}{\partial z}\right|_t \Delta z \approx F^-(r, z-\Delta z, t+\Delta t) - F^-(r,z,t), \quad (7.47)$$

  where the relation between time and spatial step $\Delta z = v_g \cdot \Delta t$ has been used. See also Section 7A.2 for detailed derivation of (7.46) and (7.47).

- Using (7.46) and (7.47), (7.44) can be expressed as

$$\begin{bmatrix} \underline{F^+} \\ \underline{F^-} \end{bmatrix}_{z\pm\Delta z, t+\Delta t} = [B]_{z,t} \begin{bmatrix} \underline{F^+} \\ \underline{F^-} \end{bmatrix}_{z,t}. \quad (7.48)$$

In (7.48), $\underline{F}^+$ and $\underline{F}^-$ are vectors with n elements of F^+ and F^-, respectively, which represent the transverse variation of the optical field from the middle of the waveguide ($r = 0$) to the side of the laser and at a particular longitudinal position (i.e., at a particular z).

- The matrix $[B]_{z,t}$ given in (7.48) can be written as

$$[B]_{z,t} = [[I] + [M] \cdot \Delta z]_{z,t}, \tag{7.49}$$

where $[M]$ is a $n \times n$ matrix of M and $[I]$ is a $n \times n$ identity matrix. In (7.49), the convergence of the calculation requires small Δz. However, plane-wave approximation on the transverse fields can be applied to improve the computing efficiency. Hence, the matrix $[B]_{z,t}$ is written as

$$\begin{aligned}[B]_{z,t} = [[I] + [M] \cdot \Delta z + \tfrac{1}{2}[M]^2 \cdot \Delta z^2 \\ + \tfrac{1}{6}[M]^3 \cdot \Delta z^3 + \tfrac{1}{24}[M]^4 \cdot \Delta z^4 + \cdots]_{z,t}\end{aligned} \tag{7.50}$$

and the matrix $[M]$ is also interpreted as a $n \times n$ tridiagonal matrix.

- The influence of DBRs along the longitudinal direction of the laser cavity is introduced into the model using an approach similar to that shown in Chapter 3. This can be done by the continuity conditions (i.e., matching phase and amplitude of the propagation fields) at the boundary between the adjacent layers [32]. The continuity conditions between layers can be calculated using the method given in Sections 3.2.1 and 3.2.3 for 1D and quasi-3D propagation of traveling waves, respectively. See Section 7A.3 for detailed derivation of the continuity conditions between layers of the quasi-3D propagation of traveling waves.

As shown above, the time-domain algorithm has the advantages of easy implementation and implicit introduction of longitudinal modes into the calculation. However, the total execution time of the time-domain traveling-wave model of VCSELs is dependent on its propagation distance Δz and the total number of layers. This is because the decrease in Δz leads to an increase in the number of computational steps due to the requirement of a spatiotemporal step relation (i.e., $\Delta z = v_g \cdot \Delta t$). Furthermore, the DBRs are calculated by applying boundary conditions at the interface of the layers so that the value of Δz used in (7.50) should be equal to the physical thickness of the layers [12,13]. Therefore, the disadvantages of using the time-domain algorithm for VCSELs are (1) the number of computational steps depends on the total number of layers, (2) Δz depends on the thickness of the layers, and (3) Δz is too small. In addition, the execution time for the quasi-3D model will be much longer than that of the 1D model of VCSELs because of the existence of the transverse dimension. However, a time-dependent transfer matrix can be utilized to speed up the calculation of the 1D and quasi-3D models, and the implementation of the time dependent transfer matrix is given in Section 7A.4.

7.5.3 Implementation of Spontaneous Emission, Optical Gain Spectrum, and Carrier Concentration inside the Active Layer of VCSELs

The implementation of spontaneous emission profile, optical gain spectrum, as well as nonuniform distribution of the carrier concentration into the traveling-wave equations are discussed in this Section [13]. For the QW active layer, the peak of optical gain, which is a function of the carrier concentration, can be approximated by

$$g(N) = a_N \log\left(\frac{N}{N_\mathrm{t}}\right). \tag{7.51}$$

If gain suppression is required in the model, the optical power dependence of the peak of the optical gain has to be written as

$$g = \frac{a_N \log(N/N_\mathrm{t})}{1 + \varepsilon[|\Psi^+|^2 + |\Psi^-|^2]}, \tag{7.52}$$

where ε is the gain suppression factor. Because of the properties of the time-domain traveling-wave equations, the dependence of gain on the frequency can be modeled using a digital filter. A possible digital filter with unit peak gain is [33]

$$y_{t+\Delta t} = By_t + (1 - B)x_t, \tag{7.53}$$

where y_t and x_t are the output and input of the digital filter at time t, respectively, and B is a complex number with $|B| < 1$. The phase of B determines the peak gain frequency, and the magnitude of B determines the bandwidth of the digital filter. It is assumed that $B = b \cdot \exp(j\omega_\mathrm{p}\,\Delta t)$, where b is a fitted parameter and ω_p is the peak frequency. The frequency response of this filter is [33]

$$g(\omega) = \frac{(1 - b)^2}{1 + b^2 - 2b\cos((\omega - \omega_\mathrm{p})\Delta t)}. \tag{7.54}$$

The parameter B can be estimated by numerically fitting the calculated gain curve with this digital filter gain. The peak frequency ω_p is also a function of the injection carrier concentration and can be approximated by

$$\omega_\mathrm{p} = \omega_\mathrm{ps} + \Delta\omega_\mathrm{p}, \tag{7.55}$$

where ω_ps is a constant representing the peak gain frequency at transparency and $\Delta\omega_\mathrm{p}$ is the change of peak gain frequency due to the variation of the carrier concentration from transparency.

Spontaneous emission fields are coupled into the forward and reverse traveling waves inside the active layer. The spontaneous emission fields $U_\mathrm{sp}^{\pm}$ and $V_\mathrm{sp}^{\pm}$ can be added to the quasi-3D and 1D traveling-wave equations, respectively, in order to take into account the influence of spontaneous emission noise on the excitation of the cavity modes. The spontaneous fields $U_{sp}^{\pm}$ and $V_{sp}^{\pm}$ coupled into the longitudinal mode can be estimated with the following assumptions [13,31]:

- The spontaneous emission fields coupled to the forward and reverse waves have equal amplitude [i.e., $U_{sp}(r,t) = U_{sp}^{+}(r,t) = U_{sp}^{-}(r,t)$ and $V_{sp}(t) = V_{sp}^{+}(t) = V_{sp}^{-}(t)$].
- The emission has a Gaussian distribution and satisfies the correlation

$$\langle U_{sp}(r,t)U_{sp}^{*}(r,t')\rangle = \frac{\beta_s \Gamma_z R_{sp}\delta(r-r')\delta(t-t')}{v_g n_w L_w}, \qquad \langle U_{sp}(r,t)U_{sp}(r,t')\rangle = 0 \tag{7.56a}$$

$$\langle V_{sp}(t)V_{sp}^{*}(t')\rangle = \frac{\beta_s \Gamma_z \langle R_{sp}\rangle \delta(t-t')}{v_g n_w L_w}, \qquad \langle V_{sp}(t)V_{sp}(t')\rangle = 0 \tag{7.56b}$$

where δ is defined as the delta function and β_s is the spontaneous emission factor of the longitudinal mode. Γ_z is the confinement factor of the spontaneous emission light and is defined as the ratio between the area enclosed by spontaneous emission (i.e., U_{sp} or V_{sp}) and the total area of the active layer.

In (7.56), the spontaneous emission is coupled only to the traveling waves inside the QWs. The enhancement of the spontaneous emission inside the microcavity can calculated with appropriate estimation of β_s [34]. R_{sp} $(=B_{sp}N^2)$ is the spontaneous emission rate generated inside the active layer. $\langle R_{sp}\rangle$ can be obtained from R_{sp} through the following relation:

$$\langle R_{sp}\rangle = \frac{\displaystyle\int_0^{\infty} R_{sp}|\Psi(r)|^2 r\,dr}{\displaystyle\int_0^{\infty} |\Psi(r)|^2 r\,dr}. \tag{7.57}$$

If $R_{sp}(\omega_p)$ represents the peak amplitude of the spontaneous emission at an injection carrier concentration, the frequency-dependent spontaneous emission profile can also be modeled by a digital filter similar to the modeling of gain spectrum shown above.

The carrier distribution $N(r,t)$ along the transverse direction of the active layer can be represented from $r_1 = 0$ to $r_n = (n-1)\cdot\Delta r$ by n elements, where $\Delta r(=W_t/n)$ is the separation between two successive points in the r direction and W_t is the radius of the laser cavity, including the core and cladding regions. The derivative of the carrier concentration at the kth element (i.e., at $r = r_k$) can also be approximated by the finite difference

$$\left.\frac{1}{r_k}\frac{\partial N}{\partial r}\right|_{r_k} = \frac{N(r_{k+1},t) - N(r_k,t)}{(k-1)\Delta r^2} \tag{7.58}$$

$$\left.\frac{\partial^2 N}{\partial r^2}\right|_{r_k} = \frac{N(r_{k+1},t) - 2N(r_k,t) + N(r_{k-1},t)}{\Delta r^2} \tag{7.59}$$

for $k > 1$. The time variation of the carrier concentration at the kth element can be approximated by

$$\left.\frac{\partial N}{\partial t}\right|_{r_k} \Delta t = N(r_k, t+\Delta t) - N(r_k, t), \tag{7.60}$$

where Δt is identical to that used in (7.46) and (7.47) and the time variation of the carrier concentration is synchronized with the traveling waves. Substituting (7.58)–(7.60) into the rate equation of the carrier concentration [i.e., (7.25)], the time variation of carrier concentration can be written as

$$N(r_k, t+\Delta t) - N(r_k, t) = \Delta t \left\{ \frac{J(r_k,t)}{qd} - R_t(r_k,t) - v_g g(r_k,t)|\Psi(r_k,t)|^2 \right.$$
$$\left. + \frac{D_n}{\Delta r^2}\left(\frac{(N(r_{k+1},t) - N(r_k,t))}{k-1} + (N(r_{k+1},t) - 2N(r_k,t) + N(r_{k-1},t))\right)\right\}, \tag{7.61}$$

where the azimuthal variation of the carrier concentration is ignored. Finally, the time-dependent carrier rate equations are solved subject to the conditions that N and its derivatives are continuous everywhere and at the element $k = 1$, the first and second derivatives take the forms of

$$\left.\frac{\partial N}{\partial r}\right|_{r_1} = 0, \tag{7.62}$$

$$\left.\frac{\partial^2 N}{\partial r^2}\right|_{r_1} = \frac{2(N(r_2,t) - N(r_1,t))}{\Delta r^2}, \tag{7.63}$$

because of the circular symmetry of the carrier concentration. It is also required that as the total element number n increases to a very large value, the carrier concentration N reduces to zero. See also Chapter 6 for the application of different boundary conditions for VCSELs with different current confinement configurations.

7.5.4 Method of Self-Consistent Calculation and Numerical Results

From the sections above, the influence of the frequency dependence of DBRs and nonuniform distribution of the carrier concentration as well as optical gain and spontaneous emission spectra have been introduced into the model. The time variation of optical field and carrier concentration can be solved self-consistently, and the corresponding procedures are shown as follows:

1. The time variation of $F^{\pm}$ is solved by the traveling-wave Equations (7.48). This is done by advancing each $F^{\pm}$ from one layer interface to another. The propagation distance of $F^{\pm}$ is Δz_i, which corresponds to a time interval of Δt. The carrier concentration is also updated at the same time step by the rate equations of the carrier concentration (7.61).
2. At the same instant, boundary conditions are applied to $F^{\pm}$ at a layer interface to calculate the reflectivity of the alternating layers. Furthermore, the optical gain and spontaneous emission spectra are applied to $F^{\pm}$ at the node corresponding to the position of the active layer.
3. Steps 1 and 2 are repeated until the required time is reached.

Figure 7.10 shows the schematic of the self-consistent calculation of the traveling-wave equations by using the numerical integration of (7.48). The nodes in Figure 7.10a indicate the position of the interface. By knowing the initial field values at time t_0, the field for forward (reverse) at the next time step $t_0 + \Delta t(\Delta t = \Delta z_i/v_g)$ can be determined at the node $z_i + \Delta z_i(z_i - \Delta z_i)$ from (7.48). The boundary conditions at the interface, optical gain, and spontaneous emission are also applied to the nodes. By knowing the initial value of the carrier concentration, we can also determine the carrier concentration at the next timestep (see Fig. 7.10b).

In the following paragraphs, the transient response of an index-guided VCSEL (i.e., an AlGaAs/GaAs QW index-guided VCSEL) is calculated using the 1D and quasi-3D models. The structure of the VCSEL is similar to that given in Figure 1.13. It is assumed that the VCSEL has a built-in index-guided structure inside the DBRs as well as the active layer. The difference in refractive index between the core and cladding layers is about 0.005. The injection current is well confined inside the core region of the active layer. The models are examined by studying the transient response of VCSELs under step–current modulation:

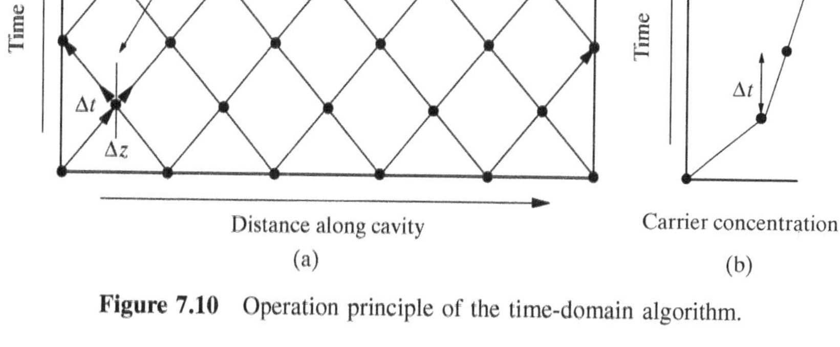

Figure 7.10 Operation principle of the time-domain algorithm.

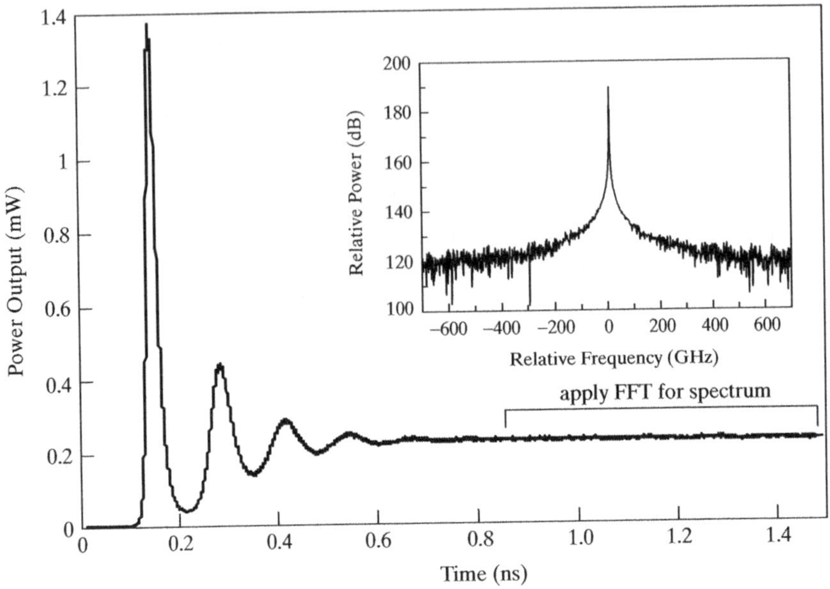

Figure 7.11 Time evolution of output power of VCSEL; inset diagram is the corresponding optical spectrum at steady state (after Ref. 13).

1. The 1D model is examined, and the transient response of VCSELs is shown in Figure 7.11. It is observed that the output power exhibits damped oscillation and settles down for more than 0.6 ns. The corresponding spectrum of the noisy optical field at steady state is also shown in Figure 7.11. The spectrum is calculated by transforming the time-dependent field using FFT with a Hamming window. Figure 7.12 plots the corresponding longitudinal field profile of the VCSEL at steady state, and the corresponding longitudinal distribution of effective refractive index is also shown in the figure. The slowly varying field distribution (i.e., $|A^+|^2 + |A^-|^2$) and standing-wave profile are also shown for comparison. It is observed that the long DBRs give high reflectivities so that the field is tightly confined within the active layer.
2. The quasi-3D model is also examined with the same longitudinal configuration of laser cavity, but no transverse confinement structure of injection carriers is assumed in the calculation. This can be modeled by (6.34) using a large magnitude of r_0. Figure 7.13a shows the switchon transient response of the transverse field distribution with r_0 equal to 4.0 μm. The corresponding time evolution of the carrier distribution is also shown in Figure 7.13b. In Figure 7.13a, a single-lobe profile in the transient response is observed during the first 1.5 ns of turnon, but a double-lobe transverse field distribution (first-order mode) dominates the remaining time interval. The switching of transverse modes occurs at the time equal to "$s1$" and is

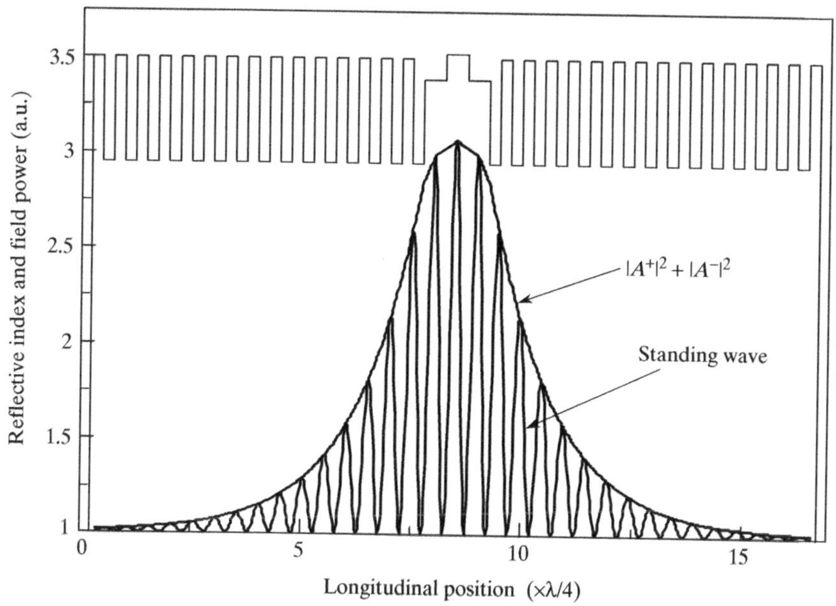

Figure 7.12 The longitudinal refractive index profile and slowly varying longitudinal optical field profile at steady state of the VCSEL.

caused by severe SHB of the carrier concentration. This is because the change of carrier distribution enhances the coupling efficiency of the first-order mode over the fundamental mode. The switchon transient response of the transverse field distribution for a device with r_0 equal to 0.02 μm is also calculated. It is noted that a single-lobe transverse field distribution (fundamental mode) is maintained during the whole time period. This is because the stimulated recombination of carrier density inside the core region is compensated for by the concentrated injection carrier distribution inside the core region and hence the SHB is not observed near the center of the core. Therefore, coupling efficiency between the carrier distribution and the first-order transverse mode is preserved at a low value during the transient period and single-transverse-mode operation is maintained at this injection profile.

In the model, the transient responses of the optical field and carrier concentration are calculated self-consistently simultaneously by solving the traveling-wave equation and rate equation of the carrier concentration. In fact, the influence of the thermal effect can be considered in the calculation by the thermal rate equation. Hence, the comprehensive model of VCSELs, including the optical field, carrier concentration, and temperature, can be developed. However, this proposed model has ignored the 3D distribution of electric potential and the flow of the current density so that the distribution of the carrier concentration and

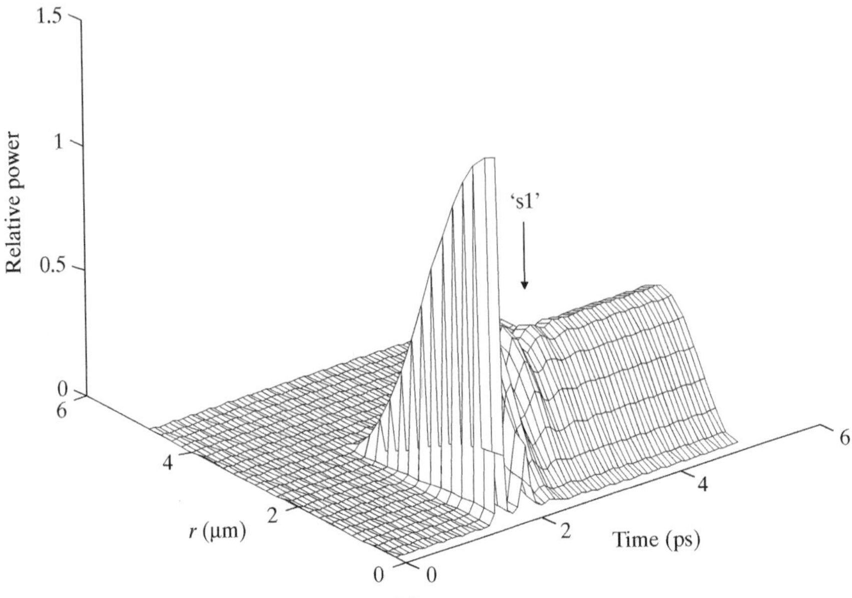

(a)

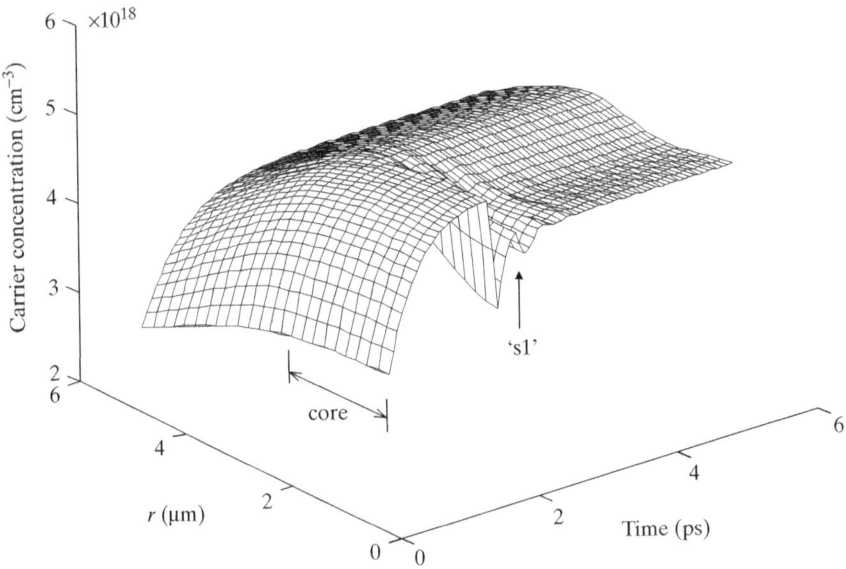

(b)

Figure 7.13 (a) Switchon transient response of the transverse field distribution and (b) carrier concentration under the modulation of step current with $D_n = 25\ s^{-1}\ cm^2$ and $r_0 = 4.0\ \mu m$. (After Ref. 12).

the temperature rise inside the laser cavity is only a first-order approximation. In order to obtain a fully comprehensive time-dependent numerical simulator of VCSELs [35], detailed electrical characteristics of the DBR as well as the active layer should also be considered (i.e., see also Chapter 6).

7.6 INFLUENCE OF CARRIER HEATING ON TRANSIENT RESPONSE OF VCSELs

In Section 7.2, the influence of the carrier transport, SHB, and self-heating on the modulation response of VCSELs has been discussed. However, the ultimate limitation of the maximum modulation bandwidth of QW lasers can be seriously affected by carrier heating (i.e., carrier scattering and relaxation) inside the QW active region. This is because the increase in carrier temperature above the cavity temperature changes the optical gain and differential gain of the active region. In fact, the phenomenon of carrier heating in VCSELs can be more serious than that of facet emitting lasers because of their small cavity size and high injection current density within the QW active region. Therefore, the influence of carrier heating on the dynamic performance of VCSELs should be determined [36].

Four major processes of the generation of hot carriers inside the QW active layer are described below [37, 38]:

- Injection heating rises from the energy difference between the barrier and QWs. The injected carriers must release their excess energy before reaching the lowest energy subband of QWs to participate in the lasing action. The excess energy will then be redistributed to the carriers inside the QWs by carrier–carrier scattering and thus inevitably cause an increase in carrier temperature.
- Stimulated recombination heating mostly occurs below the average energy of the injected carriers. Therefore, stimulated recombination deprives the carrier ensemble of its lowest-energy components and leaves the carrier system with a higher average energy and hence a higher carrier temperature.
- Nonradiative recombination heating is due to biomolecular and Auger recombination processes, which increase the energy and temperature of carriers through the excessive energy release during the recombination process.
- Free-carrier absorption describes the absorption of photon by free carriers through the nonlinear process involving phonons or impurities in order to satisfy the momentum conservation. Hence, the temperature as well as energy of free carriers increases as a result of the conversion process through the absorption of photon.

The carrier heating processes described above result in the hot carriers releasing their excess energy to phonons in order to keep the balance of energy flow. The major channel for carriers to relax their energy at room temperature is the emission of longitudinal optic (LO) phonons so that the hot LO phonons can relax their energy to the acoustic phonons (i.e., lattice phonons). Hence, the preceding explanation of the thermal dynamics of hot carriers and phonons inside the QW active region can be utilized to deduce the corresponding theoretical model, which is expected to solve the temperature of hot carriers, LO phonons, as well as acoustic phonons (i.e., lattice temperature) simultaneously.

The rate equations of hot carriers, hot phonons, and acoustic phonons can be written as [37]

$$\frac{\partial T_{\mathrm{c}}}{\partial t} = \left(\frac{\partial U_{\mathrm{c}}}{\partial T_{\mathrm{c}}}\right)^{-1} \left(\langle \Delta E_{\mathrm{c,inj}} \rangle \frac{J}{qL_{\mathrm{w}}} + \langle \Delta E_{\mathrm{c,stim}} \rangle \frac{v_{\mathrm{g}} g S}{1+\varepsilon S} + \langle \Delta E_{\mathrm{c,rec}} \rangle R_{\mathrm{t}} + \hbar\omega_0 v_{\mathrm{g}} \alpha_{\mathrm{c,ca}} S - \frac{\hbar\omega_{\mathrm{LO}}}{k_{\mathrm{B}}} \frac{N_{\mathrm{LO}}(T_{\mathrm{c}}) - N_{\mathrm{LO}}(T_{\mathrm{LO}})}{\tau_{\mathrm{c-LO}}} \right) \tag{7.64a}$$

$$\frac{\partial N_{\mathrm{LO}}(T_{\mathrm{LO}})}{\partial t} = \frac{N_{\mathrm{LO}}(T_{\mathrm{n}}) - N_{\mathrm{LO}}(T_{\mathrm{LO}})}{\tau_{\mathrm{e-LO}}} + \frac{N_{\mathrm{LO}}(T_{\mathrm{p}}) - N_{\mathrm{LO}}(T_{\mathrm{LO}})}{\tau_{\mathrm{h-LO}}} - \frac{N_{\mathrm{LO}}(T_{\mathrm{LO}}) - N_{\mathrm{LO}}(T_{\mathrm{L}})}{\tau_{\mathrm{LO}}}, \tag{7.64b}$$

$$\frac{\partial T_{\mathrm{L}}}{\partial t} = \frac{1}{C_{\mathrm{p}}} \nabla \cdot (\kappa_{\mathrm{T}} \nabla T_{\mathrm{L}}) + N \frac{\hbar\omega_{\mathrm{LO}}}{C_{\mathrm{p}}} \frac{N_{\mathrm{LO}}(T_{\mathrm{LO}}) - N_{\mathrm{LO}}(T_{\mathrm{L}})}{\tau_{\mathrm{LO}}}, \tag{7.64c}$$

where $c = n$ or p, T_{n} (T_{p}) is the temperature of electrons (holes), $\alpha_{n,ac}$ ($\alpha_{p,ac}$) is the free-carrier absorption of the electron (hole), N_{LO} is the population density of LO phonons, and ω_{LO} is the frequency of the LO phonon; $\langle \Delta E_{\mathrm{c,inj}} \rangle$, $\langle \Delta E_{\mathrm{c,stim}} \rangle$, and $\langle \Delta E_{\mathrm{c,rec}} \rangle$ are the average energy variations associated with the injection heating, stimulated recombination heating, and nonradiative recombination heating, respectively; U_{c} is the carrier energy density, T_{L} is the lattice temperature, κ_{T} is thermal conductivity, C_{p} is the specific heat capacity of the lattice, and $\tau_{\mathrm{c-LO}}$ and τ_{LO} are the relaxation time constants of the hot carriers and hot phonons, respectively.

The first four terms on the RHS of (7.64a) represent the total input power into the QW active layer due to injection heating, stimulated recombination heating, nonradiative recombination heating, and free absorption heating, with which the temperature of the carriers is increased. However, the last term on the RHS of (7.64a) represents the loss of carrier temperature as the energy of hot carriers is relaxed toward the LO phonons. Hence, (7.64a) gives the net rate of change of carrier temperature inside the QW active region. The first two terms on the RHS of (7.64b) represent the increase in phonons due to the relaxation of hot carriers. The last term on the RHS of (7.64b) represents the loss of phonons due to their relaxation toward the acoustic phonons. Hence, (7.64b) describes the rate of change of LO phonons due to the influence of hot carriers and acoustic phonons. If a single mode representation of LO phonons is used, the population distribution of LO phonons N_{LO} can be defined by the Bose–Einstein distribution function:

$$N_{\mathrm{LO}}(T) = \left(\exp\left(\frac{\hbar\omega_{\mathrm{LO}}}{k_{\mathrm{B}} T} \right) - 1 \right)^{-1}. \tag{7.65}$$

By this equation, the operational temperature of LO phonon (T_{LO}), which is not in thermal equilibrium, can be defined. Hence, the first two terms on the RHS of (7.64b) can be interpreted as the increase of phonon temperature T_{LO} due to the relaxation of hot carriers. In addition, the last term on the RHS of (7.64b) is interpreted as the loss of phonon temperature T_{LO} toward the lattice temperature T_{L}. The LO phonon lifetime τ_{LO} depends on the lattice temperature T_{L} and can be expressed as [39]

$$\tau_{\mathrm{LO}}(T_{\mathrm{L}}) = \tau_{\mathrm{LO},0}\left(1 + \left(\exp\left(\frac{0.65\hbar\omega_{\mathrm{LO}}}{k_{\mathrm{B}}T_{\mathrm{L}}}\right) - 1\right)^{-1} + \left(\exp\left(\frac{0.35\hbar\omega_{\mathrm{LO}}}{k_{\mathrm{B}}T_{\mathrm{L}}}\right) - 1\right)^{-1}\right)^{-1}, \tag{7.66}$$

where $\tau_{\mathrm{LO},0}$ is the lifetime of LO phonons at lattice temperature 0 K. Equation (7.64c) is the thermal heat equation, which is similar to that given in Chapter 5 except the heatsource is due to the relaxation of phonon energy. This equation is valid under the condition that the mean free path of the acoustic phonons is shorter than the characteristic dimension of devices. For a small-cavity VCSEL, the phonon Boltzmann equation may be necessary for accurately discussing the thermal conduction process of acoustic phonons.

The preceding rate equation model of hot carriers and hot phonons has to be solved simultaneously with the rate equations of the carrier concentration N and photon density S as shown below

$$\frac{\partial N}{\partial t} = \frac{J}{qL_{\mathrm{w}}} - R_{\mathrm{t}}(N) - \nu_{\mathrm{g}}\frac{g(N, T_{\mathrm{c}})}{1 + \varepsilon S}S, \tag{7.67a}$$

$$\frac{\partial S}{\partial t} = \left(\nu_{\mathrm{g}}\Gamma_z\frac{g(N, T_{\mathrm{c}})}{1 + \varepsilon S} - \frac{1}{\tau_{\mathrm{p}}}\right)S + \beta_{\mathrm{s}}\Gamma_z B_{\mathrm{sp}}N^2, \tag{7.67b}$$

where a single optical mode and uniform distribution of the carrier concentration rate equations are used in the analysis for simple reasons. The material gain g is also a function of the carrier temperature T_{c} and carrier Fermi level E_{Fc}, where $c = n$ or p.

This rate equation model, including the carrier heating effect, has been applied to analyze the carrier temperature of a 0.85-μm $\mathrm{Al_{0.3}Ga_{0.7}As/GaAs}$ SCH single-quantum-well (SQW) laser [37]. Using (7.64) and (7.67), it can be shown that the carrier temperature at steady state is dominated by the energy relaxation time of LO phonons. It is noted that if $\tau_{\mathrm{c-LO}}$ changes from 1 to 0.1 ps, the carrier temperature changes less than 10%. However, the slight reduction of τ_{LO} from 8 to 4 ps decreases the carrier temperature by 20% [37]. This clearly indicates that reducing the energy relaxation time of hot LO phonons will lower the energy relaxation time of hot carriers. Therefore, reducing the carrier relaxation time of LO phonons can enhance the ROF and modulation bandwidth of a QW laser. This is because ROF and the modulation bandwidth is directly proportional to the differential optical gain, which is an increasing function with the reduction of carrier temperature. In fact, there is experimental evidence that the p doping will reduce the energy relaxation time of LO phonons, and this may partially explain why the largest measured modulation bandwidth of QW lasers was achieved with p doping [40]. If the energy relaxation time of LO phonons has been optimized, the modulation response of QW lasers will be determined by the energy relaxation time of hot carriers. The influence of carrier heating on the modulation response of QW lasers has been investigated theoretically, such as AM response [6], ROF, and frequency chirp [38]. In general, the presence of hot carriers decreases the ROF but increases the damping rate of the QW lasers. Figure 7.14 shows the modulation response of the SQW laser. It is noted that the modulation response shows a significantly low frequency rolloff for a very large energy relaxation time ($\tau_{c\text{-}LO} > 20$ ps) [6]. This indicates that the carrier heating is more dominant than carrier transport and SHB on the low-frequency rolloff of the AM response.

The transient response of a GaAs/AlGaAs QW VCSEL under the influence of hot carriers and LO phonons is also analyzed experimentally [41–43]. In these studies, Wegener and co-workers used a VCSEL with a diameter of 8 μm and a threshold current of 2.9 mA. In addition, at a biased current of 5 mA, they observed a single-mode emission at 0.85 μm (i.e., ~1.459 eV photon energy) with 1 mW continuous-wave (CW) output. The experiment shows that the perturbation of an electrically pumped VCSEL at room temperature using a short (femtosecond

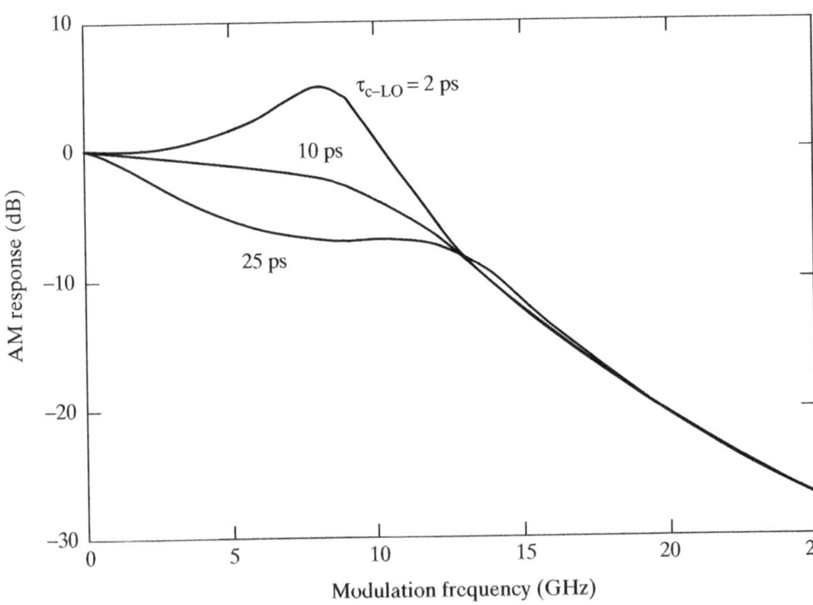

Figure 7.14 The AM modulation response of QW lasers of different $\tau_{c\text{-}LO}$ (after Ref. 6).

~120 fs FWHM) optical pulse can result in an unusually rapid switchoff of lasing emission. Figure 7.15a shows the typical transient response of VCSELs under penetration of an optical pulse with photon energy of ~1.73 eV [43]. The optical pulse is injected into the VCSEL at time $t = 0$, and a switchoff of lasing emission is observed at some biased current level. In fact, the switchoff gradually vanishes for large biased currents, and at injection bias equaling 5 mA, the switchoff mechanism disappeared. On the other hand, with decreasing pump pulse energy or decreasing photon energy, the switchoff behavior gradually disappeared. In this experiment, it is confirmed that the mechanism of switchoff is not due to the reduction of reflectivity arising from the absorption of light inside the DBRs. This mechanism is obviously distinctly different from the usual laser relaxation oscillation behavior due to the presence of carrier heating.

Figure 7.15b shows the corresponding numerical simulation obtained from (7.64) and (7.67). It is found that after the injection of the short pump pulse, the carrier concentration inside the active layer increases by ~7%. In the calculation, the carrier temperature experiences a substantial increase due to the excess energy of carriers (~275 meV >> $k_B T_c$). In this situation, the decrease in optical gain is affected mainly by the increase in carrier temperature, which overcompensates the increase in optical gain by the extra carrier concentration and the system is switched into absorption regime. Hence, this leads to a rapid switchoff of the laser emission. Hot carriers, however, quickly cool down in a timescale of τ_{c-LO} (i.e., ~1 ps) and the laser emission returns. It is noted that under the present biased conditions, the ROF is always less than 10 GHz, which is much slower than that of the initial switchoff transient response as shown in Figure 7.15a. Therefore, the experimental data cannot explain the fast transient response of VCSELs without including hot carriers. On the other hand, the increase in biased current ceases the rapid switchoff transient response of VCSELs under the penetration of optical injection. This is because the optical gain is dominated by the injection carrier concentration at a high biased level and the influence of the hot carrier is negligible. It must be noted that the transient response of VCSELs under the influence of carrier heating is quite different from that of self-heating. This is because the response time of VCSELs under the affect of carrier heating is in the picosecond range, but that of self-heating is in the microsecond range [44].

7.7 SUMMARY

This chapter explains the derivation of theoretical models, based on some known measurement results, for the analysis of the modulation response of VCSELs. Although the derivation of the laser models focuses on some particular configurations of VCSELs, other types of laser can also be modeled with slight modification.

In this chapter, the AM response and SHD of airpost VCSELs are modeled using simple rate equations. The influence of SHB, carrier transport, as well as

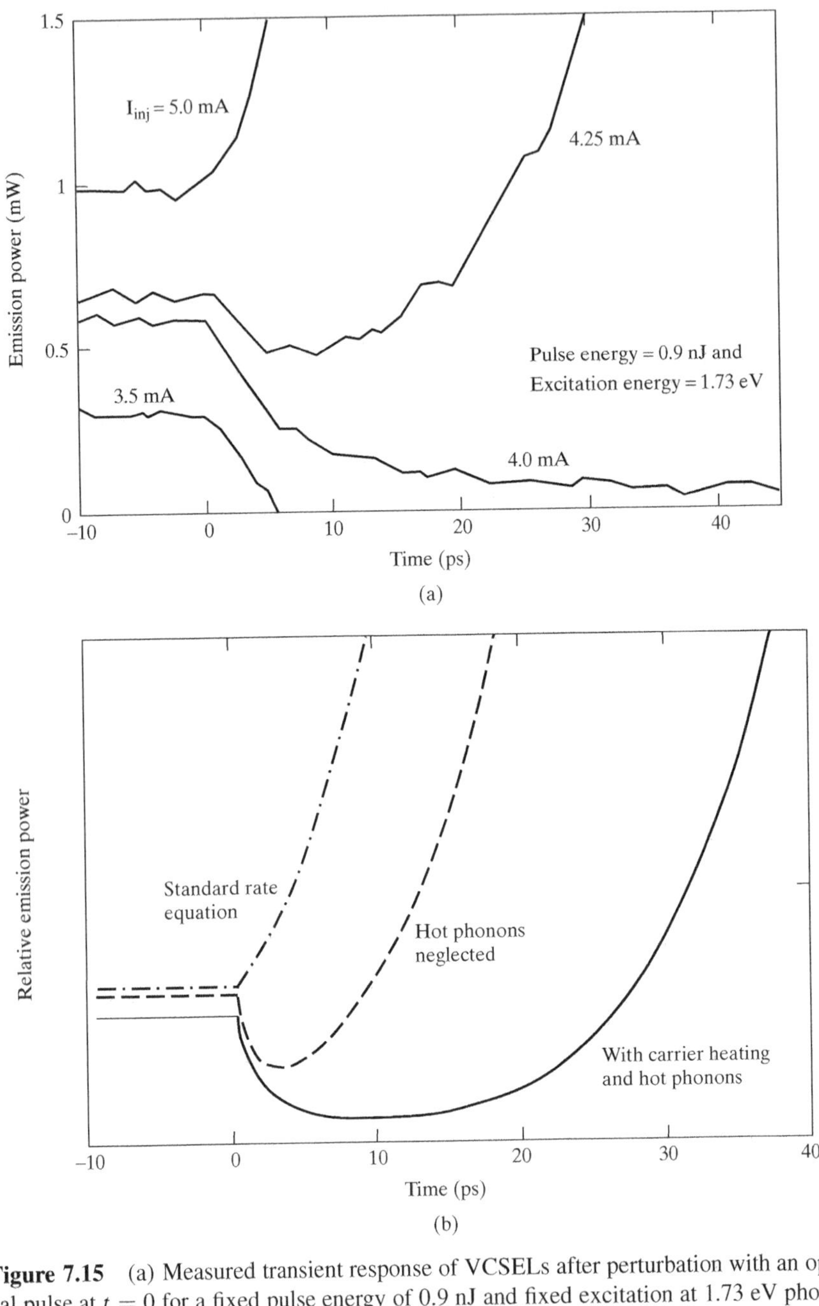

Figure 7.15 (a) Measured transient response of VCSELs after perturbation with an optical pulse at $t = 0$ for a fixed pulse energy of 0.9 nJ and fixed excitation at 1.73 eV photon energy (i.e., corresponding to an excess energy of 0.275 eV); the stationary electrical bias is a variable parameter; (b) calculated transient response of VCSEL after perturbation with a short pump pulse at $t = 0$; complete calculation includes carrier heating and hot phonons (solid line), hot phonons neglected (dashed line), and the standard rate equations (dotted line). (After Ref. 41.)

self-heating has been introduced into the model. It is found that SHB and carrier transport have less influence on the ROF, AM response, and SHD of VCSELs. However, the modulation response of VCSELs is strongly dependent on its cavity size. It is shown that VCSELs with small cavity size have better ROF, modulation bandwidth, and less SHD than do those of large cavity size. This implies that the modulation response of VCSELs is dominated by the self-heating effect. Hence, in order to improve the high-speed characteristics of VCSELs, the self-heating (i.e., thermal resistance) should be minimized.

The influence of multimode on the AM response of gain-guided VCSELs is also investigated. In this case, only the multimode effect is considered. The experimental results indicate that the spatial overlapping of transverse modes (i.e., modal gain of the transverse modes) determines the profile of the AM response of VCSELs, and this explanation has been verified by numerical analysis. It is shown that low overlapping between transverse modes gives two resonant peaks and a notch in the AM response. On the other hand, the partial overlapping of transverse modes removes the notch, but two resonant peaks are still observed in the AM response. If heavy overlapping of transverse modes is maintained in the devices, the corresponding AM response is similar to a single-mode laser. These phenomena of AM response may be due to the difference in modal gain among individual transverse modes. In the case of low overlapping, the transverse modes have different modal gains and thus exhibit different AM responses, which together give two response peaks and a notch. However, in the case of heavy overlapping, all the transverse modes receive the same modal gain so that their collective AM response appears as a response to a single-mode laser. Therefore, the AM response of multimode VCSELs is dominated by SHB and the spatial profile of transverse modes (i.e., modal gain of individual modes).

Furthermore, the large signal transient response of VCSELs is investigated. The study includes the transient response of VCSELs under current pulse modulation, sinusoidal current modulation, and external injection of optical pulses:

1. In the analysis of gain-guided VCSELs under current pulse modulation, it is shown that the delay excitation of side modes is due to the buildup time of SHB. SHB arises from the stimulation recombination of carrier concentration due to the fundamental mode. Hence, the delay time can be expressed in terms of the threshold current of individual modes and their transverse confinement factor.
2. The analysis of electrical gain switching of VCSELs using sinusoidal current modulation shows that the pulsewidth and peak power of the generated gain-switched pulses is affected by SHB. If SHB can be minimized in VCSELs, the corresponding pulsewidth and peak power can be optimized as well as the probability of the excitation of the side mode can be minimized. Hence, SHB has a more significant influence on the modulation response of multimode VCSELs rather than that of a single-mode one.
3. The gain switching of optical pulses under external optical injection in VCSELs is investigated using modified the single-mode rate equation model.

It is shown that the pump power and pulsewidth determine the pulsewidth of the gain-switched pulses. However, the minimum pulsewidth of the generated pulses is saturated at high input peak power and narrow pumped pulsewidth. This is because the minimum pulsewidth generated by optical gain switching is determined by the photon lifetime of the laser cavity. Hence, for an input pumped power and pulsewidth, the constraints of the design are the cavity length and reflectivity of DBRs.

The simple rate equation model is a powerful tool for evaluating the dynamic response of VCSELs without the requirement of extensive computing effort. However, detailed investigation of VCSELs, including the diffraction of light and the longitudinal distribution of DBRs, cannot be accounted for. Therefore, a comprehensive modeling of VCSELs including the quasi-3D distribution of traveling waves inside the laser cavity has been developed. The model is derived from the traveling-wave equations, which can be solved by the time-domain finite difference method. Using this model, it is shown that the calculation without considering light diffraction will overestimate the photon power and resonant frequency of VCSELs.

Finally, the influence of the hot carrier effect is introduced into the simple rate equation model in order to analyze the dynamic response of VCSELs. The hot carrier effect is the transfer of energy between hot carriers and phonons, with which the optical gain of the active layer is varied. It is found that the AM response exhibits low frequency rolloff due to the hot carrier effect. In addition, the short injected optical pulse can result in unusual behavior by rapid switchoff of lasing emission in VCSELs due to the heat transient dynamic between hot carriers and phonons.

7A APPENDIX

7A.1 Parameters of Single-Mode Rate Equation Model of VCSELs

Expressions for Γ_r, ξ_r, Γ_1, ξ_1, Γ_2, ξ_2, χ_1, and χ_2 are given by

$$\Gamma_r = \int_0^W r|\Psi|^2 dr \bigg/ \int_0^\infty |\Psi(r)|^2 r\,dr,$$

$$\xi_r = \frac{\int_0^W r J_0(ar)|\Psi|^2(r)dr}{\int_0^\infty |\Psi(r)|^2 r\,dr}, \tag{7A.1}$$

$$\Gamma_1 = \frac{2}{W^2}\int_0^W r|\Psi|^2 dr,$$

$$\xi_1 = \frac{2}{W^2}\int_0^W |\Psi|^2 J_0(ar) r\,dr, \tag{7A.2}$$

$$\Gamma_2 = \frac{2}{W^2}\frac{1}{J_0^2(aW)}\int_0^W rJ_0(ar)|\Psi|^2 dr,$$

$$\xi_2 = \frac{2}{W^2}\frac{1}{J_0^2(aW)}\int_0^W rJ_0^2(ar)|\Psi|^2 dr, \tag{7A.3}$$

$$\chi_1 = \frac{2}{W^2}\int_0^W \left(r\frac{\partial^2 J_0(ar)}{\partial r^2} + \frac{\partial J_0(ar)}{\partial r}\right) dr, \tag{7A.4}$$

$$\chi_2 = \frac{2}{W^2}\frac{1}{J_0^2(aW)}\int_0^W J_0(ar)\left(r\frac{\partial^2 J_0(ar)}{\partial r^2} + \frac{\partial J_0(ar)}{\partial r}\right) dr. \tag{7A.5}$$

In addition, expressions for T_{11}, T_{12}, T_{21}, and T_{22} are

$$\frac{1}{T_{11}} = \frac{v_g a_N \Gamma_r \Gamma_z \varepsilon_s \log(N_{0s}/N_t) S_s}{N_{0s}}, \tag{7A.6a}$$

$$\frac{1}{T_{12}} = v_g a_N \Gamma_r \Gamma_z \left\{\left(1 - \varepsilon_s \frac{S_s}{N_{0s}}\right)\frac{S_s}{N_{0s}} + \varepsilon_s \log\left(\frac{N_{0s}}{N_t}\right)\left(\frac{S_s}{N_{0s}}\right)^2\right\} + 2\beta_s B_{sp} N_{0s}, \tag{7A.6b}$$

$$\frac{1}{T_{21}} = v_g a_N \Gamma_p \log\left(\frac{N_{0s}}{N_t}\right)\left(\frac{1 - 2\varepsilon_s S_s}{N_{0s}}\right), \tag{7A.6c}$$

$$\frac{1}{T_{22}} = v_g a_N \Gamma_p \left\{\left(1 - \varepsilon_p \frac{S_s}{N_{0s}}\right)\frac{S_s}{N_{0s}} + \varepsilon_p \log\left(\frac{N_{0s}}{N_t}\right)\left(\frac{S_s}{N_{0s}}\right)^2\right\} - \frac{1}{\tau'_{te}}. \tag{7A.6d}$$

The expression for B_s and B_p is

$$B_i(\Omega) = \frac{S_s}{N_{0s}^2}\left(1 + \varepsilon_i\left[1 + \frac{N_{0s}}{S_s}\log\left(\frac{N_{0s}}{N_t}\right)\right]\right)\left(\frac{1}{T_{11}} + j\Omega\right) T_{12} + \frac{\varepsilon_i}{N_{0s}}\log\left(\frac{N_{0s}}{N_t}\right) - \frac{S_s^2 T_{12}^2}{N_{0s}^2}\left(\frac{1}{T_{11}} + j\Omega\right)^2 \tag{7A.7}$$

for the subscripts $i = s$ or p. The steady state characteristics of VCSELs can be calculated by using the Newton–Raphson method.

7A.2 Numerical Technique of Time-Domain Finite Difference Method

The time and spatial variations of the traveling-wave equation can be solved by a first-order difference approximation of the partial differential as follows [31]:

$$\frac{\partial F^+(t,z)}{\partial t} = \frac{F^+(t+\Delta t, z) - F^+(t,z)}{\Delta t} \quad \text{and}$$

$$\frac{\partial F^+(t,z)}{\partial z} = \frac{F^+(t,z+\Delta z) - F^+(t,z)}{\Delta z}, \tag{7A.8a}$$

$$\frac{\partial F^-(t,z)}{\partial t} = \frac{F^-(t+\Delta t,z) - F^-(t,z)}{\Delta t} \quad \text{and}$$

$$\frac{\partial F^-(t,z)}{\partial z} = \frac{F^-(t,z) - F^-(t,z-\Delta z)}{\Delta z}. \tag{7A.8b}$$

where $F^{\pm} = \Psi^{\pm}$ or $A^{\pm}$. The rate of change of F^+ with respect to time and displacement can be written as

$$\begin{aligned} &F^+(t+\Delta t, z+\Delta z) - F^+(t,z) \\ &\quad = F^+(t+\Delta t, z+\Delta z) - F^+(t, z+\Delta z) + F^+(t, z+\Delta z) - F^+(t,z) \\ &\quad = \left.\frac{\partial F^+}{\partial t}\right|_{z+\Delta z} \Delta t + \left.\frac{\partial F^+}{\partial z}\right| \Delta z, \\ &\quad = \left.\frac{\partial F^+}{\partial t}\right|_{z+\Delta z} \Delta t - \left.\frac{\partial F^+}{\partial t}\right|_{z} \Delta t + \left.\frac{\partial F^+}{\partial t}\right|_{z} \Delta t + - \left.\frac{\partial F^+}{\partial z}\right|_{t} \Delta z, \\ &\quad = \left.\frac{\partial F^+}{\partial t}\right|_{z} \Delta t + \left.\frac{\partial F^+}{\partial z}\right|_{t} \Delta z + \frac{\partial^2 F^+}{\partial t\, \partial z} \Delta z\, \Delta t, \end{aligned} \tag{7A.9}$$

where Δt and Δz are the small changes in time and propagation distances, respectively. The expression for the rate of change of F^- can also be derived in a similar way.

It is assumed that the profile of $F^{\pm}(z,t)$ is represented by $m+1$ traveling waves (forward and reverse) located at the boundary of layers, where m is the total number of layers. In addition, the thickness of the ith layer is Δz_i and the position of the boundary between the ith and $(i+1)$th layers is z_{i+1} so that z_1 and z_{m+1} represent the position of facets. If the second derivative $\partial^2 F^+(z,t)/\partial t\,\partial z$ is neglected, the traveling wave formula can be written as

$$\left.\frac{\partial F^{\pm}}{\partial t}\right|_{z_i} \Delta t \pm \left.\frac{\partial F^{\pm}}{\partial z}\right|_{t} \Delta z_i \approx F^{\pm}(z_i \pm \Delta z_i, t+\Delta t) - F^{\pm}(z_i, t). \tag{7A.10}$$

It is noted that the physical thickness of the index layer Δz_i varies with its effective refractive index so that its optical length is equal to quarter of a wavelength. If the relation between time and spatial steps in the longitudinal direction is selected as $\Delta t = \Delta z_i / v_{gi}$, where v_{gi} is the group velocity of the ith layer and Δt is a constant time step, (7.44) can be rewritten as

$$F^{\pm}(z_i \pm \Delta z_i, t+\Delta t) - F^{\pm}(z_i, t) = [M(r, z_i, t)]\Delta z_i. \tag{7A.11}$$

Now, the longitudinal distribution of traveling waves can be calculated by (7A.11). For the output power, the boundary conditions at the top and bottom surfaces of the Bragg reflectors can be written as

$$F^+(t, z_1) = r_f F^-(t, z_1) \text{ and } F^-(t, z_{m+1}) = r_r F^+(t, z_{m+1}), \tag{7A.12}$$

where r_f and r_r are the front and rear surface reflectivities, respectively.

7A.3 Determination of Light Diffraction Using Wide-Angle Beam Propagation Method

For VCSELs with diffraction of light occurring inside the laser cavity, especially for devices with small aperture such as the index-guided devices, the amplitude and phase variations of the traveling wave are nonuniformly distributed over the transverse direction within each layer. Therefore, the assumption of uniform plane wave is not valid and the term $\partial^2/\partial z^2$ should be included in the traveling-wave quasi-3D model.

The displacement and time variation of $\Psi^{\pm}(r, z, t)$ given in (7.37) can be solved by using the Padé recursion approximation [45] and time-dependent transfer matrix [31], respectively. First, rewrite (7.37) in a matrix format to describe the propagation of $\Psi^{\pm}$ for a distance Δz from z and at a time t

$$\frac{\partial}{\partial z}\left(1 - \frac{j}{2\beta}\frac{\partial}{\partial z}\right)\underline{\Psi}^+(r, z) = \frac{j}{2\beta}\left(\frac{1}{r}\frac{\partial}{\partial r}\left(r\frac{\partial}{\partial r}\right) + \Delta\varepsilon(r, z, t)k_0^2\right)\underline{\Psi}^+(r, z), \tag{7A.13}$$

where the time derivative term $\partial/\partial t$ has been ignored in the derivation. Equating the operators on Ψ^+ on the left- and right-hand sides of (7A.13) and using the Padé recursion approximation, gives

$$\frac{\partial}{\partial z} = \frac{\dfrac{j}{2\beta}[M(r, z, t)]}{1 - \dfrac{j}{2\beta}\dfrac{\partial}{\partial z}} \equiv [P], \tag{7A.14}$$

where $[P]$ treats the field variation along the transverse direction. Therefore, Ψ^+ at distance $z + \Delta z$ can be obtained by integrating (7A.14) along the longitudinal direction, and the solution is given by

$$\underline{\Psi}^+(r, z + \Delta z) = \exp([P] \cdot \Delta z)\underline{\Psi}(r, z). \tag{7A.15}$$

The expression for Ψ^- can also be determined in a similar way. Therefore, the time-independent transfer matrix can be expressed as

$$\begin{bmatrix} \underline{\Psi}^+ \\ \underline{\Psi}^- \end{bmatrix}_{r,z+\Delta z,t} = \begin{bmatrix} \exp([P] \cdot \Delta z) & 0 \\ 0 & \exp(-[P] \cdot \Delta z) \end{bmatrix} \begin{bmatrix} \underline{\Psi}^+ \\ \underline{\Psi}^- \end{bmatrix}_{r,z,t}. \tag{7A.16}$$

Then, the time and spatial displacement is chosen as $\Delta z = \Delta t \cdot v_g$. Hence, the time-dependent solution of (7.37) can be written as

$$\begin{bmatrix} s_{11}(r,z,t) & s_{12}(r,z,t) \\ s_{21}(r,z,t) & s_{22}(r,z,t) \end{bmatrix}_1 = [T_{2p}]^{-1} \begin{bmatrix} 1 & [R_{2p}] \\ [R_{2p}] & 1 \end{bmatrix} \times \begin{bmatrix} \exp([P]\cdot\Delta z) & 0 \\ 0 & \exp(-[P]\cdot\Delta z) \end{bmatrix}, \quad (7A.17)$$

where the exponential terms in (7A.17) can be simplified as follows:

$$\exp(\pm[P]\cdot\Delta z) = [I] \pm [P]\cdot\Delta z + \tfrac{1}{2}[P]^2\cdot\Delta z^2 \pm \tfrac{1}{6}[P]^3\cdot\Delta z^3 + \tfrac{1}{24}[P]^4\cdot\Delta z^4 \pm \cdots. \quad (7A.18)$$

By substituting (7A.18) into (7A.17), the displacement and time variation of $\Psi^{\pm}$ can be solved using the time-domain algorithm. It must be noted that the off-axis propagation of light at the interface of the layers has also been properly taken into account through $[R_{2p}]$ and $[T_{2p}]$ (see also Chapter 3, Section 3.4.3). Hence, a very simple but powerful quasi-3D dynamic model for VCSELs has been developed with full consideration of light diffraction inside the laser cavity.

The calculation of the phase of the propagation fields using (7A.14) can be simplified by the Padé recursion approximation on $[P]$. This is because (7A.14) suggests a recurrence relation

$$\left.\frac{\partial}{\partial z}\right|_i = \frac{\dfrac{j}{2\beta}[M]}{1 - \dfrac{j}{2\beta}\left.\dfrac{\partial}{\partial z}\right|_{i-1}}, \quad (7A.19)$$

where i is an integer. The low-order Padé operators that result from applying (7A.19) can be obtained by a recurrence relation. For example, the Padé (1,1), (2,2), and (3,3) operators can be written as

$$[P]_{(1,1)} = \frac{\dfrac{j}{2\beta}[M]}{1 + \dfrac{1}{4\beta^2}[M]}, \quad (7A.20)$$

$$[P]_{(2,2)} = \frac{\dfrac{j}{2\beta}[M] + \dfrac{j}{4\beta^3}[M]^2}{1 + \dfrac{3}{4\beta^2}[M] + \dfrac{1}{16\beta^4}[M]^2}, \quad (7A.21)$$

$$[P]_{(2,2)} = \frac{\dfrac{j}{2\beta}[M] + \dfrac{j}{3\beta^3}[M]^2 + \dfrac{3j}{32\beta^5}[M]^3}{1 + \dfrac{5}{4\beta^2}[M] + \dfrac{3}{8\beta^4}[M]^2 + \dfrac{1}{64\beta^6}[M]^3}. \quad (7A.22)$$

In fact, higher-order Padé operators can also be obtained in the same manner. It has been observed [46] that phase errors can be minimized by the Padé (1,1) operator or almost suppressed by the Padé (3,3) operator, even at a propagation angle of $\sim 30^\circ$. Therefore, the Padé operator up to the order of (3,3) is sufficient for use in the traveling-wave model of VCSELs for consideration of light diffraction.

In order to analyze the influence of light diffraction inside the resonant cavity, the laser structure used in Section 7.5 has to be changed. It is assumed that there are no built-in index-guided structures inside the spacer layers and DBRs so that the propagation fields are only transversely guided within the QW active layer in which the diffraction of propagation fields occurs. Furthermore, the core radius W and the difference of permittivity between the core and cladding regions of the active layer $\Delta\varepsilon$ is allowed to change. The influence of light diffraction on the steady-state and dynamic behaviors of VCSELs is demonstrated by equating W and $\Delta\varepsilon$ to 0.8 μm and 0.16 (i.e. strongly index-guided inside the active layer), respectively. Figure 7.16a compares the on-axis mode intensities in the longitudinal direction and mode intensities at the QW position in the transverse direction, respectively, from the models using the paraxial propagator and the Padé (3,3) operator. It is observed that the optical field is less diverged from the core region of the active layer for the case using the paraxial propagator than that using the Padé (3,3) operator. Figure 7.16b shows the corresponding dynamic response of the output power and the carrier concentration profile at steady state. As we can see, the output power and ROF are overestimated for the case using the paraxial propagator. This is because the model using the paraxial propagator underestimates the degree of light diffraction and total cavity loss.

7A.4 Increase of Propagation Distance Using Time-Dependent Transfer Matrix

In order to solve the problem of small Δz but without sacrificing the advantages of time-domain algorithm, the traveling-wave model is modified by a time-domain transfer matrix method [47]. Figure 7.17 illustrates the use of a transfer matrix $[S(r, z, t)]_1$ to represent a layer. The propagation fields can be related to $[S(r, z, t)]_i$ by

$$\begin{bmatrix} \underline{\Psi}^+(r, z+\Delta z, t+\Delta t) \\ \underline{\Psi}^-(r, z+\Delta z, t) \end{bmatrix} = [S(r, z, t)]_i \begin{bmatrix} \underline{\Psi}^+(r, z, t) \\ \underline{\Psi}^-(r, z, t+\Delta t) \end{bmatrix}$$
$$\equiv \begin{bmatrix} s_{11}(r, z, t) & s_{12}(r, z, t) \\ s_{21}(r, z, t) & s_{22}(r, z, t) \end{bmatrix}_i \begin{bmatrix} \underline{\Psi}^+(r, z, t) \\ \underline{\Psi}^-(r, z, t+\Delta t) \end{bmatrix} \tag{7A.23}$$

where the subscript i represents the position of the layer. The transfer matrix can be determined by considering the propagation of traveling waves along the semiconductor layer and both the transmission and reflection between the interface of the adjacent layers.

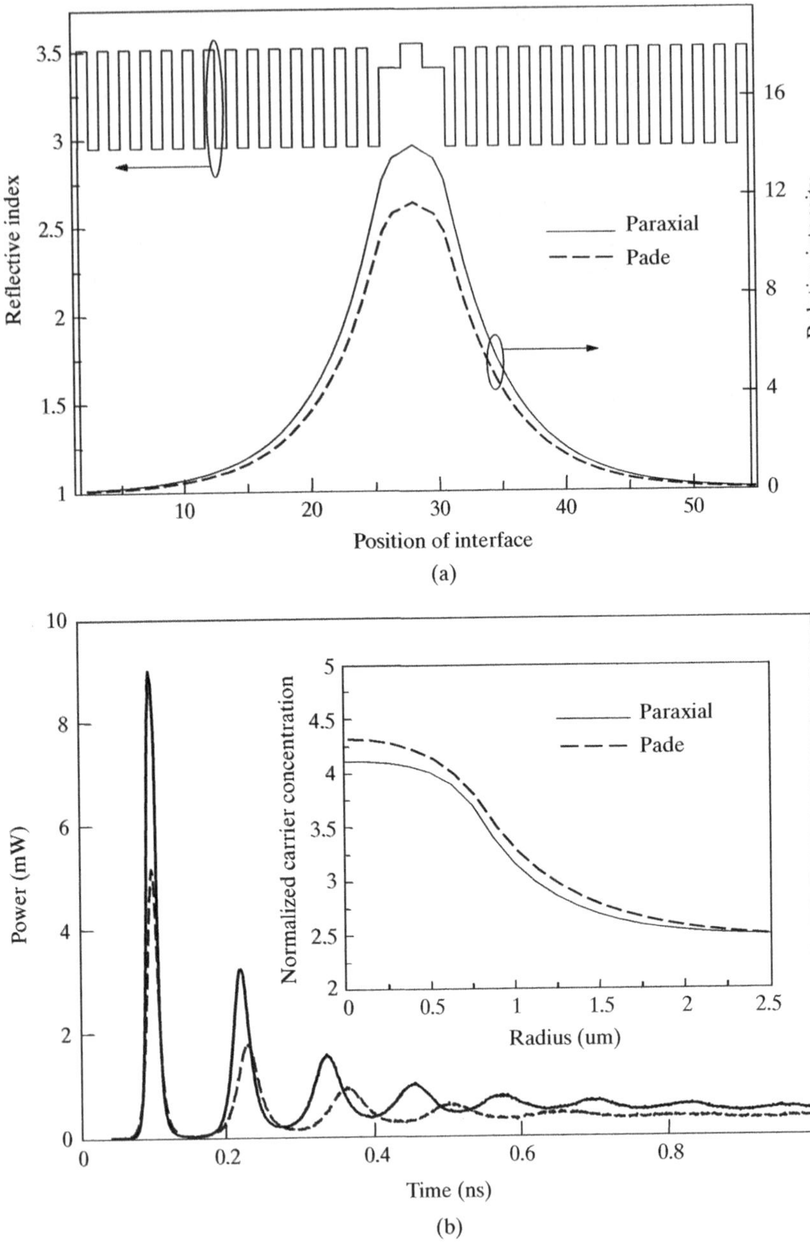

Figure 7.16 (a) On-axis mode intensities in the longitudinal direction calculated from the models using paraxial propagator and Padé (3,3) operator; (b) dynamic response of output power and inset diagram indicate the steady-state carrier concentration distribution. The results are calculated from the model using paraxial propagator (solid line) and Padé (3,3) operator (dotted line). In the calculation, W and $\Delta\varepsilon$ (i.e., inside active layer between core and cladding regions) are set to 0.8 μm and 0.16, respectively. (After Ref. 45.)

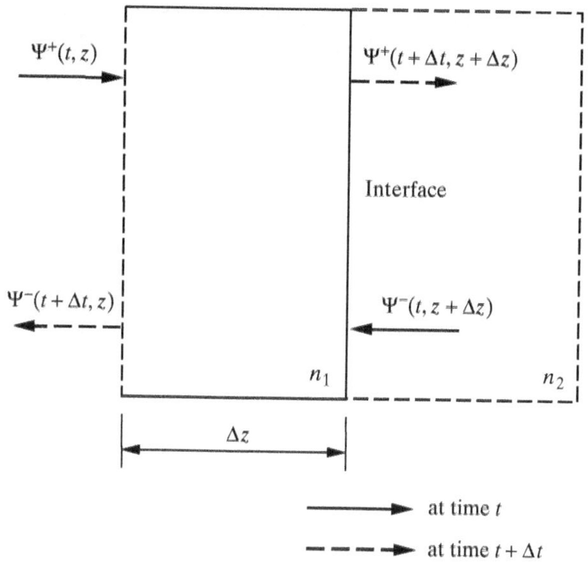

Figure 7.17 Reflection and transmission of optical fields at the boundary between layers.

From (7A.16), the propagation of forward and reverse propagating waves, Ψ^+ and Ψ^-, along the layer can be expressed as

$$\begin{bmatrix} \underline{\Psi^+} \\ \underline{\Psi} \end{bmatrix}_{r,z+\Delta z,t} = \begin{bmatrix} \exp([M]\cdot\Delta z) & 0 \\ 0 & \exp(-[M]\cdot\Delta z) \end{bmatrix} \begin{bmatrix} \underline{\Psi^+} \\ \underline{\Psi^-} \end{bmatrix}_{r,z,t}. \tag{7A.24}$$

From Section 3.4.3, the transfer matrix for a refractive index step from a layer of refractive index n_i to that of refractive index n_{i+1} can be written as

$$\begin{bmatrix} \underline{\Psi^+} \\ \underline{\Psi^-} \end{bmatrix}_{\text{at } n_{i+1}} = [T_{2p}]^{-1} \begin{bmatrix} 1 & [R_{2p}] \\ [R_{2p}] & 1 \end{bmatrix} \begin{bmatrix} \underline{\Psi^+} \\ \underline{\Psi^-} \end{bmatrix}_{\text{at } n_i}, \tag{7A.25}$$

where $[R_{2p}]$ and $[T_{2p}]$ are the matrices of reflection and transmission coefficients, respectively. Therefore, the transfer matrix $[S(r, z, t)]_i$ can be written as

$$\begin{bmatrix} s_{11}(r,z,t) & s_{12}(r,z,t) \\ s_{21}(r,z,t) & s_{22}(r,z,t) \end{bmatrix}_i = [T_{2p}]^{-1} \begin{bmatrix} 1 & [R_{2p}] \\ [R_{2p}] & 1 \end{bmatrix} \times \begin{bmatrix} \exp([M]\cdot\Delta z) & 0 \\ 0 & \exp(-[M]\cdot\Delta z) \end{bmatrix}. \tag{7A.26}$$

The exponential terms in (7A.26) can be simplified to improve the computing efficiency as

$$\exp(\pm[M]\cdot\Delta z) = [I] \pm [M]\cdot\Delta z + \tfrac{1}{2}[M]^2\cdot\Delta z^2 \pm \tfrac{1}{6}[M]^3\cdot\Delta z^3 + \tfrac{1}{24}[M]^4\cdot\Delta z^4 \pm \cdots, \tag{7A.27}$$

where $[I]$ is an identity matrix. The time-domain algorithm given in [49] can be utilized to solve (7.23). This can be done by rewriting (7A.26) in the following format:

$$\begin{bmatrix} \underline{\Psi}^{+}(r, z+\Delta z, t+\Delta t) \\ \underline{\Psi}^{-}(r, z, t+\Delta t) \end{bmatrix} = s_{22}^{-1} \begin{bmatrix} (s_{11} \cdot s_{22} - s_{12} \cdot s_{21}) & s_{12} \\ 1 & -s_{21} \end{bmatrix} \times \begin{bmatrix} \underline{\Psi}^{+}(r, z, t) \\ \underline{\Psi}^{-}(r, z+\Delta z, t) \end{bmatrix}. \tag{7A.28}$$

Now, (7A.28) can be solved by advancing $\Psi^{\pm}(r, z, t)$ from one layer to another during each time interval Δt. In addition, the boundary conditions between layers are implicitly introduced into (7A.28) so that the reflectivity of Bragg reflectors can also be evaluated in a time-dependent manner. The main differences between this model and one without the transfer matrix are that the boundary conditions are implicitly introduced and the propagating distance of $\Psi^{\pm}(r, z, t)$ is not restricted to Δz as in two earlier papers by the author [12, 13] but varies with the propagation length of the transfer matrix. Now, the total number of computational steps as well as the execution time of the traveling-wave model can be reduced by increasing the propagation length of the transfer matrix. For example, if a new transfer matrix is defined as $[S(r, z, t)]_i \times [S(r, z, t)]_{i+1}$, the corresponding propagation distance of this transfer matrix is increased to $2\Delta z$ so that the number of computational steps as well as the execution time can be reduced roughly by a factor of 2.

The computational speed of the traveling wave models using the time-dependent transfer matrix is examined. Table 7.1 shows the total execution time of the 1D and the quasi-3D time-domain traveling-wave models against the propagation distance Δz of the transfer matrices. The simulation program is written in FORTRAN 90 running on an IBM Pentium II 200-MHz PC. It is shown that the total execution time for the 1D model is reduced by more than 2.5 times when the propagation distance is increased from Δz to $2\Delta z$. This is expected because the total computational steps of (7A.28) are reduced by half through the increase of propagation distance. Hence, the total execution time for the traveling model is reduced by more than half. However, the total execution time for the quasi-3D model is increased by the application of the transfer matrix for the propagation distance of less than $4\Delta z$ (i.e., see Table 7.1). This is because the

TABLE 7.1 Total Execution Time of 1D and Quasi-3D Time-Domain Traveling-Wave Models

Propagation Distance ($\times\Delta z$)	Quasi-3D Model (Total Execution Time in minutes)		1D Model (Total Execution Time in Seconds)	
1	375	Without transfer matrix = 120	628	Without transfer matrix = 610
2	160	–	254	–
3	100	–	110	–

inverse operation of a matrix [involved in the calculation of (7A.28)] consumes half of the total execution time of the entire model. On the other hand, the 1D model requires only the multiplication and addition of complex numbers in the calculation of (7A.28) so that a significant improvement in total computational speed can be obtained.

REFERENCES

1. D. Tauber, G. Wang, R. S. Geels, J. E. Bowers, and L. A. Coldren, "Large and small signal dynamics of vertical cavity surface emitting lasers," *Appl. Phys. Lett.*, Vol. 62, no. 4, pp. 325–327, 1993.
2. K. L. Lear, A. Mar, K. D. Choquette, S. P. Kilcoyne, R. P. Schneider Jr., and K. M. Geib, "High frequency modulation of oxide-confined vertical cavity surface emitting lasers," *Electron. Lett.*, Vol. 32, no. 5, pp. 457–458, 1996.
3. U. Fiedler, G. Reiner, P. Schnitzer, and K. J. Ebeling, "Top surface emitting vertical cavity surface emitting laser diodes for 10 Gb/s data transmission," *IEEE Photon. Technol. Lett.*, Vol. 8, no. 6, pp. 746–748, 1996.
4. D. Wiedenmann, R. King, C. Jung, R. Jager, R. Michalzik, P. Schnitzer, M. Kicherer, and K. J. Ebeling, "Design and analysis of single mode oxidized VCSEL's for high speed optical interconnects," *IEEE J. Select. Topics Quantum Electron.*, Vol. 5, no. 3, pp. 503–511, 1999.
5. A. Valle, J. Sarma, and K. A. Shore, "Spatial hole burning effects on the dynamics of vertical cavity surface emitting laser diodes," *IEEE J. Quantum Electron.*, Vol. 31, no. 8, pp. 1423–1431, 1995.
6. C. Y. Tsai, Y. H. Lo, and R. M. Spencer, "Effects of spectral hole burning, carrier heating, and carrier transport on the small-signal modulation response of quantum well lasers," *Appl. Phys. Lett.*, Vol. 67, no. 21, pp. 3084–3086, 1995.
7. C. Z. Ning, R. A. Indik, and J. V. Moloney, "Self-consistent approach to thermal effects in vertical cavity surface emitting lasers," *J. Opt. Soc. Am. B*,Vol. 12, no. 10, pp. 1993–2004, 1995.
8. S. F. Yu, W. N. Wong, P. Shum, and E. H. Li, "Theoretical analysis of modulation response and second harmonic distortion in vertical cavity surface emitting lasers," *IEEE J. Quantum Electron.*, Vol. 32, no. 12, pp. 2139–2147, 1996.
9. R. R. Burton, M. S. Stern, P. C. Kendall, and P. N. Robson, "Modeling of diffraction in pillar vertical cavity surface emitting lasers with embedded Bragg layers," *Opt. Quantum Electron.*, Vol. 28, pp. 1677–1684, 1996.
10. Y. G. Zhao and J. G. McInerney, "Transverse-mode control of vertical cavity surface emitting lasers," *IEEE Quantum Electron.*, Vol. 32, no. 11, pp. 1950–1958, 1996.
11. S. F. Yu, "Analysis and design of vertical cavity surface emitting lasers for self-sustained pulsation operation," *IEEE Quantum Electron.*, Vol. 34, no. 3, pp. 497–505, 1998.
12. S. F. Yu and C. W. Lo, "Influence of transverse modes on the dynamic response of vertical cavity surface emitting lasers," *IEE Proc., Part J, Optoelectronics*, Vol. 143, no. 3, pp. 189–194, 1996.
13. S. F. Yu, "Dynamic behavior of vertical cavity surface emitting lasers," *IEEE J. Quantum Electron.*, Vol. 32, no. 7, pp. 1168–1179, 1996.

14. J. W. Scott, B. J. Thibeault, C. J. Mahon, L. A. Coldren, and F. H. Peters, "High modulation efficiency of intracavity vertical cavity lasers," *Appl. Phys. Lett.*, Vol. 65, no. 12, pp. 1483–1485, 1994.

15. R. C. Tucker and D. J. Pope, "Circuit modeling of the effect of diffusion on damping in a narrow-stripe semiconductor lasers," *IEEE J. Quantum Electron.*, Vol. 19, pp. 1179–1183, 1983.

16. W. H. Press, S. A. Teukolsky, W. T. Vetterling and B. P. Flannery, *Numerical Recipes in Fortran 90*, 2nd ed., Cambridge Univ. Press, 1996.

17. G. Morthier, F. Libbrechit, K. David, P. Vankwikelberge, and R. G. Beats, "Theoretical investigation of the second order harmonic distortion in the AM response of 1.55 μm FP and DFB lasers," *IEEE J. Quantum Electron.*, Vol. 27, no. 8, pp. 1990–2002. 1991.

18. K. H. Hahn, M. R. Tan Y. M. Houng, and S. Y. Wang, "Large area multitransverse mode VCSELs for modal noise reduction in multimode fiber systems," *Electron. Lett.*, Vol. 29, no. 16, pp. 1482–1483, 1993.

19. K. H. Hahn, M. R. Tan, and S. Y. Wang, "Intensity noise of large area vertical cavity surface emitting lasers in multimode optical fiber links," *Electron. Lett.*, Vol. 30, no. 2, pp. 139–140, 1994.

20. M. Ishikawa, R. Nagarajan, T. Fukushima, J. G. Wasserbauer, and J. E. Bower, "Long wavelength high-speed semiconductor lasers with carrier transport effects," *IEEE J. Quantum Electron.*, Vol. 28, no. 10, pp. 2230–2241, 1992.

21. Y. Satuby and M. Orenstein, "Limits of the modulation response of a single-mode proton implanted VCSEL," *IEEE Photon. Technol. Lett.*, Vol. 10, no. 6, pp. 760–762, 1998.

22. Y. Satuby and M. Orenstein, "Mode-coupling effects on the small signal modulation of multitransverse mode vertical cavity semiconductor lasers," *IEEE J. Quantum Electron.*, Vol. 35, no. 6, pp. 944–954, 1999.

23. M. Orenstein, Y. Satuby, U. Ben-Ami, and J. P. Harbison, "Transverse modes and lasing characteristics of selectively grown vertical cavity surface emitting lasers by mirror reflectivity modulation," *Appl. Phys. Lett.*, Vol. 69, pp. 1840–1842, 1996.

24. O. Buccafusca, J. L. A. Chilla, J. J. Rocca, S. Feld, C. Wilmsen, V. Morozov, and R. Leibenguth, "Transverse mode dynamics in vertical cavity surface emitting lasers excited by fast electrical pulses," *Appl. Phys. Lett.*, Vol. 68, no. 5, pp. 590–592, 1996.

25. O. Buccafusca, J. L.A. Chilla, J. J. Rocca, P. Brusenbach, and J. Martin-Regalado, "Transverse response of vertical cavity surface emitting lasers of different active region diameters," *IEEE J. Quantum Electron.*, Vol. 35, no. 4, pp. 608–615, 1999.

26. K. Tai, Y. Lai, K. F. Huang, T. C. Huang, T. D. Lee, and C. C. Wu, "Transverse mode emission characteristics of gain-guided surface emitting lasers," *Appl. Phys. Lett.*, Vol. 63, pp. 2624–2626, 1993.

27. J. Y. Law and G. P. Agrawal, "Effects of spatial hole burning on gain switching in vertical cavity surface emitting lasers," *IEEE J. Quantum Electron.*, Vol. 33, no. 3, pp. 462–468, 1997.

28. L. G. Melcer, J. R. Karin, R. Nagarajan, and J. E. Bowers, "Picosecond dynamics of optical gain switching in vertical cavity surface emitting lasers," *IEEE J. Quantum Electron.*, Vol. 27, no. 6, pp. 1417–1425, 1991.

29. P. M. Downey, J. E. Bowers, R. S. Tucker, and E. Agyekum, "Picosecond dynamics of a gain-switched InGaAsP lasers," *IEEE J. Quantum Electron.*, Vol. 12, pp. 1039–1047, 1987.

30. M. J. Adams, *Introduction to Optical Waveguide*, Wiley, New York, 1981.

31. L. M. Zhang, S. F. Yu, M. C. Nowell, D. D. Marcenac, J. E. Carroll, and R. G. S. Plumb, "Dynamic analysis of radiation and side-mode suppression in a second-order DFB laser using time domain large signal travelling wave model," *IEEE J. Quantum Electron.*, Vol. 30, no. 6, pp. 1389–1395, 1994.

32. L. A. Coldren and T. L. Koch, "Analysis and design of coupled cavity lasers-Part I: threshold gain analysis and design guidelines," *IEEE J. Quantum Electron.*, Vol. 20, pp. 657–670, 1984.

33. D. J. Jones, L. M. Zhang, J. E. Carroll, and D. D. Marcenac, "Dynamics of monolithic passively mode locked semiconductor lasers," *IEEE J. Quantum Electron.*, Vol. 31, no. 6, pp. 1051–1058, 1995.

34. T. Baba, T. Hamano, F. Doyama, and K. Iga, "Spontaneous emission factor of a microcavity DBR surface emitting laser (II)—Effects of electron quantum confinement," *IEEE J. Quantum Electron.*, Vol. 27, no. 6, pp. 1347–1358, 1991.

35. L. E. Thode, G. Csanak, L. L. So, T. J. T. Kwan, and M. Campbell, "Time dependent numerical simulation of vertical cavity lasers," *Proc. SPIE*, Vol. 2146, pp. 174–184, 1994.

36. F. Jahnke and S. W. Koch, "Many body theory for semiconductor microcavity lasers," *Phys. Rev. A*, Vol. 52, no. 2, pp. 1712–1727, 1995.

37. C. Y. Tsai, L. F. Eastman, and Y. H. Lo, "Hot carrier and hot phonon effects on high-speed quantum well lasers," *Appl. Phys. Lett.*, Vol. 63, no. 25, pp. 3408–3410, 1993.

38. C. Y. Tsai and C. Y. Tsai, "Effects of carrier heating on the frequency chirping of semiconductor lasers," *Electron. Lett.*, Vol. 33, no. 24, pp. 2043–2044, 1997.

39. M. Rieger, P. Kocevar, P. Lugli, P. Bordone, L. Reggiani, and S. M. Goodnick, "Monte Carlo studies of nonequilibrium phonon effects in polar semiconductors and quantum wells. II. Non-Ohmic transport in n-type gallium arsenide," *Phys. Rev. B*, Vol. 39, pp. 7866–7875, 1989.

40. J. A. Kash and J. C. Tsang, *Solid State Electron.*, Vol. 31, pp. 419, 1988.

41. G. Pompe, T. Rappen, and M. Wegener, "Transient response of an optically pumped short-cavity semiconductor laser," *Phys. Rev. B*, Vol. 51, no. 11, pp. 7005–7009, 1995.

42. M. Elsasser, S. G. Hense, and M. Wegener, "Subpicosecond switch-off and switch-on of a semiconductor laser due to transient hot carrier effects," *Appl. Phys. Lett.*, Vol. 70, no. 7, pp. 853–855, 1997.

43. S. G. Hense and M. Wegener, "Ultrafast switch-off of a vertical cavity semiconductor laser," *Phys. Rev. B*, Vol. 55, no. 15, pp. 9255–9258, 1997.

44. Y. G. Zhao and J. G. McInerney, "Transient temperature response of vertical cavity surface emitting lasers," *IEEE J. Quantum Electron.*, Vol. 31, no. 9, pp. 1668–1673, 1995.

45. S. F. Yu, "An improved time-domain traveling-wave model for vertical cavity surface emitting lasers," *IEEE J. Quantum Electron.*, Vol. 34, no. 10, pp. 1938–1948, 1998.

46. G. R. Hadley, "Wide-angle beam propagation using Padé approximant operators," *Opt. Lett.*, Vol. 17, pp. 1426–1428, 1991.
47. M. G. Davis and R. F. O'Dowd, "A transfer matrix method based large signal dynamic model for multielectrode DFB lasers," *IEEE J. Quantum Electron.*, Vol. 30, no. 11, pp. 2458–2466, 1994.

CHAPTER 8

Spontaneous Emission of Vertical Cavity Surface Emitting Lasers

The alternation of the spontaneous emission factor and linewidth enhancement factor in vertical cavity surface emitting lasers (VCSELs) is investigated. A simple theoretical model is used to explain the enhancement of spontaneous emission in an airposted VCSEL. In addition, the optimum configuration of a planar VCSEL for maximum spontaneous emission rate is described. It can be shown that the spontaneous emission factor and spontaneous emission rate have significant influence on the realization of thresholdless VCSEL. Finally, the approach to obtain the zero linewidth enhancement factor in VCSEL is analyzed.

8.1 INTRODUCTION

Spontaneous emission in facet emitting lasers is unlikely to be controlled because of their relatively large cavity volume [1]. However, in vertical cavity surface emitting lasers (VCSELs), spontaneous emission can be alternated by their small cavity size [2–4]. If the optical confinement in VCSELs is precisely controlled, the coupling efficiency of spontaneous emission to the lasing mode (i.e., spontaneous emission factor) as well as the spontaneous emission rate can be enhanced significantly [5–7]. If the spontaneous emission factor reaches its upper limit (i.e., 1) and the spontaneous emission rate dominates over the nonradiative recombination rate inside the laser cavity, the thresholdless operation of VCSELs can be realized [8,9]. On the other hand, the spectral linewidth, which determines the optical quality of the lasing mode, is affected by the optical confinement of VCSELs. It is shown that the linewidth of VCSELs can be narrowed below and above threshold, provided the distributed Bragg reflectors (DBRs) and the active layer are carefully constructed [10]. In addition, the linewidth is directly proportional to the linewidth enhancement factor, which is a function of differential gain and operating wavelength [11]. In order to minimize the linewidth enhancement factor as well as the linewidth of VCSELs, the use of gain materials and operating wavelength should be carefully selected [10,12]. Differential

Analysis and Design of Vertical Cavity Surface Emitting Lasers By Siu Fung Yu
ISBN 0-471-39124-7

gain, which is inversely proportional to the linewidth enhancement factor, can be increased by the reduction of the carrier concentration at threshold. Therefore, the corresponding threshold carrier concentration should also be minimized in the design of VCSELs for a small linewidth enhancement factor.

In this chapter, the spontaneous emission of VCSELs is studied through analysis of the spontaneous emission factor and spontaneous emission rate. It is shown theoretically that the maximum spontaneous emission factor of an airposted VCSEL is limited to 0.5. However, further enhancement of the spontaneous emission factor to 1 is possible if the transverse radiation of the spontaneous emission can be eliminated. Methods for measuring the spontaneous emission factor of semiconductor lasers are also described. In addition, the spontaneous emission factor of VCSELs with index- and gain-guided structures is studied. On the other hand, the obstruction of enhancing the spontaneous emission rate in planar VCSELs is investigated. It is proposed that metallic mirrors can be used to remove the angular dependence of reflectivity and oxidized aperture can be employed to confine the propagation modes, to increase the spontaneous emission rate of planar VCSELs. Furthermore, the conditions of thresholdless in VCSELs are analyzed by a simple rate equation model. In order to obtain thresholdless VCSELs, it is found that the value of spontaneous emission factor should close to 1 and the spontaneous emission rate should dominate over the nonradiative recombination rate [13]. Finally, methods to minimize the linewidth enhancement factor of semiconductor lasers are described. It is proposed that the use of modulation-doped strained QW material as the active region can force the linewidth enhancement factor to zero, and this approach has been verified by measuring the linewidth enhancement factor of facet emitting lasers. This implies that the linewidth enhancement factor of VCSELs can be reduced in a similar manner, provided the optical gain, lasing wavelength, and threshold carrier concentration are carefully selected.

8.2 SPONTANEOUS EMISSION FACTOR

The *spontaneous emission factor* β_S, is defined as the ratio of spontaneous emission coupled into the lasing mode to the total spontaneous emission. In facet emitting lasers, β_S varies between 10^{-7} and 10^{-5} and its magnitude is dependent on the corresponding optical confinement structure of lasers. This is because the lasing mode in a gain-guided laser may capture a fraction of spontaneous emission larger than that in an index-guided laser having a comparable stripe width. Hence, gain-guided facet emitting lasers usually have relatively β_S larger than those of index-guided lasers. However, this increase in β_S has no advantage in minimization of the threshold current of gain-guided facet emitting lasers, as the maximum magnitude of β_S is only 10^{-5}. The main difference between facet emitting lasers and VCSELs in terms of spontaneous emission is that the β_S of the latter case can enhance to 1 so that thresholdless and infinite modulation bandwidth can be realized in VCSELs. In the following sections, β_S

values of VCSELs with index- and gain-guided structures are measured, and the corresponding measurement techniques are also described.

8.2.1 Spontaneous Emission Factor of VCSELs

The spontaneous emission factor β_s of VCSELs has been determined experimentally [14–22]. It is found that β_s varies significantly between 10^{-5} and 10^{-1} depending on the operating temperature, optical confinement structures, as well as the cavity size of VCSELs. For a room temperature operating electrically pumped bulk GaAs VCSEL with a buried circular mesa (i.e., index-guided) structure with a diameter of 7 μm and cavity length of 5 μm, β_s is measured to be 2×10^{-5}. This small value of β_s, which is comparable with those of facet emitting lasers, is due to the relatively large cavity volume [14]. For another electrically pumped index-guided VCSEL, an oxide confined InGaAs QW VCSEL operating at around 970 nm, the corresponding β_s is found to be about 2.5×10^{-3} (4.2×10^{-2}) with 3 μm diameter (3 × 3 μm^2 area) [15,16]. For a similar oxide aperture VCSEL of area equaling 2 × 2 μm^2 but with a GaAs/AlGaAs QW active region and operating at 780 nm, the maximum value of β_s is measured to be 2.1×10^{-3} [17]. Furthermore, for an electrically pumped InGaAs QW VCSEL with airpost geometry and operating at 126 K, the value of β_s is found to be 10^{-2} in a 3-μm-diameter device [18]. This is the highest value of β_s that can be found in the literature for index-guided VCSELs. On the other hand, for an electrically pumped GaAs/AlGaAs gain-guided VCSEL with 6 μm window diameter realized by proton implantation, the maximum value of β_s measured at room temperature is found to be 2.8×10^{-2}. This value of β_s is comparable with that of an index-guided VCSEL [19]. However, further increase in window diameter (i.e., between 10 and 30 μm) causes the reduction of β_s to the order of 10^{-4} [20]. For an optically pumped GaAs VCSEL operating at 4 K, the value of β_s is found to be 10^{-2}, even with a diameter as large as 25 μm [21]. In addition, the maximum value of β_s for an optically pumped microdisk GaAs/AlGaAs QW laser operating at 80 K is found to be as large as 0.15 with a diameter of 2 μm and cavity thickness of 0.12 μm [22]. From the preceding measurement of β_s, it is noted that the value of β_s can be increased by the reduction of cavity volume.

8.2.2 Measurement of Spontaneous Emission Factor in VCSELs

Several methods to measure β_s of semiconductor lasers have been proposed [18–24]. The simplest method involves plotting the light/current curves and comparing the results with the theoretical predictions (i.e., obtained from the simple rate equation model) using β_s as the fitting parameter [14,19,21,25]. However, the laser under investigation requires having a stable single mode operation and negligible temperature rise (<5°C) near threshold. Hence, this method may be less accurate in determining the value of β_s. In fact, the influence of the thermal effect can be avoided in the analysis of β_s by using the modulation method [16,20]. The modulation method can be very accurate if

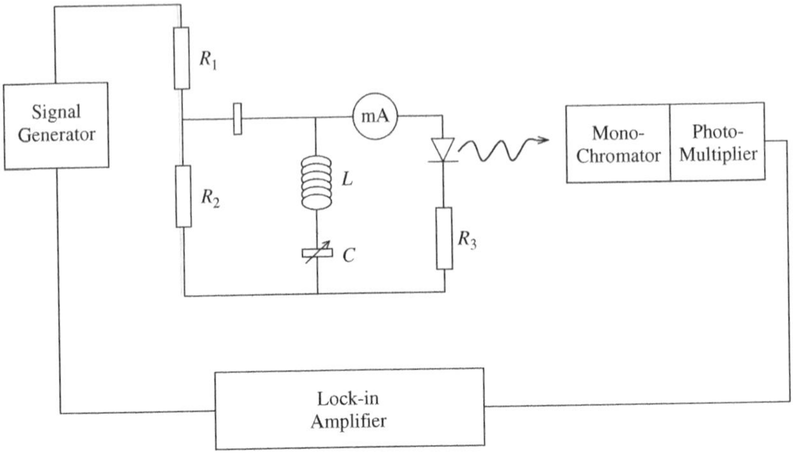

Figure 8.1 Experimental setup to measure the spontaneous emission factor of a semiconductor laser.

all the measurements are made at or near threshold, so that all current-induced carrier changes are insignificant. The accuracy of the modulation method can be further improved if the variation of the transparent current (i.e., transparent carrier density) [17] and carrier lifetime [23] are considered. In the following paragraphs, the use of the modulation method to measure β_s is explained.

Figure 8.1 shows the experimental setup for the measurement of β_s using the modulation method. In the diagram, a low-distortion modulation electrical signal is mixed with a DC bias using a standard bias-T. To avoid short-circuiting the modulation electrical signal, an inductor L and a capacitor C are used to block the DC bias. The magnitude of resistors R_1 and R_2 is carefully chosen to provide a small enough modulation signal, but large enough to be triggered by a lock-in amplifier. Using this modulated electrical signal, a modulated optical signal is generated from a laser under study. In order to analyze this optical signal, a high-resolution monochromator is used to isolate spectrally the optical signal to be studied. The first and second harmonics of the optical signal is detected by a photomultiplier. The lock-in amplifier is also used to record the amplitude of first and second harmonics from the optical signal. It must be noted that the modulation index is kept below 2% to ensure the validity of the small-signal approximation. In addition, the operation temperature is maintained by mounting the laser on a heatsink. Hence, the value of β_s can be determined from the modulation index and the amplitude ratio between the second and first harmonics at the bias current corresponding to the peak second-harmonic generation.

The theory behind the modulation method can be explained by a simple rate equation model of single-mode lasers as

$$\frac{\partial N}{\partial t} = \frac{\eta_j I}{q V_a} - R_t - G(N)S, \tag{8.1}$$

$$\frac{\partial S}{\partial t} = \left(\Gamma_z G(N) - \frac{1}{\tau_p}\right) S + \Gamma_z \beta_s R_{sp}, \tag{8.2}$$

where η_j is the injection efficiency, R_t is interpreted as the total recombination rate, and R_{sp} is the spontaneous emission rate. If the laser is modulated by an injection current with a small component of sinusoidal variation, the electrical modulation signal x can be written as

$$x = x_b + x_m \sin(\omega_m t), \tag{8.3}$$

where ω_m is the modulation frequency, x_b is the normalized bias point for current modulation, and x_m is modulation index. x_b and x_m can be expressed as [16]

$$x_b = \frac{I_b - I_{tr}}{I_{th} - I_{tr}} \quad \text{and} \quad x_m = \frac{I_m}{I_{th} - I_{tr}}, \tag{8.4}$$

where I_b is the bias current for modulation, I_{th} is the threshold current, I_{tr} is the transparency current, and I_m is the amplitude of modulation current. From the small-signal analysis of the rate equation model (i.e., see also Chapter 7), it can be shown that the second harmonic peaks very close to the threshold at $x_{peak} = 1 - 2\beta\eta_j$ and the amplitude ratio between the second and first harmonics at x_{peak} is given by [16]

$$|R_{s/f}| = \frac{1}{4} \frac{x_m}{\sqrt{4\beta_s \eta_r (1 + x_{tr})}}, \tag{8.5}$$

where

$$x_{tr} = \frac{I_{tr}}{I_{th} - I_{tr}} \quad \text{and} \quad \eta_r = \frac{R_{sp}}{R_t}. \tag{8.6}$$

Hence using (8.5), β_s can be written as [16]

$$\beta_s = \frac{1}{4\eta_r(1 + x_{tr})} \left(\frac{x_m}{4R_{s/f}}\right)^2. \tag{8.7}$$

In (8.7), the influence of the nonzero transparent current density has been included in the calculation. In order to extract β_s from the experiment using (8.7), the knowledge of three additional parameters, I_{th}, I_{tr}, and η_r, is needed. The threshold current I_{th} can be determined from the kink position by measuring the differential current/voltage characteristics (i.e., IdV/dI vs. I). The transparency current I_{tr} can be obtained from the measurement of the photoinduced current dependence on the DC bias. This can be realized by using the laser under test as a photodetector and a second identical laser as the light source [26]. The photocurrent induced in the laser is measured as a function of its belowthreshold bias. As the bias increases up to the transparency current, the detected photocurrent signal goes to zero and then changes polarity so that the transparency current can be recorded. The value

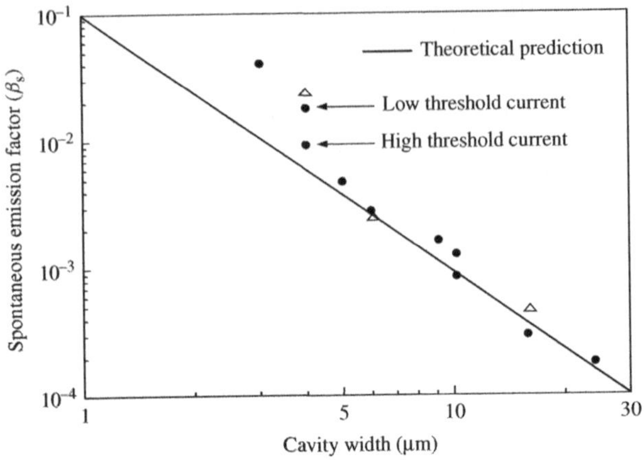

Figure 8.2 Spontaneous emission factor versus the cavity width: solid circles—experimental data obtained from the modulation method; open triangles—experimental data obtained from fitting the measured light/current curves; line—theoretical prediction using (8.8) (after Ref. 16).

of η_r can also be determined from a direct measurement of the dependence of the differential carrier lifetime on the drive current below threshold [27].

Figure 8.2 shows the variation of β_s as a function of the cavity width of an oxide aperture VCSEL with three InGaAs QWs. The laser has a square-shaped oxide aperture and operates at room temperature with a lasing wavelength of 0.98 μm [16]. If the aperture width of VCSELs varies from 3 to 25 μm, the corresponding threshold current, threshold voltage, and FWHM of spontaneous emission spectrum change from 0.27 to 3.3 mA, from 2.6 to 1.4 V, and 0.5 to 1.1 nm, respectively. In Figure 8.2, it is clearly shown that β_s scales as the inverse of the active layer volume. This type of scaling has been predicted by the classical electromagnetic theory [28] and is attributed to a decrease in the number of available cavity modes. Hence, the analytic expression of β_s can be derived from the classical electromagnetic theory as shown below

$$\beta_s = \Gamma_r \xi \frac{K\lambda_R^4}{4\pi^2 n_{eff}^3 V_a \Delta\lambda_s}, \tag{8.8}$$

where λ_R is the resonant wavelength, $\Delta\lambda_s$ is the spontaneous spectral width, V_a is the active layer volume, K is the Petermann astigmatism factor, and n_{eff} is the effective refractive index of the laser cavity. Two extra parameters, Γ_r and ξ, are the relative confinement factor and the dipole radiation enhancement in the direction perpendicular to the plane of the QWs, respectively. For typical index-guided VCSELs with QWs allocated at the center of a λ cavity, Γ_r is roughly equal to 1.8 and that of ξ is ~1.5. For an index-guided laser, K is usually set to 1 in the calculation of β_s. The calculation of β_s using (8.8) is also plotted in Figure 8.2 (solid line) as a function of the cavity width. It is observed that for cavity width greater than 5 μm, the measured data matched with the theoretical prediction. However, for devices of smaller size, the measured β_s is almost a factor of 4 higher than that predicted by the classical theory. The disagreement may be due to the error in determining the cavity size and waveguide parameters of the oxide aperture VCSELs, as the actual cavity size of VCSELs is affected by the light diffraction and self-heating, which are significant in devices with small cavity size. In addition, the reduction of the threshold current (i.e., reduction of optical loss) may increase the value of β_s in the VCSEL for a given size. It is shown in the figure that the two devices of the same size, the laser with lower threshold current as well as a lower threshold carrier concentration, exhibit higher β_s (i.e., cavity size of $4 \times 4\ \mu m^2$). In this experiment, the enhancement of β_s may not obtained from the alternation of spontaneous emission (i.e., microcavity effects) as the minimum cavity size under investigation is not small enough to achieve microcavity effects.

In the determination of β_s using (8.7), the carrier lifetime is assumed to be a constant in devices with active layers of GaAs or InGaAs QW materials. However, the assumption of constant carrier lifetime may not be applied to devices with InGaAsP gain material as β_s may overestimate by a factor of >2. Therefore, the variation of η_r should be considered using the below threshold data of output power and cavity loss. The variation of spontaneous emission lifetime with carrier concentration below threshold can be written as [17]

$$\eta_r = \frac{2B_{sp}N}{\tau_n^{-1} + 2B_{sp}N + 3C_{Aug}N^2}, \tag{8.9}$$

where B_{sp} and C_{Aug} for InGaAsP material have the typical value of 1×10^{-10} cm^3/s and 5×10^{-30} cm^6/s, respectively. The carrier concentration below threshold can be estimated from

$$\alpha - \Gamma_z G(N) = \frac{2\pi n_{eff} \Delta\lambda_s}{\lambda_R^2}, \tag{8.10}$$

where α is the total cavity loss and the FWHM of the spontaneous emission spectrum of the fundamental mode $\Delta\lambda_s$ is measured below or at threshold with the variation of the injection current.

8.3 SPONTANEOUS EMISSION IN VCSELs

The modification of spontaneous emission in VCSELs has been studied extensively using classical and quantum electrodynamical calculations [29–34]. Quantum mechanism approaches, which involve particle wavefunction and cavity modes, have been used to analyze the alternation of spontaneous emission [32–34]. In the early development of these quantum mechanism models, a two-level atom was applied to describe the electrical properties of active media [30,34]. A more realistic description of semiconductor materials using an electron–hole pair wavefunction has also been considered [31]. In order to simplify the investigation of modal density, the optical cavities of VCSELs can be grouped into two categories: (1) the fully confined 3D optical cavities, in which the optical field is confined tightly along the transverse direction [33,34,40]; and (2) the planar Fabry–Perot structure, in which the mirrors can be extended infinitely along the transverse direction [30,35]. For a 3D optical cavity such as those in airposted VCSELs, the alteration of spontaneous emission due to the existence of resonant modes, leaky modes, and propagation modes should be considered [32,38]. In addition, the dipole moment in a 3D optical cavity can be varied by electron quantum confinement, which modifies the interaction strength between the optical modes and dipole inside the active region. This should be introduced into the calculation of spontaneous emission [33]. For VCSELs with planar structure, hard mirrors are usually used to simplify the calculation of the resonant fields, but a complete description of longitudinal field profile inside a realistic VCSEL requires the consideration of DBRs. Propagation modes should also be included in the calculation; otherwise the spontaneous emission rate may be overestimated [36]. In fact, this planar structure can be used to approximate the modal fields of an oxide aperture VCSEL [32,33].

In this section, a simple 3D model for calculating the spontaneous emission energy and hence the spontaneous emission factor of an airposted VCSEL is described. The 3D model is realized by solving the electromagnetic wave equation for the radiation energy emitting from an average dipole. Using this model, the possibility to enhance the spontaneous emission factor in an airposted VCSEL with QWs active layer is explained. On the other hand, the role of the waveguide light emission in a planar VCSEL is investigated and the influence of oxide aperture on the spontaneous emission rate is also discussed.

8.3.1 Radiation Energy of Airposted VCSELs

Figure 8.3 shows a schematic of an airposted VCSEL used in the investigation. The DBRs are constructed by 15.5 pairs of AlAs ($n_L = 2.95$)/GaAs ($n_H = 3.6$) with a power reflectivity R of 99.1%. It is assumed that the DBRs are surrounded by GaAs contact layer and substrate in the longitudinal direction and air in the transverse direction. Two types of active regions, the bulk and QWs active layers, are considered in this analysis. In the case of bulk, GaAs with refractive index n_a and thickness d_a is used to construct a λ cavity. For the case of QWs, it is assumed that the InGaAs/GaAs QWs, having effective refractive index n_a and total physical thickness d_a of 3.6 and 10 nm, respectively, are sandwiched by two GaAs confinement layers. Each GaAs confinement layer ($n_c = 2.95$) with a thickness d_c is used to construct the λ cavity. The value of d_c can be determined by the following Equation [32]

$$n_a d_a + 2n_c d_c = \lambda_B, \tag{8.11}$$

where λ_B ($\sim$1 μm) is the Bragg wavelength. The schematic field distribution of the fundamental mode in the 3D microcavity is also illustrated in Figure 8.3.

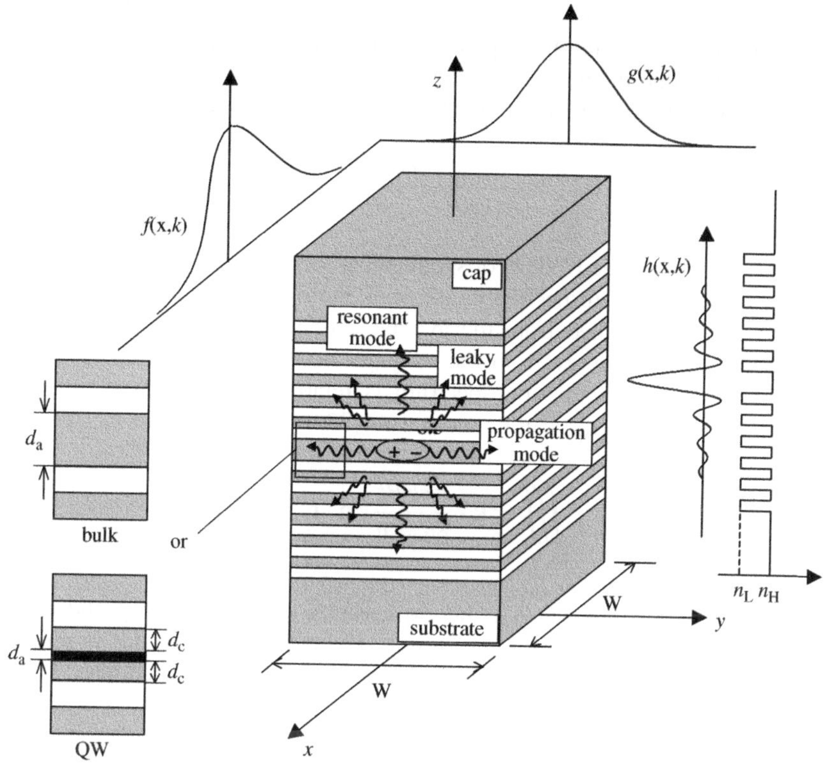

Figure 8.3 Schematic diagram of an airposted microcavity VCSEL. The longitudinal index distribution and the profile of the fundamental mode in the transverse and longitudinal directions are also shown in the figure. Bulk or QW materials can be used to construct the active layer.

The spontaneous emission rate and spontaneous emission factor of an airposted VCSEL can be evaluated with the knowledge of radiation energy coupled to the dominant mode inside the laser cavity. In fact, the radiation energy emitting from an average dipole inside a microcavity can be generally expressed in terms of the distribution of energy density over a **k** space where the **k** space corresponds to the orientation and wavelength of light. If arbitrary electric fields of polarization s and p (i.e., s wave polarizing parallel to the substrate plane and p wave polarizing inside the incident plane) are considered in the derivation, it can been shown that (see Section 8A.1) the total radiation energy, $\overline{\Re}$, from the average dipole, which couples into the unit volume of active region and the unit volume of **k** space (i.e., the unit interval of optical frequency and the unit solid angle of light) around **k**, can be written as [32,33]

$$\overline{\Re}(\mathbf{k}) = \frac{\pi c \tau_{sp} k^2 P_0^2}{8 n_a \varepsilon_o} \sum_{i=s,p} \langle |\hat{\mathbf{p}} \cdot \hat{\mathbf{e}}|^2 \rangle_i \Lambda(\mathbf{k}) \rho_i(\mathbf{k}), \tag{8.12}$$

where the subscripts s and p denote the s and p waves, respectively; τ_{sp} is the spontaneous emission lifetime; $\langle|\hat{\mathbf{p}}\cdot\hat{\mathbf{e}}|^2\rangle_i$ is the averaged dipole moment; P_0 is the initial amplitude of the polarization field; and $\Lambda(\mathbf{k})$ is the spectral function normalized so that the integration over the entire k $(=|\mathbf{k}|)$ is unity and is given by

$$\Lambda(k) = \frac{\Delta k_s/2}{\pi[(\Delta k_s/2)^2 + (k - k_p)^2]}, \tag{8.13}$$

where Δk_s and k_p are the FWHM of spontaneous emission spectrum and the peak emission wavenumber, respectively, and π is the normalization factor; $\rho_i(\mathbf{k})$ is the mode density obtained from the average mode energy in the unit volume. If the total radiation energy from the average dipole has photon energy of $\hbar\omega$ and the broadening of emission spectrum is relatively small, the spontaneous emission rate R_{sp} ($\propto 1/\tau_{sp}$) can be written as [32,33]

$$R_{sp} = \sum_{i=s,p} \iiint_{\text{all } \mathbf{k}} d^3\mathbf{k} \cdot \frac{\pi c k P_0^2}{8 n_a \varepsilon_o \hbar} \langle|\hat{\mathbf{p}}\cdot\hat{\mathbf{e}}|^2\rangle_i \Lambda(\mathbf{k}) \rho_i(\mathbf{k}). \tag{8.14}$$

In order to solve $\overline{\Re}$ and R_{sp}, the expressions of mode density and dipole moment have to be deduced. If the profile of the transverse modes is uniform over the longitudinal direction, the mode density can be decoupled into

$$\rho_i(\mathbf{k}) = \rho_{z,i}(\mathbf{k}) \times \rho_t(\mathbf{k}), \tag{8.15}$$

where $\rho_{z,i}$ is the longitudinal mode density which is identical to the mode density of a microcavity with planar DBRs and ρ_t is the transverse mode density. As discussed in Section 8A.1, $\rho_{z,i}$ is considered as a continuous function of $\mathbf{k}$ but ρ_t is discrete in the $\mathbf{k}$ space. The expressions for ρ_z and ρ_t are [32]

$$\rho_{z,i} = \sum_{\substack{\cos\\ \sin}} \frac{\Gamma_{r,i}}{2\pi A_i^2}, \tag{8.16}$$

$$\rho_t = \frac{1}{W^2} \cdot \frac{\displaystyle\iint_{-W/2}^{W/2} n_a^2 f(k_x; x) g(k_y; y) dx\, dy}{\displaystyle\iint_{-\infty}^{\infty} n_s^2 f(k_x; x) g(k_y; y) dx\, dy}, \tag{8.17}$$

where W is the cavity width, A_i is the ratio of field amplitude in the substrate to that in the active region, and $\Gamma_{r,i}$ is the relative confinement factor. The summation in (8.16) denotes the sum of two degenerate mode fields: cosine and sine types. The factor A_i for s and p polarizations can be calculated independently by using the characteristic matrix method, and the corresponding derivation is given in Section 8A.2. $\Gamma_{r,i}$, which is similar to the confinement factor defined in (2.28), represents the interaction strength between the longitudinal modes and dipoles inside the active region. For electric field of s polarization, $\Gamma_{r,s}$ can be written as [33]

$$\Gamma_{r,s} = \begin{cases} 1 + \dfrac{\sin(k_z d_a)}{k_z d_a} & \text{cosine type} \\ 1 - \dfrac{\sin(k_z d_a)}{k_z d_a} & \text{sine type} \end{cases} . \quad (8.18)$$

For the electric field of p polarization, $\Gamma_{r,p}$ is given by

$$\Gamma_{r,p} = \begin{cases} \dfrac{k_t^2}{k^2}\left[1 + \dfrac{\sin(k_z d_a)}{k_z d_a}\right] + \dfrac{k_z^2}{k^2}\left[1 - \dfrac{\sin(k_z d_a)}{k_z d_a}\right] & \text{cosine type} \\ \dfrac{k_t^2}{k^2}\left[1 - \dfrac{\sin(k_z d_a)}{k_z d_a}\right] + \dfrac{k_z^2}{k^2}\left[1 + \dfrac{\sin(k_z d_a)}{k_z d_a}\right] & \text{sine type} \end{cases} , \quad (8.19)$$

where $k_t^2 = k_x^2 + k_y^2$ and $k^2 = k_x^2 + k_y^2 + k_z^2 = (2\pi n_a/\lambda_R)^2$. It is noted that the expression of $\Gamma_{r,p}$ is the mixture of sine and cosine-type of $\Gamma_{r,s}$ because the p polarization has the parallel and normal components to the substrate plane. Figure 8.4 shows the characteristics of $\Gamma_{r,s}$ for a typical VCSEL with planar DBRs supporting fundamental longitudinal mode. It is observed that $\Gamma_{r,s}$ approaches one when $k_z d_a$ is close to infinity, which implies that the effect of $\Gamma_{r,s}$ is leveled in the large bulk. On the other hand $\Gamma_{r,s}$ varies in the range from 2 to 0 depending on whether the QW locates at the maximum or the minimum of the longitudinal mode. If the QW is arranged at the center of the λ cavity, $\Gamma_{r,s}$ can maximize to 2 and the energy coupled into the cosine-type fundamental mode almost doubles. It must be noted that for the radiation energy propagating normal to the substrate (i.e., $k_t = 0$), Figure 8.4 can also be used to describe $\Gamma_{r,p}$ but with the sine and cosine functions interchanged.

The variation of ρ_z and ρ_t as well as the total radiation energy can be easily evaluated for both polarizations provided that the dependence of $\Gamma_{r,i}$ and A_i on k is deduced. Figure 8.5 plots the variation of ρ_z in the k_z–k_t plane for devices with bulk active layer and thickness of $d_a = \lambda_B/n_a$ (i.e., λ cavity). It must be noted that for devices with bulk active region, $\Gamma_{r,i}$ for both sine and cosine types are equal to one, and only ρ_z for s and p polarizations need be included in the calculation. In the figure, the intensity of ρ_z is displayed by a shaded drawing in four levels; the largest magnitude of ρ_z, the deepest is the drawing. For s polarization as shown in Figure 8.5a, there exists a resonant mode satisfying the Bragg condition $k \sim 2\pi n_a/\lambda_B$ ($\equiv k_B$), and is indicated by a symbol A. This type of mode is usually referred to as the Bragg mode in DBR and DFB lasers [1]. Around this mode, there are spaces of very small ρ_z, which correspond to the stopband of the DBRs. These characteristics indicate that the spontaneous emission is enhanced in the direction normal to the substrate plane (i.e., $k = k_z \sim k_B$ and $k_t \sim 0$), the spontaneous emission of $k < k_B$ ($\lambda > \lambda_B$) is suppressed along the z direction, and the spontaneous emission of $k > k_B$ ($\lambda < \lambda_B$) is enhanced toward the oblique direction. In addition, the spread of

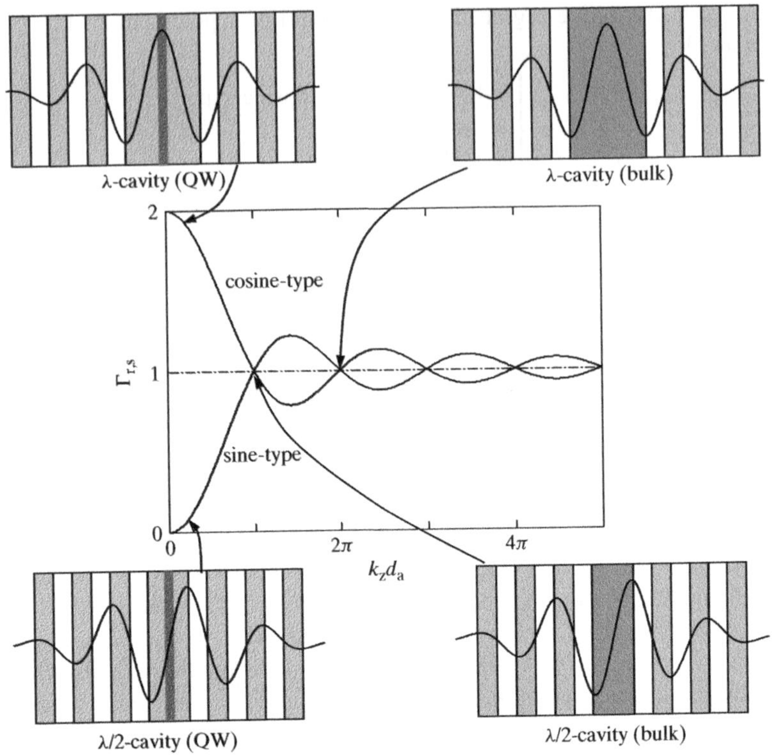

Figure 8.4 Plots of relative confinement factor Γ_r versus the normalized thickness of active layer $k_z d_a$ for both sine- and cosine-type functions. The corresponding cavity structure and field distributions in the DBRs are also shown in the figure.

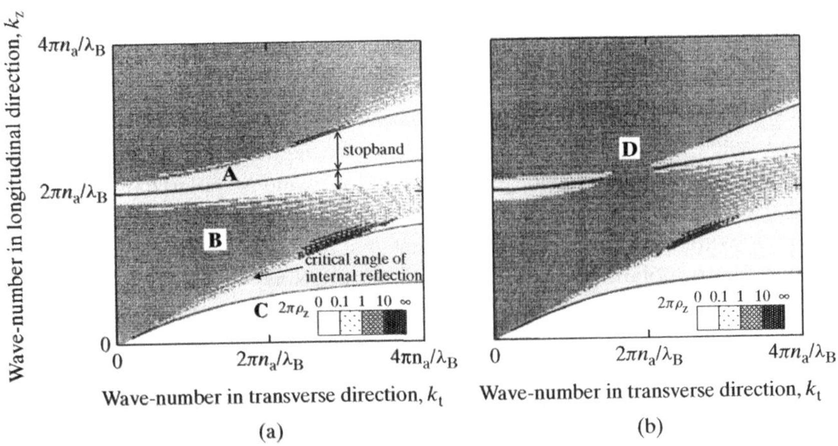

Figure 8.5 Distribution of longitudinal mode density ρ_z over the **k** space for an airposted microcavity VCSEL with λ cavity and bulk active layer: (a) *s* polarization; (b) *p* polarization. (After Ref. 32.)

stopband increases with the increase of reflectivity of the DBRs, which results in a more remarkable enhancement or suppression of spontaneous emission. In Figure 8.5a, the halftone area (indicated by symbol B) widespread around the stopband shows the weakly resonant leaky waves. The boundary of the halftone area shows the critical angle of the internal reflection due to n_a and average index of the mirrors. The curve that is indicated by the symbol C shows strongly resonant modes propagating along the transverse direction. This mode, referred to as a *propagation modes*, is discussed in more detail in Section 8.3.3. For p polarization as shown in Figure 8.5b, the Brewster angle breaks the curve of the Bragg resonant mode off, as indicated by the symbol D. However, the other characteristics of ρ_z are similar to those of the s polarization around k_B.

Figure 8.6 shows the variation of ρ_z in the k_z–k_t plane for a λ cavity laser with QW located at the center of the active region. In the calculation, $\Gamma_{r,i}$ of sine and cosine types is included. It is observed that for s wave, the cosine-type field has large mode density over all the **k** space, and maintains the Bragg resonant mode (A), leaky waves (B), and transverse propagation modes (C). This profile of ρ_z is quite close to that of a device with a bulk active region. In contrast to this, the sine-type field has quite small mode density and maintains no such special modes, caused by the difference of $\Gamma_{r,s}$ between the cosine and sine types. For the

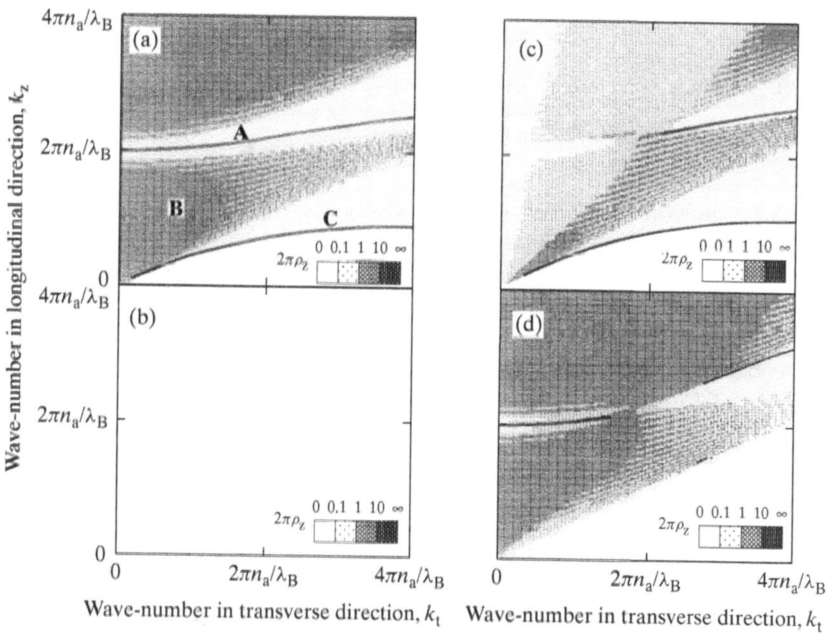

Figure 8.6 Distribution of longitudinal mode density ρ_z over the **k** space for an airposted microcavity VCSEL with λ cavity and QW active layer: (a) s-polarization cosine-type field; (b) s-polarization sine-type field; (c) p-polarization cosine-type field; (d) p-polarization sine-type field. (After Ref. 33.)

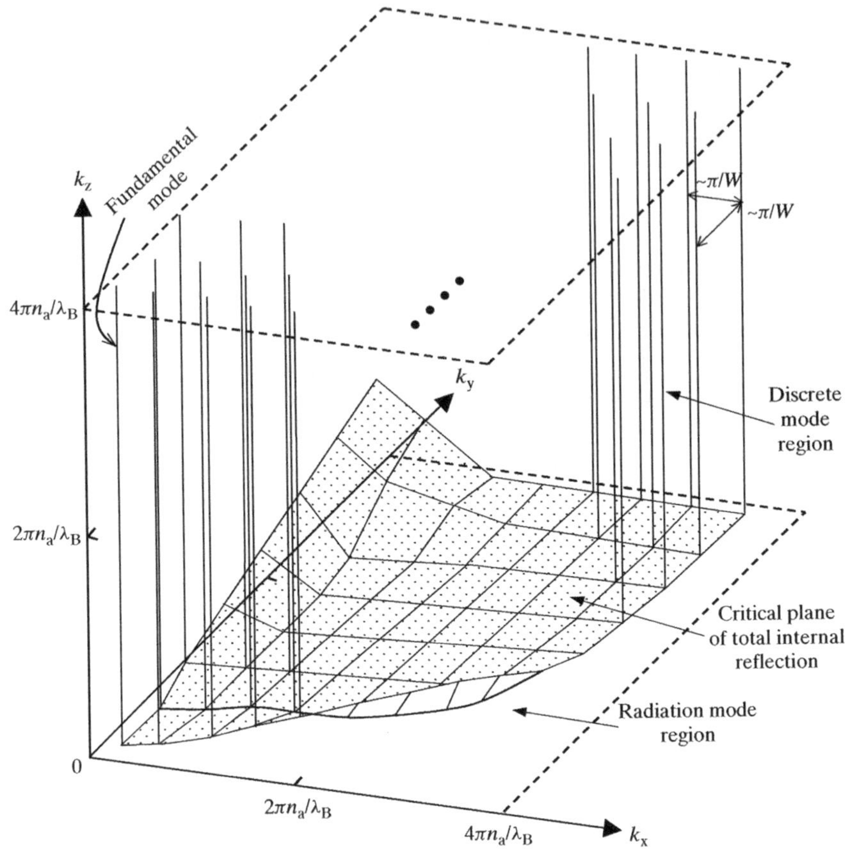

Figure 8.7 Distribution of transverse mode density ρ_t over the **k** space for an airposted microcavity VCSEL with $W = 2\lambda_B/n_a$ (~0.55 μm); not all the discrete modes are shown in the diagram (after Ref. 32).

p wave, both cosine and sine ρ_z have significant value owing to the complicated expression of $\Gamma_{r,p}$.

Figure 8.7 shows the calculated ρ_t for $W = 2\lambda_B/n_a$ (~0.55 μm) over the **k** space, but not all the discrete modes are shown. It is noted that the critical plane of total reflection separates the discrete mode region and radiation mode region of the transverse modes. The dispersion curve closest to the k_z axis denotes the fundamental transverse mode. The interval between the neighboring mode is above π/W. This figure indicates that a large portion of transverse modes suffer total internal reflection. Hence, the calculation of ρ for a 3D cavity can be obtained by overlapping the discrete transverse modes and the continuous longitudinal mode using (8.15). If the value of W increases so that there is no transverse confinement, the mode density approaches that of a free plane (i.e., $1/\pi^2$). The corresponding distribution of the mode density in this 3D cavity is obtained by multiplying the longitudinal mode density as shown in Figures 8.5 and 8.6 by $1/\pi^2$. This is identical to the mode density in a 1D microcavity with a pair of DBRs.

The dipole moment $\langle|\hat{\mathbf{p}}\cdot\hat{\mathbf{e}}|^2\rangle_i$ at the subband edge can be expressed independently for s and p polarizations by [33]

$$\langle|\hat{\mathbf{p}}\cdot\hat{\mathbf{e}}|^2\rangle_s = \begin{cases} \frac{1}{3} & \text{bulk} \\ \frac{1}{2} & \text{QW} \end{cases}, \tag{8.20}$$

$$\langle|\hat{\mathbf{p}}\cdot\hat{\mathbf{e}}|^2\rangle_p = \begin{cases} \frac{1}{3} & \text{bulk} \\ \frac{\cos^2\theta}{2} & \text{QW} \end{cases}, \tag{8.21}$$

where θ is the angle between $\mathbf{k}$ and k_z. Using (8.20) and (8.21), the influence of electron quantum confinements inside the QW active region can be illustrated by examining the radiation pattern of the spontaneous emission from a planar microcavity with DBRs. The radiation pattern is given by assuming that ρ_t is the mode density of xy free plane (i.e., $\rho_t = 1/\pi^2$) and by integrating $\overline{\Re}(\mathbf{k})$ with k. Figure 8.8 shows the calculated radiation patterns of VCSELs with QW and bulk active regions, both of $\Delta\lambda_s = 30$ nm. It is observed that the intensity of the main lobe from the QW laser is almost 3 times that from the bulk device. This is because of the difference in mode density and the effective dipole moment arises

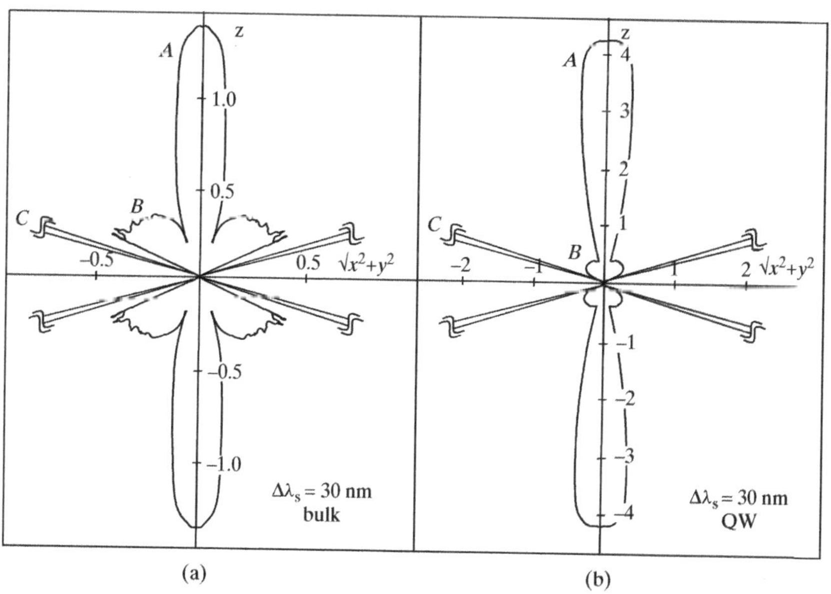

Figure 8.8 Radiation pattern from an average dipole located in (a) bulk and (b) QW active region. $\Delta\lambda_s$ for both active regions is assumed equal to 30 nm. The radiation lobes A, B, and C refer to the corresponding ρ_z over the $\mathbf{k}$ space as indicated in Figures 8.5a and 8.6a. (After Ref. 33.)

from the electron quantum confinements inside the QW active region, which improve the peak intensity and sharpness of the main lobe of the radiation profile.

8.3.2 Spontaneous Emission Factor of Airposted VCSELs

Using (8.8), the general expression of the spontaneous emission factor of ith mode $\beta_{\mathrm{s,i}}$ can be written as [32]

$$\begin{aligned}\beta_{\mathrm{s},i} &= \frac{1}{2}\left[\left(\frac{1}{2\pi}\right)^2 \times \left(\frac{n_{\mathrm{eff}}^3 V_{\mathrm{a}}}{\xi \Gamma_{\mathrm{r},i} K}\right) \times \left(4\pi k_{\mathrm{p}}^2 \frac{\pi}{2} \Delta k_{\mathrm{s}}\right)\right]^{-1}, \\ &= \tfrac{1}{2}[(\text{mode density in the free space}) \\ &\quad \times (\text{effective mode volume}) \\ &\quad \times (\text{wavenumber volume in the emission spectrum})]^{-1} \\ &= \tfrac{1}{2}[(\text{total amount of modes in the emission spectrum})]^{-1}. \qquad (8.22)\end{aligned}$$

This implies that (8.22) is derived by assuming that modes distribute uniformly so that the same radiation energy couples to every mode. This assumption is valid when the laser cavity is much larger than the lasing wavelength. However, it is not accurate to apply (8.8) to estimate β_{s} for a microcavity laser as the corresponding distribution of modal density may alternate too much from that of the uniform distribution in free space. For an accurate calculation of β_{s} in VCSELs, it is required to calculate the ratio between the radiation energy couples to the fundamental mode and that of the total radiation energy in the entire emission spectrum. In this case, the spontaneous emission factor of a linear polarized fundamental mode can be expressed as [32]

$$\beta_{\mathrm{s},0} = \mathbf{p} \cdot \frac{\displaystyle\iiint_{\text{around } k_{\mathrm{p}}} \overline{\Re}(\mathbf{k}) \cdot d^3\mathbf{k}}{\displaystyle\iiint_{\text{all over } k} \overline{\Re}(\mathbf{k}) \cdot d^3\mathbf{k}}, \qquad (8.23)$$

where $\mathbf{p}$ is the polarization probability of the polarized fundamental mode and is expressed as

$$\mathbf{p} \equiv \frac{\displaystyle\iiint_{\text{around } k_{\mathrm{p}}} \overline{\Re}_{\mathrm{s}}(\mathbf{k}) d^3\mathbf{k}}{\displaystyle\iiint_{\text{around } k_{\mathrm{p}}} \overline{\Re}(\mathbf{k}) d^3\mathbf{k}}. \qquad (8.24)$$

In the derivation of (8.24), it is assumed that the linearly polarized fundamental mode is output toward almost normal to the substrate plane and the s-polarized mode is along the x direction of the laser cavity. Furthermore, the lasing mode has a broadening of resonant wave vector $\mathbf{k}_p$ corresponding to the cavity loss at the low current injection level so that the radiation energy coupled into the polarized fundamental mode is

$$\iiint_{\text{around } k_p} \overline{\Re}_s(\mathbf{k}) d^3\mathbf{k}.$$

If a square xy cross section and a bulk active region are considered in the calculation, $\mathbf{p}$ is 0.5 for arbitrary polarization.

Figure 8.9a shows the calculated $\beta_{s,0}$ using (8.23) for the fundamental mode of VCSELs with bulk active region. $\beta_{s,0}$ is plotted against the spectral width $\Delta\lambda_s$ (i.e., $\Delta\lambda_s/\lambda_R = \Delta k_s/k_p$) with $d_a = \lambda_B/n_a$. The estimation of $\beta_{s,0}$ using (8.8) is also shown in Figure 8.9a as the dashed lines for different values of W. In order to calculate $\beta_{s,0}$, the following analytic expressions of $\Gamma_{r,0}$ and n_{eff} are used, and their derivations are shown in Section 8A.3.

$$\Gamma_{r,0} = d_a \cdot \left(d_a + \frac{\lambda_B}{2n_a} \cdot \frac{1 + (n_L/n_H)^2}{1 - (n_L/n_H)^2} \right)^{-1} \cong 2 \cdot \frac{1 - (n_L/n_H)^2}{3 - (n_L/n_H)^2}, \tag{8.25}$$

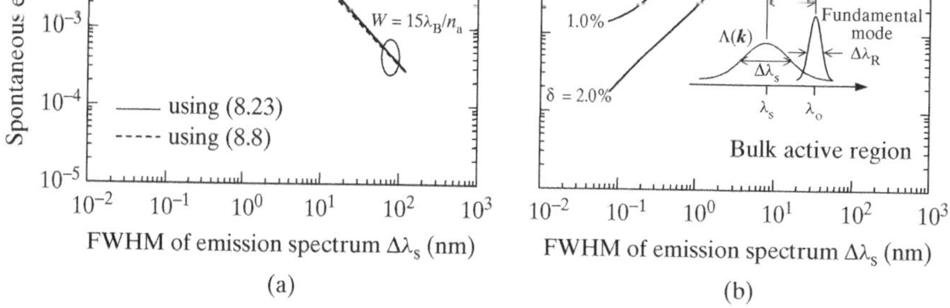

Figure 8.9 Spontaneous emission factor $\beta_{s,0}$ (fundamental mode) versus width of emission spectrum $\Delta\lambda_s$ of an airposted microcavity VCSEL with λ cavity and bulk active region: (a) peak emission wavelength λ_s tuned to the resonant wavelength of the fundamental mode λ_R; (b) λ_s detuned from λ_B by the normalized wavelength shift $\delta = (\lambda_R - \lambda_s)/\lambda_s$. (After Ref. 32.)

$$n_{\mathrm{eff}} = n_{\mathrm{a}} \sqrt{\frac{d_{\mathrm{a}} + \left(\dfrac{n_{\mathrm{H}}}{n_{\mathrm{a}}}\right)^2 \dfrac{\lambda_{\mathrm{B}}}{2n_{\mathrm{a}}} \cdot \dfrac{2(n_{\mathrm{L}}/n_{\mathrm{H}})^2}{1-(n_{\mathrm{L}}/n_{\mathrm{H}})^2}}{d_{\mathrm{a}} + \dfrac{\lambda_{\mathrm{B}}}{2n_{\mathrm{a}}} \cdot \dfrac{1+(n_{\mathrm{L}}/n_{\mathrm{H}})^2}{1-(n_{\mathrm{L}}/n_{\mathrm{H}})^2}}} \cong \sqrt{\frac{2n_{\mathrm{H}}^4}{3n_{\mathrm{H}}^2 - n_{\mathrm{L}}^2}}, \tag{8.26}$$

where n_L and n_H are the refractive indices of the lower and higher index layers of the DBRs, respectively, and d_{a} is assumed to be $\lambda_{\mathrm{B}}/n_{\mathrm{a}}$. In addition, ξ is assumed equals to 1 in the calculation of $\beta_{\mathrm{s},0}$ for devices with bulk active region. From the figure, it is noted that when $\Delta\lambda_{\mathrm{s}}$ is greater than 10 nm, $\beta_{\mathrm{s},0}$ is inversely proportional to $\Delta\lambda_{\mathrm{s}}$ and W^2. This agrees with the calculation of (8.22) as indicated by the dashed curves. For $W = 2\lambda_{\mathrm{B}}/n_{\mathrm{a}}$, $\beta_{\mathrm{s},0}$ exceeds 0.1, although $\Delta\lambda_{\mathrm{s}}$ is larger than 30 nm. If $\Delta\lambda_{\mathrm{s}}$ reduces to less than 30 nm, $\beta_{\mathrm{s},0}$ approaches 0.5 because of its upper limit determined by the polarization probability ($\beta_{\mathrm{s},0} = 0.11$, 0.19, and 0.28, with $\Delta\lambda_{\mathrm{s}} = 20$, 10, and 5 nm, respectively).

Although the dependence of $\beta_{\mathrm{s},0}$ on only $\Delta\lambda_{\mathrm{s}}$ and W is shown in Figure 8.9, $\beta_{\mathrm{s},0}$ is also a function of d_{a} and the reflectivity of DBRs R. It is noted that $\beta_{\mathrm{s},0}$ is inversely proportional to d_{a} when d_{a} is much larger than the wavelength in the active region. However, when d_{a} becomes as small as the wavelength, the increase of $\beta_{\mathrm{s},0}$ is not linear against d_{a} because of the light penetration outside the active region. It can be shown that for $d_{\mathrm{a}} = \lambda_{\mathrm{B}}/2n_{\mathrm{a}}$, $\beta_{\mathrm{s},0}$ is found to be 1.7 times larger than that for the case $d_{\mathrm{a}} = \lambda_{\mathrm{B}}/n_{\mathrm{a}}$. If d_{a} is much smaller than the resonant wavelength, the relation between $\beta_{\mathrm{s},0}$ and d_{a} will change according to the position for the active region against the maximum of the mode field due to the presence of Γ_{r}. The dependence of R is obtained from the relationship between the spontaneous spectral width $\Delta\lambda_{\mathrm{s}}$ and resonant spectral width $\Delta\lambda_{\mathrm{R}}$. As shown in the curve for $W = 15\lambda_{\mathrm{B}}/n_{\mathrm{a}}$ in Figure 8.9a, $\beta_{\mathrm{s},0}$ saturates below the upper limit 0.5 around $\Delta\lambda_{\mathrm{s}} = \Delta\lambda_{\mathrm{R}}$ (~0.16 nm). This is because the amount of energy coupled to the fundamental mode no longer increases when the emission spectrum is completely included into the resonant bandwidth. Conversely, the results of Figure 8.9 are valid as long as the bandwidth $\Delta\lambda_{\mathrm{R}}$ is smaller than the critical $\Delta\lambda_{\mathrm{s}}$ at which $\beta_{\mathrm{s},0}$ approaches the upper limit 0.5. In Figure 8.9a, the critical $\Delta\lambda_{\mathrm{s}}$ for $W = 2\lambda_{\mathrm{B}}/n_{\mathrm{a}}$ is 6–7 nm. Thus, the enhancement of $\beta_{\mathrm{s},0}$ is possible by $\Delta\lambda_{\mathrm{R}} < 5$ nm, which is equivalent for devices with $R > 90\%$.

Figure 8.10 shows the calculated characteristics of $\beta_{\mathrm{s},0}$ for lasers with bulk and QW active regions. $\beta_{\mathrm{s},0}$ is plotted against the transverse size W of the cavity. The spectral widths $\Delta\lambda_{\mathrm{s}}$ of the bulk and QW devices are assumed equal to 30 and 15 nm, respectively, in the calculation. Furthermore, the resonant wavelength is tuned to the peak emission wavelength. It is observed that $\beta_{\mathrm{s},0}$ of devices with QW active regions exceeds 10^{-2} even with $W = 3\ \mu$m. When W is as small as $2\lambda_{\mathrm{B}}/n_{\mathrm{a}}$, $\beta_{\mathrm{s},0}$ of the QW device can be as large as 0.15. The enhancement of $\beta_{\mathrm{s},0}$ in VCSELs with QW active regions is due to the fact that (1) the increase of mode density $\rho(\mathbf{k})$ through $\Gamma_{\mathrm{r},i}$ with the suitable allocation of QW layers inside the active region, (2) the anisotropic emission intensity and polarization increase the effective dipole moment $\langle|\hat{\mathbf{p}} \cdot \hat{\mathbf{e}}|^2\rangle_i$ of QW active regions over the bulk ones, and

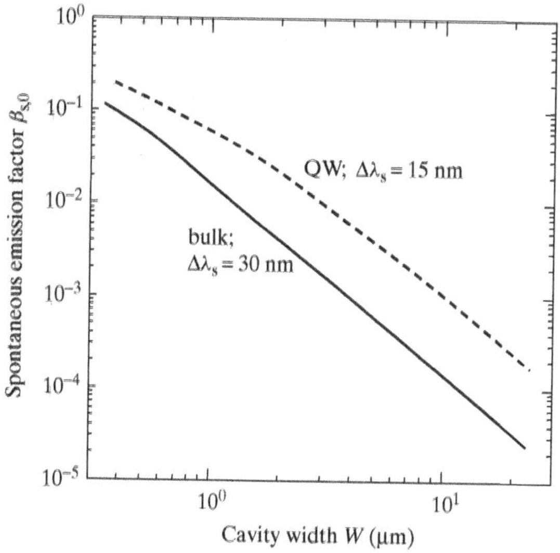

Figure 8.10 Spontaneous emission factor $\beta_{s,0}$ (fundamental mode) of airposted microcavity VCSELs with bulk (solid line) and QW (dashed line) versus the cavity width W (after Ref. 33).

(3) the spectral width $\Delta\lambda_s$ of QW material is reduced by the electron quantum confinement. However, the upper limit of $\beta_{s,0}$ for lasers with both bulk and QW active regions is 0.5. In general, the value of $\beta_{s,0}$ is much larger than that of conventional facet emitting lasers.

In the preceding sections, it is shown that the enhancement of β_s is possible if (1) a λ cavity is used, (2) the QW active layer is allocated at the peak of longitudinal mode, and (3) the transverse size of the laser cavity is reduced with the appropriated selection of $\Delta\lambda_s$ and R. However, the maximum value of β_s is saturated at 0.5 because of the polarization probability of a linear polarized mode. It is believed that the remaining polarization light will radiate into the direction other than that from the surface of the laser cavity. This postulation can be verified by measuring β_s of an airposted VCSEL along the transverse direction [31]. The gain region of the airposted VCSEL emitting near 0.85 μm from the surface is constructed by five 70-Å GaAs QWs allocated at the center of a 1λ-long AlGaAs cavity. The laser cavity is sandwiched by two DBRs in which the lower n-type mirror has 29 periods and the upper p-type mirror has 20 periods. Each period is comprised of quarter-wavelength-thick AlAs/AlGaAs semiconductor layers. A monolithic photodetector, which completely encircles the laser cavity, is used to measure the spontaneous emission emitted from the transverse direction. It is demonstrated that the spontaneous emission from the cavity perimeter (i.e., transverse direction) is overwhelmingly larger than the same emitting along the lasing direction (i.e., vertical direction) [31]. Hence, it can be understood that the upper limit of β_s in the airposted VCSEL (see also Section 8.4.2) is due to the radiation of polarized light along the transverse direction. Therefore, it is expected that VCSELs with metal-clad optical waveguide can enhance the value of β_s above the 0.5 limitations as well as reduce the spontaneous emission lifetime τ_{sp}. It is estimated that the use of metal clad optical waveguide in airposted VCSELs can increase β_s to nearly 1 and reduce τ_{sp} by 100 times when compared to facet emitting lasers [30].

8.3.3 Spontaneous Emission Rate in Planar VCSELs

In Section 8.3.1, the spontaneous emission rate R_{sp} of an airposted VCSEL has been derived using a radiation energy approach. It can be shown that R_{sp} is dependent on the spontaneous emission linewidth $\Delta\lambda_s$ and the cavity size and R, the reflectivity of the DBRs. R_{sp} increases with increase of $\Delta\lambda_s$ and saturates for a large value of $\Delta\lambda_s$. However, the variation of R_{sp} is not directly proportional to the cavity sizes because of the resonant conditions of the transverse modes. Furthermore, R_{sp} increases with increase of R and is maximized as R approaches one.

The spontaneous emission rate R_{sp} can also be evaluated from Fermi's golden rule with the assumption that the transition of an atom in a two-level system (i.e., dipole model) generates an optical field coupled to the cavity modes. Hence, the modal density of the quantum-mechanical model can assume constant within the transition linewidth. This approximation is reasonable because a typical semiconductor has a homogeneously broadened transition line with a width of a few percent of transition center frequency. In this case $\Delta\lambda_s$ is much wider than $\Delta\lambda_R$ so that the dependence of R_{sp} on $\Delta\lambda_s$ can be omitted in the calculation. Therefore, this approach allows us to concentrate on the optimization of R_{sp} with different mirrors in planar VCSELs [29,30,36,37]. In fact, it can be shown that if the planar microcavity VCSELs have (1) cavity length of $\lambda/2$, (2) dipole sheet placed in the center of the cavity, and (3) reflectivity of hard mirrors equal to 100%, R_{sp} can be enhanced by a factor of 3 (i.e., compared to the case without cavity) [29]. However, other investigations have shown that for any dissipative cavity (i.e., $R < 100\%$), the actual cavity length should be slightly longer than $\lambda/2$ in order to maximize R_{sp} [36]. Figure 8.11 plots the normalized R_{sp} (normalized to the case without cavity) versus normalized cavity length (normalized to lasing wavelength) for a symmetric planar VCSEL with hard mirrors. It is assumed that the microcavities have three different reflectivities (i.e., $R = 90\%$, 95%, 99%) and using $R = 0$ as a comparison. It is seen that for a dissipative cavity ($R < 100\%$), the spontaneous emission rate is about 1.4 times larger than that of the case with $R = 0$ for $\lambda/2$ cavity. Radiation patterns of three different cavity length with $R = 95\%$ (i.e., positions A, B, and C) are also plotted in the figure. For cases A and B with the normalized $R_{sp} \leq 1.4$, a single-lobe emission is observed. But for case C with normalized $R_{sp} \sim 2.2$, a double-lobe shape is excited. The double-lobe shape, which corresponds to one of the higher-order transverse modes in a 3D cavity, is undesirable in many applications. Therefore, there is a tradeoff between the maximum R_{sp} and the desired radiation pattern of spontaneous emission.

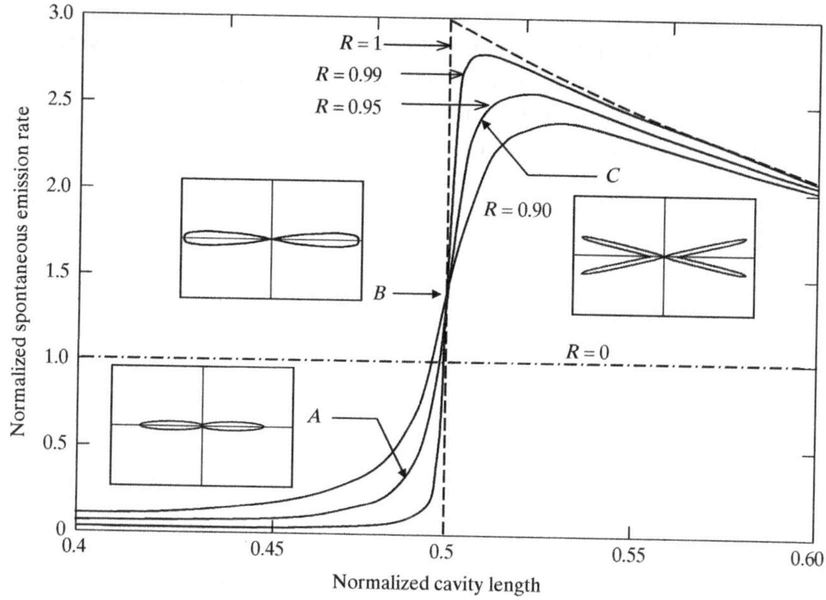

Figure 8.11 Normalized spontaneous emission rate versus normalized cavity length of planar microcavity VCSELs with dipole sheet placed in the cavity center but with different reflectivities. The insert diagrams are the radiation patterns of VCSEL with $R = 0.95$ but with different cavity lengths. Symbols A, B, and C represent normalized cavity length equal to 0.49, 0.5, and 5.1, respectively. (After Ref. 36.)

It is noted that the DBRs are commonly used to realize planar VCSELs and should be considered in the analysis of R_{sp} [37]. In addition, the reflectivity of DBRs is dependent on the operating wavelength as well as the incident angle of optical fields so that the calculated R_{sp} of planar VCSELs with DBRs may be less than that with hard mirrors. The diffraction of resonant modes, which is dependent on the design of the waveguide geometry of planar VCSELs, also affects the value of R_{sp}. The investigation on R_{sp} of planar VCSELs with different types of DBRs (e.g., AlAs/GaAs, SiO_2/Si, GaAs/AlO_x) has been performed using the quantum-mechanical model in which the influence of propagation modes is accounted for [37,39]. Consider a VCSEL with 13 periods of n- and p-GaAs/AlAs DBRs (i.e., contrast ratio between the refractive indices of GaAs and AlA is 1.2) and the laser cavity is constructed by two types of spacer layers: (1) a high-refractive-index (RI) spacer (i.e., GaAs) where AlAs is the first layer of mirrors and (2) a low-RI spacer (i.e., AlAs) where GaAs is the first layer of mirrors. It can be shown that the numerical evaluation of these configurations of VCSELs (i.e., with $\lambda/2$ cavity length) gives a weak normalized R_{sp} between 0.8 and 1.2 for high- and low-RI spacers, respectively, relative to free-space value [39]. This is because the absolute value and the phase of the reflectivity of GaAs/AlAs DBRs have a strong angular dependence. The reflectivity of the DBRs is weak over a very large fraction of solid angle (i.e., 55% and 90% of the total reflectivity of the high- and low-RI spacer, respectively, is weak). This indicates that along the direction of solid angle of weak reflectivity, the mirror produces very little modification on spontaneous emission. Hence, the overall spontaneous emission rate is not very different from that of the free-space value.

It is expected that the reflectivity of DBRs with high contrast ratio can be maximized over a wider solid angle. For a VCSEL with GaAs/AlO_x DBRs (i.e., with contrast ratio as large as 2.28) and AlAs spacer layers, the corresponding reflectivity of the mirrors is nearly unity over 75% of all solid angles. However, the normalized R_{sp} is found to equal 0.7 and 1.4 of the free-space value at the cavity length of 0.49λ and 0.51λ, respectively. It seems the high contrast ratio of GaAs/AlO_x mirrors has less enhancement or inhibition of the spontaneous emission, and the reason for this poor performance of GaAs/AlO_x mirrors lies in the high refractive index of the semiconductor cavity spacer. This is because the propagation modes can be supported along the spacer layers, which trap the spontaneous emission generated from the active region (i.e., the refractive index of AlAs is higher than that of AlO_x). As a result, the spacer layer reduces any enhancement or inhibition of the emission rate significantly. It should be noted that if the spacer itself were made of AlO_x or a lower-RI material, the modification of spontaneous emission would be similar to that of an ideal nonguiding cavity.

Spontaneous emission control in VCSELs can be effectively optimized if metallic mirrors are used instead of DBRs [39]. Gold, with a complex RI of $0.239 + j6.521$, can be used to form metallic mirrors. Good-quality metallic mirrors can be realized by depositing gold on a GaAs semiconductor substrate to achieve high reflectivity of 95% and the remaining 5% is the absorption of metal. The absorption will degenerate the quality factor of the laser cavity, but the reflectivity of the metallic mirror is essentially uniform over all the angles of incidence with a small amount of optical loss. In the design of VCSELs using metallic mirrors, one must be aware that the complex refractive index of metal introduces a phase change of field in addition to the π phase shift assumed in the ideal mirror model. It can be shown that the additional phase shift of a GaAs/Au mirror interface is roughly 1 radian, which corresponds to an optical thickness of $\sim\lambda/2\pi$. Therefore, the required optical thickness of spacer layers should be geometrically shorter by $\sim\lambda/2\pi$ if metallic mirrors are used. Figure 8.12 shows the calculation of normalized R_{sp} from a planar microcavity with a point dipole. Maximum enhancement is found to be 4.4 at the cavity length equals $\sim$0.35λ. It should be remembered that the maximum enhancement from an ideal planar microcavity (i.e., $R = 100\%$) is only 3.

R_{sp} in planar VCSELs with DBRs can be further enhanced if the propagation modes are strongly confined along the transverse direction. It has been demonstrated that R_{sp} of an airposted VCSEL is 5 times larger than that of planar VCSEL because of the strong transverse confinement of propagation modes [40]. In fact, an oxide aperture can be used to improve the transverse optical confinement of planar VCSELs. In addition, quantum dots (QDs) can be employed to enhance the electronic confinement as well as minimize the surface

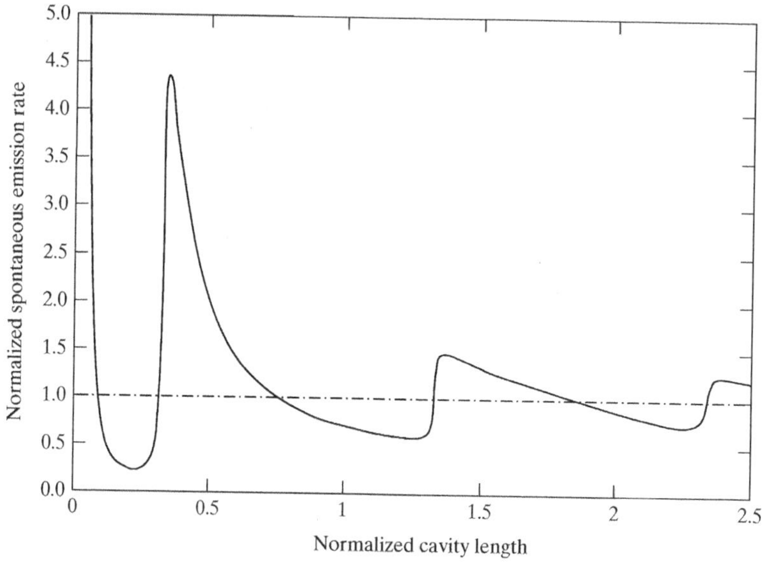

Figure 8.12 Spontaneous emission rate versus the normalized cavity length of a planner microcavity VCSEL with Au mirrors and GaAs spacer; the dashed line indicates the free-space value for a horizontal dipole placed in the center of the planar cavity (after Ref. 39).

recombination in such a small oxide aperture [41]. Hence, significant improvement in R_{sp} can be achieved in oxide aperture VCSELs with QD light emitters. Figure 8.13 shows the schematic of an apertured VCSEL using two oxide layers, which are sandwiched by two DBRs. The transverse confinement of optical modes in oxide aperture VCSELs is obtained by blueshifting the longitudinal resonance in the oxidized region relative to that of the unoxidized region [42]. Although the propagation mode is supported along the transverse direction of the high-RI layers, the oxidized region blocks the propagation modes inside the active layer so that an aperture mode is formed. Spontaneous emission, which is generated by the excited QDs light emitters, is coupled to the lowest-order cavity mode confined by the aperture. In this case, R_{sp} can be increased by a factor of 10 when compared with the case without the cavity effect. The other advantage of using QDs as the light emitters is that they can provide a spontaneous lifetime shorter than that of QWs. This is because QDs can still obtain the required electronic confinement in a laser cavity of very small aperture.

Theoretical prediction indicates that the value of R_{sp} can be increased provided that narrowband QDs, which are coupled directly into the aperture mode, are placed at the mode center. Hence, the combination of oxide aperture VCSEL and QDs can generate sizable microcavity effects. In fact, it can be shown that the Purcell factor [43] of an oxide aperture VCSEL (with the emitter placed at the mode center) increases from 2.2 to 8.8 for a mode diameter change from 2 to 1 μm, respectively [41]. This is because the reduction of aperture size restricts

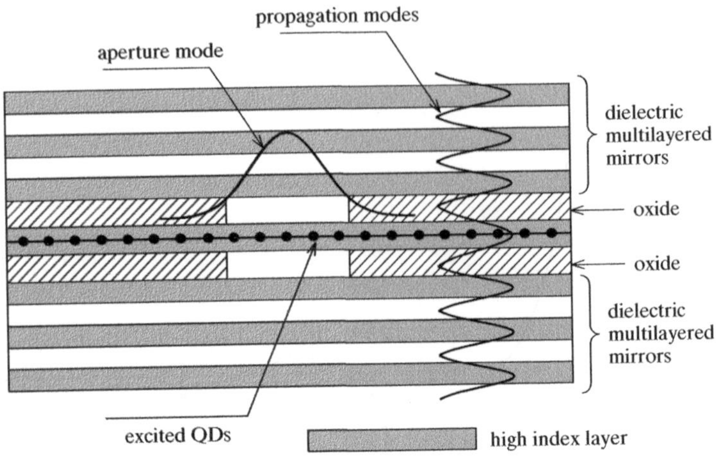

Figure 8.13 Schematic illustration of an oxide aperture VCSEL with a uniformly excited sheet of quantum dots. Spontaneous emission is coupled to the aperture modes that radiate normal to the surface and propagation modes that propagate along the high index layers of the cavity.

the number of transverse modes to be supported so that the spontaneous emission power distributed to the transverse modes is increased. However, further reduction of aperture size beyond the cutoff condition of the fundamental transverse mode is not desired as the corresponding field profile as well as Purcell factor will reduce. In order to obtain a large value of R_{sp}, it is necessary to obtain a low-level inhomogeneous broadening in the distribution of QDs, realize a submicrometer-size mode, and obtain high-quality factors at these small modal sizes.

8.4 RATE EQUATION ANALYSIS OF VCSELs

In this section, threshold current and modulation bandwidth of VCSELs are analyzed using the simple rate equation model. The alteration of spontaneous emission is also calculated through β_s and R_{sp}. Hence, the possibility of achieving thresholdless and unlimited modulation bandwidth in VCSELs is investigated. On the other hand, the degeneration of modulation bandwidth in VCSELs due to the nonlinear gain (i.e., arising from the heat effects) is also discussed.

8.4.1 Threshold Current and Modulation Bandwidth

Threshold current and modulation bandwidth of VCSELs can be analyzed by the rate equation model as shown below

$$\frac{\partial N}{\partial t} = \frac{I}{qV_a} - \left(\frac{1}{\tau_{sp}} + \frac{1}{\tau_n}\right) N - v_g g S, \tag{8.27}$$

$$\frac{\partial S}{\partial t} = \left(\nu_g g - \frac{1}{\tau_p}\right) S + \beta_s \frac{N}{\tau_{sp}}, \tag{8.28}$$

where $V_a = A_{eff} \cdot d$, τ_n is the nonradiative recombination lifetime of carrier concentration, τ_p is the photon lifetime, and ν_g is the group velocity. In (8.27), it is assumed that the injection efficiency is equal to one and S represents the photon density inside the active region (i.e., $\Gamma_z = 1$). The enhancement of spontaneous emission is introduced into the rate equation model through (1) β_s, which can be as large as one; and (2) the spontaneous emission lifetime τ_{sp}, which can be much smaller than τ_n. The term $(1/\tau_{sp} + 1/\tau_n)N$ in (8.27) represents R_t in (8.1), and N/τ_{sp} in (8.28); represents R_{sp} in (8.2); g is the optical gain, which has a linear relation with the injected carrier concentration into the active region, and can be expressed as

$$g = g_N(N - N_t), \tag{8.29}$$

where N_t is the transparency carrier concentration. In order to analyze the influence of size effect on the threshold conditions of microcavity lasers, the differential gain constant g_N is expressed as [44]

$$g_N = \frac{\beta_s V_a}{\nu_g \tau_{sp}}. \tag{8.30}$$

If the threshold current is defined as the balance between net stimulated gain and cavity loss, additional pumping will generate net stimulated emission so that lasing occurs. Using this idea, the threshold current density N_{th} can be expressed as

$$N_{th} = N_t + \frac{\tau_{sp}}{\beta_s V_a \tau_p} \equiv N_t\left(1 + \frac{1}{\zeta}\right), \tag{8.31}$$

where ζ is a dimensionless parameter defined by

$$\zeta = \beta_s V_a N_t \frac{\tau_p}{\tau_{sp}}. \tag{8.32}$$

The corresponding threshold current I_{th} can be deduced from (8.28) using (8.31) with S and $\partial S/\partial t$ set to zero, gives

$$I_{th} = \frac{q}{\beta_s \tau_p}(1 + \zeta)\left(1 + \frac{\tau_{sp}}{\tau_n}\right). \tag{8.33}$$

where it is noted that there will always be a threshold current penalty as soon as τ_n becomes smaller than τ_{sp} (i.e., dominant by nonradiative recombination or large ζ). If nonradiative recombination is neglected (i.e., $\tau_n \to \infty$), the expression for the threshold current will consist of the summation of two terms: the left term, which depends mainly on the properties of the cavity (i.e., β_s and τ_p); and the right term, which depends mainly on the properties of the gain medium (i.e., N_t, V_a, and τ_{sp}). If ζ becomes sufficiently small, (8.33) reduces to

$$I_{th} \approx \frac{q}{\beta_s \tau_p}\left(1 + \frac{\tau_{sp}}{\tau_n}\right). \tag{8.34}$$

On the other hand, if $\zeta \gg 1$, the threshold current can be approximated by

$$I_{th} \approx \frac{qN_t V_a}{\tau_{sp}}\left(1 + \frac{\tau_{sp}}{\tau_n}\right). \tag{8.35}$$

If ζ reduces with the reduction of V_a, the threshold current will decrease until the domination of the left term in (8.33). In this case, (8.34) is only proportional to $1/\beta_s$, which sets the fundamental limit for the threshold gain of a given cavity. Therefore, the ultimate threshold current is independent on τ_{sp}, N_t, and V_a, provided the nonradiative recombination is negligible and further reduction of threshold current can be achieved only by increasing β_s. Figure 8.14 plots the threshold current as a function of β_s of a typical VCSEL using (8.33) in the calculation. It is shown that as long as $\beta_s < 10^{-3}$ and $\zeta < 1$, the threshold current can be approximated by (8.34) for a laser with larger V_a (i.e., 10^{-12} cm^{-3}). When $\beta_s > 10^{-3}$, the threshold current is roughly given by (8.35). For a laser with smaller V_a (i.e., 10^{-15} cm^{-3}), the threshold current follows (8.34) as long as $\beta_s < 0.1$. It must be noted that for lasers with negligible nonradiative

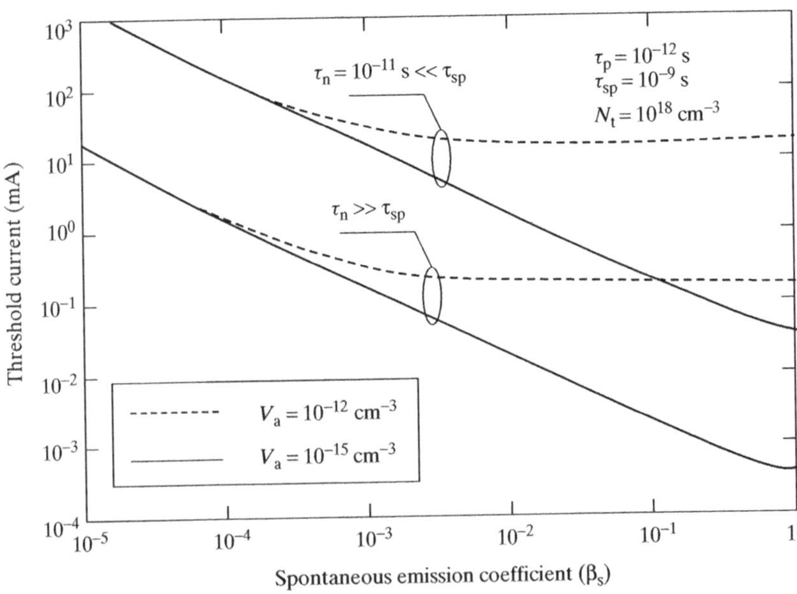

Figure 8.14 Threshold current versus spontaneous emission factor (after Ref. 44).

recombination, the corresponding threshold current is always lower than that of the domination of nonradiative recombination.

If the microcavity laser supports only one mode within its gain bandwidth and nonradiative recombination is negligible (i.e., $\tau_n \gg \tau_{sp}$), the photon emission is the only means of power dissipation and the quantum efficiency of the device must be unity both below and above threshold. This observation can be verified using (8.27)–(8.30) to express the injection current I versus the photon number p, which is given by

$$I = \frac{q}{\beta_s \tau_p}\left(\frac{p}{1+p}(1+\zeta)\left(1+\beta_s p+\frac{\tau_{sp}}{\tau_n}\right)-\zeta\beta_s p\right), \tag{8.36}$$

where $p = SV_a$. Figure 8.15 shows the injection current/photon power curves of three different combinations of parameters. It must be noted that except for the case $\beta_s \sim 1$ and $\tau_n \gg \tau_{sp}$, there is a sharp increase in the photon number as soon

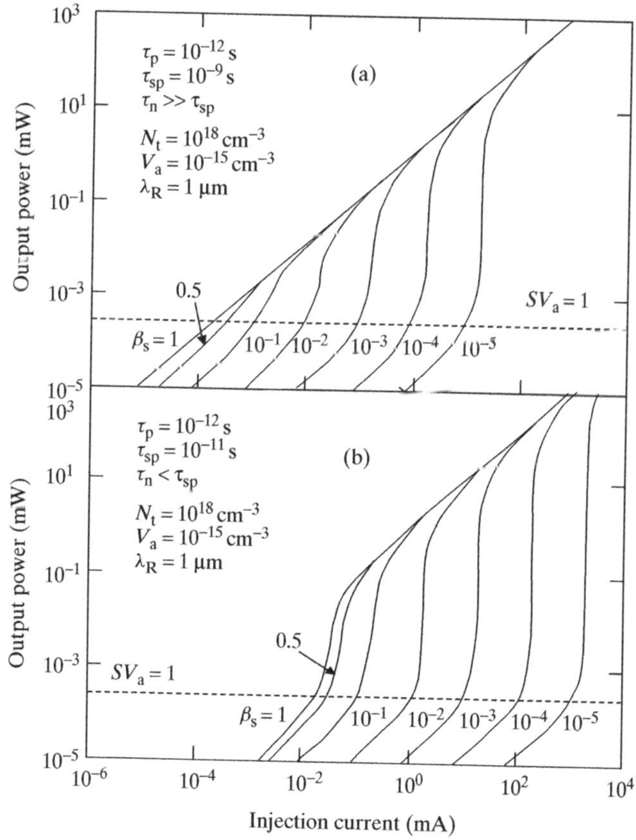

Figure 8.15 Output power (at $\lambda_R = 1\ \mu m$) versus the pump current (after Ref. 44).

as the photon number exceeds unity (i.e., threshold point). It must be noted that for an ideal VCSEL (i.e., $\zeta \sim 0$ and $\beta_s = 1$), (8.34) can be simplified to

$$I_{th} = \frac{q}{\tau_p}, \tag{8.37}$$

which implies that the only loss mechanism is the photon coupled into the lasing mode. The photon emission rate exactly equals the injection rate of carriers for the compensation of loss. Hence, the photon loss rate in this case is p/τ_p so that the corresponding mean photon number at threshold is unity. In the light/current curves, the kinks at the photon number equals one are the threshold point of the lasers. If $\beta_s \sim 1$ and the nonradiative recombination is negligible, the transition from below to above threshold is smooth. As a result, there is no threshold point (i.e., thresholdless operation) that can be extracted from the light/current curve given in Figure 8.15a. It can be concluded that thresholdless VCSELs can be realized if $\beta_s \to 1$ (i.e., V_a is smaller than $\tau_{sp}/N_t\beta_s\tau_p$) and $\tau_n \gg \tau_{sp}$ (i.e., negligible of nonradiative recombination).

The modulation response of VCSELs can also be deduced from (8.27) and (8.28). This can be done by assuming that the injection current is modulated by a small sinusoidal fluctuation at a steady-state bias of I_b, that is, $I = I_b + \Delta I \exp(j\omega t)$, where ΔI is the small perturbation of injection current and ω is the angular frequency of modulation. The small fluctuation of carrier concentration and photon density can then be written as $N = N_s + \Delta N \exp(j\omega t)$ and $S = S_s + \Delta S \exp(j\omega t)$, respectively. Inserting these into the rate equations and linearized by ignoring the higher-order terms, it can be shown that the relaxation oscillation frequency ω_n and modulation response can be written as [45]

$$\omega_n = \sqrt{\left(\tau_p^{-1} + v_g \frac{\partial g}{\partial S} S\right) \frac{1}{\tau_{st}} - \frac{1}{4}\left(\omega_F + v_g \frac{\partial g}{\partial S} S\right)^2}, \tag{8.38}$$

$$\left|\frac{\Delta S}{\Delta I}\right|^2 = \frac{1/(qV_a)^2}{\tau_{st}^2[(\omega^2 - \omega_n^2 - \omega_D^2)^2 + 4\omega^2\omega_D^2]}, \tag{8.39}$$

where $\omega_F = 1/\tau_{sp} + 1/\tau_n + 1/\tau_{st}$, $\tau_{st} = (v_g S g_N)^{-1}$ is the stimulated lifetime and $\omega_D = \frac{1}{2}(\omega_F - v_g S\, \partial g/\partial S)$ is the damping rate. In the derivation of (8.38), the spontaneous emission, which is the last term on the RHS of (8.28), has been ignored for a simple reason. Inserting the values of S_s and N_s obtained from the state-steady analysis, the modulation response as a function of biased current I_b and angular modulation frequency ω can be obtained. The modulation response can be normalized by multiplying (8.39) by I_b/S_s to obtain a normalized response function equal to unity. This implies that for a 10% current modulation there will result in 10% photon density modulation. Figure 8.16 plots the absolute square of the normalized intensity response function of VCSELs with $\beta_s = 1$ at different biased current levels. It is noted that for the bias below and near threshold, the cutoff value of the modulation response is around the inverse of the spontaneous

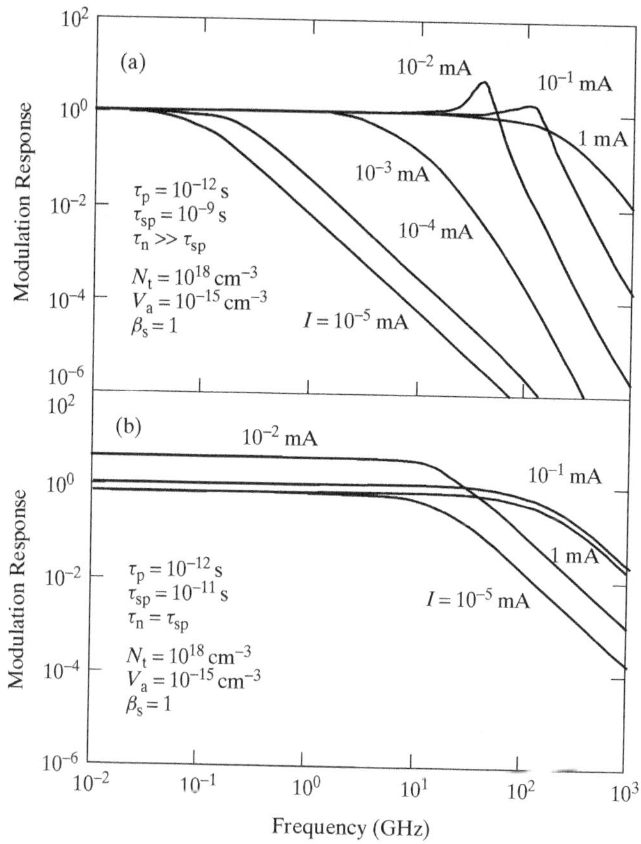

Figure 8.16 Normalized intensity modulation response versus modulation frequency of direct-current modulation without (a) and with (b) enhancement of spontaneous emission rate (after Ref. 44).

emission lifetime. If the bias current increases further, the modulation bandwidth increases until its cutoff value reaches the inverse of cavity lifetime (i.e., τ_p), which is the ultimate limit. This analysis shows that the intensity modulation response is not much different from the facet emitting lasers but the bandwidth of the modulation response can be improved significantly by increasing the value of β_s to one.

8.4.2 Influence of Nonlinear Gain on Modulation Bandwidth of VCSELs

This section analyzes the modulation bandwidth of VCSELs under the constraints of nonlinear gain, which may have resulted from the self-heating of the laser cavity. It can be shown that there is an optimum allowed injection current for maximum modulation bandwidths. Hence, under the influence of nonlinear gain, the modulation bandwidth of VCSELs is limited even with the presence of microcavity effects. The modulation bandwidth can be calculated using (8.39) but with optical gain modified to

$$g = \frac{g_N(N - N_t)}{1 + \varepsilon S}, \tag{8.40}$$

where ε is the gain suppression factor. The denominator of (8.40) induces gain compression due to the high optical density inside the active region. Figure 8.17 shows the modulation bandwidth of VCSELs with three different diameters for the cases with and without nonlinear gain. In order to identify the influence of nonlinear gain from microcavity effect on the modulation bandwidth, it is assumed that the corresponding spontaneous emission lifetime for the three cavities are equal. In fact, it can be shown that the enhancement of spontaneous emission has a less significant influence on the modulation bandwidth if nonlinear gain is considered. It is observed from Figure 8.17 that the lasers rapidly achieve very high photon densities and broad bandwidths, but the modulation bandwidth becomes restricted by the influence of nonlinear gain. This can be a good explanation for the reason that the measured results on the modulation bandwidth of VCSELs with small cavity is limited to several tens of gigahertz [13]. In the design of VCSELs, the maintenance of gain linearity is as important as the enhancement of spontaneous emission inside the laser cavity.

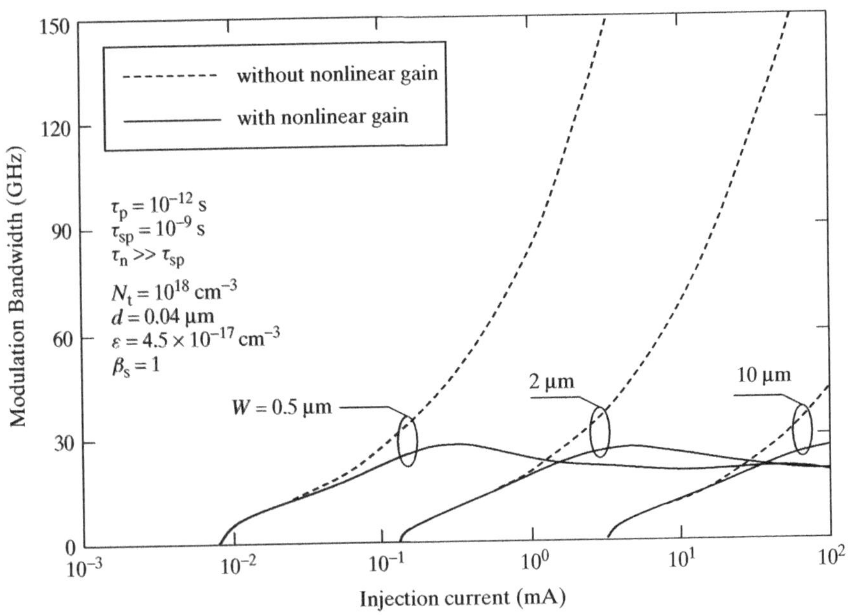

Figure 8.17 Modulation bandwidth of InGaAsP QW VCSELs as a function of current density versus cavity diameter $2W$: solid lines—nonlinear gain model; dashed lines—without consideration of nonlinear gain (after Ref. 45).

8.5 LINEWIDTH ENHANCEMENT FACTOR AND LINEWIDTH

The linewidth enhancement factor α_H (Henry alpha factor) is a material parameter, and its value is dependent on both laser frequency and threshold carrier concentration of the lasers. In semiconductor lasers, it is desired to reduce α_H to zero so that the corresponding linewidth can be minimized. It is noted that using QWs with high differential gain as the active layer, the value of α_H can be reduced. In fact, α_H can reduce to zero if strained QWs with modulation doping are used as the gain material of semiconductor lasers. In the following sections, methods to achieve minimum linewidth enhancement factor are explained. In addition, methods to estimate linewidth enhancement factor in single- and multiple-mode semiconductor lasers are described. The design consideration to minimize linewidth enhancement factor of VCSELs is also discussed.

8.5.1 Possibility of Zero Linewidth Enhancement Factor

The broadening of laser linewidth $\Delta\nu$ is attributed to the fluctuations in a phase of the optical field. These fluctuations have resulted from the spontaneous emission events, which discontinuously alter the phase and intensity of the laser field. The theoretical model, which is developed by Schawlow–Towns using the assumption of fluctuations in a phase of the optical field, has successfully explained the linewidth phenomena of gas lasers [46] such as the Lorentzian shape of linewidth spectrum and the inverse relation between $\Delta\nu$ and the output power of lasers. However, this Schawlow–Towns formula has underestimated $\Delta\nu$ of semiconductor lasers by a factor of 30, due to ignorance of the variation of the refractive index with the carrier concentration inside the semiconductor materials [47,48]. This variation of the refractive index comes from the spontaneous-emission-induced phase and intensity change in the laser field. The restoration of the laser to its steady-state intensity results in changes in the imaginary part of the refractive index. These changes are accompanied by the changes in the real part of the refractive index, which cause additional phase fluctuation and line broadening. It must be noted that the change in refractive index due to the injection carrier is negligible in gas lasers. In order to take into account the influence of carriers induced index changed in the analysis of the linewidth of semiconductor lasers, a linewidth enhancement factor α_H (or Henry alpha factor) is defined and can be expressed as [47]

$$\alpha_H = -\frac{4\pi}{\lambda_R}\frac{\partial n/\partial N}{\partial g/\partial N}. \tag{8.41}$$

The nominator and denominator on the RHS of (8.41) represent the differential refractive index and gain with respect to the carrier concentration inside the laser cavity. It must be noted that α_H has strong wavelength dependence, reflecting the spectrum shape of the differential gain. From (8.41), it is clear that α_H is dependent mainly on the physical properties of material. Hence, the modified $\Delta\nu$

of semiconductor lasers can be expressed as

$$\Delta\nu = \Delta\nu_{ST}(1 + \alpha_H^2), \tag{8.42}$$

where $\Delta\nu_{ST}$ is the linewidth obtained from the Schawlow–Towns formula and the detailed derivation of $\Delta\nu$ can be found in a paper by Henry [47]. In (8.42), it is shown that for a single-mode semiconductor laser, the coupling of phase to intensity fluctuations increases the laser linewidth by $1 + \alpha_H^2$. It can be seen from the definition that α_H is a material parameter, and its value is dependent on both the lasing wavelength and the threshold carrier concentration of the lasers. However, the threshold carrier concentration of semiconductor lasers is also affected by the laser's cavity length, waveguide absorption loss, and mirror reflectivities. Hence, the linewidth enhancement factor and linewidth are dependent on both the material properties and well as the waveguide design of the semiconductor lasers. It must be noted that α_H plays a key role in optical communication in semiconductor lasers as the spectral chirping width is broadened by the factor $\sqrt{1 + \alpha_H^2}$ under direct modulation [49], and the spectral linewidth in CW operation exceeds Schawlow–Towns formula by the factor $1 + \alpha_H^2$ [47]. Therefore, very low-chirped modulation, which is highly desired, could be obtained by reducing the α_H factor to zero.

There are many possible methods to reduce α_H in semiconductor lasers. It is believed that the replacement of bulk by QWs as the active region of semiconductor lasers can reduce the magnitude of α_H. This is because the differential gain of QWs materials is relatively higher than that of bulk materials [50,51]. Furthermore, semiconductor lasers with SQWs are expected to be superior to their MQW counterparts in the spectral linewidth because of the smaller spontaneous emission factor in SQW lasers [52]. However, detailed investigation of QW semiconductor lasers has shown that the state filling effect (i.e., high threshold current density due to the unavoidable thermal population of injection carriers into the optical confining region) suppresses the enhancement of differential gain in SQW lasers so that the magnitude of α_H has not been reduced [53]. In addition, α_H is also an increasing function of electrical pumping as the differential gain is inversely proportional to carrier concentration at the gain peak wavelength (i.e., $\partial g/\partial N = a_N/N$) [54]. Hence, there are an optimal number of QWs for achieving the narrowest spectral linewidth in QWs semiconductor lasers. However, in the design of facet emitting lasers, SQW is preferred because of its internal loss is lower than that of the MQW. In order to minimize α_H in semiconductor lasers, the injection current density into the QWs should also be limited. In fact, the application of graded-index separate-confinement heterostructure (GRINSCH structure) in SQW semiconductor lasers can minimize the injection current density into the QW so that the influence of the filling effect and the magnitude of α_H can be reduced [55]. Furthermore, it is shown that utilizing the delicate interplay between structure- and gain-material-dependent carrier nonlinearities, it is possible to design GRINSCH SQW lasers with α_H clamped at a range of injection current [56]. This is because the density of states at the band edge can be properly tailored so that the strong blueshift of gain peak, the possibility of occupation of off-resonant states, and the variation of refractive index of QW materials can be realized simultaneously. In fact, in the measurement of a GRINSCH SQW (5-nm $In_{0.14}Ga_{0.86}As$ QW in 150-nm $Al_{0.2}Ga_{0.8}As$ barrier with $Al_{0.6}Ga_{0.4}As$ cladding layers), a facet emitting laser lasing at gain peak wavelength around 0.9–1.0 μm, α_H is measured to be clamped near 2.1 for a range of injection carrier concentration (i.e., between 2 and 5×10^{12} cm^{-2}). However, for the same device but with a well width of 10 nm, α_H is measured to be a linear function of injection carrier concentration and is roughly equal to 6 at carrier concentration of 5×10^{12} cm^{-2} [56]. Furthermore, it is also shown that α_H of strained QW lasers is dependent on the crystal orientation of substrate. In the theoretical calculation, it is found that the value of α_H in the non(001)-oriented strained QW lasers, especially in the (111)-strained QW laser, is less than that in conventional (001)-oriented lasers by 1.4 times [57].

It must be noted that the magnitude of α_H can be reduced to zero, provided the QW structure is modified appropriately. It is noted that α_H is related to the differential gain via Kramers–Kronig relation so that (8.41) can be rewritten in terms of the differential gain [50]

$$\alpha_H = \frac{-\displaystyle\int_{-\infty}^{+\infty} \frac{\partial g(E,N)/\partial N}{E - \hbar\omega} dE}{\partial g(\hbar\omega, N)/\partial N}, \tag{8.43}$$

where $\hbar\omega$ is the photon energy. This equation (8.43) implies that if a differential gain spectrum $\partial g(E,N)/\partial N$ has a symmetric peak, α_H takes zero at the gain peak. This can be verified by calculating α_H of a typical lattice-matched QW laser such as a 90-Å InGaAs/InGaAsP SQW laser operating near 1.55 μm [50,58]. Figure 8.18 shows the calculated spectrum of material gain, differential gain, and α_H of the SQW material. It is observed that the differential gain spectrum has a peak at about 1.33 μm. On the other hand, the magnitude of α_H monotonically decreases to a shorter wavelength and crosses over zero at about 1.34 μm. As expected from (8.43), the wavelength that gives zero α_H exists near the peak of the differential gain spectrum. However, the wavelength at which $\alpha_H = 0$ lies within the absorption region, which is about 150 nm away from the gain peak, so that lasing oscillation may not be achieved in this wavelength. For lasing at the wavelength with zero magnitude of α_H, it is necessary to boost the optical gain (dashed line in Figure 8.18a) around the peak of the differential gain.

It is possible to broaden the gain spectrum so that the optical gain is obtained near the peak of the differential gain. It is noted that the peak wavelength of the differential gain is determined mainly by the quasi-Fermi energy of electrons, while the spectral width of the optical gain is affected by the quasi-Fermi energy of both electrons and holes. Therefore, increasing the quasi-Fermi energy of holes broadens the gain spectrum while the peak wavelength of the differential gain remains unchanged. This can be illustrated by the subband structure versus the in-plane wavevector **k** of a QW material as given in Figure 8.19. The corresponding

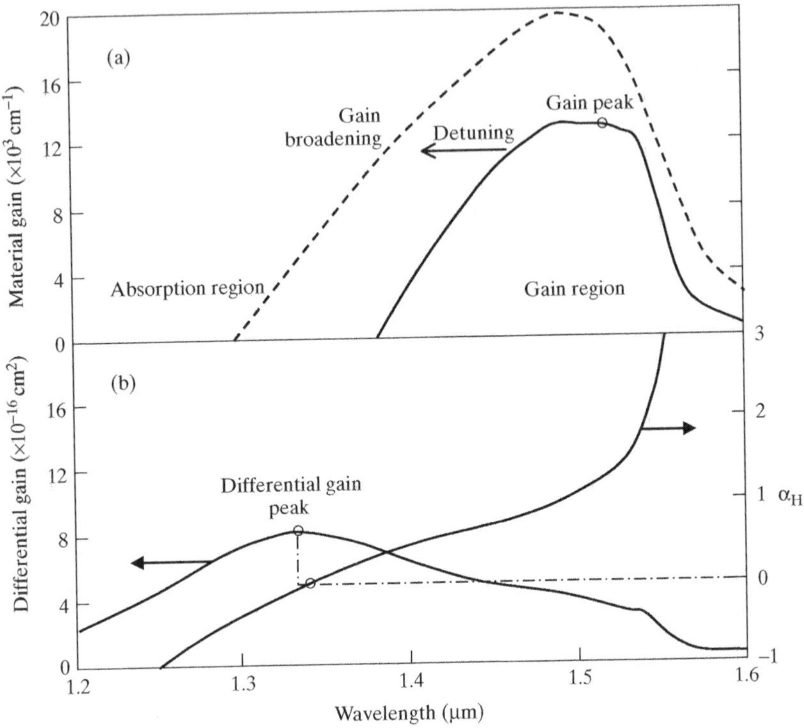

Figure 8.18 Plots of (a) material gain and (b) differential gain and linewidth enhancement factor versus wavelength for lattice-matched QWs under an injection of carrier concentration (after Ref. 50).

carrier density distribution at a certain injection level is also shown in the diagram. The QW material is assumed to have a conduction band (C) and two decoupled parabolic valence subbands, which are labeled HH (heavy hole) and LH (light-hole) with constant effective masses m_{hh} and m_{lh}, respectively. The split of the two lowest-valence subbands (i.e., HH and LH subbands) is due to the quantum size effect by energy of δE. In the diagram, E_{g} is the energy bandgap and E_{Fn} (E_{Fp}) is the quasi-Fermi energy for the conduction (valence) band. The curvature in the dispersion curve of each band determines the density of states from which effective masses are defined as a function of **k**.

The dependence of $\partial g(E, N)/\partial N$ on the quasi-Fermi energy of the electron can be proved by expressing $\partial g(E, N)/\partial N$ in terms of the Fermi–Dirac carrier distribution functions in the conduction and valence bands, $f_{\mathrm{c}}(E, N)$ and $f_{\mathrm{v}}(E, N)$, respectively, as shown below:

$$\begin{aligned} \frac{\partial g(E, N)}{\partial N} &\propto \frac{\partial}{\partial N}(f_{\mathrm{c}} - f_{\mathrm{v}}) \\ &\propto f_{\mathrm{c}}(1 - f_{\mathrm{c}})\frac{\partial E_{\mathrm{Fn}}}{\partial N} + f_{\mathrm{v}}(1 - f_{\mathrm{v}})\frac{\partial E_{\mathrm{Fp}}}{\partial N}. \end{aligned} \tag{8.44}$$

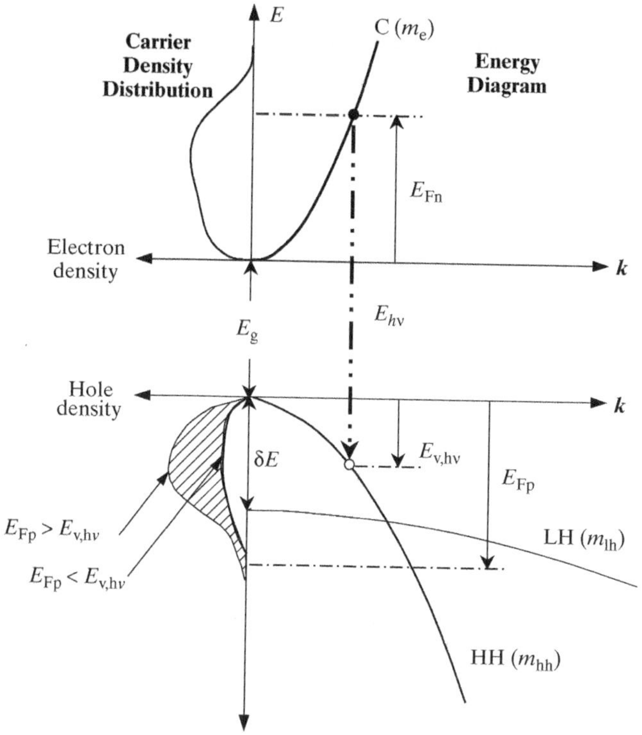

Figure 8.19 Schematic illustration of band structures of the lowest conduction and lowest valence subbands in the parabolic approximation and of the carrier density distributions. Carrier energy is measured from the bottom of each band. It is assumed that an electron at the electron quasi-Fermi energy E_{Fn} can recombine with a hole with energy $E_{\mathrm{v,peak}}$ and emit a photon with energy $E_{h\nu}$. The peak of the differential gain $E_{h\nu}$ is located around the electron quasi-Fermi energy. In the crucial situation of $E_{\mathrm{Fn}} > E_{\mathrm{v,peak}}$, hole density becomes high energy to hold the inversion condition at $E_{\mathrm{v,peak}}$. (After Ref. 50.)

In general, $\partial E_{\mathrm{Fn}}/\partial N \gg \partial E_{\mathrm{Fp}}/\partial N$ and the function $\partial f_c/\partial N$ dominates the differential gain spectrum through the contribution of $\partial E_{\mathrm{Fn}}/\partial N$ in (8.44). As the function $f_c(1 - f_c)$ has a peak at E_{Fn}, the function $\partial f_c/\partial N$ also takes a maximum value at E_{Fn}, which corresponds to the photon energy $E_{h\nu}$ given by

$$E_{h\nu} = \frac{m_e}{m_r} E_{\mathrm{Fn}} + E_g \tag{8.45}$$

where m_r is the reduced mass defined by $m_e m_h/(m_e + m_h)$, m_e is the electron mass, and m_h is the total mass of hole. This indicates that the peak wavelength of $\partial g/\partial N$ is determined mainly by the quasi-Fermi energy of the electrons E_{Fn} and the corresponding peak value of $\partial g/\partial N$ appears at $E_{h\nu}$.

The broadening of the gain spectrum by increasing the quasi-Fermi energy of holes can be verified by relating $\partial g(E, N)/\partial N$ in (8.43) to f_c and f_v. In order to obtain positive gain (i.e., $f_c - f_v > 0$ at the peak of $\partial g/\partial N$) at the photon energy of $E_{h\nu}$, the following conditions for the transition of photons with $E_{h\nu}$, should be satisfied:

$$f_v(E_{v,h\nu}) < \tfrac{1}{2}(\equiv f_c(E_{Fn})), \tag{8.46}$$

$$E_{v,h\nu} \equiv (E_{h\nu} - E_g)\frac{m_r}{m_h}. \tag{8.47}$$

It can be shown that these conditions can be realized provided that E_{Fp}, which gives $f_v(E_{Fp}) = \frac{1}{2}$, must penetrate deeper into the valence band than the hole energy $E_{v,h\nu}$:

$$E_{Fp} > E_{v,h\nu}. \tag{8.48}$$

Once E_{Fp} of QW materials increases and is larger than that of the hole energy level $E_{v,h\nu}$, the peak of $\partial g/\partial N$ moves to the positive gain region.

In the preceding paragraph, the condition of positive optical gain at the peak of the differential gain peak has been deduced. The remaining problem is to make (8.48) hold for QW materials. It must be noted that in QWs under laser condition, the energy level of E_{Fp} is very low because of the large density of states in the valence band. This suggests that the population of holes in the valence band have to be boosted in order to satisfy (8.48). It is known that p-type modulation doping increases the quasi-Fermi energy as well as the population of holes but without reducing the effective mass of valence band [59]. However, this effect of doping is spoiled by the increase in the density of states. In the case of lattice-matched QWs (i.e., Fig. 8.19), energy splitting δE is about 80 meV, which is as small as the thermal energy $k_B T$ at room temperature. The density of states near the zone center is determined by the effective mass of the LH band (m_{lh}) rather than that of the HH band (m_{hh}). If compressive strain is introduced to increase δE by pushing the HH band upward and LH band downward, the density of states decreases to the value determined mostly by m_{hh}. In this case, E_{Fp} is easy to penetrate into the valence band and the condition $m_{lh} \gg m_{hh} > m_e$ is still held even in the compressive strained band. This is because compressive strain reduces the density of states as well as the effective mass near the zone center of valence band. Therefore, the influence of modulation doping on the value of the density of states will be eliminated by the strain effect so that the requirement of (8.48) can be maintained.

Experiments have verified α_H of a strained MQW facet emitting laser with modulation doping are much lower than that with bulk or strained MQW active layers [60,61]. In the experiment, α_H of 1.3-μm InGaAsP/InP strained MQW DFB lasers and conventional InGaAsP/InP bulk DFB lasers are measured. The active layer of the strained MQW DFB laser consists of eight InGaAsP 1.1% compressive strained wells and InGaAsP barriers with a bandgap wavelength of 1.1 μm. The wells and barriers are of 30 and 100 Å thick, respectively. The thickness of both guide layers are 1100 and 500 Å. Because of the single-mode behavior of DFB structure, α_H is measured above threshold by biasing the lasers at

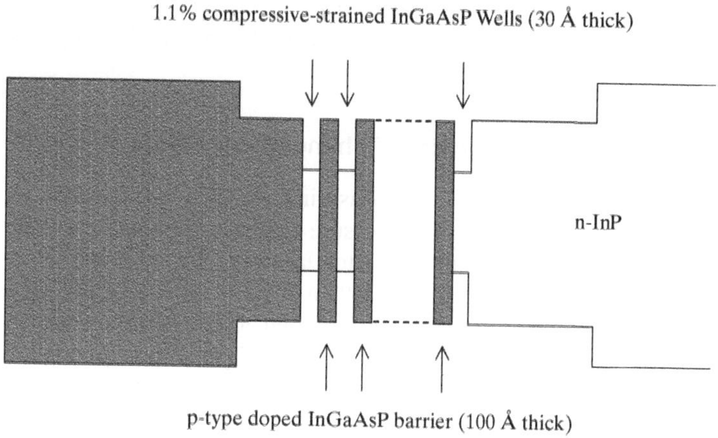

Figure 8.20 Schematic diagram of a modulation-doped strained MQW configuration.

1.5 times its threshold value and directly modulating with sinusoidal current of 4 GHz modulation frequency. It is shown that α_H of the strained MQW DFB lasers is measured to be about 2 and that of the bulk DFB lasers is greater than 5 [60]. The same measurement is applied to the strained MQW DFB lasers with modulation doping. Figure 8.20 shows a detailed configuration of the modulation-doped strained MQW lasers; p-type doping is applied to the barrier and p-side optical guide layer with doping concentrations of 0.5 and 1×10^{18} cm^{-3}, respectively. It can be shown that α_H can decrease to less than 1 if the lasing wavelength is detuned toward the shorter wavelength but with positive optical gain maintained [60]. In this measurement, it is shown that if lasing wavelength and optical gain are suitably selected, α_H of a strained QW laser with modulation doping can be reduced to zero. Modulation-doped strained MQW lasers of 1.55-μm InGaAsP/InP composition are also fabricated for small α_H. The MQW active layer consists of four InGaAsP wells with 1.4% compressive strain and InGaAsP barriers with a bandgap wavelength of 1.1 μm. The wells and barriers have thicknesses of 70 Å and 200 Å, respectively. The active layer is sandwiched by 1.1 μm composition InGaAsP optical guide layers, whose thickness are 1100 and 500 Å, respectively. For modulation doping, the barriers and the p-side optical guide layer have p-type doping with a concentration of 1.5×10^{18} cm^{-3}. Devices with DFB and edge emitting FP structure are fabricated so that α_H is measured above and below threshold. It can be shown that the value of α_H can reduce to less than 1 by detuning of −50 nm (i.e., lasing wavelength 1.5 μm) but with positive optical gain maintained [50]. In addition, the value of α_H at gain peak decreases with the level of doping concentration [60]. From the preceding measurement of 1.3- and 1.55-μm InP-based MQW lasers, a reduction of α_H is demonstrated in the modulation-doped strained MQW lasers. A small α_H of value less than 1 is obtained, not at wavelengths for the gain peak but within certain wavelength ranges (i.e., detuning toward the shorter wavelength) where the gain is positive in both 1.3- and 1.55-μm modulation-doped strained MQW lasers. These results suggest that it is very likely to fabricate high-performance lasers, such as low-chirp lasers or narrow-linewidth lasers, based on the modulation-doped strained MQW structure.

8.5.2 Measurement of Linewidth Enhancement Factor

There are two possible methods for measuring α_H of a single-mode semiconductor laser. The first approach is to measure the linewidth/output power curve. Linewidth of a single-mode semiconductor laser can be measured by using a scanning Fabry–Perot interferometer with a narrow free spectral range (i.e., <6 GHz). Stable temperature control is required to maintain a uniform lasing frequency with the variation of injection current. If the plot of linewidth versus the inverse of the output power is a straight line, α_H can be evaluated from (8.42) as $\Delta\nu_{ST}$ is inversely proportional to the output power. Alternatively, α_H can be measured by sinusoidally modulating the single-mode semiconductor laser at reasonably high frequencies [60,62,63]. This type of measurement has been performed using a gain-guiding GaAs QW VCSEL [64]. The lasing wavelength of the VCSEL is about 0.85 μm, with a threshold current of 2.8 mA and a maximum output power of 0.8 mW for single-mode operation (at 4.7 mA). Current confinement of the laser is realized by double-proton implantation so that the corresponding diameter of the output window is 8 μm. Figure 8.21 shows the

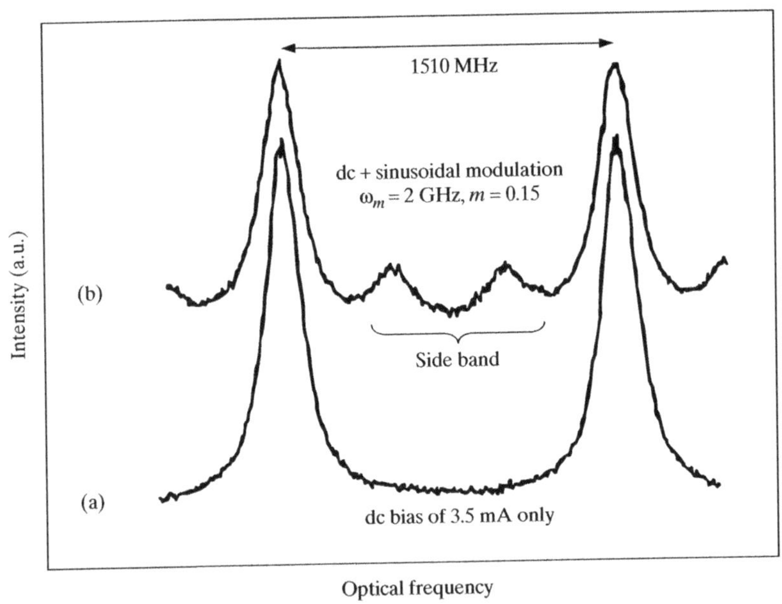

Figure 8.21 Fabry–Perot interferometer spectra: (a) DC bias with $I_b = 3.5$ mA; (b) DC bias and sinusoidal modulation at a frequency ω_m of 2 GHz and a modulation depth m of 0.15; at wavelength 0.85 μm, the interferometer has a free spectral range of 1500 MHz and a resolution of 15 MHz (after Ref. 64).

measured optical spectra under DC bias (i.e., DC bias of 3.5 mA) and that under small signal sinusoidal modulation (i.e., with amplitude modulation depth and modulation frequency equal to 0.15 and 2 GHz, respectively) superimposed on the DC bias [64]. In this experiment, amplitude modulation depth and modulation frequency are allowed to vary from 0.1 to 0.3 and from 0.5 to 2.0 GHz, respectively.

The spectra illustrated in Figure 8.21 can be used to estimate α_H. Consider an electric field $E(t)$, describing a sinusoidally modulated cavity mode, is written as [64]

$$E(t) = E_0(1 + \tfrac{1}{2}A_m \cos(2\pi\omega_m t)) \cos[2\pi\omega_R t + x_m \cos(2\pi\omega_m t)] \qquad (8.49)$$

where ω_R is the lasing frequency, ω_m is the modulation frequency, A_m is the amplitude modulation depth, and x_m is the frequency modulation index, which can be determined from the relative sideband strength. The sideband strength, including the intensity modulation component, is given by

Centerline at ω_R $\qquad I_0 = E_0^2(J_0^2(x_m) + m^2 J_1^2(x_m)), \qquad (8.50a)$

First sidebands at $\omega_R + \omega_m$ $\qquad I_1 = E_0^2(J_1^2(x_m) + (\tfrac{1}{2}A_m\{J_0(x_m) - J_2(x_m)\})^2), \qquad (8.50b)$

where $J_n(x_m)$ are the nth-order Bessel functions. Hence, under small-signal modulation (linear case), α_H can be written as

$$\alpha_H = \frac{2x_m}{A_m}. \qquad (8.51)$$

The frequency modulation index x_m is determined from the relative sideband strength. Hence, using (8.51), the magnitude of α_H is determined from the relation between the amplitude modulation index A_m and the relative sideband strength I_1/I_0.

For semiconductor lasers with multiple-mode operation, α_H can be measured from its amplitude spontaneous emission spectra below threshold [60]. It is noted from (8.43) that α_H is defined as the variation of real and imaginary parts of refractive index due to the small change of carrier concentration. Hence, α_H can be determined from the analysis of emission spectra between different currents below the threshold. The change of refractive index Δn can be obtained by the shift of wavelength $\Delta\lambda$ of the Fabry–Perot mode through the injection current (i.e., below threshold) according to the relation

$$\Delta n = \frac{\lambda_R}{2L\, d\lambda}\Delta\lambda, \qquad (8.52)$$

where $d\lambda$ is the Fabry–Perot mode spacing, L is the cavity length, and $\Delta\lambda$ is the shift of the wavelength due to the small variation of injection current. The mode gain g_m can be deduced from the depth of the modulation in the spontaneous emission intensity, caused by the Fabry–Perot resonance, using the relation

$$g_m = \frac{1}{L}\left(\log\left(\frac{\sqrt{P_r}-1}{\sqrt{P_r}+1}\right) - \log(R)\right), \tag{8.53}$$

where $P_r = P_{max}/P_{min}$, (P_{max} and P_{min} are the intensities at the adjacent peak and valley in the emission spectra) and R is the power reflectivity at the laser facets. Substituting (8.52) and (8.53) into (8.41) and canceling the term of carrier concentration gives

$$\alpha_H = \frac{2\pi}{L\,d\lambda}\frac{\Delta\lambda}{\Delta g_m}, \tag{8.54}$$

where Δg_m is the change in gain of the optical mode as the current is changed slightly. In this subthreshold measurement of α_H, lasers with Fabry–Perot structure should be used in order to obtain the P_{max} and P_{min} in the subthreshold emission spectra.

8.5.3 Linewidth Enhancement Factor and Linewidth of VCSELs

The linewidth enhancement factor and the linewidth of semiconductor lasers deteriorate if the laser cavity is not optimized, especially for small-cavity devices. This is because small-cavity devices such as VCSELs have high threshold carrier concentration, which reduces the differential gain and increases the value of α_H significantly. As discussed Section 8.5.1, the differential gain of QW materials can be enhanced by blueshifting the lasing wavelength from the peak of the optical gain so that α_H as well as linewidth can be reduced. This approach can be applied to optimize α_H of VCSELs as their lasing wavelength can be tuned with a suitable design of cavity length and DBRs [10,60]. On the other hand, using strained QWs with modulation doped as the active layer can enhance the differential gain but without scarifying the optical gain of the lasers. However, this is difficult to realize in VCSELs because of their vertical configuration. In this section, the value of α_H and linewidth in the most recent development of VCSELs is reviewed.

In the current investigation of VCSELs, the value of α_H has been measured to vary between 1.7 and 6.0, which is compatible to that of facet emitting lasers [12,26,58,65–72]. For a 4QW GaAs/AlGaAs gain-guided VCSEL (where transverse current confinement is achieved by ion implantation) with a diameter of 15 μm and lasing wavelength of 0.85 μm, the corresponding α_H and linewidth measured near room temperature are found to be 5.4 and 65 MHz, respectively at 400 μW [65]. Furthermore, another similar GaAs/AlGaAs MQW gain-guided VCSEL (where transverse current confinement is realized by double-proton implantation) with a diameter of 8 μm and lasing wavelength of 0.85 μm has α_H varying between 3.7 and 4.5 at room temperature [64,71]. In these experiments, the variation of α_H is due mainly to the difference in threshold current as well as threshold carrier concentration of the lasers. On the other hand, the measurement of α_H for InGaAs/GaAs MQW gain-guided VCSELs also has been reported [67,68]. For a 5QW InGaAs/GaAs VCSEL (in which mesa structure is used to confine transverse current spreading) lasing at 0.98 μm and with a diameter of 40 μm, the corresponding value of α_H is found to be equal 4.0 at 1.1 times its threshold [67]. Another 3QW InGaAs/GaAs VCSEL (where transverse current confinement is achieved by proton implantation) with a diameter of 12 μm has linewidth and α_H of 70 MHz and 3.7, respectively, at the lasing wavelength of 0.996 μm and optical power of 370 μW [68]. In these measurements of GaAs/AlGaAs and InGaAs/GaAs QW VCSELs, the lasing wavelength of VCSELs is maintained near the gain peak. In addition, the value of α_H measured from VCSELs is comparable to that of facet emitting lasers using bulk material. This is because the enhancement of differential gain in QW material is suppressed by the high carrier concentration inside the VCSELs cavity. Furthermore, the microcavity effects have no advantage in the reduction of α_H in VCSELs. As a result, the value of α_H for QW VCSELs is close to that of facet emitting lasers with bulk materials.

In Section 8.5.1, it has been shown that α_H can be minimized if the lasing wavelength is detuned to the shorter wavelength of the gain peak with the appropriate control of the threshold current. This approach has been adopted to fabricate a 3QW InGaAs/GaAs gain-guided VCSEL with an active layer diameter equaling 34 μm [69]. The device is designed to lase at 0.966 μm at room temperature, and the subthreshold ($\sim 0.25 I_{th}$) α_H is found to be around 0.7 for the device operating close to the gain peak. This result verifies the requirement of low threshold gain (i.e., threshold carrier concentration) in achieving high differential gain so that the value of α_H can be reduced. Further investigation has also explained the dependence of α_H on the operation temperature [70]. It can be shown that the value of α_H reduces monotonically with increase of temperature (i.e., increases from 220 to 420 K) under conditions where the lasing mode coincides with the peak gain wavelength at 290 K. This is because at low temperature, there is a redshift of resonant wavelength relative to the peak gain wavelength (i.e., see also Fig. 5.2). The increase in temperature from 220 to 300 K reduces the separation between the resonant wavelength and peak gain wavelength so that α_H is reduced. The decrease in α_H is particularly steep for temperature varying between 220 and 300 K as there is an additional decrease in threshold current concentration as the wavelength of the peak gain is closer to the resonant mode. However, further increase in temperature over 300 K produces a blueshift in the resonant wavelength relative to the peak gain wavelength so that the threshold carrier concentration increases. In this case, the value of α_H remains unchanged or increases slightly relative to the threshold gain of the VCSEL. Hence, the closer the resonant mode to the peak wavelength of the differential gain reduces α_H, whereas operations away from the peak gain wavelength increase the threshold carrier concentration as well as α_H [70]. From the preceding analysis, the value of α_H in VCSELs is dependent on the differential gain and threshold carrier concentration, which are functions of temperature. In order to reduce the value of α_H, the differential gain (threshold carrier concentration) at resonant wavelength should be enhanced (reduced), which can be achieved by slightly blueshifting the resonant wavelength with respect to the peak gain wavelength. Therefore, in the design of VCSELs with narrow linewidth, the resonant wavelength should be blueshifted at an appropriated magnitude and operating at a particular temperature.

8.6 SUMMARY

In this chapter, the characteristics of spontaneous emission in VCSELs are investigated through the study of the spontaneous emission factor and spontaneous emission rate. Theoretical calculation has indicated that the spontaneous emission factor of an airposted VCSEL with λ cavity can be as large as 0.5. In addition, the value of the spontaneous emission factor is dependent on the size of laser cavity and the location of the QW layer (or dipole). The modulation method has been applied to measure the spontaneous emission factor of an airposted VCSEL. It is found that the value of the spontaneous emission factor varies with the size of the laser cavity, and the maximum value is found to be 10^{-2} [18]. Further enhancement of the spontaneous emission factor to 1 is possible, provided the spontaneous emission radiating in the transverse direction can be eliminated. In fact, the spontaneous emission of an airposted VCSEL emitting from both the vertical and transverse directions has been measured. It is found that the radiation power in the transverse direction is overwhelmingly greater than that in the vertical direction [31]. Hence, VCSELs with metal-clad optical waveguides are proposed to confine the spontaneous emission radiating along the transverse direction [30].

The dependence of spontaneous radiation rate (i.e., a rate inversely proportional to the spontaneous emission lifetime) on the reflectivity of mirrors and cavity length is also investigated. It is well known that for microcavity lasers with $\lambda/2$ cavity and hard mirrors, the spontaneous emission rate can has a three-fold enhancement over the free-space value [5,6]. However, the spontaneous emission rate deteriorates if DBRs are used instead of hard mirrors. This is because the reflectivity of the DBRs varies over the solid angle (i.e., 25% of the total solid angle has low reflectivity for 13 pairs of $GaAs/AlO_x$ DBR). Therefore, the use of metallic mirrors is proposed to improve the spontaneous emission rate of a planar VCSEL. It can be shown that the maximum enhancement in spontaneous emission rate relative to free-space value can be as large as 4.4 at a cavity length of 0.35λ. Another limitation on the enhancement of spontaneous emission rate in a planar VCSEL is the lack of transverse confinement of propagation modes. Hence, a planar VCSEL with oxidized aperture is suggested to improve the transverse optical confinement. In addition, narrowband quantum dots are used as the light emitters to increase the electronic confinement in a very small aperture.

The influence of the spontaneous emission factor and spontaneous emission rate on the threshold current and modulation bandwidth of VCSELs is also studied. It is found that thresholdless operation can be realized in VCSELs if (1) the spontaneous emission factor is one, (2) the volume of the active region is small, and (3) the spontaneous emission rate dominates over the nonradiative recombination rate. However, it is noted that the spontaneous emission factor and spontaneous emission rate cannot be maximized simultaneously using the same cavity length. This is because the spontaneous emission factor is maximized using λ cavity design but the spontaneous emission rate is maximized using $\lambda/2$ (or 0.35λ) cavity design with DBRs (or metallic mirrors). In order to achieve thresholdless operation in VCSELs, λ cavity design should be used to maximize the spontaneous emission factor. In addition, nonradiative recombination rate should be suppressed instead of maximizing the spontaneous emission rate. On the other hand, it is found that the wide modulation bandwidth of VCSELs can deteriorate significantly with the influence of nonlinear gain.

Linewidth enhancement factor, which is a function of operating wavelength and injection carrier concentration, is a material-dependent parameter. It is shown that the linewidth enhancement factor can be reduced to zero if the operating wavelength is at the differential gain peak of QWs. However, the corresponding optical gain is a negative value (i.e., absorption) so that the lasing action cannot be realized at this operating wavelength. Therefore, modulation-doped QWs with strain is proposed to boost the optical gain at the differential gain peak. Experimental results have shown that the linewidth enhancement factor of a modulation doped QWs DFB laser with strain introduced can be less than one. However, the use of modulation doped QWs with strain is not the only requirement for realizing small linewidth enhancement factor in VCSELs. This is because extensive heat generation and high injection of carrier concentration inside the laser cavity significantly influence the linewidth enhancement factor. Therefore, careful design of the operating wavelength, heat generation rate, as well as injection carrier concentration is also required to minimize the linewidth enhancement factor in VCSELs.

8A APPENDIX

8A.1 Derivation of Radiation Energy Density

Radiation energy density from an average dipole can be derived in the form of energy density distribution over a **k** space, which corresponds to the orientation and wavelength of the optical field. Hence, the radiation pattern of the spontaneous emission of various wavelengths can be represented by the distribution of energy density. First, the distribution of radiation energy inside the laser cavity is evaluated from the classical wave equation with respect to an electric field vector **E** as shown below [32]

$$\nabla^2\mathbf{E} - 2\tau_d^{-1}\varepsilon_0\mu_0 n^2\frac{\partial\mathbf{E}}{\partial t} - \varepsilon_0\mu_0 n^2\frac{\partial^2\mathbf{E}}{\partial t^2} = \mu_0\frac{\partial^2\mathbf{P}}{\partial t^2}, \tag{8A.1}$$

where τ_d is the decay time constant of the mode energy by the absorption loss and n is the refractive index, ε_0 and μ_0 are the free-space permittivity and permeability, respectively; $\mathbf{P}$ is the microscopic polarization characterized by an electric dipole locating at $\mathbf{r}_{\mathscr{P}} = (x_{\mathscr{P}}, y_{\mathscr{P}}, z_{\mathscr{P}})$ and is given by

$$\mathbf{P} = \mathscr{P}_o(\hat{\mathbf{p}} \cdot \hat{\mathbf{e}}) \exp\left(-\frac{1}{\tau_{\text{sp}}} + j\omega_{\mathscr{P}} t\right) \delta(\mathbf{r} - \mathbf{r}_{\mathscr{P}}), \tag{8A.2}$$

where $\mathscr{P}_0$ is the initial amplitude of polarization; $\hat{\mathbf{p}}$ and $\hat{\mathbf{e}}$ are unit vectors of dipole moment and electric field, respectively; τ_{sp} is the spontaneous emission lifetime; $\omega_{\mathscr{P}}$ is the angular optical frequency of an oscillating dipole; δ is the Dirac delta function; and $\mathbf{r}$ is an arbitrary position vector.

The calculation of $\mathbf{E}$ using (8A.1) can be simplified if the transverse components of $\mathbf{E}$ are strongly confined inside the laser cavity. Hence, the longitudinal components of $\mathbf{E}$ propagating along the laser cavity can be separated from the transverse components of $\mathbf{E}$. The longitudinal distribution of the $\mathbf{E}$ field is determined by the longitudinal allocation of DBRs as well as the substrate and cladding layers. It is noted that the refractive index of both the substrate and cladding layer is relatively higher than that of the DBRs so that the longitudinal modes of $\mathbf{E}$ suffer radiation loss into the substrate and cladding layer. Any mode having loss is a continuous mode in the classification of orthogonal normal modes, although the mode shows some resonant properties. Therefore, the longitudinal mode is not considered as some discrete quasimode as it usually is in the analysis of conventional lasers, but treats the longitudinal modes specified by a continuous eigenvalue. Such treatment is convenient to analyze resonant modes and leaky waves uniformly. On the other hand, the transverse modes are discrete modes specified by the discrete eigenvalues owing to the total reflection at the boundaries between the semiconductor and air. In addition, the assumption of a large transverse confinement allows an arbitrary polarization of the electric field of modes. This means that (8A.1) can be reduced to the scalar wave equation with respect to an amplitude of electric field E expressed as

$$\begin{aligned} E(\mathbf{r}, t) &= \int\int_{-\infty}^{\infty}\int S(\mathbf{k}, t) f(k_x; r_t) g(k_y; r_t) h(k_z; z) \exp(j\omega t) d^3\mathbf{k} \\ &= \int\int_{-\infty}^{\infty}\int E(\mathbf{r}, \mathbf{k}, t) d^3\mathbf{k}, \end{aligned} \tag{8A.3}$$

where $\mathbf{k}$ is a wavevector given by $\mathbf{k} = (k_x, k_y, k_z)$; k_x and k_y are discrete wavenumbers, and k_z is the continuous one, and they are related by $|\mathbf{k}|^2 = k_x^2 + k_y^2 + k_z^2 = k^2 = (2\pi n_{\text{a}}/\lambda_{\text{R}})^2$; S is a function varying slowly with time; and f, g, and h are functions of spatial distribution of E as shown in Figure 8.3. The general solution of the transverse components of E along the x direction, f, is $\cos(k_z z)$ or $\sin(k_z z)$ in the active region when the center of the active region is chosen as the origin of the coordinate system. g and h are given similarly.

The vector wave equation can be rewritten in a scalar description by substituting (8A.2) and (8A.3) into (8A.1). Multiplying both sides of the scalar description of the vector wave equation by $f(k_x';x)\cdot g(k_y';y)\cdot h(k_z';z)$ and integrating over both space and **k** space, gives [32]

$$\frac{\partial^2 S}{\partial t^2}+2\tau_{\mathrm{d}}^{-1}\frac{\partial S}{\partial t}+\omega^2 S$$
$$=\frac{\omega_{\mathscr{P}}^2}{\varepsilon_0}\mathscr{P}_0(\hat{\mathbf{p}}\cdot\hat{\mathbf{e}})e^{(-1/\tau_{\mathrm{sp}}+j\omega_{\mathscr{P}})t}$$
$$\times\frac{f(k_x;x)g(k_y;y)h(k_z;z)|_{r=r_P}}{\iint_{-\infty}^{\infty}\int d^3\mathbf{r}\iint_{-\infty}^{\infty}\int d^3\mathbf{k}\cdot n^2 f(k_x;x)f(k_x';x)g(k_y;y)g(k_y';y)h(k_z;z)h(k_z';z)}. \tag{8A.4}$$

The general solution of the differential equation (8A.4) for arbitrary polarization is given by

$$S=\frac{\omega_{\mathrm{p}}P_0(\hat{\mathbf{p}}\cdot\hat{\mathbf{e}})}{j2\pi\varepsilon_0}\cdot\frac{e^{(-1/\tau_{\mathrm{d}}+j\omega_{\mathscr{P}})t}-e^{(-1/\tau_{\mathrm{sp}}+j\omega_{\mathscr{P}})t}}{(\tau_{\mathrm{sp}}^{-1}-\tau_{\mathrm{d}}^{-1})+j(\omega-\omega_{\mathscr{P}})}$$
$$\times\frac{f(k_x;x)g(k_y;y)h(k_z;z)|_{r=r_{\mathscr{P}}}}{A^2\int\int_{-\infty}^{\infty}n_{\mathrm{s}}^2 f^2(k_x;x)g^2(k_y;y)dx\,dy}, \tag{8A.5}$$

where A is an amplitude of h in the substrate (and in the cap layer). In the derivation of (8A.5), the formula for the continuous mode [32]

$$\int_{-\infty}^{\infty}n^2 h(k_z;z)h(k_z';z)dz=\pi n_{\mathrm{s}}^2A^2\delta(k_z-k_z'), \tag{8A.6}$$

has been used. Applying the Poynting theory and using (8A.6), the radiation energy density $\Re$, at a wavevector **k** can be expressed as

$$\Re(\mathbf{k})=\frac{1}{2}\int_0^{\infty}dt\cdot 2\tau_{\mathrm{d}}^{-1}\int\int_{-\infty}^{\infty}\int d^3\mathbf{r}$$
$$\times\int\int_{-\infty}^{\infty}\int d^3\mathbf{k}'\cdot\varepsilon_0 n^2 E(\mathbf{r},\mathbf{k}',t)\cdot E^*(r,k',t)$$
$$=\frac{\mathscr{P}_0^2|\hat{\mathbf{p}}\cdot\hat{\mathbf{e}}|^2}{8\pi\varepsilon_0}\cdot\frac{\tau_{\mathrm{sp}}(\tau_{\mathrm{sp}}^{-1}-\tau_{\mathrm{d}}^{-1})\omega_{\mathscr{P}}^2}{(\tau_{\mathrm{sp}}^{-1}-\tau_{\mathrm{d}}^{-1})^2+(\omega-\omega_{\mathscr{P}})^2}$$
$$\times\frac{f^2(k_x;x)g^2(k_y;y)h^2(k_z;z)|_{r=r_{\mathscr{P}}}}{A^2\int\int_{-\infty}^{\infty}n_{\mathrm{s}}^2 f^2(k_x;x)g^2(k_y;y)\,dx\,dy}. \tag{8A.7}$$

In practical situations, the radiation energy from the laser is of average measurement by the photodetector so that $\Re(\mathbf{k})$ is also required to be average with respect to the position of dipoles and the emission frequency [32]. In addition, the arbitrary field is assumed to be constructed by s- and p-polarization waves where s wave polarizes parallel to the substrate plane and the p wave polarizes inside the incident plane. Hence, the average of $\mathfrak{R}(\mathbf{k})$ with respect to the position $\mathbf{r}_{\mathscr{P}}$ can be obtained by integrating $\mathfrak{R}(\mathbf{k})$ over the active region and dividing by the volume of active region $V_a(=\pi W^2 d_a)$. The average of $\mathfrak{R}(\mathbf{k})$ with respect to the emission frequency $\omega_{\mathscr{P}}$ can also be calculated simultaneously by multiplying $\mathfrak{R}(\mathbf{k})$ by the emission spectrum function $\Lambda(k)$, which is normalized so that the integration over $\mathbf{k}$ should be unity. As a result, the averaged radiation energy density of s and p polarizations, $\overline{\mathfrak{R}}_{s,p}(\mathbf{k})$, can be obtained from (8A.7) and is given by

$$\begin{aligned}\overline{\mathfrak{R}}_i(\mathbf{k}) &= \Lambda(k)\cdot\frac{1}{V_a}\int_{\substack{\text{active}\\ \text{region}}}\mathfrak{R}(\mathbf{k})d^3 r_{\mathscr{P}} \\ &= \frac{c\tau_{sp}\mathscr{P}_0^2\langle|\hat{\mathbf{p}}\cdot\hat{\mathbf{e}}|^2\rangle}{16\varepsilon_0 n_a W^2}k^2\Lambda(k)\cdot\left(\frac{(\tau_{sp}^{-1}+\tau_d^{-1})\sqrt{\varepsilon_0\mu_o}n_a}{\pi(\tau_{sp}^{-1}+\tau_d^{-1})^2\varepsilon_0\mu_o n_a^2+(k-k'_{\mathscr{P}})^2}\right) \\ &\quad\times\frac{\displaystyle\iint_{-W/2}^{W/2} n_a^2 f^2(k_x;x)g^2(k_y;y)dx\,dy}{\displaystyle\iint_{-\infty}^{\infty} n_s^2 f^2(k_x;x)g^2(k_y;y)dxdy}\cdot\sum_{\substack{\cos\\ \sin}}\frac{\Gamma_{r,i}}{A_i^2},\end{aligned} \tag{8A.8}$$

where $i=s$ or p and $\omega_{\mathscr{P}}$ was converted to $k'_{\mathscr{P}}$ by the relation $k'_{\mathscr{P}}=n_a\omega_{\mathscr{P}}\sqrt{\varepsilon_0\mu_o}$. In order to simplify the analysis, the absorption loss and the decay rate due to relaxation can be assumed much smaller than those of the mirror loss [i.e., $(\tau_{sp}^{-1}-\tau_d^{-1})\sqrt{\varepsilon_0\mu_o}\ll k-k'_{\mathscr{P}}$], so that the term sandwiched between the brackets in (8A.8) can be eliminated, and $\overline{\mathfrak{R}}_{s,p}(\mathbf{k})$ is reduced to

$$\overline{\mathfrak{R}}_{s,p}(\mathbf{k})=\frac{\pi c\tau_{sp}\mathscr{P}_o^2}{8n_a\varepsilon_0}k^2\langle|\hat{\mathbf{p}}\cdot\hat{\mathbf{e}}|^2\rangle_{s,p}\Lambda(k)\rho_{s,p}(\mathbf{k}), \tag{8A.9}$$

where $\rho_{s,p}(\mathbf{k})$ is the mode density defined by the time-average mode power in the unit volume of the active region and the unit wavenumber volume. $\rho_{s,p}(\mathbf{k})$ can be deduced from the product of the mode densities of longitudinal mode ρ_z and the transverse mode ρ_t, each of which is given by [33]

$$\rho_{z,i}=\sum_{\substack{\cos\\ \sin}}\frac{\Gamma_{r,i}}{2\pi A_i^2}, \tag{8A.10}$$

$$\rho_t=\frac{1}{W^2}\cdot\frac{\displaystyle\iint_{-W/2}^{W/2} n_a^2 f^2(k_x;x)g^2(k_y;y)dx\,dy}{\displaystyle\iint_{-\infty}^{\infty} n_s^2 f^2(k_x;x)g^2(k_y;y)dx\,dy}, \tag{8A.11}$$

where the summation in (8A.10) indicates the sum of two degenerate functions, cosine and sine. For s polarization, $\Gamma_{r,s}$ can be written as

$$\Gamma_{r,s} = 1 \pm \frac{\sin(k_z d_a)}{k_z d_a}, \tag{8A.12}$$

where the plus and minus signs respectively correspond to the cosine and sine functions in (8A.10) and d_a is the thickness of the active layer. On the other hand, since the electric field of p polarization has the parallel and the normal components to the substrate plane, the final expression of $\Gamma_{r,p}$ is given by the mixture of $\Gamma_{r,s}$. For these two components, they are

$$\Gamma_{r,p} = \frac{k_t^2}{k^2}\left[1 \pm \frac{\sin(k_z d)}{k_z d}\right] + \frac{k_z^2}{k^2}\left[1 \mp \frac{\sin(k_z d)}{k_z d}\right], \tag{8A.13}$$

where the upper and lower signs respectively correspond to the cosine and sine functions in (A8.10), $k_t^2 = k_x^2 + k_y^2$ and $k^2 = k_x^2 + k_y^2 + k_z^2 = (2\pi n_a/\lambda_R)^2$.

8A.2 Derivation of Optical Field Distribution inside VCSELs

The amplitude of the longitudinal field at substrate $A_{s,i}$, is given by using the characteristics matrix $M_{m'}$, as follows [32]

$$A_{s,i}^2 = \left(\frac{\chi_{s,i}\Psi'_l}{k_{s,i}}\right)^2 + \Phi_i^2, \tag{8A.14}$$

$$k_{s,i}^2 = \left(\frac{n_{s,i}}{n_{a,i}}\right)^2 k^2 - k_{t,l}^2, \tag{8A.15}$$

where $i = s$ or p polarization and the subscripts a and s denote the parameters of the active and substrate regions, respectively. The propagation field in the longitudinal direction can be written in a matrix format as shown below:

$$\begin{bmatrix}\Phi_i \\ \Psi_i\end{bmatrix} = \prod_{m'=1}^{m} M_{m',i} \times \begin{cases} \begin{bmatrix} \cos\left(\dfrac{k_z d_a}{2}\right) \\ \dfrac{\mp k_z}{\chi_{a,i}} \sin\left(\dfrac{k_z d_a}{2}\right) \end{bmatrix} & \text{cosine type} \\ \begin{bmatrix} \sin\left(\dfrac{k_z d_a}{2}\right) \\ \dfrac{k_z}{\chi_{a,i}} \cos\left(\dfrac{k_z d_a}{2}\right) \end{bmatrix} & \text{sine type,} \end{cases} \tag{8A.16}$$

where m is the total number of dielectric layers and

$$\bullet M_{m',i} = \begin{cases} \begin{bmatrix} \cos(|k_{m',i}|d_{m'}) & \left(\dfrac{\chi_{m',i}}{|k_{m',i}|}\right)\sin(|k_{m',i}|d_{m'}) \\ \left(-\dfrac{|k_{m',i}|}{\chi_{m',i}}\right)\sin(|k_{m',i}|d_{m'}) & \cos(|k_{m',i}|d_{m'}) \end{bmatrix} & k^2_{m',\mathrm{i}} > 0 \\ \begin{bmatrix} \cosh(|k_{m',i}|d_{m'}) & \dfrac{\chi_{m',i}}{|k_{m',i}|}\sinh(|k_{m',i}|d_{m'}) \\ (|k_{m',i}|/\chi_{m,i})\sin h(|k_{m',i}|d_{m'}) & \cosh(|k_{m',i}|d_{m'}) \end{bmatrix} & k^2_{m',\mathrm{i}} < 0 \end{cases} . \tag{8A.17}$$

$$k^2_{m',i} = \left(\frac{n_{m',i}}{n_{a,i}}\right)^2 k^2 - k^2_{\mathrm{t},i}, \tag{8A.18}$$

The subscript m' $(= 1, \ldots, m)$ denotes the number of the dielectric layers of the VCSEL. $\chi_{m',i}$ is the polarization dependent factor defined by

$$\begin{aligned} \chi_{m',s} &= 1 && s \text{ polarization} \\ \chi_{m',p} &= n^2_{m',p} && p \text{ polarization} \end{aligned} . \tag{8A.19}$$

8A.3 Analytic Expression of Spontaneous Emission Coefficient of Fundamental Mode

Figure 8.3 shows the light power distribution inside the laser cavity. The normalized power confined inside the active region U_{a} is given by

$$U_{\mathrm{a}} = \tfrac{1}{2} n_{\mathrm{a}} d_{\mathrm{a}}. \tag{8A.20}$$

If the number of pairs of DBRs is large enough that the power penetration into the substrate is sufficiently small, the power outside the active region U_0 can be approximately given by

$$\begin{aligned} U_0 &= 2\left[\frac{1}{2} n_{\mathrm{L}} \frac{\lambda_{\mathrm{B}}}{4n_{\mathrm{L}}} + \sum_{i=1}^{\infty} \left(\frac{1}{2} n_{\mathrm{H}} \frac{\lambda_{\mathrm{B}}}{4n_{\mathrm{H}}} + \frac{1}{2} n_{\mathrm{L}} \frac{\lambda_{\mathrm{B}}}{4n_{\mathrm{L}}}\right)\left(\frac{n_{\mathrm{L}}}{n_{\mathrm{H}}}\right)^{2i}\right] \\ &= \frac{\lambda_{\mathrm{B}}}{4} \cdot \frac{1 + (n_{\mathrm{L}}/n_{\mathrm{H}})^2}{1 - (n_{\mathrm{L}}/n_{\mathrm{H}})^2}. \end{aligned} \tag{8A.21}$$

From (8A.20) and (8A.21), the confinement factor of the light power in the longitudinal direction is given by

$$\Gamma_{\mathrm{r},0} = \frac{d_0}{d_{\mathrm{eff}}} = d_0 \cdot \left(d_{\mathrm{a}} + \frac{\lambda_{\mathrm{B}}}{4} \cdot \frac{1 + (n_{\mathrm{L}}/n_{\mathrm{H}})^2}{1 - (n_{\mathrm{L}}/n_{\mathrm{H}})^2}\right)^{-1}, \tag{8A.22}$$

where $d_0 = \lambda_B/n_a$ and $n_a = n_H$ are assumed. If light confinement toward the xy direction is large enough, $\Gamma_{r,0}$ becomes

$$\Gamma_{r,0} = 2 \cdot \frac{1 - (n_L/n_H)^2}{3 - (n_L/n_H)^2}. \tag{8A.23}$$

Similarly, the effective index n_{eff} is obtained by integrating the product of n^2 and the power distribution U divided by the total power:

$$n_{eff}^2 = \frac{\int_{-\infty}^{\infty} n^2 U \, dz}{\int_{-\infty}^{\infty} U \, dz}$$

$$= n_a^2 \frac{d_a + \left(\dfrac{n_H}{n_a}\right)^2 \dfrac{\lambda_B}{2n_a} \cdot \dfrac{2(n_L/n_H)^2}{1 - (n_L/n_H)^2}}{d_a + \dfrac{\lambda_B}{2n_a} \cdot \dfrac{1 + (n_L/n_H)^2}{1 - (n_L/n_H)^2}}. \tag{8A.24}$$

If $d_a = \lambda_B/n_a$ and $n_a = n_H$, n_{eff} can be written as

$$n_{eff} \cong n_H \sqrt{\frac{2}{3 - (n_L/n_H)^2}} = n_H^2 \sqrt{\frac{2}{3n_H^2 - n_L^2}}. \tag{8A.25}$$

REFERENCES

1. G. P. Agrawal and N. K. Dutta, *Long Wavelength Semiconductor Lasers*, Van Nostrand Reinhold, 1986.
2. E. Yablonovitch, T. J. Gmitter, and R. Bhat, "Inhibited and enhanced spontaneous emission from optical thin AlGaAs/GaAs double heterostructures," *Phys. Rev. Lett.*, Vol. 61, no. 22, pp. 2546–2549, 1984.
3. A. M. Vredenberg, N. E. J. Hunt, E. F. Schubert, D. C. Jacobson, J. M. Poate, and G. J. Zydzik, "Controlled atomic spontaneous emission from Er^{3+} in a transparent Si/SiO_2 microcavity," *Phys. Rev. Lett.*, Vol. 71, no. 4, pp. 517–520, 1993.
4. I. Abram, "Semiconductor microcavities for spontaneous emission control," *J. Luminesc.*, Vols. 72–74, pp. 29–33, 1997.
5. H. Yokoyama, K. Nishi, T. Anan, Y. Nambu, S. D. Brorson, and E. P. Ippen, "Controlling spontaneous emission and threshold-less laser oscillation with optical microcavities," *Opt. Quantum Electron.*, Vol. 24, pp. S245–S272, 1992.
6. F. De Martini, M. Marrocco, P. Mataloni, L. Crescentini, and R. Loudon, "Spontaneous emission in the optical microscopic cavity," *Phys. Rev. A*, Vol. 43, no. 5, pp. 2480–2497, 1991.

7. G. Bjork, S. Machida, Y. Yamamoto, and K. Igeta, "Modification of spontaneous emission rate in planar dielectric microcavity structures," *Phys. Rev. A*, Vol. 44, pp. 669–681, 1991.
8. H. Yokoyama and S. D. Brorson, "Rate equation analysis of microcavity lasers," *J. Appl. Phys.*, Vol. 66, no. 10, pp. 4801–4805, 1989.
9. D. G. Deppe, D. L. Huffaker, T. B. Oh, H. Deng, and Q. Deng, "Low threshold vertical cavity surface emitting lasers based on oxide confinement and high contrast distributed bragg reflectors," *IEEE J. Select. Topics Quantum Electron.*, Vol. 3, no. 3, pp. 893–904, 1997.
10. T. Makino, "Spontaneous emission model of surface emitting DFB semiconductor lasers," *IEEE J. Quantum Electron.*, Vol. 29, no.1, pp. 14–22, 1993.
11. J. Singh, *Semiconductor Optoelectronics—Physics and Technology*, McGraw-Hill, New York, 1995.
12. G. Bjork, A. Karlsson, and Y. Yamamoto, "On the linewidth of microcavity lasers," *Appl. Phys. Lett.*, Vol. 60, no. 3, pp. 304–306, 1992.
13. D. Tauber, G. Wang, R. S. Geels, J. E. Bowers, and L. A. Coldren, "Large and small signal dynamics of vertical cavity surface emitting lasers," *Appl. Phys. Lett.*, Vol. 62, no. 4, pp. 325–327, 1993.
14. F. Koyama, K. Morito, and K. Iga, "Intensity noise and polarization stability of GaAlAs/GaAs surface emitting lasers," *IEEE J. Quantum Electron.*, Vol. 27, no. 6, pp. 1410–1416, 1991.
15. D. G. Deppe, D. L. Huffaker, J. Shin, and Q. Deng, "Very low threshold index confined planar microcavity lasers," *IEEE Photon. Technol. Lett.*, Vol. 7, no. 9, pp. 965–967, 1995.
16. D. V. Kuksenkov, H. Temkin, K. L. Lear, and H. Q. Hou, "Spontaneous emission factor in oxide confined vertical cavity surface lasers," *Appl. Phys. Lett.*, Vol. 70, no. 1, pp. 13–15, 1997.
17. J. H. Shin, Y. G. Ju, H. E. Shin, and Y. H. Lee, "Spontaneous emission factor of oxidized vertical cavity surface emitting lasers from the measured below threshold cavity loss," *Appl. Phys. Lett.*, Vol. 70, no. 18, pp. 2344–2346, 1997.
18. R. J. Ram, E. Goobar, M. G. Peters, L. A. Coldren, and J. E. Bowers, "Spontaneous emission factor in post microcavity lasers," *IEEE Photon. Technol. Lett.*, Vol. 8, no. 5, pp. 599–561, 1996.
19. G. Shtengel, H. Temkin, T. Uchida, M. Kim, P. Brusenbach, and C. Parsons, "Spontaneous emission factor and its scaling in vertical cavity surface emitting lasers," *Appl. Phys. Lett.*, Vol. 64, no. 9, pp. 1062–1064, 1994.
20. Y. G. Zhao, J. G. McInerney, and R. A. Morgan, "Measurement of spontaneous emission factor for vertical cavity surface emitting semiconductor lasers," *IEEE Photon. Technol. Lett.*, Vol. 7, no. 11, pp. 1231–1233, 1995.
21. R. J. Norowicz, H. Heitmann, Y. Kadota, and Y. Yamamoto, "GaAs microcavity quantum well laser with enhanced coupling of spontaneous emission to the lasing mode," *Appl. Phys. Lett.*, Vol. 61, no. 4, pp. 393–395, 1992.
22. U. Mohideen, W. S. Hobson, S. J. Pearton, F. Ren, and R. E. Slusher, "GaAs/AlGaAs microdisk lasers," *Appl. Phys. Lett.*, Vol. 64, no. 15, pp. 1911–1913, 1994.
23. Y. Z. Huang, "On the rate equation of semiconductor lasers for measuring spontaneous emission factor," *IEEE Photon. Technol. Lett.*, Vol. 7, no. 9, pp. 977–979, 1995.

24. J. C. Goodwin, and B. K. Garside, "Measurement of spontaneous emission factor for injection lasers," *IEEE J. Quantum Electron.*, Vol. 18, pp. 1264–1271, 1982.

25. Y. Suematsu, S. Akiba, and T. Hong, "Measurement of spontaneous emission factor AlGaAs double heterostructure semiconductor lasers," *IEEE J. Quantum Electron.*, Vol. 13, pp. 596–600, 1977.

26. D. Kuksenkov, H. Temkin, and S. Swirthun, "Measurement of internal quantum efficiency and losses in vertical cavity surface emitting lasers," *Appl. Phys. Lett.*, Vol. 66, no. 14, pp. 1720–1722, 1995.

27. J. W. Bae, G. Shtengel, D. Kuksenkov, and H. Temkin, "Threshold carrier density in vertical cavity surface emitting lasers," *Appl. Phys. Lett.*, Vol. 66, no. 16, pp. 2031–2033, 1995.

28. K. Petermann, "Calculated spontaneous emission factor for double heterostructure injection lasers with gain-induced waveguiding," *IEEE J. Quantum Electron.*, Vol. 15, pp. 566–570, 1979.

29. S. D. Brorson, H. Yokoyama, and E. P. Ippen, "Spontaneous emission rate alteration in optical waveguide structures," *IEEE J. Quantum Electron.*, Vol. 26, no. 9, pp. 1492–1499, 1990.

30. Y. Yamamoto, S. Machida, and G. Bjork, "Microcavity semiconductor laser with enhanced spontaneous emission," *Phys. Rev. A*, Vol. 44, no. 1, pp. 657–668, 1991.

31. T. Zhang, J. G. Wohlbier, K. D. Choquette, and N. Tabatabale, "Microcavity vacuum field configuration and the spontaneous emission power," *IEEE J. Select. Topics Quantum Electron.*, Vol. 1, no. 2, pp. 606–615, 1995.

32. T. Baba, T. Hamano, F. Koyama, and K. Iga, "Spontaneous emission factor of a microcavity DBR surface emitting laser," *IEEE J. Quantum Electron.*, Vol. 27, no. 6, pp. 1347–1358, 1991.

33. T. Baba, T. Hamano, F. Koyama, and K. Iga, "Spontaneous emission factor of a microcavity DBR surface emitting laser (II)—effects of electron quantum confinement," *IEEE J. Quantum Electron.*, Vol. 28, no. 5, pp. 1310–1319, 1992.

34. R. J. Ram, D. I. Babic, R. A. York, and J. E. Bowers, "Spontaneous emission in microcavities with distributed mirrors," *IEEE J. Quantum Electron.*, Vol. 31, no. 2, pp. 399–410, 1995.

35. F. Jahnke, and S. W. Koch, "Many body theory for semiconductor microcavity lasers," *Phys. Rev. A*, Vol. 52, no. 2, pp. 1712–1727, 1995.

36. G. Bjork, "On the spontaneous lifetime change in an ideal planar microcavity—transition from a mode continuum to quantized modes," *IEEE J. Quantum Electron.*, Vol. 30, no. 10, pp. 2314–2318, 1994.

37. C. C. Lin, D. G. Deppe, and C. Lei, "Role of waveguide light emission in planar microcavities," *IEEE J. Quantum Electron.*, Vol. 30, no. 10, pp. 2304–2313, 1994.

38. A. Onischenko, and J. Sarma, "Analysis of vertical and in-plane spontaneous emission in large diameter VCSELs," *IEE Proc. Optoelectron.*, Vol. 144, no. 1, pp. 39–43, 1997.

39. I. Abram, I. Robert, and R. Kuszelewicz, "Spontaneous emission control in semiconductor microcavities with metallic or bragg mirrors," *IEEE J. Quantum Electron.*, Vol. 34, no. 1, pp. 71–76, 1998.

40. T. Tezuka, S. Nunoue, H. Hoshida, and T. Noda, "Spontaneous emission enhancement in pillar-type microcavities," *Jpn. J. Appl. Phys.*, Vol. 32, pp. L54–L56, 1993.

41. D. G. Deppe, L. A. Grahama, and D. L. Huffaker, "Enhanced spontaneous emission using quantum dots and an apertured microcavity," *IEEE J. Quantum Electron.*, Vol. 35, no. 10, pp. 1502–1508, 1999.
42. L. A. Graham, D. L. Huffaker, and D. G. Deppe, "Spontaneous lifetime control in a native oxide aperture microcavity," *Appl. Phys. Lett.*, Vol. 74, no. 17, pp. 2408–2410, 1999.
43. E. M. Purcell, "Spontaneous emission probabilities at radio frequencies," *Phys. Rev.*, Vol. 69, p. 681, 1946.
44. G. Bjork, and Y. Yamamoto, "Analysis of semiconductor microcavity lasers using rate equations," *IEEE J. Quantum Electron.*, Vol. 27, no. 11, pp. 2386–2396, 1991.
45. A. Karlsson, R. Schatz, and G. Bjork, "On the modulation bandwidth of semiconductor microcavity lasers," *IEEE Photon. Technol. Lett.*, Vol. 6, no. 11, pp. 1312–1314, 1994.
46. A. Yariv, *Quantum Electronics*, 3rd ed., New York, Wiley, 1989, Appendix A.
47. C. H. Henry, "Theory of the linewidth of semiconductor lasers," *IEEE J. Quantum Electron.*, Vol. 18, no. 2, pp. 259–264, 1982.
48. M. Osinksi, and J. Buus, "Linewidth broadening factor in semiconductor lasers—an overview," *IEEE J. Quantum Electron.*, Vol. 23, no. 1, pp. 9–29, 1987.
49. T. L. Koch, and J. E. Bowers, "Nature of wavelength chirping in directly modulated semiconductor lasers," *Electron. Lett.*, Vol. 20, pp. 1038–1040, 1984.
50. T. Yamanaka, Y. Yoshikuni, K. Yokoyama, W. Lui, and S. Seki, "Theoretical study on enhancement differential gain and extremely reduced linewidth enhancement factor in quantum well lasers," *IEEE J. Quantum Electron.*, Vol. 29, no. 6, pp. 1609–1616, 1993.
51. B. Zhao, T. R. Chen, and A. Yariv, "A comparison of amplitude-phase coupling and linewidth enhancement in semiconductor QW and bulk lasers," *IEEE J. Quantum Electron.*, Vol. 29, no. 4, pp. 1027–1030, 1993.
52. Y. Arakawa, and A. Yariv, "Theory of gain, modulation response and spectral linewidth in AlGaAs quantum well lasers," *IEEE J. Quantum Electron.*, Vol. 21, pp. 1666–1674, 1985.
53. B. Zhao, T. R. Chen, and A. Yariv, "The extra differential gain enhancement in multiple quantum well lasers," *IEEE Photon. Technol. Lett.*, Vol. 4, no. 2, pp. 124–126, 1992.
54. T. A. DeTemple, and C. M. Herzinger, "On the semiconductor laser logarithmic gain-current density relation," *IEEE J. Quantum Electron.*, Vol. 29, no. 5, pp. 1246–1252, 1993.
55. M. Hochholzer, and V. Jordan, "Discussion of the linewidth enhancement factor α of GaAs/GaAlAs quantum well lasers," *IEE Proc-Optoelectron.*, Vol. 141, no. 5, pp. 311–315, 1994.
56. J. Hader, D. Bossert, J. Stohs, W. W. Chow, and J. V. Moloney, "Clamping of the linewidth enhancement factor in narrow quantum well semiconductor lasers," *Appl. Phys. Lett.*, Vol. 74, no. 16, pp. 2277–2279, 1999.
57. T. Ohtoshi, T. Kuroda, A. Niwa, and S. Tsuji, "Dependence of linewidth enhancement factor on crystal orientation in strained quantum well lasers," *IEEE Photon. Technol. Lett.*, Vol. 6, no. 12, pp. 1424–1426, 1994.

58. M. Asada, A. Kameyama, and Y. Suematsu, "Gain and intervalence absorption in quantum-well lasers," *IEEE J. Quantum Electron.*, Vol. 20, pp. 745–753, 1984.

59. K. Uomi, "Modulation doped multiple quantum well (MD-MQW) lasers. I. Theory," *Jpn. J. Appl. Phys.*, Vol. 29, pp. 88–94, 1990.

60. F. Kano, T. Yamanaka, N. Yamamoto, Y. Yoshikuni, H. Mawatari, Y. Tohmori, M. Yamamoto, and K. Yokoyama, "Reduction of linewidth enhancement factor in InGaAsP/InP modulation doped strained multiple quantum well lasers," *IEEE J. Quantum Electron.*, Vol. 29, no. 6, pp. 1553–1559, 1993.

61. F. Kano, T. Yamanaka, N. Yamamoto, H. Mawatari, and Y. Yoshikuni, "Linewidth enhancement factor in InGaAsP/InP modulation doped strained multiple quantum well lasers," *IEEE J. Quantum Electron.*, Vol. 30, no. 2, pp. 533–537, 1994.

62. C. Harder, K. Vahala, and A. Yariv, "Measurement of the linewidth enhancement factor α of semiconductor lasers," *Appl. Phys. Lett.*, Vol. 42, pp. 328–330, 1983.

63. I. D. Henning, and J. V. Collines, "Measurement of the semiconductor laser linewidth broadening factor," *Electron. Lett.*, Vol. 19, pp. 927–929, 1983.

64. D. V. Kuksenkov, H. Temkin, and S. Swirthun, "Frequency modulation characteristics of gain-guided AlGaAs/GaAs vertical cavity surface emitting lasers," *Appl. Phys. Lett.*, Vol. 66, no. 24, pp. 3239–3241, 1995.

65. G. R. Olbright, R. P. Bryan, W. S. Fu, R. Apte, D. M. Bloom, and Y. H. Lee, "Linewidth, tunability, and VHF-millimeter wave frequency synthesis of vertical cavity GaAs quantum well surface emitting laser diode arrays," *IEEE Photon. Technol. Lett.*, Vol. 3, no. 9, pp. 779–781, 1991.

66. R. Jin, D. Boggavarapu, G. Khitrova, H. M. Gibbs, Y. Z. Hu, S. W. Kock, and N. Peyghambarian, "Linewidth broadening factor of a microcavity semiconductor laser," *Appl. Phys. Lett.*, Vol. 61, no. 16, pp. 1883–1885, 1992.

67. T. Mukaihara, M. Araki, H. Maekawa, Y. Hayashi, N. Ohnoki, F. Koyama, and K. Iga, "Dynamic wavelength shift of vertical cavity surface emitting lasers," *Electron. Lett.*, Vol. 30, no. 20, pp. 1677–1678, 1994.

68. B. Moller, E. Zeeb, U. Fiedler, T. Hackbarth, and K. J. Ebeling, "Linewidth enhancement factor of vertical cavity surface emitting laser diodes," *IEEE Photon. Technol. Lett.*, Vol. 6, no. 8, pp. 921–923, 1994.

69. P. Dowd, H. D. Sumers, I. H. White, M. R. T. Tan, Y. M. Houng, and S. Y. Wang, "Measurement of differential gain and linewidth enhancement factor of InGaAs vertical cavity surface emitting lasers," *Electron. Lett.*, Vol. 31, no. 7, pp. 557–558, 1995.

70. H. D. Summer, P. Dowd, I. H. White, and M. R. T. Tan, "Calculation of differential gain and linewidth enhancement factor in 980 nm InGaAs VCSELs," *IEEE Photon. Technol. Lett.*, Vol. 7, no. 7, pp. 736–738, 1995.

71. D. Kuksenkov, S. Feld, C. Wilmsen, H. Temkin, S. Swirthun, and R. Leibenguth, "Linewidth and α factor in AlGaAs/GaAs vertical cavity surface emitting lasers," *Appl. Phys. Lett.*, Vol. 66, no. 3, pp. 277–279, 1995.

72. P. Michler, W. W. Ruhle, G. Reiner, J. K. Ebeling, and A. Moritz, "Time bandwidth product of gain switched $In_{0.2}Ga_{0.8}As/GaAs$ microcavity lasers," *Appl. Phys. Lett.*, Vol. 67, no. 10, pp. 1363–1365, 1995.

CHAPTER 9

Nonlinear Characteristics of Vertical Cavity Surface Emitting Lasers

The nonlinear characteristics of single-cavity VCSELs, including self-sustained pulsation and bistability, are studied with the influence of self-focusing and diffraction loss taken into consideration. The effect of diffraction loss on the modulation response of single-cavity VCSELs is also discussed. On the other hand, the conditions of self-sustained pulsation and bistability of coupled cavity VCSELs are investigated. The dual-wavelength operation of coupled-cavity VCSELs is also analyzed. Polarization switching and bistability in single-cavity VCSELs are also discussed. The possibility of using polarization bistability in high-speed optical digital systems is investigated. Methods for achieving wavelength tunability in VCSELs are also studied.

9.1 INTRODUCTION

Vertical cavity surface emitting lasers (VCSELs) are rapidly emerging as a stronger competitor over the facet emitting semiconductor laser in becoming the light source in high-speed optical fiber communication systems [1]. This is because of the three obvious advantages of VCSELs: (1) low diverged circular output beam, which enhances the coupling efficiency into an optical fiber even without the use of an objective lens [2]; (2) low production cost due to the possibility of monolithic fabrication processes and wafer scalar testing [2]; and (3) high direct modulation speed [3]. For the application of VCSELs in digital switching systems (i.e., all-optical switching and memory systems) [4], optical disk (CD and DVD) readout devices as well as optical wavelength division-multiplexed networks [5], nonlinear characteristics such as self-sustained pulsation (SSP) [6], bistability [7,8], and multiple and tunability wavelength operations [9] are required.

In split-contact facet emitting lasers, the generation of SSP and optical bistability has been demonstrated [10]. The excitation of SSP and optical bistability is reproducible and is dependent on the negative differential resistance of the

Analysis and Design of Vertical Cavity Surface Emitting Lasers By Siu Fung Yu
ISBN 0-471-39124-7

saturable absorber. In addition, it is found that facet emitting lasers with narrow stripe width support SSP, due to an unpumped saturable absorption region [11]. Therefore, it is expected that VCSELs with similar configurations can realize SSP and optical bistability. In fact, it is shown that coupled cavity VCSELs with an intracavity absorber demonstrate SSP and optical bistability [12]. These nonlinear characteristics are due to the negative differential resistance of the saturable absorber, which can be controlled by an electrical bias. On the other hand, it is believed that single-cavity VCSELs of small cavity size (e.g., those with selective oxidization, proton, or ion implantation configuration) support SSP because of the excessive saturable absorption of the unpumped region. However, no strong signal of SSP has been observed experimentally [6]. Therefore, it is usually recognized that output characteristics of single-cavity VCSELs are more stable than those of facet emitting lasers, especially under the influence of external optical feedback [13,14]. The suppression of SSP in single-cavity VCSELs has been explained by the influence of diffraction loss, which dominates over the saturable absorption [15]. Hence, large saturable absorption is not the sufficient requirement to generate SSP in single-cavity VCSELs with small cavity size. In order to utilize VCSELs with single and coupled cavities in high-speed digital systems, the generation conditions of SSP and optical bistability must be determined.

Dual-wavelength laser sources are desirable in several applications such as two-wavelength interferometry for distance measurement, terahertz difference signal generation, and frequency mixing. An optimal source for these applications would be a single-diode laser capable of simultaneous coaxial emission at two different wavelengths. It is noted that cleaved coupled cavity (C^3) facet emitting semiconductor lasers could give rise to dual-wavelength emission [16]. On the other hand, two monolithic cavities, one grown on top of the other and sharing a common mirror form coupled cavity VCSELs, which are believed to give better performance than C^3 facet emitting lasers. This is because the coupled dual-wavelength emissions of the coupled cavity VCSELs are widely spaced (i.e., ~13 nm), and have the same threshold under optical pumping [17]. Therefore, it is interesting to further investigate the lasing characteristics of coupled cavity VCSELs with various configurations.

In Chapter 4, the polarization stability of single-cavity VCSELs under the influence of an injection current has been discussed. In the analysis, the influence of gain anisotropy and birefringence on the selection mechanism of the two orthogonal polarized modes has been studied. In this chapter, the switching mechanisms of the two orthogonal polarized modes under the trigger of external optical injection are analyzed [18]. The influence of self-heating and longitudinal distribution of optical field is also considered. It is interesting to study the fast switching characteristic of the two orthogonal polarized modes by external optical triggering as this behavior has significant potential in the application of all-optical switching and optical memory [4].

Tunable lasers are used in spectroscopy, beam steering, wavelength-division multiplexing, interferometry, and a wide variety of other applications [4]. In facet emitting semiconductor lasers, the continuous tuning of the wavelength is realized by using diffraction grating and bulk optics to form an external cavity [19]. However, an integrated and continuously tunable laser is far more desirable and potentially cost-effective. In this chapter, the possibility of utilizing VCSELs as continuously tunable wavelength lasers is discussed.

This chapter is organized as follows. First, SSP and optical bistability of single-cavity VCSELs are discussed and analyzed. The conditions of SSP and optical bistability are derived, including calculation of self-focusing and diffraction loss. It is found that the conditions of optical bistability, which are similar to those of facet emitting lasers, are dependent on the carrier lifetime and differential gain between the regions of gain and saturable absorption. However, the conditions of SSP in VCSELs are different from those of facet emitting lasers because of the presence of diffraction loss. In addition, the influence of diffraction loss on the modulation response of VCSELs is studied. Then, SSP and optical bistability of coupled cavity VCSELs with an intracavity absorber are investigated. The corresponding conditions of SSP and optical bistability are also derived. On the other hand, the dual wavelength emission of coupled cavity VCSELs with two active cavities is discussed and analyzed. Different approaches to realizing polarization switching and bistable operation in single-cavity VCSELs are then studied. A theoretical model is developed to analyze the switching of the two orthogonal polarized modes under the trigger of external optical injection. Finally, design considerations of tunable wavelength single-cavity VCSELs are discussed.

9.2 INFLUENCE OF SELF-FOCUSING AND DIFFRACTION LOSS ON THE NONLINEAR DYNAMIC OF SINGLE-CAVITY VCSELs

In this section, the influence of self-focusing and diffraction loss on the generation mechanism of SSP in VCSELs is studied. In the investigation, self-focusing and diffraction loss are expressed in terms of two parameters—differential confinement factor and differential cavity loss—with which the analytic model is deduced. The corresponding conditions of SSP in single-cavity VCSELs is obtained, also accounting for self-focusing and diffraction loss. Using the model, it can be shown that the presence of diffraction loss suppresses SSP in VCSELs significantly. In addition, the second-harmonic distortion can be increased in low-frequency range because of the diffraction loss. In the following paragraphs, the phenomenon of self-focusing and diffraction loss in VCSELs is explained. Derivation of the analytic model for single-cavity VCSELs using differential confinement factor and differential cavity loss is also described. Hence, the conditions of SSP in VCSELs with different optical confinement structures are calculated numerically. Furthermore, the influence of diffraction loss on the small-signal modulation response of VCSELs is investigated.

9.2.1 Self-Focusing and Diffraction Loss in VCSELs

Figure 9.1 shows the phenomenon of self-focusing in VCSELs. The mechanism of self-focusing can be explained as the increase in focusing of the transverse

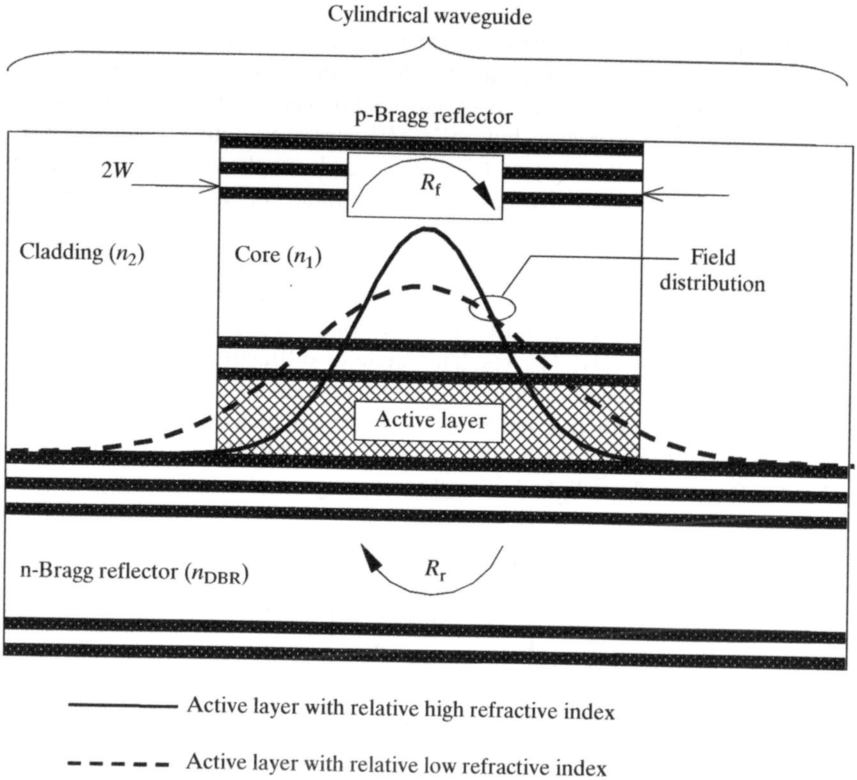

Figure 9.1 Transverse distribution of optical fields inside the active layer of a VCSEL; the solid line indicates the focus of transverse field into the core region due to the increase of refractive index inside the active layer (after Ref. 29).

field into the core region as the refractive index inside the active layer increases. As a result, the transverse optical confinement factor Γ_t of VCSELs is increased. The optical confinement factor is defined as

$$\Gamma_t = \frac{\int_0^W |\Psi(r)|^2 r\,dr}{\int_0^\infty |\Psi(r)|^2 r\,dr}. \tag{9.1}$$

For lasers under small-signal modulation, the variation of the refractive index inside the active layer is due mainly to the change of carrier concentration. Hence, the change of confinement factor due to self-focusing is related to carrier concentration N and can be approximated by

$$\Gamma_t(N_s + \Delta N(t)) = \Gamma_{t,s} + \frac{\partial \Gamma_t}{\partial N}\Delta N(t), \tag{9.2}$$

where ΔN is the small change of carrier concentration and $\Gamma_{t,s}$ and N_s are the confinement factor and carrier concentration, respectively, at steady state. The parameter $\partial\Gamma/\partial N$ is defined as the *differential confinement factor*, which describes the modulation response of the confinement factor. The author has shown [15] that $\Gamma_{t,s}$ decreases but $|\partial\Gamma_t/\partial N|$ increases with the reduction of core radius. However, the sign of $\partial\Gamma_t/\partial N$ depends on the optical confinement structure of lasers and $\partial\Gamma_t/\partial N$ has a negative value for VCSELs with index-guided structures.

Figure 9.2a shows the diffraction of light from a cylindrical waveguide into a distributed Bragg reflector (DBR). The core radius of the device is W; the effective refractive indices of core and cladding regions of the cylindrical waveguide are n_1 and n_2, respectively. The resultant angle of diffraction θ_r from the cylindrical waveguide (core and cladding) of effective refractive index n_{WG} into the DBR with effective refractive index n_{DBR} can be calculated from the theory of Fraunhofer diffraction and Snell's law of refraction [20]. Figure 9.2b shows the angle of diffraction θ_f as light leaves the core region of a circular aperture of

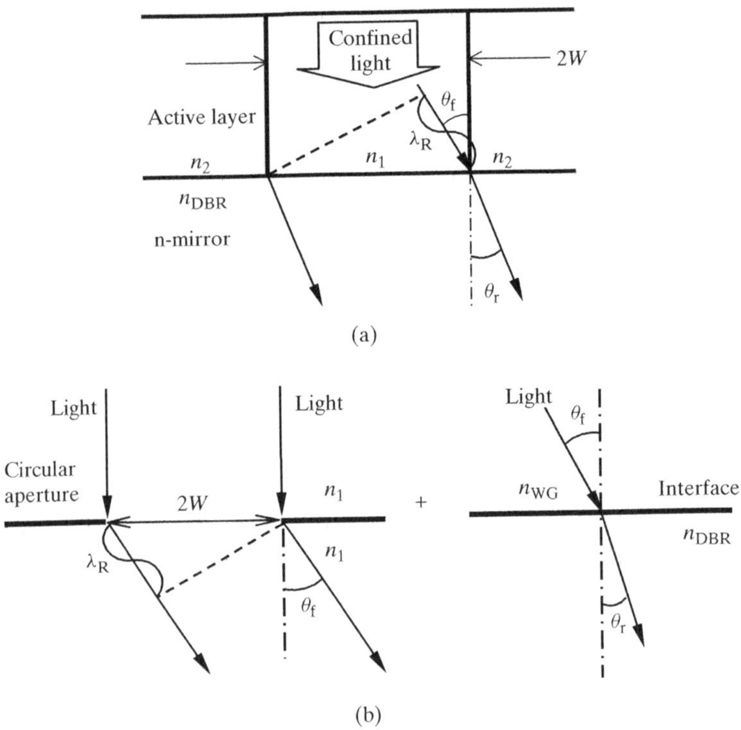

Figure 9.2 The mechanism of light diffraction from a cylindrical waveguide: (a) light diffraction through a circular aperture and refraction of light between two dielectric materials of refractive indices n_{eff} and n_{DBR}, respectively; (b) explanation of light diffraction from a cylindrical waveguide using the theory of Fraunhofer diffraction and Snell's law of refraction. (After Ref. 29.)

radius W. From Fraunhofer diffraction theory, we obtain

$$\theta_f = \sin^{-1}\left(\frac{\lambda_R}{2W}\right), \tag{9.3}$$

where λ_R is the wavelength in the core region. Light is further refracted to θ_r in the mirror. By Snell's law of refraction

$$n_{WG}\sin(\theta_f) = n_{DBR}\sin(\theta_r). \tag{9.4}$$

Substituting (9.4) into (9.3) gives

$$\theta_r = \sin^{-1}\left(\frac{\lambda_R n_{WG}}{2n_{DBR}W}\right). \tag{9.5}$$

Since the reflectivity of DBRs is a function of θ_r [21], (9.5) implies that the reflectivity is also dependent on n_{WG}. In general, the total cavity loss of VCSELs can be written as

$$\alpha = \alpha_m + \frac{1}{L_{eff}}\log\left(\frac{1}{|R_r||R_f|}\right), \tag{9.6}$$

where α_m is the effective material loss and L_{eff} is the effective cavity length. $|R_r|$ and $|R_f|$ are the field reflectivity of the n- and p-type mirrors, respectively. Hence, α is also dependent on n_1 as $|R_r|$ and $|R_f|$ are both functions of n_{WG}.

For devices under small-signal modulation, any variation of α with a small change of n_1 can be related to ΔN by

$$\alpha(N_s + \Delta N(\mathrm{t})) = \alpha_s + \frac{\partial\alpha}{\partial N}\Delta N(t), \tag{9.7}$$

where α_s is the total cavity loss at steady state and $\partial\alpha/\partial N$ is the differential cavity loss. The value of α_s can be deduced from the steady-state value of $|R_r|$ and $|R_f|$ through (9.6). $\partial\alpha/\partial N$ can be obtained by differentiating (9.6) with respect to the carrier concentration

$$\frac{\partial\alpha}{\partial N} = -\frac{1}{L_{eff}}\left(\frac{1}{|R_r|}\frac{\partial|R_r|}{\partial N} + \frac{1}{|R_f|}\frac{\partial|R_f|}{\partial N}\right), \tag{9.8}$$

where α_m and L_{eff} are assumed to be independent of the carrier concentration. $\partial|R_r|/\partial N$ and $\partial|R_f|/\partial N$ are the rates of change of $|R_r|$ and $|R_f|$ with respect to the carrier concentration, respectively. The value of $\partial\alpha/\partial N$ can thus be calculated from (9.8), provided $|R_r|$ and $|R_f|$ are determined. The reflectivity of DBRs under the influence of diffraction loss can be evaluated using the mode-matching technique as described in Section 3.3. In fact, the field reflectivity R_r can be simplified to

$$R_r \approx \int_0^\infty \frac{\beta_0 - \beta_D}{\beta_0 + \beta_D}\Psi_0\Psi_0^*\,ds. \tag{9.9}$$

It must be noted that β_D, β_0, and Ψ_0 are dependent on n_{WG}. R_f can be calculated in a similar manner.

It is expected that $|R_r|$ (as well as $|R_f|$) decreases with the reduction of W due to the increase of diffraction. Hence, the value of α_s increases with the reduction of W as α_s is inversely proportional to $|R_r|$. $|\partial\alpha/\partial N|$ would also increase with the reduction of W, but the sign of $\partial\alpha/\partial N$ depends on the optical confinement structure of VCSELs. For devices with an index-guided structure, $\partial\alpha/\partial N$ is negative. This can be shown by expressing (9.9) into a Fresnel reflectivity formula as given by [22]

$$|R_r| \approx \frac{n_{DBR} - n_{WG}}{n_{DBR} + n_{WG}}. \tag{9.10}$$

$|R_r|$ must be greater than zero, and this implies $n_{DBR} > n_{WG}$. For devices with an index-guided structure, the imaginary part of n_{WG} is negligible so that $\partial|R_r|/\partial N$ can be expressed as

$$\frac{\partial|R_r|}{\partial N} = \frac{-2n_{DBR}}{(n_{DBR} + n_{WG})^2}\frac{\partial n_{WG}}{\partial N}, \tag{9.11}$$

where $\partial n_{WG}/\partial N$ ($\approx\partial n_1/\partial N < 0$) is the effective change of the refractive index with carrier concentration. From (9.11), $\partial|R_r|/\partial N$ is positive, and this also applies for $\partial|R_f|/\partial N$. Alternatively, it can be shown that (1) $\partial|R_{r,f}|/\partial\theta_r$ is negative as $\theta_r = 0$ is the global maximum of $|R_{r,f}|$ [8] and (2) $\partial\theta_r/\partial N$ is also negative as indicated in (9.5); thus $\partial|R_{r,f}|/\partial N = \partial|R_{r,f}|/\partial\theta_r \times \partial\theta_r/\partial N$ is positive. Both of these arguments show that $\partial\alpha/\partial N$ has a negative value for index-guided lasers. In the paragraphs above, the purpose of defining the differential confinement factor and differential cavity loss is to simplify the derivation of the conditions of SSP in single-cavity VCSELs using the simple rate equation approach so that self-focusing and diffraction loss can be factored in easily.

9.2.2 Condition of Self-Sustained Pulsation in VCSELs

The schematic of VCSELs as shown in Figure 9.3a shows that the condition of SSP in VCSELs, including self-focusing and diffraction loss, can be derived using the following modified rate equation model [15]

$$\frac{\partial S(t)}{\partial t} = v_g(\Gamma_{t,1}(N_1, N_2)g_1(N_1) + \Gamma_{t,2}(N_1, N_2)g_2(N_2) - \alpha(N_1, N_2))S(t) + R_{sp}(N_1, N_2), \tag{9.12a}$$

$$\frac{\partial N_1(t)}{\partial t} = \frac{J(t)}{qd} - \frac{N_1(t)}{\tau_{n1}} - v_g\Gamma_{t,1}(N_1, N_2)g_1(N_1)S(t), \tag{9.12b}$$

$$\frac{\partial N_2(t)}{\partial t} = -\frac{N_2(t)}{\tau_{n2}} - v_g\Gamma_{t,2}(N_1, N_2)g_2(N_2)S(t), \tag{9.12c}$$

where S is the effective photon density; N_1 and N_2 are the equivalent carrier concentration inside the core (gain) and cladding (absorption) regions of the laser,

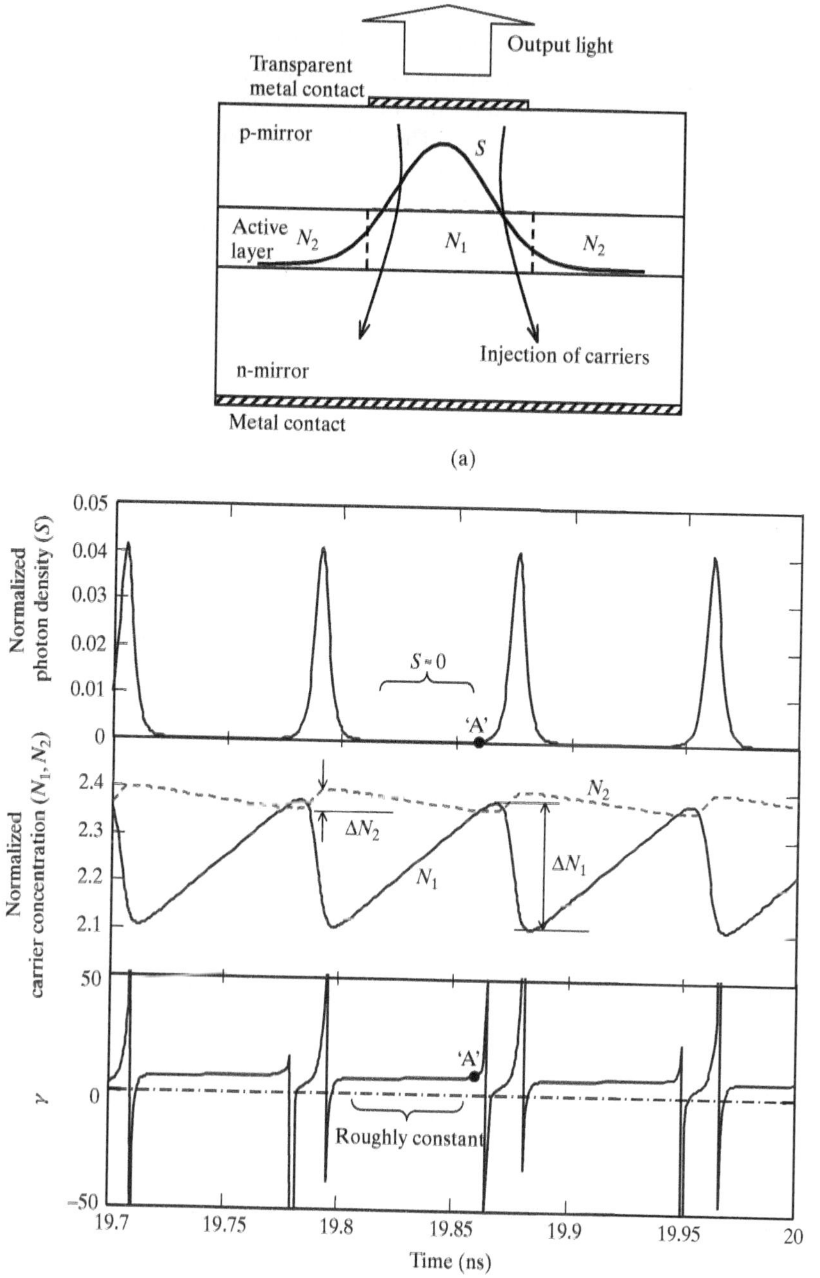

Figure 9.3 (a) Schematic used to derive the conditions of self-sustained pulsation operation of VCSELs with current confinement structure similar to narrow-stripe wide-facet emitting lasers; (b) transient response of photon density S, carrier concentration inside the core N_1 and cladding N_2 regions, and γ. (After Ref. 25.)

respectively; g_1 and g_2 are the corresponding optical gain; and τ_{n1} and τ_{n2} are the corresponding carrier lifetimes. The influence of self-focusing is introduced into the model through the carrier-dependent confinement factors $\Gamma_{t,1}$ and $\Gamma_{t,2}$ [23]. This is because the transverse field profile varies as a result of the carrier-induced index change inside the active layer. The influence of diffraction loss is also included in the model through the carrier-dependent cavity loss α. This is because the reflectivity of DBRs, which is inversely proportional to the cavity loss, varies with the penetration depth (i.e., effective guidewidth) of the transverse field [24].

For lasers in SSP, the carrier concentrations inside the gain region (N_1) and absorption region (N_2) are varied out of phase [25] (see Fig. 9.3b). This is because as N_1 approaches its threshold, photon density starts to generate and is absorbed in the absorption region. Hence, N_2 in the absorption region is increased relative to the absorption. Afterward, N_1 reduces with the stimulated recombination due to the high power of light. Therefore, it is possible to simplify our analysis by relating the net change of carrier concentration inside the active (ΔN_1) and absorption (ΔN_2) regions to the net change of equivalent carrier concentration ΔN as below

$$\Delta N \equiv \Delta N_1 \approx -\gamma \Delta N_2, \tag{9.13}$$

where $\gamma \geq 1$ and is assumed to be time-independent. The change of optical gain Δg_1 and Δg_2 in the active and absorption regions, respectively, can be expressed as

$$\Delta g_1 = \frac{\partial g_1}{\partial N_1}\Delta N_1 \equiv \frac{\partial g_1}{\partial N}\Delta N \qquad \text{and} \qquad \Delta g_2 = \frac{\partial g_2}{\partial N_2}\Delta N_2 \equiv -\gamma^{-1}\frac{\partial g_2}{\partial N}\Delta N, \tag{9.14}$$

where $\partial g_1/\partial N$ and $\partial g_2/\partial N$ are the corresponding differential gain. The carrier induced change in confinement factors are given by

$$\Delta\Gamma_{t,1} = \frac{\partial \Gamma_{t,1}}{\partial N_1}\Delta N_1 + \frac{\partial \Gamma_{t,1}}{\partial N_2}\Delta N_2 \equiv \frac{\partial \Gamma_t}{\partial N}\Delta N \tag{9.15a}$$

$$\Delta\Gamma_{t,2} = \frac{\partial \Gamma_{t,2}}{\partial N_1}\Delta N_1 + \frac{\partial \Gamma_{t,2}}{\partial N_2}\Delta N_2 \equiv -\frac{\partial \Gamma_t}{\partial N}\Delta N, \tag{9.15b}$$

where it is assumed that $\Gamma_{t,1} + \Gamma_{t,2} = 1$ and $\partial\Gamma_t/\partial N$ is defined as the equivalent differential confinement factor. Furthermore, the change of cavity loss can be written as

$$\Delta\alpha = \frac{\partial \alpha}{\partial N_1}\Delta N_1 + \frac{\partial \alpha}{\partial N_2}\Delta N_2 \equiv \frac{\partial \alpha}{\partial N}\Delta N, \tag{9.16}$$

where $\partial\alpha/\partial N$ is defined as the equivalent differential cavity loss. Substituting (9.13)–(9.16) into (9.12) gives [23]

$$\frac{\partial \Delta S}{\partial t} = v_g S_s \left((g_{1s} - g_{2s})\frac{\partial \Gamma_t}{\partial N} + \left(\Gamma_{t,1s}\frac{\partial g_1}{\partial N} - \gamma^{-1}\Gamma_{t,2s}\frac{\partial g_2}{\partial N} \right) - \frac{\partial \alpha}{\partial N} \right) \Delta N, \tag{9.17}$$

$$\frac{\partial \Delta N}{\partial t} = -\left(\frac{1}{\tau_{n1}} + v_g S_s\{g_{1s}\frac{\partial \Gamma_t}{\partial N} + \Gamma_{t,1s}\frac{\partial g_1}{\partial N}\}\right)\Delta N - v_g \Gamma_{t,1s} g_{1s}\,\Delta S, \quad (9.18)$$

$$-\frac{\partial \Delta N}{\partial t} = \left(\frac{1}{\tau_{n2}} + v_g S_s\{g_{2s}\gamma\frac{\partial \Gamma_t}{\partial N} + \Gamma_{t,2s}\frac{\partial g_2}{\partial N}\}\right)\Delta N - v_g \Gamma_{t,2s} g_{2s}\gamma\,\Delta S, \quad (9.19)$$

where the subscript "s" stands for steady state and ΔS is the small change in photon density. In the derivation of (9.17)–(9.19), the spontaneous emission term R_{sp} is ignored in the calculation.

The term ΔN can be eliminated from (9.17) by adding (9.18) and (9.19), then substituting the expression of ΔN back into (9.17), giving

$$\frac{\partial \Delta S}{\partial t} = -v_g \alpha'_s \frac{v_g S_s\left((g_{1s} - g_{2s})\frac{\partial \Gamma_t}{\partial N} + \left(\Gamma_{t,1s}\frac{\partial g_1}{\partial N} - \gamma^{-1}\Gamma_{t,2s}\frac{\partial g_2}{\partial N}\right) - \frac{\partial \alpha}{\partial N}\right)}{\frac{1}{\tau_{n1}} - \frac{1}{\tau_{n2}} + v_g S_s\left((g_{1s} - g_{2s}\gamma)\frac{\partial \Gamma_t}{\partial N} + \left(\Gamma_{t,1s}\frac{\partial g_1}{\partial N} - \Gamma_{t,2s}\frac{\partial g_2}{\partial N}\right)\right)}\Delta S, \quad (9.20)$$

where $\alpha'_s = \Gamma_{t,1s} g_{1s} + \Gamma_{t,2s} g_{2s}\gamma > 0$. For unstable operations (i.e., SSP), ΔS should increase with time and the factor in front of ΔS must be positive. The carrier lifetime inside the active region should be much shorter than that inside the absorption region, due to the high injection of carrier concentration, so that $1/\tau_{n1} - 1/\tau_{n2} > 0$. For devices with weak optical confinement (i.e., $\Gamma_{t,2} > \Gamma_{t,1}$), $\Gamma_{t,1s}\partial g_1/\partial N - \Gamma_{t,2s}\partial g_2/\partial N < 0$ is satisfied, as the differential gain of the absorption region is relatively higher than that of the gain region. However, the denominator of the factor can be assumed to be greater than zero. This is because SSP occurs only for the device biased near threshold, and this implies that the steady-state value of S_s is very small. Therefore, the condition of SSP is reduced to

$$\eta \equiv \underbrace{(g_{1s} - g_{2s})\frac{\partial \Gamma_t}{\partial N}}_{(1)} + \underbrace{\left(\Gamma_{t,1s}\frac{\partial g_1}{\partial N} - \gamma^{-1}\Gamma_{t,2s}\frac{\partial g_2}{\partial N}\right)}_{(2)} - \underbrace{\frac{\partial \alpha}{\partial N}}_{(3)} < 0, \quad (9.21)$$

where η is the instability constant. In (9.21), $g_{1s} - g_{2s} > 0$ because g_{2s} is the optical loss in the absorption region. Now, the condition of SSP is determined by the signs of (1) the differential confinement factor arising from the self-focusing effect, (2) the differential gain difference between the gain and absorption regions, and (3) the differential cavity loss arising from the diffraction loss.

The validity of (9.21) depends on whether the assumptions given in (9.13)–(9.16) are realistic. One of the main assumptions describes the rate of change of the carrier concentration inside the core and cladding regions (i.e., $\Delta N_1/\Delta N_2 = -\gamma$) may be in debate as γ is also a function of time. Furthermore, this assumption also infers $\partial\Delta N_1/\partial t/\partial\Delta N_2/\partial t = -\gamma$. If γ is not a constant at the position of linearization of the rate equation model, the derivation given in (9.21) is absolutely meaningless. In fact, it can be shown that the linearization of the rate equation model can be performed at some particular values of S, N_1, and N_2 so that γ can be maintained as a constant. Figure 9.3b shows the time variation of S, N_1, N_2, and γ under SSP obtained from the model given in [25]. It is noted that during the turnon/off time of S, the value of γ varies significantly and its maximum absolute magnitude can be as large as 10^4. On the other hand, γ remains constant during the time interval when S is negligible. If assumption (9.13) is applied for the linearization of the rate equation model at the time interval when S is negligible, γ will be a constant so that the stability condition of S can be examined by (9.20), which describes the growth of S from spontaneous emission. The meaning of γ can be recognized by comparing (9.21) with the SSP condition of edge emitting lasers. If the influence of self-focusing and diffraction loss are ignored in the derivation of (9.21), the condition of SSP reduces to $\eta_a = \Gamma_{t,1s}\partial g_1/\partial N - \gamma^{-1}\Gamma_{t,2s}\partial g_2/\partial N$. The condition of SSP given by Kawaguchi [26] is equal to $\eta_b = \Gamma_{t,1s}|g_{1s}|\partial g_1/\partial N - \Gamma_{t,2s}|g_{2s}|\partial g_2/\partial N$, where two extra terms, g_{1s} and g_{2s}, are observed in the expression of η_b. Therefore, γ in (9.21) can be interpreted as the mixing effect of the carrier concentration between the core and cladding regions so that $\gamma = |g_{1s}/g_{2s}|$ can be assumed.

The waveguide structure of lasers determines the sign of $\partial\Gamma_t/\partial N$. It is shown in [23] that $\partial\Gamma_t/\partial N$ has a negative value for index-guided lasers but a positive value for gain or anti-index-guided devices so that devices with index-guided structure enhance SSP. In fact, Chen et al. have shown [13] that edge emitting devices with strong index-guided structure enhance SSP through the presence of the self-focusing effect (i.e., $\partial\Gamma_t/\partial N$). In Figure 2 of the Yuri et al. paper [11], it is observed that the reduction of mode gain (i.e., reduction of carrier concentration) decreases the FWHM of the optical field (i.e., increases the optical confinement of the optical mode), which implies that $\partial\Gamma_t/\partial N$ has a negative value. Therefore, their numerical analysis [11] verifies (9.21) in explaining the generation mechanism of SSP in narrow stripe facet emitting lasers. It seems that oxide confined VCSELs have much stronger index-guided structures than do facet emitting lasers, and SSP should be easily generated through the negative value of $\partial\Gamma_t/\partial N$. However, SSP similar to that occurring in facet emitting lasers [11] has not been observed in VCSELs [6], and this can be explained by the presence of $\partial\alpha/\partial N$ in (9.21). It has been shown in Section 9.2.1 that $\partial\alpha/\partial N$ has a negative value. Therefore, the diffraction loss works against the generation of SSP in VCSELs as indicated in (9.21).

9.2.3 Calculation of $\partial\alpha/\partial N$ and $\partial\Gamma_t/\partial N$ and the Condition of SSP

In the preceding analytical analysis of SSP, the influence of thermal lensing and SHB has been ignored in order to simplify our derivation. In the following development of the numerical model of VCSELs, the influence of SHB (as well as self-focusing and diffraction loss) is considered. However, the thermal lensing effect is ignored in our analysis because the response time of heat flow is much longer than the time variation of the carrier concentration and photon density, and the thermal lensing effect is less dominant for the device biased close to its threshold.

The modified rate equation model of VCSELs is given by [27]

$$\frac{\partial N(r,t)}{\partial t} = \frac{J(r,t)}{qd} - \frac{N(r,t)}{\tau_n} - v_g \Gamma_z g(N)|\Psi(r,t)|^2 S(t) + D_n \frac{1}{r}\frac{\partial}{\partial r}\left(r\frac{\partial N(r,t)}{\partial r}\right), \tag{9.22}$$

$$\frac{\partial S(t)}{\partial t} = v_g(\Gamma_z \langle g(t)\rangle - \langle \alpha_m(t)\rangle - \alpha(t))S(t) + \langle R_{sp}(t)\rangle, \tag{9.23}$$

where Γ_z is the longitudinal confinement factor and g is the optical gain of the QW active layer to be determined. The symbol Ψ represents the normalized transverse field (i.e., $1 = 2\int_0^\infty |\Psi|^2 r\,dr/W^2$) of the fundamental mode (i.e., LP_{01} mode) with S as the corresponding photon density and W as the radius of the core region. In (9.22) and (9.23), the self-focusing effect is directly introduced into the computation through the time dependence of Ψ. The time variation of Ψ can be calculated by solving the complex scalar wave equation (see Chapter 3) with the influence of nonuniformly transverse distribution of refractive index included in the calculation. On the other hand, the parameter $\alpha(t)$ is introduced into (9.23) to include the influence of diffraction loss.

The current spreading from a circular-disk contact along the active region can be expressed as [28]

$$J(r,t) = \begin{cases} J_0(t) & r \le W \\ J_0(t)\exp\left(\dfrac{-(r-W)}{r_0}\right) & r > W \end{cases}, \tag{9.24}$$

where J_0 denotes the current density at the edge within the contact area ($r < W$) and r_0 is the effective diffusion length of the injection carrier. In the photon rate equation, the effective optical gain $\langle g \rangle$ is given by

$$\langle g(t)\rangle = \frac{2}{W^2}\int_0^\infty g(N)|\Psi(r,t)|^2 r\,dr. \tag{9.25}$$

The effective material loss in the active layer α_m and the spontaneous emission rate R_{sp} are given by

$$\langle \alpha_m(t)\rangle = \frac{2}{W^2}\int_0^\infty \alpha_m(r)|\Psi(r,t)|^2 r\,dr, \tag{9.26a}$$

$$\langle R_{sp}(t)\rangle = \frac{2}{W^2}\beta_s \Gamma_z B_{sp}\int_0^W N^2(r,t)|\Psi(r,t)|^2 r\,dr, \tag{9.26b}$$

where α_m ($\sim$20 cm^{-1}) is the transverse distribution of material loss of the QW active layer. The transient behavior of N and S is calculated from the coupled differential equations, (9.22) and (9.23), by using the fourth-order Runge–Kutta numerical integration method. The variation of the carrier concentration is obtained by using the finite difference method subjected to the corresponding continuity and boundary conditions [27] so that the influence of SHB can be computed. The transverse distribution of Ψ is also solved simultaneously with the variation of the carrier concentration so that both self-focusing and diffraction loss [i.e., arising from the time-dependent cavity loss $\alpha(t)$] can be considered.

From the numerical model of VCSELs given above, the corresponding condition of SSP, including SHB, self-focusing, and diffraction loss, is given by

$$\eta = (\langle g_1\rangle - \langle g_2\rangle)\frac{\partial \Gamma_t}{\partial\langle N\rangle} + \left(\Gamma_{t,1}\frac{\partial\langle g_1\rangle}{\partial\langle N_1\rangle} - \langle\gamma\rangle^{-1}\Gamma_{t,2}\frac{\partial\langle g_2\rangle}{\partial\langle N_2\rangle}\right) - \frac{\partial\alpha}{\partial\langle N\rangle} < 0, \tag{9.27}$$

where $\langle g\rangle = \langle g_1\rangle + \langle g_2\rangle$, $\partial\langle g\rangle/\partial\langle N\rangle$ is the effective differential gain, $\langle\gamma\rangle^{-1} = \langle g_2\rangle/\langle g_1\rangle$, $\partial\Gamma_t/\partial\langle N\rangle$, and $\partial\alpha/\partial\langle N\rangle$ are the effective differential confinement factor and cavity loss, respectively. The differential confinement factor $\partial\Gamma_t/\partial\langle N\rangle$ and the differential cavity loss $\partial\alpha/\partial\langle N\rangle$ can be approximated as

$$\frac{\partial\Gamma_t}{\partial\langle\Delta N\rangle} \approx \frac{\Delta\Gamma_t}{\langle\Delta N\rangle} \approx \frac{2\Delta\Gamma_t}{\langle\Delta N_1\rangle - \langle\Delta N_2\rangle} \quad \text{and} \quad \frac{\partial\alpha}{\partial\langle N\rangle} \approx \frac{\Delta\alpha}{\langle\Delta N\rangle} \approx \frac{2\Delta\alpha}{\langle\Delta N_1\rangle - \langle\Delta N_2\rangle}, \tag{9.28}$$

where $\langle\Delta N_1\rangle$ and $\langle\Delta N_2\rangle$ are the average change of carrier concentration inside the core and cladding regions, respectively.

For the calculation of differential confinement factor and differential cavity loss in VCSELs as shown in Figure 9.1, it is assumed that the DBR mirror has a circular symmetry with 30-pair GaAs/AlAs layers of half-wavelength each. The fundamental transverse field (LP_{01} mode) has a resonant wavelength of 0.85 μm confined inside the active layer but diverged into the DBR. Using this optical confinement structure of VCSELs, the confinement factor and cavity loss as well as their derivatives can be evaluated. Figure 9.4 shows the variation of $\partial\Gamma_t/\partial\langle N\rangle$ and $\partial\alpha/\partial\langle N\rangle$ of VCSELs with a weakly index-guided structure (i.e., $\Delta n = 0.0225$) and an anti-index-guided structure (i.e., $\Delta n = -0.0225 + j0.012$) for different values of W. In the calculation, the variation of Δn is related to the carrier concentration by $\partial n/\partial N \sim 1 \times 10^{-20}$ cm^3 for GaAs/AlGaAs QW material. In Figure 9.4a, the variation of $-\partial\Gamma_t/\partial\langle N\rangle$ for the index-guided device and $\partial\Gamma_t/\partial\langle N\rangle$ for the antiguiding device are plotted by varying W. It is shown that the device with an index-guiding structure gives a negative value of $\partial\Gamma_t/\partial\langle N\rangle$ but that with an antiguided structure gives a positive value of $\partial\Gamma_t/\partial\langle N\rangle$. In addition, the absolute magnitude of $\partial\Gamma_t/\partial\langle N\rangle$ for both types of transverse guiding structure increases dramatically with the reduction of W. On the other hand, the variation of $-\partial\alpha/\partial\langle N\rangle$ for the index-guided device and $\partial\alpha/\partial\langle N\rangle$ for the anti-index-guided device are plotted in Figure 9.4b by varying W. It is noted that the sign of $\partial\alpha/\partial\langle N\rangle$ is similar to that of $\partial\Gamma_t/\partial\langle N\rangle$ and the absolute magnitude of $\partial\alpha/\partial\langle N\rangle$ is increased with reduction of W. Hence, the condition of SSP can be enhanced by $\partial\Gamma_t/\partial\langle N\rangle$ for an index-guided structure or $\partial\alpha/\partial\langle N\rangle$ for an anti-index-guided structure. However, optimizing the magnitude of $\partial\Gamma_t/\partial\langle N\rangle$ or $\partial\alpha/\partial\langle N\rangle$ to enhance SSP in VCSELs is not recommended because

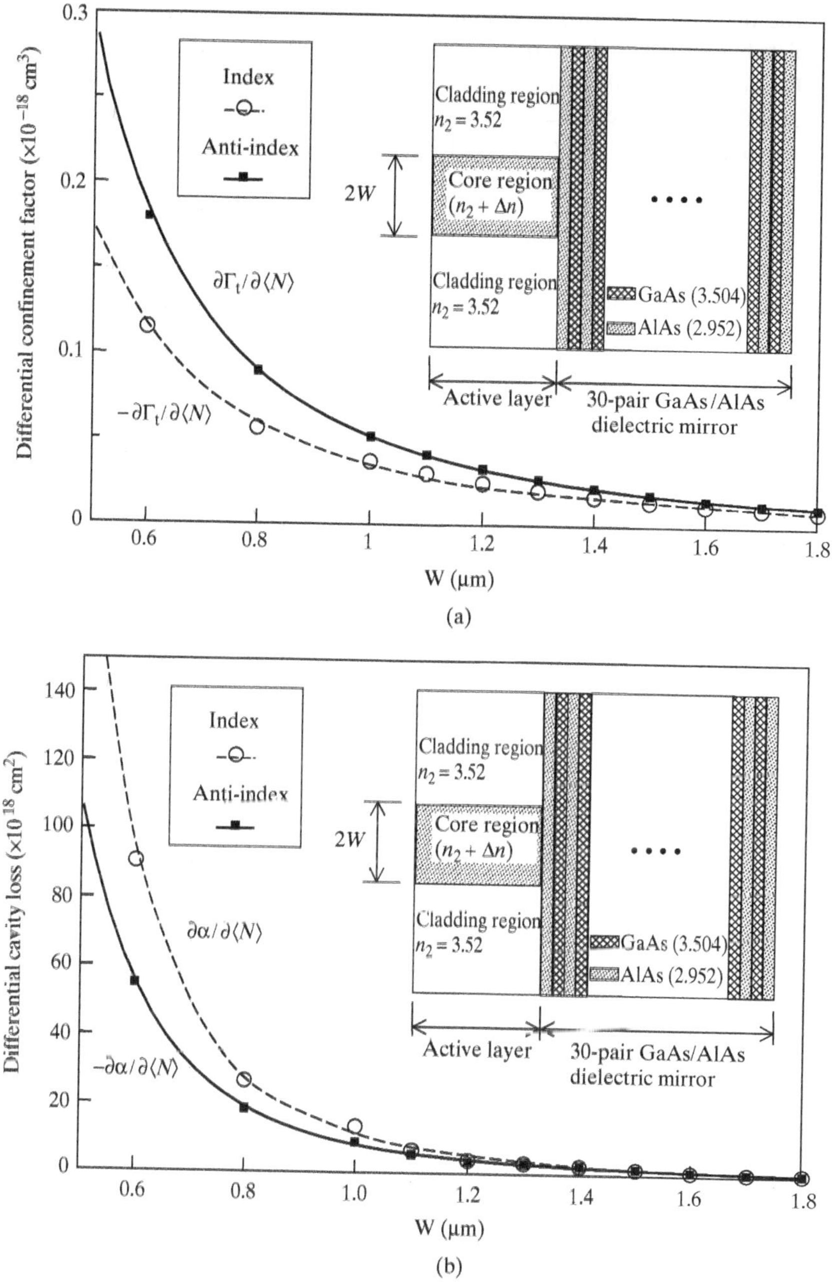

Figure 9.4 Variation of (a) $\partial\Gamma_t/\partial\langle N\rangle$ and (b) $\partial\alpha/\partial\langle N\rangle$ with W, for devices of index-guided (dotted line) and anti-index-guided (solid line) structures; insert diagram is schematic of DBR used to calculate the reflectivity under the influence of diffraction; Δn is difference of complex refractive index between core and cladding regions. (After Ref. 15.)

- The magnitude of $\partial\Gamma_t/\partial\langle N\rangle$ and $\partial\alpha/\partial\langle N\rangle$ is also dependent on the carrier distribution inside the active layer.
- The influence of $\partial\Gamma_t/\partial\langle N\rangle$ is also affected by the magnitude of the optical gain difference [see (9.27)].
- The magnitude of $\partial\Gamma_t/\partial\langle N\rangle$ and $\partial\alpha/\partial\langle N\rangle$ is significant only for small W (i.e. <1 μm).

In fact, it is more practical to control the transverse overlap effects to enhance SSP in VCSELs, provided the diffraction loss effects can be minimized. This is because the diffraction loss effects are difficult to control and can deteriorate the SSP of VCSELs. Therefore, a novel structure of VCSELs, with immunity to self-focusing and diffraction loss as well as adjustable overlap volume between the saturable absorption region and the lasing mode is required.

In fact, using the interdiffusion technique to enhance the transverse overlap effects in VCSELs for SSP has been proposed [15]. Figure 9.5a plots the variation of instability constant η as a function of W and Δn. The change of Δn is due to the application of interdiffusion into the core region of the GaAs QWs active layer (i.e., well width of 100 Å), and the corresponding optical gain will also be affected. If the optical gain of the core region of the active layer is expressed in terms of the logarithm function [i.e., $g = a_N \log(N/N_t)$, where a_N is the gain parameter and N_t is the carrier concentration at transparency], it can be shown that $a_N = 1780$, 1507, and 468 cm^{-1}, and $N_t = 2.1996 \times 10^{18}, 2.0515 \times 10^{18}$, and 2.4902×10^{18} cm^{-3} for Δn equals 0, −0.012, and −0.045, respectively. The injection current of the devices is set to about 1.1 times its threshold value. It is shown that

- The range of W (for $\eta < 0$) increases with $|\Delta n|$ because of the enhancement of absorption loss inside the cladding region.
- The parameters $\partial\Gamma_t/\partial\langle N\rangle$ and $\partial\alpha/\partial\langle N\rangle$ have significant influence on the instability constant η only for small W.
- For devices with large W, the instability constant η is affected mostly by the saturable absorption inside the cladding region.

The current dependence of pulse repetition frequency and peak pulse power is also shown in Figure 9.5b. It is assumed that the core radius of the devices is equal to 1.3 μm with Δn varying between 0 and −0.045. It is observed that the pulse repetition frequency (for both SSP and ROF) as well as the peak output power increase with $|\Delta n|$. It is noted that the conditions to get SSP with high oscillation frequency and peak power are low optical confinement and small differential gain inside the core region of the active layer.

9.2.4 Influence of Diffraction Loss on Modulation Response of VCSELs

The modulation response of VCSELs under the influence of self-focusing and diffraction loss can also be analyzed using the rate equation model given in

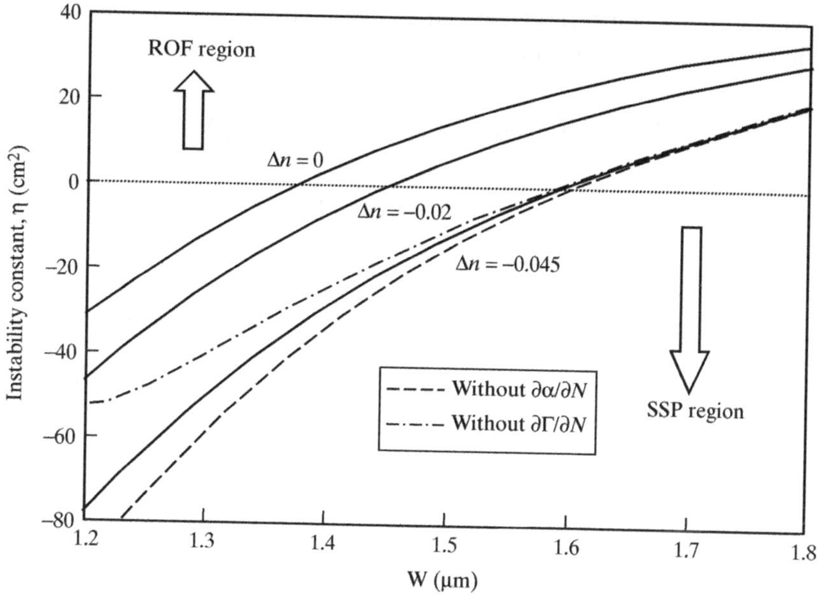

(a)

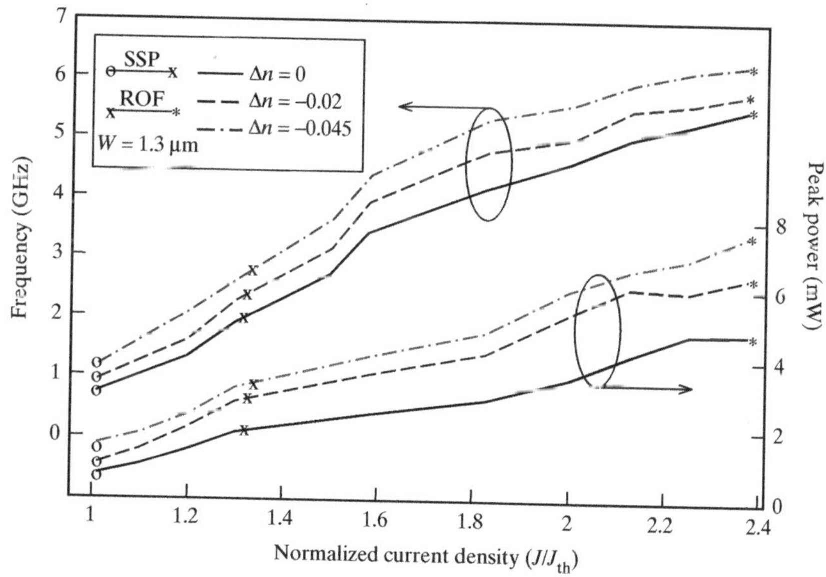

(b)

Figure 9.5 (a) Variation of instability constant η with W for different optical confinement structures, where the lasers are set to 1.1 times their threshold values; (b) current dependence of pulse repetition frequency and output peak power of VCSEL with $W =$ 1.3 μm and different optical confinement structures. (After Ref. 15.)

Section 7.2. If the influence of the carrier transport effect is ignored, the rate equation model can be written as [29]

$$\frac{\partial S(t)}{\partial t} = v_g(\Gamma_t(t)g(t) - \alpha(t))S(t) + \beta_s\Gamma_z B_{sp}N^2(t), \tag{9.29a}$$

$$\frac{\partial N(t)}{\partial t} = \frac{I(t)}{qV_a} - \frac{N(t)}{\tau_n} - v_g\Gamma_t(t)g(t)S(t), \tag{9.29b}$$

where I is the injection current and V_a is the volume of the active layer; α and Γ_t here take $\Gamma_{t,s}$ and α_s as well as $\partial\Gamma_t/\partial N$ and $\partial\alpha/\partial N$ into account for determining the self-focusing and diffraction loss. The thermal effect can be factored in using the auxiliary thermal equation. Furthermore, the influence of nonlinear gain is considered in the model through the carrier and photon dependence of g defined as

$$g = \frac{g_0}{1 + \varepsilon S}, \tag{9.30}$$

where ε is the gain compression factor. g_0 is the unsaturated optical gain that is dependent only on the carrier concentration and can be expressed as

$$g_0 = \Gamma_z a_N \log\left(\frac{N}{N_t}\right). \tag{9.31}$$

where the parameters a_N and N_t vary with temperature. In addition, the Arrhenius-type relation can describe the temperature dependence of the threshold current density. Chapter 5 describes the implementation of temperature dependence of optical gain and threshold current density.

The small-signal modulation response of VCSELs can be obtained using the perturbation method [30]. This can be done by equating I, N, and S with

$$I(t) = I_s + \Delta I(t) = I_s + I_1 e^{j\omega t}, \tag{9.32a}$$

$$N(t) = N_s + \Delta N(t) = N_s + N_1 e^{j\omega t} + N_2 e^{j2\omega t}, \tag{9.32b}$$

$$S(t) = S_s + \Delta S(t) = S_s + S_1 e^{j\omega t} + S_2 e^{j2\omega t}, \tag{9.32c}$$

where I_s is the bias current and I_1 and ω are the amplitude and frequency of the sinusoidal modulation current, respectively. The subscript "s" stands for the steady state. By substituting (9.32) into the rate equations, and after some manipulations, the AM response and SHD of VCSELs with diffraction loss effects are given by

$$\frac{S_1}{I_1} = \frac{\dfrac{v_g S_s}{qV_a}\Gamma_{t,s}\dfrac{\partial g_{eff}}{\partial N}}{(\omega_n^2 - \omega^2) + j\omega v_g\left(\kappa_s - \Gamma_{t,s}\dfrac{\partial g}{\partial S}S_s\right)}, \tag{9.33}$$

$$\frac{S_2}{S_1} = \frac{(2j\omega + \nu_g\kappa_s)F(\omega) - \nu_g S_s \Gamma_{t,s} \dfrac{\partial g_{\text{eff}}}{\partial N} E(\omega)}{(\omega_n^2 - 4\omega^2) + 2j\omega\nu_g\left(\kappa_s - \Gamma_{t,s}\dfrac{\partial g}{\partial S} S_s\right)} S_1. \tag{9.34}$$

The relaxation oscillation frequency (ROF) ω_n is defined as

$$\omega_n^2 = \nu_g^2 \Gamma_{t,s}\left(\alpha_{\text{eff}}\frac{\partial g_{\text{eff}}}{\partial N} - \kappa_s \frac{\partial g}{\partial S}\right) S_s, \tag{9.35}$$

where

$$\frac{\partial g_{\text{eff}}}{\partial N} = \frac{\partial g}{\partial N} + \frac{1}{\Gamma_{t,s}}\left(g_s \frac{\partial \Gamma_t}{\partial N} - \frac{\partial \alpha}{\partial N}\right), \qquad \alpha_{\text{eff}} = \alpha_s\left(1 - \frac{\varepsilon S_s}{(1+\varepsilon S_s)}\right),$$

$$\kappa_s = \frac{1}{\nu_g \tau_n} + \left(g_s \frac{\partial \Gamma_t}{\partial N} + \Gamma_{t,s}\frac{\partial g}{\partial N}\right) S_s,$$

$$\frac{\partial g}{\partial S} = -\frac{\varepsilon \alpha_s}{\Gamma_{t,s}(1+\varepsilon S_s)}, \text{ and } \frac{\partial g}{\partial N} = \frac{\Gamma_z a_N}{N_t}\exp\left\{-\frac{\alpha_s}{\Gamma_{t,s}\Gamma_z a_N}\right\}.$$

In (9.34), $F(\omega)$ and $E(\omega)$ are defined as

$$E(\omega) = \nu_g S_s \frac{\partial \Gamma_s}{\partial N} K(\omega)\left[\frac{\partial g}{\partial S} + \frac{\partial g}{\partial N}K(\omega)\right] + \nu_g \Gamma_{t,s}\left[\frac{\partial g}{\partial S} + \left(\frac{\partial g_{\text{eff}}}{\partial N} + \frac{1}{\Gamma_{t,s}}\frac{\partial \alpha}{\partial N}\right)K(\omega)\right], \tag{9.36a}$$

$$F(\omega) = \nu_g S_s \frac{\partial \Gamma_t}{\partial N} K(\omega)\left[\frac{\partial g}{\partial S} + \frac{\partial g}{\partial N}K(\omega)\right] + \nu_g \Gamma_{t,s}\left[\frac{\partial g}{\partial S} + \frac{\partial g_{\text{eff}}}{\partial N}K(\omega)\right], \tag{9.36b}$$

where

$$K(\omega) = \left(\nu_g S_s \Gamma_{t,s}\frac{\partial g_{\text{eff}}}{\partial N}\right)^{-1} \times \left(j\omega - \nu_g \Gamma_{t,s} S_s \frac{\partial g}{\partial S}\right).$$

In (9.36), $E(\omega)$ and $F(\omega)$ differ only by an extra term, $\nu_g(\partial\alpha/\partial N)K(\omega)$. In this derivation, the spontaneous emission term is ignored.

From (9.33), it is observed that the self-focusing and diffraction loss have no relation to the AM response of VCSELs. Figure 9.6 shows the calculated SHD of a strongly index-guided VCSEL using (9.34) for devices with $W = 3.5$ and 5 μm. The device is biased around 4 mA so that the steady-state output power of 1 mW undergoes a small-signal current modulation ($\sim$0.5 mA) of frequency ω. In the calculations, the values of $\Gamma_{t,s}$, α_s, and R_{TH} are found to be 0.993 (0.999), 64 cm^{-1} (50.3 cm^{-1}), and 4200 cm^{-1} (2900 cm^{-1}), respectively, for devices of

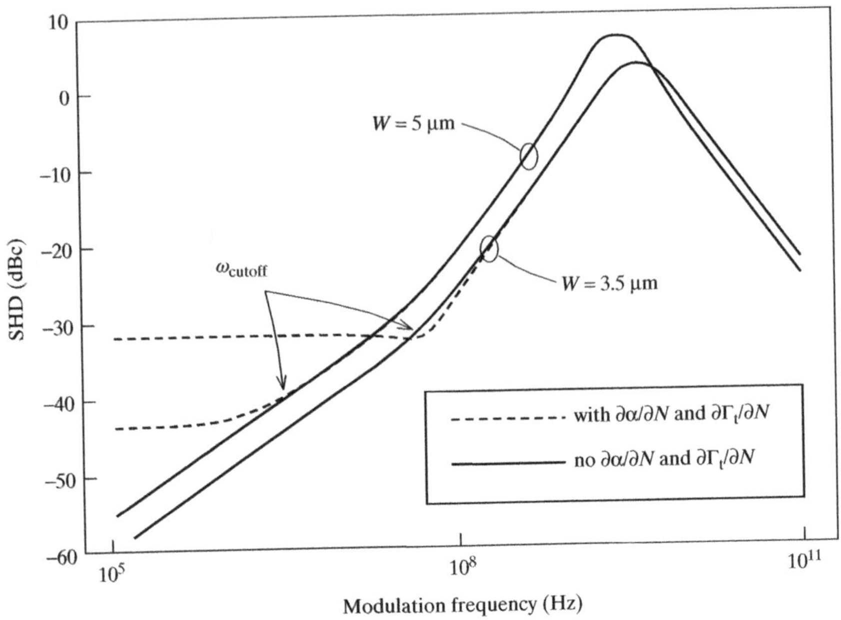

Figure 9.6 Second-harmonic distortion of VCSELs with core radius of 3.5 and 5 μm; the steady-state output power of VCSELs is set to 1 mW (after Ref. 29).

$W = 3.5$ μm (5 μm). The influence of cavity size and thermal effect has been factored into the analysis of SHD (i.e., see Section 7.2). In the first calculation, both $\partial\Gamma_t/\partial N$ and $\partial\alpha/\partial N$ are set to zero to simulate the SHD response without self-focusing and diffraction loss; in the second one, $\partial\Gamma_t/\partial N$ and $\partial\alpha/\partial N$ are assigned their most probable values (i.e., at $W = 3.5$ μm, $\partial\Gamma_t/\partial N$ and $\partial\alpha/\partial N$ are equal to -0.4×10^{-20} cm^3 and -0.1×10^{-18} cm^2; at $W = 5$ μm, $\partial\Gamma_t/\partial N$ and $\partial\alpha/\partial N$ are equal to -0.02×10^{-20} cm^3 and -0.03×10^{-18} cm^2). It is observed that, at low modulation frequencies, SHD is high for the case with $\partial\Gamma_t/\partial N$ and $\partial\alpha/\partial N$ factored in. For modulation above the cutoff (i.e., ω_{cutoff}), $\partial\Gamma_t/\partial N$ and $\partial\alpha/\partial N$ have little effect on the SHD. Furthermore, when the core radius W is increased from 3.5 to 5 μm, the influence of $\partial\Gamma_t/\partial N$ and $\partial\alpha/\partial N$ as well as SHD are reduced.

From this numerical analysis, it is clear that either (or both) of the two diffraction loss parameters, $\partial\Gamma_t/\partial N$ or (and) $\partial\alpha/\partial N$, has (have) a profound effect on SHD only at low modulation frequencies. It is interesting, then, to further investigate how each nonlinear parameter affects SHD. This can be done analytically by simplifying the expression for SHD in (9.34) at low modulation frequencies. Hence, for $\omega \ll \omega_n$, (9.33) and (9.34) are reduced to

$$\frac{S_1}{I_1} \approx \frac{\dfrac{v_g S_s}{q V_a}\Gamma_{t,s}\dfrac{\partial g_{\text{eff}}}{\partial N}}{\omega_n^2}, \tag{9.37a}$$

$$\frac{S_2}{S_1} \approx \frac{v_g \kappa_s F(\omega) - v_g S_s \Gamma_{t,s} \dfrac{\partial g_{eff}}{\partial N} E(\omega)}{\omega_n^2} S_1. \tag{9.37b}$$

For large S_s (say, 5×10^{15} cm^{-3}), the value of $\Gamma_{t,s}(\partial g/\partial N)S_s$ is roughly equal to 0.25 cm^{-1}, which is larger than $1/v_g\tau_n$ (=0.04 cm^{-1}). Thus $(1/v_g\tau_n) + S_s g_s(\partial g_{eff}/\partial N) \ll S_s\Gamma_{t,s}(\partial g/\partial N)$; therefore $\kappa_s \approx S_s\Gamma_{t,s}(\partial g_{eff}/\partial N)$. If the numerator of (9.37b) becomes $v_g^2 S_s \Gamma_{t,s}(\partial g_{eff}/\partial N)(\partial\alpha/\partial N)K(\omega)$, where $K(\omega) \approx (\partial g/\partial S)(\partial g_{eff}/\partial N)^{-1}$ at low modulation frequencies, (9.37b) can then be reduced to

$$\frac{S_2}{S_1} \approx v_g^2 S_s \Gamma_{t,s} \frac{\partial\alpha}{\partial N}\frac{\partial g}{\partial S}\frac{S_1}{\omega_n^2}. \tag{9.38}$$

Thus the expression for SHD is reduced to a form independent of ω, conforming to the leveling SHD region in Figure 9.6 for modulation frequencies below ω_{cutoff}. From (9.33) and (9.35), S_1 and ω_n are related to $\partial\Gamma_t/\partial N$ and $\partial\alpha/\partial N$ through

$$\frac{\partial g_{eff}}{\partial N} = \frac{\partial g}{\partial N} + \frac{1}{\Gamma_{t,s}}\left(g_s\frac{\partial\Gamma_t}{\partial N} - \frac{\partial\alpha}{\partial N}\right).$$

In a typical VCSEL, the true values of $g_s\, \partial\Gamma_t/\partial N$ and $\partial\alpha/\partial N$ are 100 times less than $\partial g/\partial N$ so that $(\partial g_{eff}/\partial N) \approx (\partial g/\partial N)$ (see Fig. 3 of Ref. 29), indicating that S_1 and ω_n are essentially independent of $\partial\Gamma/\partial N$ and $\partial\alpha/\partial N$. The term $(\partial g/\partial P) = -\{\varepsilon\alpha_s/[\Gamma_{t,s}(1+\varepsilon S_s)]\}$ does not involve $\partial\Gamma_t/\partial N$ and $\partial\alpha/\partial N$. Thus (9.38) expresses the direct proportionality of SHD to $|\partial\alpha/\partial N|$ at low modulation frequencies but indicates that $\partial\Gamma_t/\partial N$ has no effect here. In fact, the numerical calculation can verify that the presence of $\partial\Gamma_t/\partial N$ has no influence on the SHD of VCSELs.

In the preceding paragraph, the increase in SHD at low frequency in VCSELs due to diffraction loss through the parameter $\partial\alpha/\partial N$ is explained. It is found that the diffraction loss kicks in at modulation frequencies below ω_{cutoff} (e.g., 600 MHz in the device with $W = 3.5$ μm). It must be noted that SHD is another type of noise source, which reduces the usable bandwidth of the high-speed analog communication systems [31]. A device with $W = 5$ μm exhibits about −45 dB SHD, and one with $W = 3.5$ μm exhibits a little over −30 dB SHD. Such levels of intensity noise are pretty much in line with those found in edge emitting lasers [32]. However, with careful consideration of the influence of diffraction loss, VCSELs can offer more promising solutions than edge emitting lasers to problems that exist in analog communication systems. It is noted that small devices are desired for application in high-speed communication systems because of their high modulation bandwidth. However, the increase in SHD in small-cavity VCSELs due to the influence of diffraction loss is unavoidable. In fact, the increase in SHD is due to a combination of increase in $\Gamma_{t,s}$, α_s, and $\partial\alpha/\partial N$ as W decreases. Thus it becomes more critical to consider the influence of diffraction loss and means of minimizing SHD, when designing VCSELs with small core size.

9.3 OPTICAL BISTABILITY OF SINGLE-CAVITY VCSELs

In the analysis of SSP in semiconductor lasers, it is noted that bistability is accompanied by SSP but at slightly different biased conditions [10,26,33,34]. There are two main types of bistable operation in semiconductor lasers: (1) optical (i.e., amplitude) bistability due to the presence of saturable absorption [10,26] and (2) wavelength/polarization bistability under the influence of nonlinear variation of refractive index and optical gain [26,35] inside the active layer of lasers. The cavity of VCSELs is quite different from that of facet emitting lasers because VCSELs have (1) short gain length of several QWs thick, (2) short cavity length in order of the wavelength, and (3) small mirror loss due to the high reflectivity (>90%) of reflectors. However, the corresponding optical bistability characteristics of VCSELs will be similar to those of facet emitting lasers, provided the saturable absorption is introduced into the active layer such as those using (1) parasitic p-n-p-n structure adjacent to the active layer [7] or (2) proton implantation gain-guided devices [8]. This is because the differential confinement factor and differential cavity loss have no influence on the steady-state behavior of VCSELs.

In this section, the optical bistability characteristics of a GaAs QW gain-guided VCSEL are studied [8]. The current confinement structure of VCSELs is realized by proton implantation and a hollow metallization with a 15-μm-diameter window formed for the emission of lasing light (i.e., see Fig. 1.10). Figure 9.7 shows the typical light/current and current/voltage curves, in which optical bistability is demonstrated in both curves. It is observed in the light/current curve that the

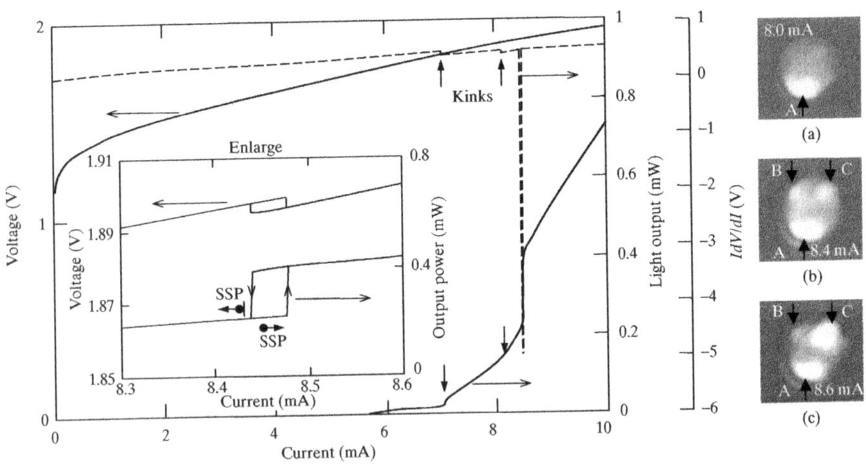

Figure 9.7 Light/current and current/voltage curves (solid lines) of VCSEL measured at 20°C. The dashed line shows the $I(dV/dI)$ dependence versus current as the current is increased. The inset shows the hysteresis in the light/current and current/voltage characteristics. The photos inset on the right-hand side of the figure measure the near-field emission pattern of VCSEL at (a) 8, (b) 8.4, and (c) 8.6 mA. (After Ref. 8.)

laser starts to lase at ~6.9 mA. In addition, the slope of the light/current curve increases for the injection current varies between 8.1 and 8.5 mA. This indicates that the corresponding differential quantum efficiency is also an increasing function of the injection current. At injection near 8.5 mA, a step change in light output is observed within a very small transition of the injection current. The inset diagram in Figure 9.7 shows the corresponding enlarged light/current and current/voltage curves, which demonstrate bistable loops. It is found that the current hysteresis width is ~36 μA and the output power discontinuity at the switching point is ~0.18 mW. On the other hand, the current/voltage curve shows a voltage discontinuity of ~2.5 mV at the switching point. At higher currents (i.e., >8.5 mA), the light intensity continues to increase essentially linearly with the same differential quantum efficiency as within the range of the injection current between 8.1 and 8.5 mA. The dashed curve in Figure 9.7 shows the measured current times the derivative of the voltage with respect to current, $I(dV/dI)$, versus increasing current. Kinks occur at current 6.9 and 8.1 mA corresponding to a step change of differential quantum efficiency at those points. The total $I(dV/dI)$ decrease at these two kinks is 0.054 V. On the other hand, a large negative differential resistance is present at the onset of the hysteresis loop. A similar negative differential resistance is also observed at the turnoff transition of the hysteresis loop with decreasing injection current.

The differential quantum efficiency and the origin of bistability behavior in the light/current curves can be clarified from the near-field emission profile of VCSELs [8]. Figure 9.7 also shows the measured near field of VCSELs at (a) 8.0, (b) 8.4, and (c) 8.6 mA, corresponding to the range of (a) 6.9–8.1 mA, (b) 8.1–8.5 mA, and (c) >8.5 mA, respectively, in the light/current curve. There is only one emission spot (spot A) observed within the current range between 6.9 and 8.1 mA, but two more spots (spots B and C) are excited within the current range between 8.1 and 8.5 mA. The emission profile can be considered as the filamentary of an ideal fundamental transverse-mode pattern. This is because the three spots are found to have the same wavelength and polarization. The nonuniform filamentary laser emission can be attributed to the small nonuniform distribution of injection carriers across the active and nonlinear absorption regions. The nonlinear absorption may be due to (1) the deep level traps caused by proton implantation [36], (2) the unionized donors, or (3) the residual H^+ from the MOCVD process. For injection current greater than 8.6 mA, three emission spots still occur, but the emission intensity of spot C increases dramatically, corresponding to the discontinuity of the light/current curves. This can be explained as the bleaching of saturable absorber in the region around spot C. The low differential quantum efficiency observed in the range between 6.9 and 8.1 mA is due to a single filament spot, which takes up only a small fraction of the laser aperture. However, higher differential quantum efficiency is shown for the range between 8.1 and 8.5 mA. This is because additional filaments begin to lase at different parts of the aperture, indicating that the threshold condition has been reached in these locations. The onset of the new spots also results in the second kink in the $I(dV/dI)/I$ curve. The large negative differential resistance as shown in Figure 9.7 can also be attributed to saturable absorption existing in the laser cavity [37]. The sudden increase of light output depletes the carriers in the active region, which in turn decreases the Fermi energy. Since the voltage across the laser diode equals the separation of the quasi-Fermi levels in the conduction and valence bands, a change in external voltage occurs. Therefore, the bistability in the current/voltage curve that accompanies the bistability in light/current curve of the VCSEL is due to change in the carrier concentration.

Furthermore, the bistable behavior of VCSELs is accompanied by random fluctuation and SSP near the transition point of the hysteresis loops. It is observed that SSP occurs only in a narrow current range just below the switching point. When the laser operates in the high-current region, no noise spectrum or SSP spectrum is detected. When the current is driven downward to the switching point, the random fluctuation and SSP spectrum are not observed until the laser operates in the turnoff region. It must be noted that the noise spectrum and SSP are emitted only from spot C. This confirms that local saturable absorption under spot C is responsible for the bistable behavior of the laser.

The bistability theory developed for the split-contact facet emitting lasers can be utilized to study the optical bistability characteristics of VCSELs [26,33]. It is noted that the bistable conditions of split-contact facet emitting lasers are dependent on ratios of (1) differential gain between the gain region and absorption region, (2) the carrier lifetime between the gain region and the absorption region, and (3) the absorption magnitude in the absorption region to the cavity loss. On the basis of the observation of the bistable operation of VCSEL given above, the region of bistability can be written as [33]

$$\frac{\tau_{n1}\partial g_1/\partial N}{\tau_{n2}\partial g_2/\partial N} < \frac{-\xi}{1-\xi}, \tag{9.39}$$

where $\xi = \Gamma_2 g_2/\alpha$ and Γ_2 is interpreted as the portion of light (i.e., spot C) inside the absorption region of the active layer.

9.4 NONLINEAR DYNAMIC OF COUPLED CAVITY VCSELs

In analogy with facet emitting lasers, VCSELs can also exhibit SSP by introducing a saturable absorption layer within the multilayer structure along the longitudinal direction, which is not at the position of the active layer. This VCSEL configuration can be realized using coupled cavity structure so that one of the cavities provides gain and the other one gives sufficient saturable absorption to support SSP. In fact, the idea of using coupled cavity VCSELs to generate SSP has been proposed theoretically [38] as well as verified experimentally [12,39]. In the following, the design requirement of SSP of VCSELs is studied. In addition, the practical application of coupled cavity VCSELs to generate SSP is also discussed. On the other hand, the dual-wavelength operation in coupled cavity VCSELs is also investigated in this section.

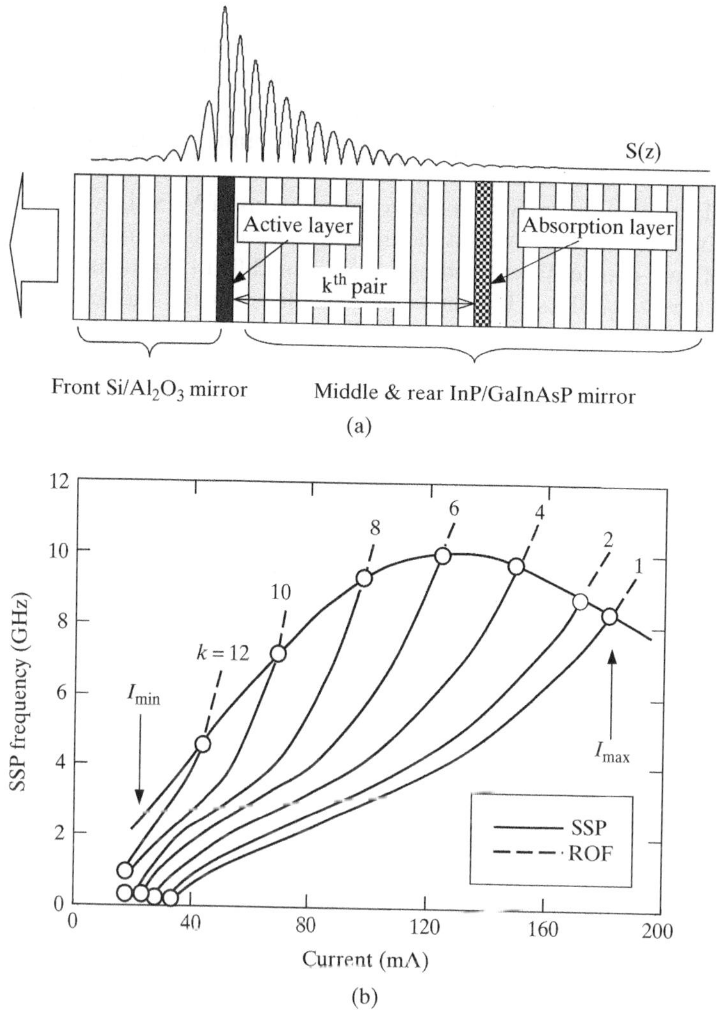

Figure 9.8 (a) Photon density distribution across device and schematic representation of coupled cavity VCSELs with intracavity absorber; (b) pulsation frequency against injection current. (After Ref. 38.)

9.4.1 Design of Coupled Cavity VCSELs for Self-Sustained Pulsation Operation

Figure 9.8a shows a typical configuration of VCSELs with two-cavity design for SSP [12]. It is assumed that the two cavities, consisting of absorption and active layers, respectively, are sandwiched between two set of DBRs. The two laser cavities are weakly coupled together by light through the middle DBR. The active layer supports the laser cavity mode, which is the only one longitudinal mode inside the coupled cavity as another longitudinal mode due to the absorber cavity is highly suppressed. Hence, the production of SSP is due to the unstable interaction of the gain region and the saturable absorption region, via the photon distribution along the longitudinal direction of the device. It must be noted that the position of the saturable absorption layer and the profile of the longitudinal optical field are the important parameters in determining the SSP behavior of coupled cavity VCSELs [26,33].

In order to analyze the SSP in such configurations of VCSELs with a saturable absorption layer allocated within the multilayer structure, a simple rate equation model of split-contact facet emitting lasers is modified [33] to study the SSP in coupled cavity VCSELs. The three rate equations for the carrier concentration inside the active and saturable absorption layers, N_1 and N_2, and the total number of photons inside the laser cavity P are given by [38]

$$\frac{dP}{dt} = v_g(\Gamma_1 g_1(N_1) + \Gamma_2 g_2(N_2) - \alpha)P + V_{a1}\beta_s B_{sp} N_1^2, \tag{9.40}$$

$$\frac{dN_1}{dt} = \frac{I_1}{qV_{a1}} - \frac{N_1}{\tau_{n1}} - v_g \Gamma_1 g_1(N_1)\frac{P}{V_{a1}}, \tag{9.41}$$

$$\frac{dN_2}{dt} = \frac{I_2}{qV_{a2}} - \frac{N_2}{\tau_{n2}} - v_g \Gamma_2 g_2(N_2)\frac{P}{V_{a2}}, \tag{9.42}$$

where I_1 and I_2 are the injection current and V_{a1} and V_{a2} are the volume of active and saturable absorption layers, respectively. The confinement factors, Γ_1 and Γ_2, are defined as the ratio of photon number in the active and saturable absorption layers, respectively, to that within the entire laser cavity, which can be written as

$$\Gamma_1 = \frac{\int_{d_a} S(z)dz}{\int_{\text{all}} S(z)dz} \quad \text{and} \quad \Gamma_2 = \frac{\int_{d_s} S(z)dz}{\int_{\text{all}} S(z)dz}, \tag{9.43}$$

where d_a and d_s are the thickness of active and saturable absorption layers, respectively, and $S(z)$ is the longitudinal distribution of photon density to be deduced with the information of the longitudinal configuration of VCSEL. If the coupled cavity of VCSELs is designed to confine the photon density strongly inside the cavity of the active layer, the position of the saturable absorption layer has less influence on the longitudinal distribution of the photon density and Γ_1 can be assumed to be a constant. This has been verified that in such configurations of dielectric structure, the longitudinal distribution of the lasing mode is independent of the saturable absorber. Therefore, the influence of the absorption cavity on the phase information of the lasing mode can be ignored and a model of split-contact laser can be applied to analyze the nonlinear dynamic behavior of coupled cavity VCSELs.

The rate equation model given above can analyze the SSP of coupled cavity VCSELs shown in Figure 9.8a. In addition, the longitudinal distribution of photon density $S(z)$ inside the laser cavity can be calculated using the transfer matrix method as described in Chapter 3. It is assumed that the total length of the rear and middle mirrors has 40 pairs of InP/GaInAsP layers and that of the front mirror has 4 pairs of Si/Al_2O_3 layers. Each pair of layers is a half-wavelength thick. The purpose of using this model is to ascertain how the allocation of the saturable absorption layer within the multilayered structure affects the frequency and peak power of SSP with different levels of bias. Figure 9.8b shows the pulsation frequency of the coupled cavity VCSELs with different levels of current and different lengths of the middle multilayered structure (i.e., number of pairs, k). The corresponding SSP performance can be explained as follows:

- SSP is bounded between upper and lower currents. This is because below the lower bounded current, the laser is subthreshold so no pulses can occur. Above the upper bounded current, the device is driven so hard as to permanently bleach the absorbing layer so that ROF is supported.
- There is an optimum position (i.e., $k = 6$) for the saturable absorption layer within the bottom multilayer structure to achieve the maximum range of SSP. This is because if the saturable absorption layer is far from the active layer, the range of SSP is limited because of the weak coupling with the active layer. However, as the saturable absorption layer is moved closer to the active layer, it will become more susceptible to permanent saturation, and damped oscillation will ensure.

The stability condition of coupled cavity VCSELs can be calculated from (9.40)–(9.42) using an approach similar to that described in Section 9.2. The parameters P, N_1, and N_2 are written as the sum of a time-independent term and a small deviated term [15]

$$P = P_s + \Delta P,\ N_1 = N_{1s} + \Delta N_1,\ \text{and}\ N_2 = N_{2s} + \Delta N_2, \tag{9.44}$$

where the subscript "s" stands for steady state. Substituting (9.44) into (9.40)–(9.42) and ignoring the higher-order term of ΔP, ΔN_1, and ΔN_2 gives

$$\frac{d}{dt}\begin{bmatrix}\Delta P\\ \Delta N_1\\ \Delta N_2\end{bmatrix} = \begin{bmatrix} 0 & v_g P_s \Gamma_1 \dfrac{\partial G_1}{\partial N} & v_g P_s \Gamma_2 \dfrac{\partial G_2}{\partial N} \\ \dfrac{-v_g \Gamma_1 G_{1s}}{V_1} & -\left(\dfrac{1}{\tau_1} + v_g P_s \Gamma_1 \dfrac{\partial G_1}{\partial N}\right) & 0 \\ \dfrac{-v_g \Gamma_2 G_{2s}}{V_2} & 0 & -\left(\dfrac{1}{\tau_2} + v_g P_s \Gamma_2 \dfrac{\partial G_2}{\partial N}\right) \end{bmatrix} \times \begin{bmatrix}\Delta P\\ \Delta N_1\\ \Delta N_2\end{bmatrix}, \tag{9.45}$$

where the spontaneous emission is ignored in the derivation. In addition, it is assumed that Γ_1 and Γ_2 are time-independent. Hence, setting the time-dependent components of ΔP, ΔN_1, and ΔN_2 as exp(λt) and substituting them into (9.45) gives [26]

$$\begin{bmatrix} -\lambda & v_g P_s \Gamma_1 \dfrac{\partial g_1}{\partial N} & v_g P_s \Gamma_2 \dfrac{\partial g_2}{\partial N} \\ \dfrac{-v_g \Gamma_1 g_{1s}}{V_{a1}} & -\left(\dfrac{1}{\tau_{n1}} + v_g P_s \Gamma_1 \dfrac{\partial g_1}{\partial N}\right) - \lambda & 0 \\ \dfrac{-v_g \Gamma_2 g_{2s}}{V_{a2}} & 0 & -\left(\dfrac{1}{\tau_{n2}} + v_g P_s \Gamma_2 \dfrac{\partial g_2}{\partial N}\right) - \lambda \end{bmatrix} \times \begin{bmatrix} \Delta P \\ \Delta N_1 \\ \Delta N_2 \end{bmatrix} = 0. \qquad (9.46)$$

Setting the determinant of the 3×3 matrix in (9.46) to zero gives the third-order polynomial equation of λ. If the real part of λ is positive, the coupled cavity VCSELs are operated at SSP. This condition of SSP can be obtained from the coefficient of λ of order zero in the third-order polynomial equation, which can be written as

$$\eta = \frac{1}{\tau_{n2}} \frac{\Gamma_1^2}{V_{a1}} \frac{\partial g_1}{\partial N} g_{1s} + \frac{1}{\tau_{n1}} \frac{\Gamma_2^2}{V_{a2}} \frac{\partial g_2}{\partial N} g_{2s} < 0, \qquad (9.47)$$

where it is assumed that P_s is small in the derivation of (9.47). Comparing (9.47) with (9.21), apart from the differential confinement factor and differential cavity loss, the conditions of SSP for the two devices are almost identical. If the total volume of the coupled cavity is assumed to be equal to one (i.e., $\Gamma_1/V_{a1} = \Gamma_2/V_{a2} = 1$) and τ_{n1}/τ_{n2} is set to one, (9.47) can be transformed to (9.21). In this analysis, the assumption of constant Γ_1 may not be valid, as it is a function of time due to the carrier-induced index change inside the active layer so that the profile of longitudinal photon density will be affected. In addition, self-focusing and diffraction loss may have some influence on the generation mechanism of SSP if these nonlinearities can be properly introduced into the rate equation model.

9.4.2 Implementation of VCSELs with an Intracavity Absorber

VCSELs with an intracavity absorber can be realized by integrating an additional QW layer into the upper DBR of a conventional VCSELs (see Fig. 9.9a) [12,39]. In this case, the negative differential resistance, which occurs in the current/voltage (I_b/V_b) curves of the saturable absorber, generates SSP. In the experiment, the VCSEL is designed so that the lasing wavelength ($\lambda_R \sim 959$ nm) at $2I_{th}$ of the laser section is 11 nm longer than the absorption peak ($\lambda_a \sim 948$ nm) of the saturable absorber. The gain layer consists of two 80-Å $In_{0.15}Ga_{0.85}As$ QWs with a gain peak at 940 nm. The absorber is a single 80-Å $In_{0.15}Ga_{0.85}As$ QW [12,39]. The VCSEL is essentially a coupled-cavity laser with a npn configuration. The

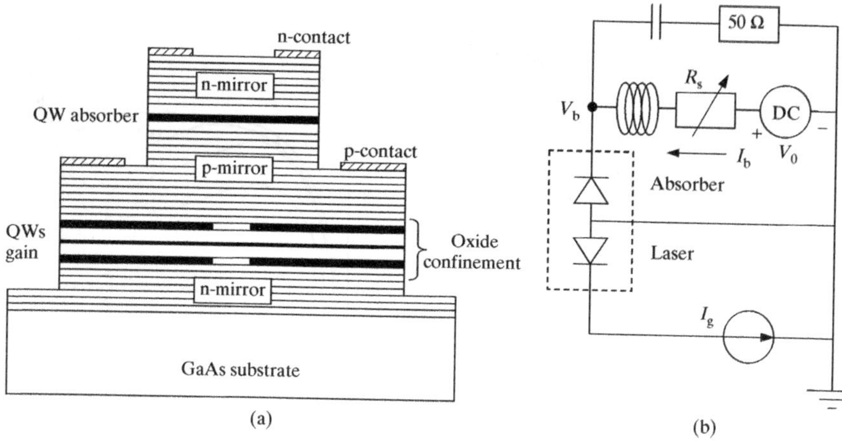

Figure 9.9 (a) Schematic diagram of an oxidized confined coupled cavity VCSEL with an intracavity absorber; (b) biasing circuitry used to obtain optical bistability and self-sustained pulsation. (After Ref. 12.)

forward biased bottom pn junction serves as the active layer and the top np junction within the top mirror stack serves as the voltage-controlled saturable absorber. The single QW saturable absorber is biased with a voltage source V_0 in series with a variable resistor R_s, and the corresponding controlling circuit is shown in Figure 9.9b. The applied voltage and current into the saturable absorber are V_b and I_b, respectively. Hence, using V_0 and R_s, the differential resistance of the saturable absorber can be adjusted so that it is possible to generate SSP.

The typical light/current (L/I_g) curves from the laser section are shown in Figure 9.10a for different magnitudes of R_s. A step change in light output and a hysteresis loop are observed, indicating the possibility of bistable operation. On the other hand, the hysteresis loop can be made closer with the variation of bias voltage and resistance so that SSP can begin. The transition between optical bistability and SSP can be explained by the I_b/V_b curves of the saturable absorber as shown in Figure 9.10b for different levels of current to the gain layer I_g. If R_s and V_b are adjusted so that the load line crosses the absorber I_b/V_b curve 3 times, then bistability is observed. However, if the load line is tilted so that it matches the negative slope of the I_b/V_b curves more closely (i.e., negative differential resistance), the system enters a state of instability and SSP commences [10].

In Figure 9.10b, the absorber reverse bias increases but the laser bias remains unchanged. It is observed that the negative differential resistance appears at high reverse bias (i.e., ~ -2 to -12 V) for a range of laser bias current (i.e., $I_g \sim$ 8–9.5 mA). This characteristic of I_b/V_b curves is similar to that of facet emitting lasers using bulk material as the saturable absorber [10]. If the absorber current I_b is proportional to the applied electric field across the absorber junction and the optical power, the negative differential resistance can be explained as follows:

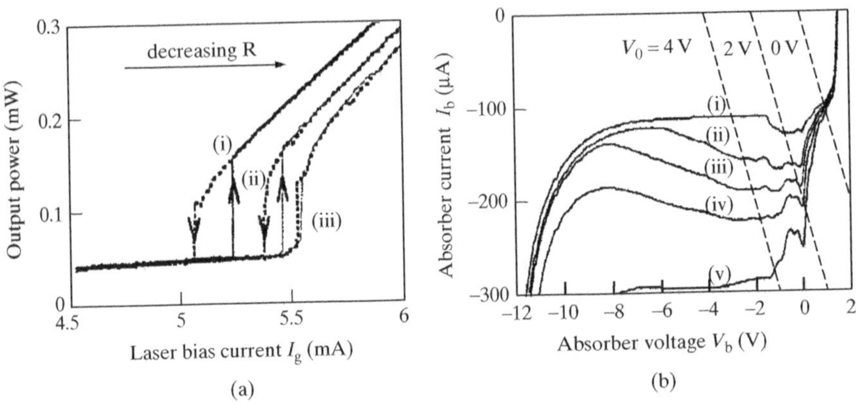

Figure 9.10 (a) Light/current (L/I_g) curves for a coupled cavity VCSEL with intra-cavity absorber at $V_0 = 40$ V [(i) $R_s = 450$ kΩ; (ii) $R_s = 350$ kΩ; (iii) $R_s = 300$ kΩ]; (b) current/voltage (I_b/V_b) curves for the absorber at different levels of current [(i) 8.0 mA; (ii) 8.5 mA; (iii) 8.75 mA; (iv) 9.0 mA; (v) 9.5 mA]; load lines (dashed lines) of 10 kΩ for $V_0 = 4$, 2, and 0 V. (After Ref. 12.)

- The increase in electric field causes the increase in absorber current I_b, and this effect predominates for small V_b ($V_b \sim -2$ to -3 V).
- The increased absorption reducing the laser power causes the absorber current I_b to decrease, and this characteristic is predominates for large V_b ($V_b \sim -3$ to -8 V).
- When the reverse bias voltage increases further, the absorption of the absorber junction is too great and the VCSELs stop lasing. The resultant drop in the optical power causes the absorber current to drop and approach the value determined solely by the applied electric field ($V_b \sim -8$ to -12 V).

Therefore, the negative differential resistance appears in the region where the absorber current decreases with increasing reverse bias ($V_b \sim -3$ to -8 V).

On the other hand, a sharp negative differential resistance also appears between $V_b = -0.5$ and 0.25 V in the I_b/V_b curves. This is because of the redshift of the band edge of the QW absorber as the increase in reverse bias voltage V_b. The mechanisms of redshift of the band edge of the QW absorber are explained as follows [40]:

- For $I_g = 0$, if the applied reverse voltage bias to the absorber is increased, the absorber band edge redshifts by ~5 nm as a result of the quantum confined Stark effect.
- For $I_g \sim I_{th}$, the absorption band edge redshifts by more than 9 nm relative to the spectrum measured with zero I_g, due to the heating of laser cavity and carrier screening. It is estimated that the heating of laser cavity at

$I_g \sim 6$ mA redshifts the absorber band edge by 2 nm. In addition, the redshift in absorption spectrum due to bandgap shrinkage is about 7.5 nm.

Figure 9.11 shows the resulting absorption spectra and CW lasing wavelength of the device with $I_g = 6$ mA. In this figure, the absorption spectrum, which was measured with zero bias, has been redshifted by 9 nm, due to the influence of cavity heating and carrier screening. The figure shows the following:

- At $V_b = 0.25$ V, the absorption at the lasing wavelength is relatively low, which results in a low absorber current I_b (i.e., point *a* of Fig. 9.11).
- As V_b varies from 0.25 to −0.1 V, the absorption and hence the absorber current increase, reaching a maximum when the excitonic peak of the absorber spectrum coincides with the lasing wavelength at $V_b \sim -0.1$ V (i.e., between points *b* and *c*).
- Further reduction in V_b ($V_b < -0.1$ V) moves the redshift of excitonic peak away from the lasing wavelength and consequently the absorber current decreases again (i.e., point *d*).

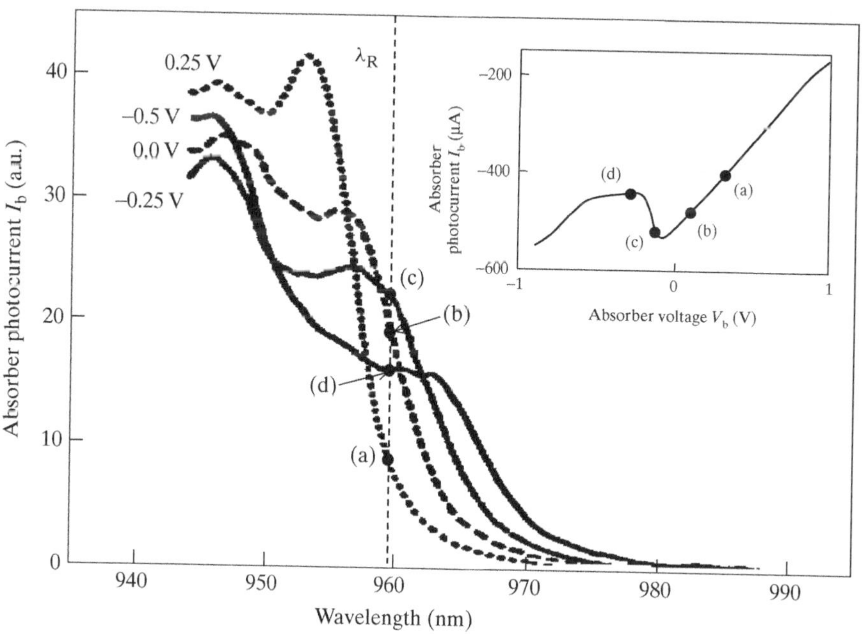

Figure 9.11 Origin of the observed sharp negative differential resistance in the absorber I_b/V_b trace. All curves are for a gain bias current I_g of 6 mA. Dashed curves show absorption spectra with varying applied absorber bias voltage V_b. The vertical line shows the position of the CW lasing wavelength. The inset shows the measured current/voltage (I_b/V_b) curve response for the absorber. (After Ref. 40.)

The Figure 9.11 inset shows that the measured absorber I_b/V_b curves for the device at I_g equals 6 mA, with sharp negative differential resistance occurring in the absorber I_b/V_b curve trace near $V_b \sim -0.1$ V. Therefore, it is clear that the sharp negative differential resistance is due to the redshift of the absorption band edge of the QW absorber, which is dependent on the lasing wavelength. In the explanation above, the bleaching of absorption at various V_b values under the influence of lasing power has been ignored. It must be noted that if the lasing wavelength is shorter than the band edge wavelength of the absorber, negative differential resistance will be suppressed in the I_b/V_b curves of the absorber layer. The importance of the strategic placement of the lasing wavelength and the QW absorption edge significantly enhances the negative differential resistance, providing flexibility in the design for SSP or bistability.

The SSP of coupled cavity VCSELs is also measured using a high-speed detector and an RF spectrum analyzer. It is found that for laser-biased current of 9 mA, $V_0 = 2$ V and $R_s = 10$ kΩ, SSP centered at 1.86 GHz with RF linewidth <10 MHz is observed. In fact, SSP of frequency range varying between 1.1 and 2 GHz can be easily obtained using different combinations of V_b (between −2 and −10 V), R_s (between 1 and 100 kΩ), and I_g (between 5 and 9 mA). In this case, the saturable absorber has negative differential resistance, which also leads to the generation of SSP.

The rate equation model (9.40)–(9.42) can be utilized to analyze the steady-state characteristics of coupled cavity VCSELs. The currents I_1 and I_2 in the model can be interpreted as the injection of carriers into the gain and absorption layers. However, under the influence of injection carriers, the dependence of wavelength on the gain and absorption spectra should also be introduced into the calculation in order to model the sharp negative differential resistance of the absorber. In addition, the external voltage V_b across the absorption layer can be roughly approximated by [41]

$$N = N_c \sum_i \log\left\{1 + \exp\left[-\frac{E_{ci} - E_{Fn}}{k_B T}\right]\right\}, \tag{9.48a}$$

$$P \approx N_v \sum_i \log\left\{1 + \exp\left[\frac{E_{Fp} - E_{vi}}{k_B T}\right]\right\}, \tag{9.48b}$$

$$V_b = E_{Fn} - E_{Fp}, \tag{9.49}$$

where P is also defined as the carrier concentration of holes and i is the number of quantized states. The quasineutrality (i.e., $N = P$) is required to solve the dependence of N on V_b. In the derivation of (9.48), the subband structure of the conduction and valence bands of the QWs are assumed to have a parabolic profile.

The rate equation model (9.40)–(9.42) has assumed that the gain and absorption layers are electrically insulated and that they are interacting only optically. However, there is an additional electrical coupling, which consists of the following two parts [10]:

- A constant parasitic resistance is due to finite conductance of the layers in between the gain and absorption layers. However, the multilayer structure is only slightly doped or nondoped so that its contribution to the leakage current can be neglected.
- The generation of photocurrent inside the absorber due to the lasing power, referred to as the *photoconductive effect*, should be added to (9.42) as the extra I_2. This is because the internal resistance of the saturable absorber can be very large (i.e., >200 kΩ) with the absence of lasing light but can be reduced to a small value (i.e., ~2 kΩ) because of the absorption of lasing light.

Hence, if the I_b/V_b curves need to be deduced from the rate equation model, the influence of photoconductive effect should also be considered. However, this is the most difficult part of the model because it is not easy to calculate the absorption spectrum of the QW absorber in order to estimate the absorber photocurrent of VCSELs.

9.4.3 Condition of Optical Bistability in Coupled Cavity VCSELs

If the absorber bias I_2 is considered as a control parameter in the laser, the model (9.40)–(9.42) can be solved for steady-state photon number versus current I_1 dependence. In order to simplify the derivation of the bistability condition of VCSELs, it is assumed that the optical gains g_1 and g_2 are linearly proportional to their corresponding carrier concentration:

$$g_1 = g_{N1}(N_1 - N_{t1}) \text{ and } g_2 = g_{N2}(N_2 - N_{t2}). \tag{9.50}$$

Solving (9.40) and (9.42) for the carrier concentrations with the consideration of (4.50) in the coupled cavity VCSELs, (9.41) yields [42]

$$P = \frac{a^{-1}P_1}{1 + b(1 + P/P_1)^{-1}}\left[\frac{I_1}{I_{\text{th}}} - \left(1 + \frac{ab}{1 + P/P_2}\right)\right], \tag{9.51}$$

where $P_1 = V_1(\upsilon_g g_1 g_{N1}\tau_{n1})^{-1}$ and $P_2 = V_2(\upsilon_g g_2 g_{N2}\tau_{n2})^{-1}$ are the saturation photon inside the two cavities. Furthermore, the effective cavity losses of the active and absorber regions are defined as $\alpha_{\text{eff1}} = \Gamma_1 g_{N1} N_{t1}$ and $\alpha_{\text{eff2}} = \Gamma_2 g_{N2}(N_{t2} - I_2\tau_{n2}/qV_{a2})$, respectively. The parameters I_0, a, and b are given by

$$I_{\text{th}} = \frac{qV_{a1}}{\tau_{n1}} N_{t1}\left(1 + \frac{\alpha}{\alpha_{\text{eff1}}}\right), \tag{9.52}$$

$$a = \frac{\alpha}{\alpha + \alpha_{\text{eff1}}} \quad \text{and} \quad b = \frac{\alpha_{\text{eff2}}}{\alpha}, \tag{9.53}$$

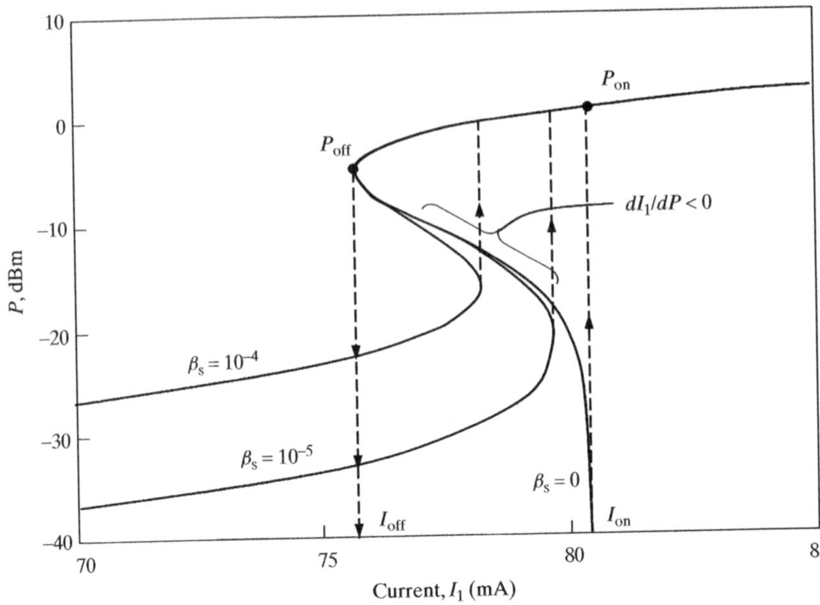

Figure 9.12 Typical light/current curves for a two-segment laser with saturable absorber with different values of spontaneous emission factor β_s. (After Ref. 43.)

where I_{th} is interpreted as the threshold current in the absence of the saturable absorber. The analysis becomes simpler if the single-valued function $I_1(P)$ are used. Observing the shape of I_1/P curves as shown in Figure 9.12, it is easy to show that the laser becomes bistable when $dI_1/dP|_{P=0} < 0$, the laser turns on from the $P = 0$ point as the spontaneous emission is neglected in the calculation. Hence, the condition of bistability can be deduced and is given by [42]

$$\eta \equiv b\left(\frac{P_1}{P_2} - 1\right) > 1. \tag{9.54}$$

The bistability condition implies two simultaneous requirements:

- The saturation photon of the absorber must be smaller than that of the gain section:

$$P_2 < P_1 \qquad \text{or} \qquad \frac{\Gamma_1 g_{N1} \tau_{n1}}{V_{a1}} < \frac{\Gamma_2 g_{N2} \tau_{n2}}{V_{a2}}. \tag{9.55}$$

- The portion of cavity loss of the total cavity loss must be sufficiently large:

$$\frac{\alpha_{eff2}}{\alpha + \alpha_{eff2}} < \frac{P_2}{P_1}. \tag{9.56}$$

The corresponding switchon and switchoff photon powers, P_{on} and P_{off}, respectively, of the hysteresis loop can be obtained from (9.51) and are given by [42]

$$P_{\text{on}} = P_2(\eta - 1) \quad \text{and} \quad P_{\text{off}} = P_2(\sqrt{\eta} - 1). \tag{9.57}$$

The switchon and switchoff currents $[P_{\text{on}}(I_{\text{on}})$ and $P_{\text{off}}(I_{\text{off}})]$ can be obtained from (9.51) and (9.54). In the derivation of (9.54) and (9.57), the influence of the spontaneous emission has been ignored in the calculation. In fact, it has been shown that the widths of the hysteresis loops in the bistable lasers depend strongly on the value of the spontaneous emission factor, β_s [43]. Figure 9.12 gives some typical results of the calculation on the light/current curves of a bistable laser with and without the spontaneous emission factor included in the calculation. It is noted that the output power is no longer zero at low biased level of the laser section due to the presence of spontaneous emission. The condition $dI_1/dP|_{P=0} < 0$ is still observed in the light/current curves, provided the laser is in bistable operation and this assumption is valid in analyzing the bistable operation of coupled cavity VCSELs. However, it is observed that the increase in the spontaneous emission factor reduces the hysteresis loops. It must be noted that the spontaneous emission factor is larger than that in facet emitting lasers so that the hysteresis loops calculated from the model without spontaneous emission included in the calculation may be overestimated.

9.4.4 Dual-Wavelength Operation in Coupled Cavity VCSELs

In the preceding paragraphs, the coupled cavity VCSEL with an intracavity absorber under investigation works like a two-section device with one longitudinal mode controlled mainly by the active cavity. Therefore, the theories developed for split-contact facet emitting single-cavity lasers are applied to analyze the nonlinear dynamic of coupled cavity VCSELs (Sections 9.3.1 and 9.3.2) [15,33,42]. However, the models have ignored the phase information of the lasing mode under the influence of the saturable absorption cavity. This approximation is satisfied as the lasing mode, which is less dependent on the saturable absorption cavity of VCSELs, is tightly confined inside the laser cavity. In coupled cavity VCSELs with twin active cavities, the two cavities have well-separated modes so that the mode interaction in such a coupled microcavity leads to strong mode pushing and splitting [44–46]. This phenomenon can be explained as when the two longitudinal modes of the individual cavities are at resonance, the modes of the coupled system are shifted to either side of the resonance wavelength [17]. The amount of pushing and splitting depends on the degree of coupling between the two cavities. Hence, this detailed information of the modal behavior of coupled cavity VCSELs should be calculated by the scattering matrix method discussed in Chapter 2.

The modal characteristics of coupled cavity VCSELs can be studied by investigating the corresponding reflectivity spectrum. Figure 9.13a shows a schematic of a coupled cavity VCSEL used in the experiment [17,44]. The device, which

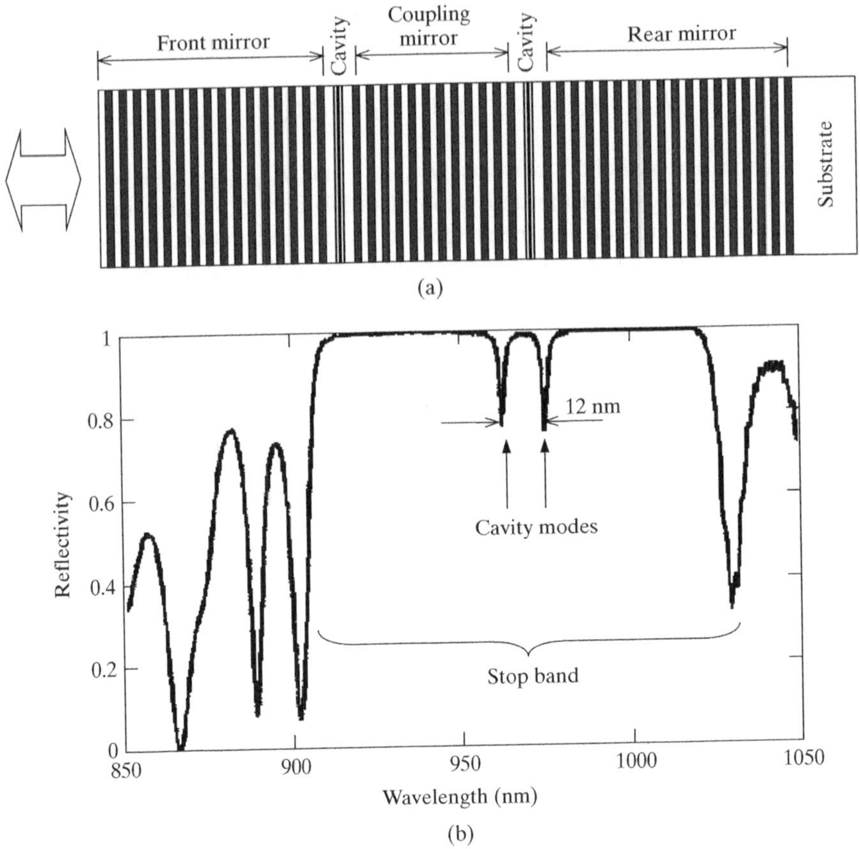

Figure 9.13 (a) Schematic diagram of a coupled cavity VCSEL consisting of a GaAs substrate, a rear DBR of 18.5 pairs, a GaAs λ cavity with three 75-Å $In_{0.24}Ga_{0.76}As$ quantum wells, a middle DBR mirror of 10.5 pairs, a second GaAs λ cavity with three 75-Å $In_{0.24}Ga_{0.76}As$ quantum wells, and a front DBR mirror of 15-pair DBR (the mirrors are made of $\lambda/4$ layers of AlAs and $Al_{0.1}Ga_{0.9}As$ pseudoalloy); (b) a broadband reflectivity spectrum at 300 K of the coupled cavity VCSEL at resonance. (After Ref. 44.)

is fabricated by molecular beam epitaxy (MBE), consists of one each of rear, middle (coupling), and front $Al_{0.1}Ga_{0.9}As$/AlAs quarter-wavelength (i.e., $\lambda_R/4$) DBRs, and their corresponding number of pairs is 18.5, 10.5, and 15, respectively. Two GaAs cavities, which contain three 75-Å $In_{0.23}Ga_{0.77}As$ QWs as the active layer, are of one-wavelength-layer thickness (i.e., $L_f = L_r = \lambda_R$). The structure is considered symmetric as the front and rear reflectivities are almost equal ($R_f \cong R_r = 99.0\%$) while the coupling mirror has a lower reflectivity ($R_c \sim 88.6\%$). The reflectivity of the coupled cavity VCSEL is measured at room temperature by a white light reflectometer, which has a spatial resolution <25 μm and a spectral resolution <1 Å. Figure 9.13b shows a wide spectrum of the coupled cavity VCSEL at room temperature. Two notches in the reflectivity spectrum represent the two resonant modes (i.e., the two cavities are of identical length) with wavelengths approximately equaling 963.5 and 974.5 nm. This is because at room temperature, the QW band edge has a wavelength longer than that of cavity modes, leading to absorption in both cavities.

Under the resonant condition (i.e., $L_f \approx L_r$), the phase separation between the two lasing modes can be written as [44]

$$\Delta\phi = \sqrt{2 - \sqrt{R_m R_c}(1 + R_m^{-1})}, \tag{9.58}$$

where $R_m(\equiv R_f = R_r) = 0.99$ and $R_c = 0.886$ so that $\Delta\phi \sim 0.3427$. If $\Delta\phi$ is related to the split of wavelength Λ through the round-trip oscillation of the resonant mode λ_R, the expression of Λ can be approximated by

$$\Lambda \approx \frac{\lambda_R^2}{2\pi n_{eff} L_{eff}} \Delta\phi, \tag{9.59}$$

where λ_R ($\sim$969 nm) is the resonant wavelength of the individual cavity; n_{eff} ($\sim$3.4) and L_{eff} ($\sim 3.5\lambda_R$) are the effective refractive index and cavity length of the couple cavity VCSELs, respectively; and L_{eff} can be estimated under the assumption that it should be greater than the total cavity length of the two cavities plus the portion of optical field penetrated into the DBRs. It can be shown that $\Lambda \sim 15$ nm, which is closer to the measured Λ (i.e., $\sim$12 nm) from Figure 9.13b.

In order to analyze the lasing characteristics of coupled cavity VCSELs, the device is optically pumped at room temperature under quasi-CW conditions (i.e., 1 μs pulse, 5 kHz) above the GaAs bandgap using a tunable titanium–sapphire laser at a reflectivity minimum outside of the mirror stopband (820–865 nm) [45]. The pump beam is incident on the front side, and the light emission from both the front and rear of the structure is measured as a function of the incident pump power. The emission is analyzed using a 1-m spectrometer and charge-coupled device (CCD) detector. On the other hand, the influence of off-resonance lasing characteristic of the coupled cavity VCSELs is studied from a systematic wedge of a sample wafer. The sample wafer can be fabricated by stopping the rotation motion of the substrate during the growth of each cavity (i.e., after the growth of QWs). In this case, the lengths L_f (and L_r) of the front (and rear) cavity are varied as a function of the position on the sample. Hence, the lasing characteristics of different structures can be investigated at different points of the wafer corresponding to different cavity detunings, $\delta = (\lambda_f - \lambda_r)/\Lambda$, where the mode splitting Λ is assumed to be equal to 13 nm. λ_f and λ_r are the isolated resonant cavity modes of the front and rear cavities, respectively.

The input/output curves of the coupled-cavity VCSELs measured from the front mirror are shown in Figure 9.14. It is noted that if the device is in lasing condition, each individual cavity can support a single longitudinal mode (i.e., of wavelength λ_f or λ_r), and the coupling between the cavities gives rise to a pair of optical modes for the system, namely, λ_L (for long-wavelength) and

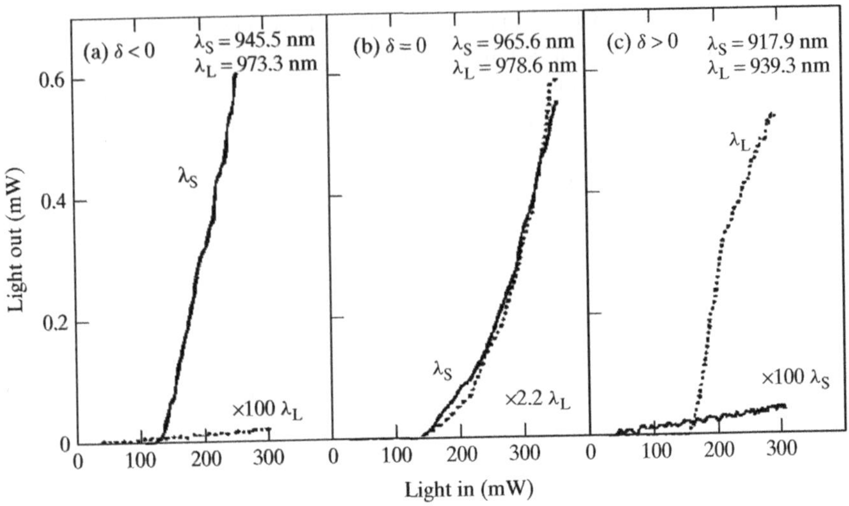

Figure 9.14 Optical output power in the top direction versus incident power for the two-cavity modes, λ_S (solid line) and λ_L (dashed line)—each plot is from a different point on the wafer, corresponding to a different cavity detuning: (a) negative detuning ($L_f < L_r$); (b) resonant cavities ($L_f = L_r$); (c) positive detuning ($L_f > L_r$). (After Ref. 45.)

λ_S (for short-wavelength) modes. At resonance (i.e., $\delta = 0$), dual lasing occurs and the light/light curves for λ_S and λ_L are identical except for a scaling factor of ~2. The structure lases simultaneously at two different wavelengths with the same threshold and no apparent competition; neither kinks nor instabilities between the modes are observed as the pump power increases. The values of λ_L and λ_S obtained from the lasing measurement are close to those obtained in the reflectivity spectrum at resonance. Far from resonance (i.e., $\delta < 0$ or $\delta > 0$), only one of the two modes lases and the lasing mode is always the one that associates with the top cavity.

The single-wavelength lasing behavior of the coupled cavity VCSELs at off-resonance condition can be explained as follows [17]:

- The cavities are uncoupled, and each mode is spatially localized in one or other cavity. This implies that there is almost no overlap between the mode localized in one cavity and the gain in the other cavity. Therefore, the system behaves like two separate lasers, a front laser and a rear laser, each one with its own gain medium, threshold power, and resonant wavelength.
- Front and rear lasers of the coupled cavity VCSELs should have a lower threshold than that of single-cavity VCSELs with the same reflectivities (i.e., R_f and R_r). Therefore, the emission of lasing light can be generated from the top or bottom cavities.
- As a result of the asymmetric pumping, the incident pump beam is absorbed mainly in the front cavity, leaving only a smaller fraction to pump the rear

cavity. Hence, the front cavity always reaches threshold before the rear cavity, and only the front cavity mode lases.

- For different detunings, the spatial asymmetry of the cavity modes changes while the pump asymmetry remains unchanged. However, if the coupled cavity laser is pumped extensively, both widely spaced modes can be made to lase independently.

On the other hand, a coupled-cavity VCSEL can no longer be considered as two separate lasers with independent gains at the condition of resonance. This is because the optical field is uniformly distributed over the entire laser cavity, and the two modes see both gain media equally but the two gain media are unequally pumped. This asymmetry in gain profiles supports dual-wavelength lasing in coupled cavity VCSELs, and both lasing modes with the same threshold maintain stability with the increase in external pumped power. These characteristics of coupled cavity VCSELs at resonance can be explained by the mechanism of self-regulation [17] as described below:

1. It is assumed that there is gain at λ_L in both cavities, but it is not sufficient to reach threshold, while at λ_S there is gain in the front cavity and absorption in the rear cavity.
2. If light lases at λ_S from the front gain medium is efficiently absorbed by the QWs located in the rear cavity, λ_S acts as an additional pump for the rear gain medium.
3. Through this internal pump mechanism, the rear cavity gain increases until its value at λ_L is large enough, when combined with the contribution from the front gain, to initiate laser oscillation at λ_L.
4. However, the stimulated emission at λ_L will deplete the number of carriers in the front gain medium. As a result, the lasing emission at λ_S will be reduced through the competition of the lasing mode at λ_L so that the internal pump mechanism will be weakened and there will be in turn a decrease in lasing power at λ_L.

Through procedures of self-regulation 2–4, the onset of laser oscillation at λ_L from tuning off laser oscillation at λ_S is prevented. In this case, the simultaneous laser emission at both wavelengths is obtained even at high external pumped power. In addition, both lasing modes have the same threshold condition. In fact, the observation of soft thresholds and identical shapes of the light/light curves in Figure 9.14 for both lasing modes support the explanation given above.

In summary, it is shown that only one wavelength is lased at off-resonance condition. This is because the resonators uncoupled and the front cavity is pumped more efficiently than the rear cavity. However, dual-wavelength emission, with two widely spaced wavelengths ($\sim$13 nm), with the same threshold and dependence on pump power, is observed when the laser is in resonance condition. This unusual situation of coupled dual-laser emission requires a system that has (1) two Fabry–Perot modes, which occurs only for coupled cavity lases at resonance; (2) independent and unequally pumped gain media, with which only one mode will lase at one wavelength; and (3) band edge position so that lasing at the lower energy mode does not start first in order to realize the mechanism of self-regulation.

9.5 POLARIZATION SWITCHING AND BISTABILITY IN SINGLE-CAVITY VCSELs

The stabilization mechanisms of single polarization operation in single-cavity VCSELs have been described in Chapter 4. In this section, the methods to control the switching of two orthogonal polarized modes in VCSELs are studied. The characteristics of polarization bistable in single-cavity VCSELs are also investigated. A simple model of single-cavity VCSELs with polarization bistability is developed to analyze their switching characteristics under the trigger of external optical injection. This is because bistable VCSELs have potential application in high-speed switching systems.

9.5.1 Polarization Switching in VCSELs

The polarization switching behavior of VCSELs is not attributed to a single event but is dependent on several nonlinearities of the laser cavity [47–55]. It has been explained that self-heating of gain-guided VCSELs by the injection current induces polarization switching of two nondegenerated orthogonal modes [49–51]. There are at least two conditions for the switching of polarization modes to be realized in VCSELs:

- The resonant modes are of nondegeneration type so that they have a finite wavelength difference (i.e., ~0.3 Å).
- The two cavity modes and peak laser gain are optimally aligned at operating temperature so that the gain is blue- or redshifted to the cavity resonance.

Therefore, due to the spectral splitting of the polarizations of fundamental mode, under the condition of blueshifted gain, the shorter-wavelength polarized mode will experience enhanced gain overlap, while for redshifted gain, the longer wavelength polarization will dominate. This explains the mechanism of polarization switching under the influence of self-heating. It should be noted that this phenomenon is of thermal origin (i.e., thermal shift of the bandgap), and thus the modulation frequencies are found to be less than 80 kHz [51].

In gain-guided QW VCSELs, if self-heating due to the change of injection current can be avoided, the switching mechanism of orthogonal polarization modes can be affected by carrier dynamics associated with the different magnetic sublevels of the conduction and heavy hole valence bands in QWs. The population difference is annihilated by spin–flip relaxation processes in a timescale of tens of picoseconds [53], which is comparable to the photon lifetime and thus slow enough to have an important effect on the dynamics of VCSELs. The four-level spin–flip model, including the vector nature of the laser field, saturable dispersion, and VCSEL anisotropies (birefringence and dichroism) as well as the spin–flip relaxation process, has been described in Chapter 4. The mechanism of polarization switching can be explained as follows [53]:

1. The nonzero value of spin sublevel population difference is coupled to the linearly polarized orthogonal modes through the amplitude phase coupling, which is associated with the linewidth enhancement factor.
2. Energy of the population difference and the weakly polarized mode is relaxed through the damped relaxation oscillation, and the corresponding relaxation frequency is dependent on the injection current.
3. When this oscillatory decay comes in resonance with the birefringent splitting of the mode frequencies, the phase relations become favorable for energy transfer between the polarization modes and the damped oscillations become undamped.

Hence, in processes 1–3, the dominant mode becomes unstable and polarization switching is achieved. This switching mechanism of orthogonally polarized modes is studied experimentally by plotting the injection current at the position of polarization switching against the temperature of the active layer of a gain-guided VCSEL [53]. In order to verify the influence of the spin–flip relaxation process on the switching mechanism, the influence of self-heating is avoided in the experiment. The gain difference between the two orthogonal polarized modes at different temperatures can be extracted from the plot using the theoretical model given in Section 4.4 [i.e., (4.102) and (4.103)]. If the slope of gain difference versus temperature is found to be constant, the idea of polarization switching under the influence of the spin–flip relaxation process can be verified. In fact, it has been found that the slope of gain difference versus temperature is a constant over a range of operation temperatures.

In index-guided VCSELs, polarization switching is dominated by birefringence and gain anisotropy of the laser cavities. It is shown experimentally that gain anisotropy of the two polarization modes can be realized by thermal lensing [54]. However, the requirements of polarization switching using thermal lensing are linearly polarized modes coexisting in the slightly anisotropic laser resonator and cavity resonance wavelength at the short-wavelength side of the gain maximum. Hence, subsequent competition of modal gains for the two orthogonal modes is realized by thermal lensing. Although self-heating is responsible for the switching mechanism, the modulation speed of the switching can be up to 50 MHz with a polarization contrast ratio of 20 : 1, which can be observed when the device has a prebias close to the switching point [54]. On the other hand, birefringence inside a weakly index-guided VCSEL can also be realized through SHB [52]. In Section 4.2.4, the influence of SHB on the excitation of the two orthogonally polarized modes has been investigated. It must be noted that the influence of SHB on the polarization switching is highly dependent on the detailed optical confinement structure of VCSELs. The corresponding switching mechanism is less effective as it is achieved only when the device is in high output power. On the other hand, it has been shown that the use of injection current control on anisotropic gain and birefringence in airpost cruciform VCSELs can achieve small-signal modulation responses up to 50 MHz and large-signal modulation responses up to 10 MHz for switching of the two orthogonally polarized modes [49].

Self-modulation of the two orthogonally polarized modes in VCSELs is a promising technique for generating ultra-high-frequency optical pulses without the requirement of high-speed electronics [56–58]. In addition, the corresponding switching mechanism is less dependent on anisotropic gain and birefringence of laser cavity resulting from self-heating, thermal lensing, as well as SHB, so the modulation bandwidth can be as high as several gigahertz. The self-oscillation in VCSELs can be obtained when a portion of the output power is injected back into the laser after having rotated its polarization by 90° with respect to the initial laser polarization state. The experimental setup can be easily realized using a VCSEL with an extended cavity, which is formed by a partial reflector with a zeroth-order quarter-wave plate inserted in between [57]. The reflectivity of the partial reflector is of 65% for both vertical and horizontal polarizations. The optical axis of the quarter-wave plate is oriented at 45° with respect to the polarization directions of both polarization modes. After one complete round trip through the extended cavity, the quarter-wave plate changes the polarization state of the reflected light by 90°. Hence, this optical feedback forces the polarization of the laser to switch, and the corresponding mechanism is called *polarization injection locking*. It must be noted that the principle behind polarization switching in VCSELs is different from that in facet emitting lasers as the front mirror of VCSELs has a very high reflectivity, so that slight variation of the position of the quarter-wave plate will not affect the self-oscillation of the system. In the experiment, it is shown that the polarization self-modulation frequency obtained can be up to 6 GHz [57].

Polarization switching can also be controlled through polarization optical feedback. In gain-guided VCSELs with weak anisotropies, optical feedback for a high-reflectivity mirror (>99%) can be used to control the polarization state of the system, and the amount of feedback is controlled by an attenuator sitting between the laser and the external mirror. Because of the high reflectivity of the feedback mirror, the location of the mirror can have a significant influence on the modal gain of the laser, in which the piezoelectric ceramic is utilized to control the displacement of the mirror. In the experiment, the mirror is placed 3–10 cm away from the VCSELs [58]. Hence, the polarization switching under external optical feedback can be explained in terms of the anisotropic losses of the equivalent mirror. If the equivalent reflectivity of the vertical mode is r_v, the reflectivity of the horizontal mode is $r_h = r_v + (1 - r_v^2) r_{ext} e^{j\phi}$, where r_v (>99%) is assumed equal to the reflectivity of the front mirror of the VCSEL and r_{ext} is the effective reflectivity of the external cavity. Hence, the losses of the two polarization modes can be expressed as $\Delta r_v^2 = 2 r_v r_{ext}(1 - r_v^2)\cos(\phi)$, where ϕ

is changed with the external cavity length L_{ext} by moving the external mirror so that $\phi = 4\kappa L_{eff}$, where κ is the coupling coefficient. The polarization switching can be achieved by modulating the position of the external mirror, but this is not desired as the time response of the piezoelectric ceramic is very low, of the order of tens of kilohertz. Alternatively, polarization switching can also be obtained in the light/current curves by applying a small sinusoidal voltage. This is because the variation of the carrier concentration changes the optical length of the laser cavity, which is equivalent to the external cavity length of the system. In fact, it can be shown that the modulation bandwidth up to 80 MHz can be realized in such an application [58].

In the discussions on the switching mechanism of the two orthogonal polarized modes presented above, the controllable modulation speed of VCSELs is limited to 80 MHz, which is much slower than what can be achieved by electrical devices. In order to use the polarization properties of VCSELs in future optical communication and switching systems, switching speed should be enhanced to the gigahertz range. In the following sections, bistable VCSELs are used to increase the switching speed of the two orthogonal polarization modes to within that range. The switching speed of the bistable device is expected to be high because the change in carrier population is minimal during the polarization switching process, and because stimulated emission leads to an effective shortening of the carrier lifetime [59].

9.5.2 Polarization Bistability in Single-Cavity VCSELs

The polarization bistability of gain-guided VCSELs under the influence of optical injection has been analyzed experimentally [60]. In the experiment, the free running VCSELs (with horizontally polarized light emitted at 835 nm) are perturbed by an external optical injection with light polarized perpendicular to that of the free-running VCSELs. Polarization bistability can be achieved only under the following conditions:

- Polarization switching of the two orthogonal modes of VCSELs should be observed at a value of injection current–switching current.
- The laser should bias at a current slightly above the level of the switching current.
- The wavelength of the triggering injection light should be longer than that of the free-running mode of VCSELs.
- The wavelength of the output power of VCSELs should be locked to the injected signal when the polarization is switched.

If these requirements are satisfied, the polarization hysteresis loops of the optical output power versus optical input power can be observed. It is noted that the polarization bistability of VCSELs is due to the difference in self- and cross-gain saturation (i.e., dispersive bistability), which has been explained in Chapter 4. The switching of the two bistable states is achieved through the injection locking where the free-running mode of VCSELs is suppressed by optical injection via gain saturation. The locking mechanism can be explained as the injected optical signal stimulates carrier concentration recombination, with the effect of increasing the refractive index of the active layer and moving the resonance frequency of the VCSELs toward the incident laser frequency (i.e., increasing the free-running wavelength toward the wavelength of incident optical power). However, if the injection optical wavelength is shorter than that of free-running VCSELs, no bistable switching can be obtained. In this case, only gradual transition from the free-running VCSEL mode to the injected polarization mode is recorded.

Polarization bistability is also observed in gain-guided VCSELs with uniform gain spectra [61]. However, this bistability behavior is different from that described in the preceding paragraph. In their experiment, Kawaguchi et al. observed polarization bistability in the light/current curves of the VCSELs under the variation of the injection current [62]. This bistability behavior of VCSELs is also dependent on temperature and is obtained only at 90°C. On the other hand, polarization switching and bistability are observed in the light/current curves at 18.5°C under the influence of polarized trigger light pulse, where polarization is perpendicular and with a wavelength slightly longer than that of the free-running lasing mode of VCSELs. As the memory effect is observed in the light/current curves, the mechanism of polarization switching and bistability is not due to injection locking. In addition, the experiment shows that the wavelength of VCSELs is not locked to the wavelength of the injection light. In this situation, polarization bistability can be explained as the extra saturable absorption arisen from the proton implanted cladding region due to the small core area (i.e., 6 μm^2) [62], which dominates over the anisotropic gain and birefringence so that the memory effect is observed in the light/current curves. Using these types of lasers, the 1.2-GHz all-optical flip-flop operation of polarization bistable VCSELs by injecting optical trigger pulses with two orthogonal polarizations is demonstrated [63]. The high switching speed is due mainly to the polarization bistable operation, which is highly desirable in realizing VCSELs. However, one major drawback of using such polarization bistable VCSELs in the application of all-optical logic circuits is dependence of bistable operations on the operation temperature of lasers.

In Chapter 4, the conditions of polarization bistability under the influence of the relaxation process of magnetic sublevels carriers as well as anisotropic gain and birefringence have been derived. On the other hand, if only anisotropic gain is considered in the calculation, the corresponding conditions of polarization bistability, determined by the saturation gain coefficient of the two orthogonal polarizations, will also be deduced [64]. Therefore, the derivation of polarization bistability conditions in VCSELs is not repeated in this section.

9.5.3 Polarization Bistability in Single-Cavity VCSELs under External Optical Injection

In the preceding discussion of two possible mechanisms of polarization bistability in VCSELs, it is believed that the former case (i.e., injection locking and dispersive bistability) is more promising than the latter case (i.e., saturable absorption). This is because, in comparison to the latter device, the performance of the former is less dependent on the operation temperature. In this section, the polarization bistability of VCSELs under external optical injection is analyzed with self-heating included in the calculation [18]. It is assumed that the two-nondegenerated orthogonal polarization states, h- and v-modes, are weakly guided inside the laser cavity with the profile of LP_{01} mode. The free-running output light is linearly polarized h-mode with a lasing wavelength of 850 nm, which is shorter than that of the v mode as well as the external injected polarized light parallel to the v mode. The difference in the lasing thresholds of h and v modes is due to the anisotropic gain and birefringence of the laser cavity.

9.5.3.1 *Modeling the Polarization Bistability of VCSELs*

It is assumed that simultaneous v-polarized light amplification takes place when a v-polarized light is injected into a laser diode. The term of carrier consumption due to the v-polarized light injection can be introduced into the rate equation of carrier concentration as

$$\frac{\partial N}{\partial t} = \frac{J}{qL_w n_w} - \frac{N}{\tau_n} - \nu_h \Gamma_{t,h} g_h S_h - \nu_v \Gamma_{t,v} g_v S_v, \tag{9.60}$$

where $S_{h(v)}$ is the average photon density inside the active layer, $\Gamma_{t,h(v)}$ is the transverse confinement factor, and $\nu_{h(v)}$ is the group velocity of the two nearly degenerate orthogonal fundamental LP polarization modes. The optical gain $g_{h(v)}$ is dependent on the temperature and the resonance wavelength of h- and v-polarized light. In the calculation, it is assumed that $\nu_h = \nu_v = 0.86 \times 10^{10}$ cm/s, $L_w = 100$ Å, $n_w = 3$, and $\tau_n = 3 \times 10^{-9}$ s. The calculation of optical gain and its method of implementation in (9.60) can be found in Section 9A.1.

The photon density of the h-polarized light can be described by

$$\frac{\partial S_h}{\partial t} = \left(\nu_h \Gamma_{t,h} g_h - \frac{1}{\tau_{p,h}} \right) S_h + \beta_s \Gamma_z B_{sp} N^2, \tag{9.61}$$

where $\tau_{p,h}$ is the photon lifetime of the h mode and β_s ($\sim 1 \times 10^{-4}$) is the spontaneous emission factor. The output power of h-polarized light from the rear surface $P_{h,out}$ can be expressed as

$$P_{h,out} = \tfrac{1}{2} \hbar \omega A_h \Gamma_z \Gamma_{t,h}^{-1} (1 - r_{r,h}^2) S_h, \tag{9.62}$$

where $r_{r,h}$ is the facet reflectivity of the rear surface of h-polarized light, Γ_z is the longitudinal coupling factor, and A_h ($\sim \pi W^2$, $W = 10$ μm) is the area of the core region. In addition, the influence of self-heating can be included in the calculation using the auxiliary thermal equation (i.e., $T = T_0 + R_{TH}(P_{IV} - P_{hv})/\kappa_T$) with normalized effective thermal resistance R_{TH} of value 1200 cm^{-1} (see Chapter 5). The resonant conditions of the h mode can be evaluated using the transfer matrix method. The threshold condition requires the optical fields to reproduce itself after each round trip under steady-state operating conditions. This leads to the condition

$$mh_{21}r_{\mathrm{f},h}e^{2\gamma_{h,1}\Delta L_1} + mh_{22} - r_{\mathrm{r},h}(mh_{11}r_{\mathrm{f},h}e^{2\gamma_{h,1}\Delta L_1} + mh_{12}) = 0, \tag{9.63}$$

where the derivation of (9.63) can be found in Section 9A.2. Solving (9.63) gives the threshold gain, resonant frequency, as well as the intensity distribution profile of h-polarized light. The parameters $\tau_{\mathrm{p},h}$ and Γ_z in (9.61) can also be determined from the transfer matrix method as they are dependent on the configuration and materials used in VCSELs.

In order to describe the polarization bistability in VCSELs, the wavelength of the injected v-polarized light and its dependence on the DBRs have to be considered. In addition, the matrix method is adopted to describe the injected fields inside the cavity of the VCSEL. Hence, the input v-polarized propagation field F_{in} and output v-polarized propagation field F_{out} are related by

$$\frac{F_{\mathrm{out}}}{F_{\mathrm{in}}} = \frac{t_{\mathrm{r},v}t_{\mathrm{f},v}e^{\gamma_{v,1}\Delta L_1}(mv_{11}\, mv_{22} - mv_{12}\, mv_{21})}{mv_{22} + mv_{21}r_{\mathrm{f},v}e^{2\gamma_{v,1}\Delta L_1} - r_{\mathrm{r},v}(mv_{12} + mv_{11}r_{\mathrm{f},v}e^{2\gamma_{v,1}\Delta L_1})}. \tag{9.64}$$

The v-polarized field distribution inside the laser cavity is given by

$$F_k^+ = (r_{\mathrm{f},v}e^{2\gamma_{v,1}\Delta L_1}p_{11} + p_{12})F_1^- + p_{11}t_{\mathrm{f},v}e^{\gamma_{v,1}\Delta L_1}F_{\mathrm{in}}, \tag{9.65a}$$

$$F_k^- = (r_{\mathrm{f},v}e^{2\gamma_{v,1}\Delta L_1}p_{21} + p_{22})F_1^- + p_{21}t_{\mathrm{f},v}e^{\gamma_{v,1}\Delta L_1}F_{\mathrm{in}}, \tag{9.65b}$$

where k is an integer and F_k^+ and F_k^- are the forward and reverse propagation fields. The electric field F_1^- given in (9.65) is expressed as

$$\frac{F_1^-}{F_{\mathrm{in}}} = \frac{(mv_{11}r_{\mathrm{r},v} - mv_{21})t_{\mathrm{f},v}e^{\gamma_{v,1}\Delta L_1}}{mv_{22} - r_{\mathrm{r},v}mv_{21} + (mv_{21} - r_{\mathrm{r},v}mv_{11})r_{\mathrm{f},v}e^{2\gamma_{v,1}\Delta L_1}}. \tag{9.66}$$

Derivations of (9.64)–(9.66) are shown in Section 9A.3.

The optical power of the input v-polarized light $P_{\mathrm{in},v}$ is defined as

$$P_{\mathrm{in},v} = \tfrac{1}{2}\hbar\omega A_v\Gamma_{\mathrm{t},v}^{-1}S_{\mathrm{in},v}, \tag{9.67}$$

where $S_{\mathrm{in},v}$ is the injected photon density. The corresponding output power of v-polarized light $P_{\mathrm{out},v}$ can be obtained from (9.64) and is given by

$$P_{\mathrm{out},v} = (1 - r_{\mathrm{r},v}^2)\left(\frac{|F_{\mathrm{out}}|^2}{|F_{\mathrm{in}}|^2}\right)P_{\mathrm{in},v}. \tag{9.68}$$

The corresponding average photon density S_v inside the active layer is given by

$$\frac{S_v}{S_{\mathrm{in},v}} = \frac{\sum_i [(1 + e^{-2\mathrm{Re}[\gamma_{v,i}\Delta L_i]})|F_i^+|^2 + (1 + e^{2\mathrm{Re}[\gamma_{v,i}\Delta L_i]})|F_i^-|^2]\Delta L_i}{|F_{\mathrm{in}}|^2 L_{\mathrm{w}}n_{\mathrm{w}}}, \tag{9.69}$$

where the RHS represents the average photon power of v polarization inside the active layer. As the input power of the v-polarized light and its wavelength are defined, the values of S_h and N can be solved self-consistently by using (9.60), (9.61), (9.69), and the auxiliary thermal equation.

9.5.3.2 Numerical Analysis of Bistability Behavior of VCSELs under Optical Injection

The configuration of VCSELs used in this analysis is similar to that in Figure 7.12. It is assumed that the number of n- and p-doped DBRs, in which antireflection is coated on both front and rear surfaces (i.e., $r_f = r_r = 0$), is equal to 42. Furthermore, the resonance wavelength of the v-polarized light is 0.1 Å longer than that of h-polarized light so that the two polarized modes, with the transverse distribution of LP_{01} mode, are spectrally nondegenerated as a result of polarization gain dependence. The effective refractive index of the active layer for the h-polarized mode is 3.524, and that of the v-polarized mode is 3.528. The refractive indices of the spacer layer as well as the high and low indices of the DBRs are set to 3.393, 3.503, and 2.950, respectively, for both polarization modes. The device is biased at 1.1 times its threshold current value and an external v-polarized light is injected onto the front surface of the VCSEL with a wavelength 1.1 Å longer than that of h-polarized light.

As a result of the spectral splitting of the h- and v-polarization modes, the polarized mode with a shorter wavelength will experience a slightly higher modal gain. However, the mode with a longer wavelength will dominate under the condition of red-shifted gain, whereby the switching of polarization modes is possible through the thermal effects as the optical gain spectrum can be red- or blueshifted depending on the cavity temperature (see Fig. 9.15) [51]. Therefore,

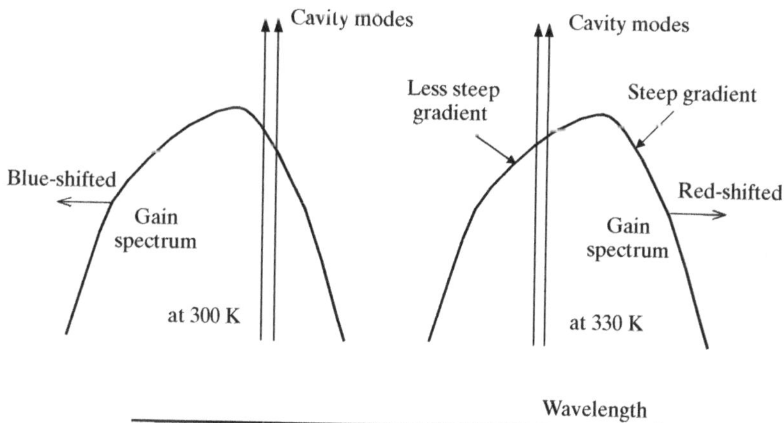

Figure 9.15 Schematics for the alignment of optical gain spectrum and cavity modes of VCSELs at background temperatures 300 K (where the optical gain is blueshifted and the optical gain is redshifted to the cavity mode) and 330 K.

in the design of VCSELs for polarization bistable operation, the choice of resonance wavelength is critical because of the high thermal resistance of VCSELs. In this investigation, the bistable characteristics of VCSELs with optical gain blue- and redshifting relative to the resonance wavelength are all studied. This can be calculated by assuming that the corresponding operation temperature is 300 and 330 K but keeping the resonance wavelength at 0.85001 μm. At 300 K the gain peak wavelength is 0.84407 μm, and at 330 K the gain peak wavelength is 0.851573 μm, so that polarization switching is avoided for temperature rises of less than 15 K inside the laser cavity.

Figure 9.16 illustrates the bistability characteristics of VCSELs with different thermal conductivities $\kappa_T(A, \kappa_T = \infty; B, \kappa_T = 1.0; C, \kappa_T = 0.9; D, \kappa_T = 0.45\ \mathrm{W\,cm^{-1}\,K^{-1}})$. The background temperature of the devices is set to 300 K, and it is expected that temperature rise inside the laser cavity is due to the

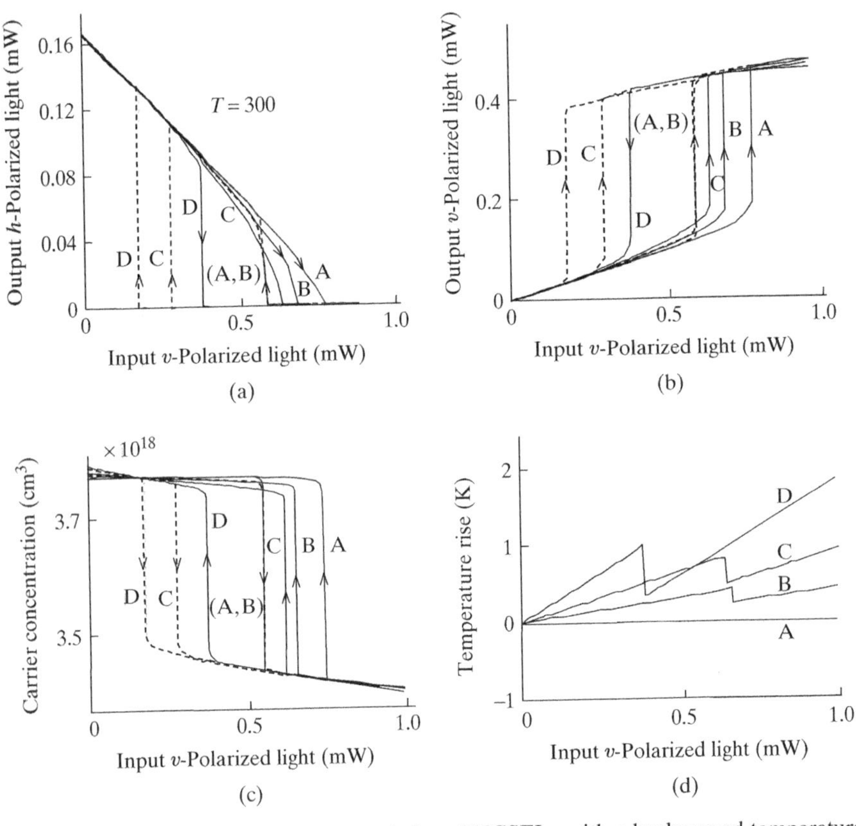

Figure 9.16 The steady-state characteristics of VCSELs with a background temperature of 300 K: (a) h-polarized light output versus v-polarized light input; (b) v-polarized light output versus v-polarized light input; (c) carrier concentration versus v-polarized light input; (d) temperature rise versus v-polarized light input (solid line—forward process; dashed line—reverse process). (After Ref. 18.)

absorption of input v-polarized optical power. Figures 9.16a and 9.16b show the output power (h- and v-polarization light) as a function of input v polarization light. It can be seen that the output h- and v-polarization light exhibits bistability with the input h-polarization light. In general, the hysteresis loop becomes narrower with the decrease of κ_T and starts to shift toward the left (low-power side of v-polarized light) as κ_T is less than 0.9 $\mathrm{W\,cm^{-1}\,K^{-1}}$. But the hysteresis loop vanishes for κ_T 0.2 $\mathrm{W\,cm^{-1}\,K^{-1}}$. Figure 9.16c illustrates the carrier concentration as a function of v-polarization input power. When the v-polarized light is injected and increased by a small amount, for case A, the carrier concentration is clamped at the level for stimulated emission of h-polarized light. However, for cases B, C, and D, the carrier concentration is slightly reduced with the input v-polarized light but still weakly clamped at the upper level. This is due to the redshift of the optical gain spectrum with the increase of cavity temperature, which enhances the optical gain of the h-polarization light. A sudden drop of carrier concentration from its clamped level is observed for further increase of input v-polarization light. The refractive index change induced by the carrier concentration would result in a shift of the resonance wavelength, which will increase the mirror loss and destroy the condition of laser emission. As a consequence, the h-polarized light reduces to spontaneous emission while the v-polarized light amplification takes place. Figure 9.16d shows the change in the temperature of the devices under the influence of input v-polarized light. Temperature increases linearly with the input power and kinks (instead of hysteresis loops) are observed at the point where h-polarized light is quenched lasing. This calculation is close to the experimental results described in Section 9.4.2. The analysis given above is also repeated when the background temperature of the devices is set at 330 K. In general, the polarization bistability of VCSELs under this situation is similar to that at a background temperature of 300 K except that higher input power is required for the bistable operation so that the operation temperature is relatively high.

The influence of the input wavelength of v-polarized light on the bistable mechanism of VCSELs is also analyzed. First, it is stated that the laser has $\kappa_T = 0.9\ \mathrm{W\,cm^{-1}\,K^{-1}}$ and the background temperature is set to 300 K. It is shown that VCSELs exhibit polarization bistability for the wavelength detuned toward the longer resonance wavelength by not more than 9 Å; however, no polarization bistability is recorded for the wavelength detuned by more than 1 Å toward the shorter resonance wavelength. Similar behavior of VCSELs is observed with the background temperature of 330 K. The influence of reflectivity of the DBRs on the bistable behavior of VCSELs is also analyzed. It is found that for devices with low reflectivity (i.e., by reducing the number of layers), the required switching power of the two orthogonally polarized modes can be reduced. However, the coupling strength between the profile of injected light and the active layer is reduced so that the maximum width of the hysteresis loops is also reduced. On the other hand, for mirrors of VCSELs with high reflectivity, the corresponding coupling strength is improved so that the maximum width of the hysteresis loops is increased but high external injection power is required to achieve polarization switching.

From the preceding analysis, the polarization bistable characteristics of VCSELs can be summarized as follows:

- The hysteresis loop shift toward left or right side on the input/output plots is dependent on the alignment between the optical gain and cavity mode. For the gain, which is blueshifted relative to the cavity mode, hysteresis loops exhibit left shift to the low-input-power side. However, the opposite is observed for the redshifted gain design.
- For devices with blueshifted gain design, power dissipated in terms of heat is relatively low. This is because polarization switching is achieved through low triggered input-power so that the influence of self-heating effects at the active region can be minimized.
- In order to avoid the influence of self-heating on the polarization bistable operation of VCSELs, thermal conductivity as well as thermal resistivity of VCSELs should be limited to a certain range of magnitude.

Therefore, in order to utilize VCSELs as optical switch or optical memory, the gain peak should be blueshifted relative to the resonance wavelength for reasons of low input power and less heat dissipation inside the laser cavity.

9.6 WAVELENGTH TUNABILITY IN VCSELs

VCSELs with wavelength tunability are highly attractive for application in wavelength division multiplexed transmission systems and free-space optical interconnects in massively parallel computers [4]. In the realization of wavelength tunable VCSELs, two possible approaches, the thermal tuning technique (i.e., thermoelectric effect) [65–69] and mechanical tuning method [70–81], are commonly used. The thermal tuning technique, which has the advantage of a simple fabrication process, is first employed to develop wavelength tunable VCSELs. For example, wavelength tunable VCSELs with InGaAs/GaAs QW active layers emitting between 940 and 990 nm have been fabricated using the thermal tuning technique. The maximum tuning range of $\sim$10.1 nm achieved in single-cavity VCSELs has been reported [66]. On the other hand, the lowest threshold current of tunable VCSELs is measured to be 650 μA [67], and the maximum output power is found to be 300 mW [69]. Figure 9.17 shows a schematic of a wavelength tunable VCSELs using the thermal tuning technique [66]. The laser structure is similar to a typical gain-guided VCSEL with ion implantation for the confinement of injection carrier and optical field inside the active layer. A stripe thin-film heater, containing 80-nm AuBe and 40-nm Au, is wrapped around the VCSEL to increase the total length and resistance. The heater is 144 μm long and 8 μm wide, and has a resistance of 34 Ω. The wavelength is tuned by applying a DC current through the thin-film metal

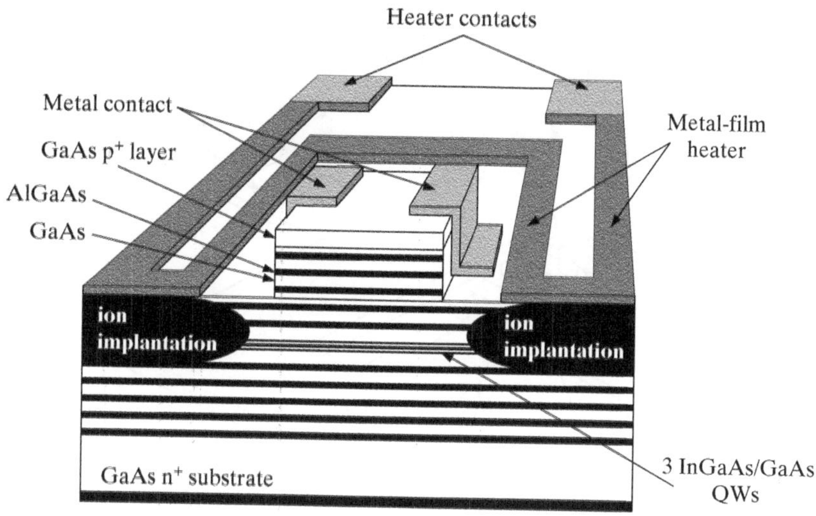

Figure 9.17 Schematic diagram of wavelength tunable VCSEL using thermoelectric effect (after Ref. 66).

for the generation of heat. The tuning mechanism achieved through the thermal effect (see also Chapter 5) causes the redshift of lasing wavelength. However, the corresponding tuning time of VCSELs is about several hundred nanoseconds or even longer if limited by the thermal effect.

Continuous-wavelength tuning of 40 Å has been demonstrated in GaInAsP/InP QW VCSELs emitting at 1440 nm using an external mirror [70]. In the experiment, a VCSEL is mounted on a heatsink, and an external mirror with reflectivity greater than 99% is put on the laser chip and fixed by a plastic plate. In the initial state, the external mirror is set in contact with the laser chip surface. The external cavity length can be slightly changed mechanically by tuning the spacing between the laser cavity and the external mirror so that the lasing wavelength can be changed continuously. Figure 9.18 shows a corresponding schematic of the wavelength tunable VCSEL with an external mirror [70]. However, the main disadvantage of this tunable wavelength VCSEL is that the displacement of the external mirror is dependent on the mechanically controlled tuning rod.

In order to avoid the use of mechanically moved parts, VCSELs with a microelectromechanically moved mirror is applied to modulate the effective cavity length of the laser cavity. It is shown that the maximum tuning range for a 1440-nm InP MQW vertical cavity LED using a surface-micromachined tunable top DBR is 75 nm [77]. On the other hand, the maximum tuning ranges for 950-nm InGaAs MQW and 850-nm GaAs/AlGaAs QW tunable VCSELs are found to be 44 nm [81] and 30 nm [79], respectively, using a microelectromechanically movable front mirror. In addition, the minimum threshold current of 0.34 mA can be obtained in VCSELs emitting at 950 nm [78], but

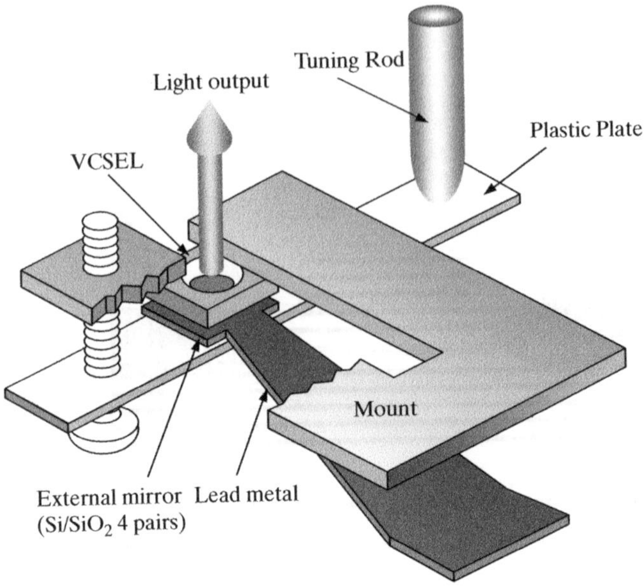

Figure 9.18 Wavelength tuning VCSELs using an external movable mirror (after Ref. 70).

maximum output power of 1.6 mW [75] is measured in electrically pumped tunable VCSELs. There are two typical types of microelectromechanically front mirrors: a freely suspended cantilever [71–75] and a deformable membrane [76–81], as shown in Figure 9.19. It is noted that the devices are of a semiconductor-coupled cavity design, where the QW active layer is located inside the semiconductor cavity where the thickness is a multiple of $\lambda/2$, and the gap with the thickness, an odd multiple of $\lambda/4$, is a part of the front mirror. The front mirror reflectance of this structure is rather high (>99%), aided largely by the two semiconductor–air interfaces. The movable mirror, which is suspended inside the air, is employed to modulate the effective cavity length. The wavelength shift $\Delta\lambda$ due to the change in mirror position Δd can be expressed as [80]

$$\frac{\Delta\lambda}{\lambda_R} = \kappa_\phi \frac{\Delta d}{L_{eff}} \tag{9.70}$$

where κ_ϕ is the phase coupling factor between the airgap and the semiconductor cavity (defined as round-trip phase change of the structure to round-trip phase change of the airgap) and L_{eff} is the effective cavity length, taking into account the penetration of optical waves into the mirror. Although the phase coupling factor (i.e., $\kappa_\phi \sim -0.1$) of this structure is rather small (i.e., resulting in a small shift of cavity resonance per change in airgap thickness), the corresponding threshold gain required for lasing is low and compatible with tunable VCSELs using the thermoelectric effect.

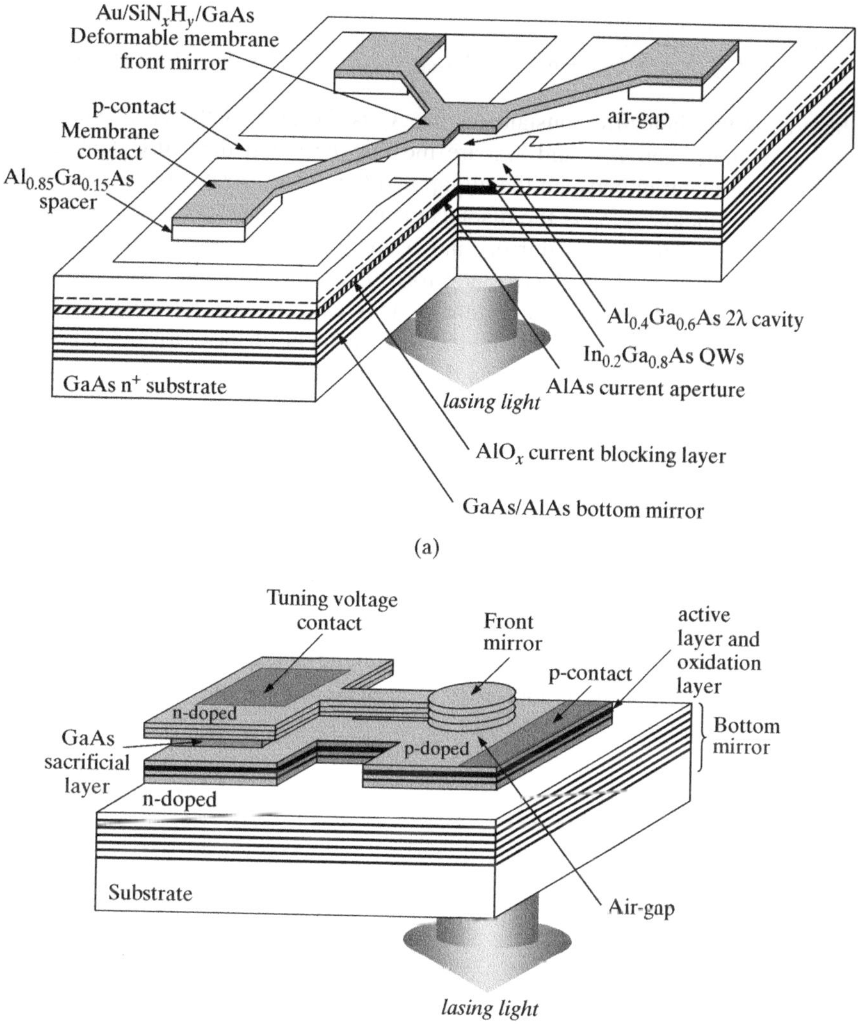

Figure 9.19 Schematic diagrams of micromachined tunable VCSELs with (a) a four-leg configuration of micromachined deformable-membrane top mirror and (b) an n-type DBR etched into the shape of a cantilever over an airgap. (After Refs. 72 and 80.)

The vertical displacement of the front mirror can be realized by applying an electrostatic bias between the mirror and cavity region so that the airgap thickness between the mirror and the semiconductor cavity can be modulated. For small mirror displacements, a parabolic relationship between the vertical displacement of the mirror and the applied voltage is expected because the mirror and substrate form a parallel-plate capacitor. The electrostatic force F on the moving mirror is then given by [80]

$$F = \frac{\varepsilon_0 A_c V_b^2}{2(L_{mi}/\varepsilon_{mi} + L_{g0} - \delta_c)^2} + \frac{n_{leg}\varepsilon_0 b L_b V_b^2}{2(L_{mi}/\varepsilon_{mi} + L_{g0} - \delta_c)(L_{mi}/\varepsilon_{mi} + L_{g0})}, \tag{9.71}$$

where ε_0 is the dielectric constant of air, A_c is the area of the electrode, V_b is the bias voltage, and L_{mi} and ε_{mi} are the thickness and dielectric constant of any insulting material between the airgap and the mirror, respectively. L_{g0} is the initial airgap thickness, δ_c is the displacement of the central reflector, n_{leg} is the number of legs holding the mirror, and b and L_b are the width and length of the legs, respectively. This electrostatic force is balanced by the elastic restoring force of the legs. The spring constant of legs can then be extracted from the plot of the electrostatic force versus the displacement of the mirror. In fact, the maximum airgap change is roughly a third of the initial airgap thickness, beyond which the electrostatic force is always greater than the elastic restoring force of the membrane legs, causing the mirror to be pulled into contact with the substrate. The advantage of the four-leg design (shown in Fig. 9.19a) over the cantilever approach (Fig. 9.19b) is that the deformable front mirror maintains a position parallel to the bottom mirror. Figure 9.20 shows the typical CW lasing spectra of the multiple-transverse-mode tunable VCSEL with the four legs deformable front mirror with injection current of 1.2 times threshold. The membrane bias is increased from 0 to 21.1 V, the corresponding resonance frequency is estimated to be 0.5–1 MHz, and the wavelength switching speed is in the microsecond range. It is noted that the higher-order transverse modes can be removed if the radius of the core region is reduced.

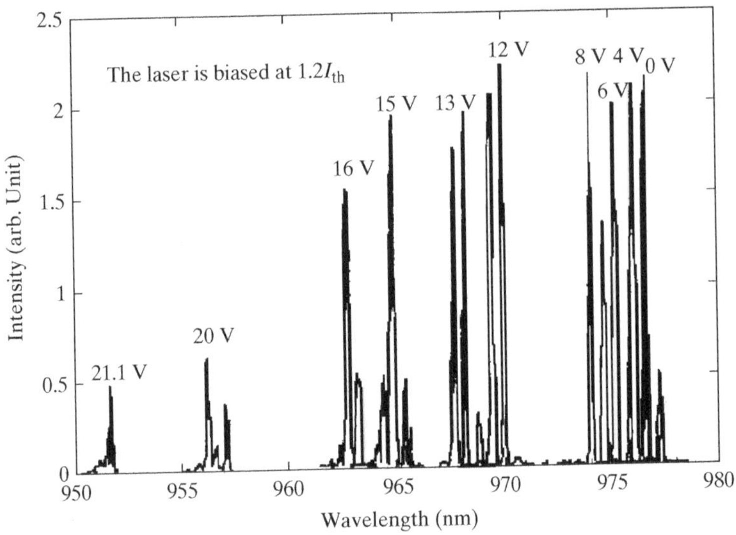

Figure 9.20 Lasing wavelength versus tuning voltage at 1.2I_{th} for a tunable VCSEL with four-leg configuration of micromachined deformable membrane top mirror (after Ref. 80).

9.7 SUMMARY

In single-cavity VCSELs with small cavity size, it is shown that the suppression of SSP is due to the presence of diffraction loss (i.e., differential diffraction loss dominates over the excessive saturable absorption of the unpumped region), which stabilizes the dynamic response of VCSELs. Hence, this explains the reason why VCSELs are more stable than facet emitting lasers and no strong signal of SSP is reported experimentally. The condition of SSP in VCSELs is different from facet emitting lasers mainly because of the domination of diffraction loss, which is negligible in facet emitting lasers. Furthermore, it is found that the bistable operation of single-cavity VCSELs is similar to that of facet emitting lasers as the diffraction loss has negligible influence on the steady-state response of VCSELs. On the other hand, the modulation response of single-cavity VCSELs is investigated in terms of self-focusing and diffraction loss. It is found that for single-cavity VCSELs with small cavity size, extensive SHD is induced at low modulation frequencies due to the presence of differential diffraction loss.

The nonlinear dynamic of coupled cavity VCSELs is also analyzed. It is shown that SSP and optical bistability can be realized in coupled cavity VCSELs in the presence of an intracavity absorber. Negative differential resistance, which induces SSP and optical bistability in coupled cavity VCSELs, is measured in the light/current curves of the intracavity absorber. Suitable biased voltage and current onto the intracavity absorber can control the value of negative differential resistance. Hence, the conditions of SSP and optical instability of coupled cavity VCSELs are similar to those of facet emitting lasers with split contact. On the other hand, the dual-wavelength lasing behavior of coupled cavity VCSELs is studied. It is shown that dual-lasing behavior of coupled cavity VCSELs occurs only when both laser cavities have the same length (i.e., resonance condition) and under asymmetric pumping of the top and bottom cavities through external optical injection. This dual-wavelength lasing mechanism of coupled cavity VCSELs is referred to as *self-regulation*.

The mechanisms of polarization switching and self-modulation of two orthogonal polarized modes in single-cavity VCSELs are also discussed. The studies of the dynamic response of VCSELs include the influence of thermal effects, SHB, and the spin–flip relaxation process as well as injection current and optical feedback. It is found that the maximum self-modulation frequency is 6 GHz for single-cavity VCSELs under appropriate optical feedback. However, the switching speed of VCSELs can be limited to 80 MHz because of thermal effects. Hence, in order to make use of the polarization properties of single-cavity VCSELs in high-speed optical communication and switching systems, the bistable behavior of the two polarized modes should be adopted to enhance the corresponding switching speed to the gigahertz range. This is because the change in carrier population is minimal during the switching process, especially under the trigger of external optical injection, so that the speed of polarization switching can be enhanced to the gigahertz range. In this chapter, the influence of spectral mismatch between the gain and cavity mode, self-heating effects, and the reflectivity of DBRs on the polarization bistability of VCSELs is also analyzed, and the results can be summarized as follows:

- The required input power for achieving polarization bistability is dependent on the alignment between gain and cavity mode. When the gain is blueshifted relative to the cavity mode, the hysteresis loops shift toward the low-power side of input v-polarized light, especially for devices with low κ_T. However, the opposite is observed for the redshift alignment.
- The influence of self-heating is minimized for the blueshifted alignment, due to low injection power required for the bistable operation.
- The reflectivity of the DBR also affects the bistable characteristics of VCSELs.

Therefore, in order to use VCSELs in all-optical switching and optical memory, it is recommended that the gain be blueshifted relative to the cavity mode for the requirement of low input power and minimal self-heating effects. Reflectivity of DBRs should also be optimized for suitable hysteresis loop width.

Finally, the methods used to realize continuous-wavelength tunable single-cavity VCSELs are discussed. It is noted that thermal tuning and mechanical tuning techniques (i.e., using a movable front mirror) are two typical methods employed to achieve wavelength tunability in VCSELs. Although both methods have low tuning speed of in the kilohertz range, the maximum tuning range of 75 nm can be achieved with stable output power by using a microelectromechanically moving front mirror, so the mechanical tuning method is preferred to the thermal tuning method.

9A APPENDIX

9A.1 Approximation of Optical Gain

The gain spectrum of the $GaAs/Al_{0.3}Ga_{0.7}As$ QW active layer can be calculated by using the $k \cdot p$ method, and the detailed calculation can be found in Zory's text [41]. However, direct implementation of the calculated data into our model is not practical, as it requires a huge database. Therefore, the optical gain is approximated as follows.

- The dependence of peak gain on the injection carrier concentration and temperature is approximated by

$$g_p = a_N \log \frac{N}{N_t}, \tag{9A.1}$$

where the temperature dependence of a_N and N_t can be expressed as

$$\begin{aligned} a_N(T) &= a_0 + a_1 T + a_2 T^2, \\ N_t(T) &= b_0 + b_1 T, \end{aligned} \tag{9A.2}$$

where $a_0 = 2478.3\ \text{cm}^{-1}$, $a_1 = -7.6369\ \text{cm}^{-1}\,\text{K}^{-1}$, $a_2 = 1.3177 \times 10^{-2}\ \text{cm}^{-1}\,\text{K}^{-2}$, $b_0 = 8.4226 \times 10^{17}\ \text{cm}^{-3}$, and $b_1 = 5.0679 \times 10^{15}\ \text{cm}^{-3}\,\text{K}^{-1}$.

- For the injection carrier concentration larger than $3 \times 10^{18}\ \text{cm}^{-3}$, the gain peak is almost independent of the injection concentration. Hence, the variation of gain peak wavelength λ_p with temperature can be approximated by

$$\lambda_p = \lambda_0 + \lambda_1 T, \tag{9A.3}$$

where $\lambda_0 = 0.76904\ \mu\text{m}$ and $\lambda_1 = 2.501 \times 10^{-4}\ \mu\text{m/K}$.

- If the operation wavelength of the VCSELs $\lambda_{h(\nu)}$ is shifted less than or equal to ± 10 nm away from the gain peak, the gain spectrum can be approximated by

$$g_{h(\nu)} = g_P - g_\lambda (\lambda_{h(\nu)} - \lambda_p)^2, \tag{9A.4}$$

where g_λ is a constant and is equal to $6.25 \times 10^6\ \text{cm}^{-1}\,\mu\text{m}^{-2}$ for $\lambda_{h(\nu)} > \lambda_p$ but to $1.254 \times 10^6\ \text{cm}^{-1}\,\mu\text{m}^{-2}$ for $\lambda_{h(\nu)} < \lambda_p$ in order to take into account the asymmetric gain profile.

9A.2 Modeling of Field Distribution within Multilayered Mirrors

The forward and reverse propagation fields of the h-polarized light, F_n^+ and F_n^- at the rear surface, can be expressed as [20,21] (see also Fig. 9.21):

$$\begin{bmatrix} F_n^+ \\ F_n^- \end{bmatrix} = \prod_{i=1}^{n-1} \begin{bmatrix} e^{\gamma_{h,i+1}\Delta L_{i+1}} & 0 \\ 0 & e^{-\gamma_{h,i+1}\Delta L_{i+1}} \end{bmatrix} \begin{bmatrix} \dfrac{nh_{i+1} + nh_i}{2nh_{i+1}} & \dfrac{nh_{i+1} - nh_i}{2nh_{i+1}} \\ \dfrac{nh_{i+1} - nh_i}{2nh_{i+1}} & \dfrac{nh_{i+1} + nh_i}{2nh_{i+1}} \end{bmatrix}$$

$$\begin{bmatrix} F_1^+ \\ F_1^- \end{bmatrix} = \begin{bmatrix} mh_{11} & mh_{12} \\ mh_{21} & mh_{22} \end{bmatrix} \begin{bmatrix} F_1^+ \\ F_1^- \end{bmatrix}, \tag{9A.5}$$

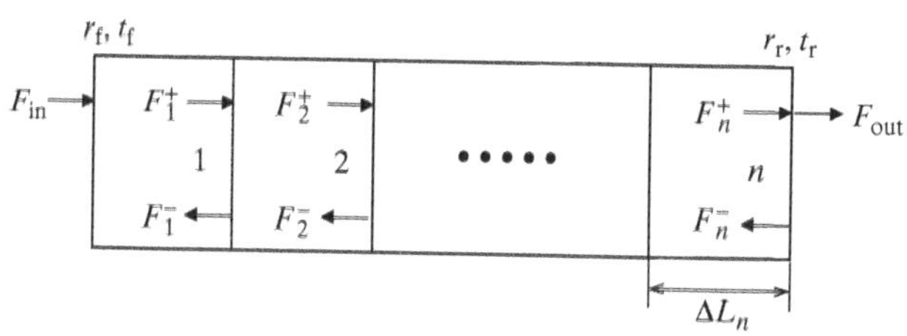

Figure 9.21 Schematic of a VCSEL with DBRs, where F_n^+ and F_n^- are the forward and reverse optical fields, respectively, propagating along the longitudinal direction and ΔL_n is the thickness of the nth layer; F_{in} is the injection input field and F_{out} is the output field; r_f and t_f (r_r and t_r) are the reflection and transmission coefficients on the front (rear) surface.

where F_1^+ and F_1^- are the propagation fields at the interface between the first and second matrix sections respectively; $nh_{i(i+1)}$ is the effective refractive index of the ith [$(i+1)$th] layer; ΔL_{i+1} is the thickness of the $(i+1)$th layer; and $\gamma_{h,i+1}$ is defined as

$$\gamma_{h,i+1} = \begin{cases} \frac{1}{2}(\Gamma_{t,h} g_{h,i+1} - \alpha_h) + j\dfrac{2\pi nh_{i+1}}{\lambda_h} & \text{active layer} \\ -\frac{1}{2}\alpha_h + j\dfrac{2\pi nh_{i+1}}{\lambda_h} & \text{elsewhere} \end{cases}, \tag{9A.6}$$

where λ_h (=0.85 μm) is the resonant wavelength, α_h (=20 cm^{-1}) is the free-carrier absorption loss, and $\Gamma_{t,h}$ (=0.998) is the transverse confinement factor of h-polarized light. The boundary conditions for the h-polarized light at the right and left surfaces are given as

$$F_n^- = r_{r,h} F_n^+ \quad \text{and} \quad F_1^+ = r_{f,h} e^{2\gamma_{h,1}\Delta L_1} F_1^-, \tag{9A.7}$$

where $r_{f,h}$ and $r_{r,h}$ are the reflectivity of the rear and front surfaces, respectively. Therefore, using (9A.5) and (9A.7), the eigenvalue equations for F_n^+ and F_1^- can be obtained and are given by

$$\begin{bmatrix} mh_{11} r_{f,h} e^{2\gamma_{h,1}\Delta L_1} + mh_{12} & -1 \\ mh_{21} r_{f,h} e^{2\gamma_{h,1}\Delta L_1} + mh_{22} & -r_{r,h} \end{bmatrix} \begin{bmatrix} F_1^- \\ F_n^+ \end{bmatrix} = \begin{bmatrix} 0 \\ 0 \end{bmatrix}. \tag{9A.8}$$

The threshold gain and the resonance frequency of VCSELs can be determined from the root of the determinant of (9A.8), which is

$$mh_{21} r_{f,h} e^{2\gamma_{h,1}\Delta L_1} + mh_{22} - r_{r,h}(mh_{11} r_{f,h} e^{2\gamma_{h,1}\Delta L_1} + mh_{12}) = 0, \tag{9A.9}$$

so that the value of the effective photon lifetime $\tau_{p,h}$ and the coupling factor Γ_z can also be calculated.

9A.3 Longitudinal Distribution of v-Polarization Light

The forward and reverse propagation fields F_n^+ and F_n^- at the rear surface can be expressed as [20,21] (see also Fig. 9.20)

$$\begin{bmatrix} F_n^+ \\ F_n^- \end{bmatrix} = \prod_{i=1}^{n-1} \begin{bmatrix} e^{\gamma_{v,i+1}\Delta L_{j+1}} & 0 \\ 0 & e^{-\gamma_{v,i+1}\Delta L_{i+1}} \end{bmatrix} \begin{bmatrix} \dfrac{nv_{i+1} + nv_i}{2nv_{i+1}} & \dfrac{nv_{i+1} - nv_i}{2ny_{j+1}} \\ \dfrac{nv_{i+1} - nv_i}{2nv_{i+1}} & \dfrac{nv_{i+1} + nv_i}{2nv_{i+1}} \end{bmatrix}$$

$$\begin{bmatrix} F_1^+ \\ F_1^- \end{bmatrix} = \begin{bmatrix} mv_{11} & mv_{12} \\ mv_{21} & mv_{22} \end{bmatrix} \begin{bmatrix} F_1^+ \\ F_1^- \end{bmatrix}, \tag{9A.10}$$

where $nv_{i(i+1)}$ is the effective refractive index of the ith [$(i+1)$th] layer. ΔL_{i+1} is the thickness of the $(i+1)$th layer and $\gamma_{v,i+1}$ is defined as

$$\gamma_{v,i+1} = \begin{cases} \frac{1}{2}(\Gamma_{t,v} g_{v,i+1} - \alpha_v) + j\dfrac{2\pi n v_{i+1}}{\lambda_v} & \text{active layer} \\ -\frac{1}{2}\alpha_v + j\dfrac{2\pi n v_{i+1}}{\lambda_v} & \text{elsewhere} \end{cases}, \tag{9A.11}$$

where λ_v (=0.85001 μm) is the optical wavelength of the injection light, α_v (=20 cm^{-1}) is the free-carrier absorption loss, and $\Gamma_{t,v}$ (=0.997) is the transverse confinement factor of v-polarized light. Inside the active layer, the carrier-induced index change Δn is approximated by

$$\Delta n = -\Gamma_{t,v}\frac{\alpha_H \lambda_v}{4\pi}\frac{\partial g}{\partial N}\Delta N + \frac{\partial n}{\partial T}\Delta T, \tag{9A.12}$$

where α_H (=3.0) is the linewidth enhancement factor, $\partial g/\partial N$ (=4×10^{-16} cm^2) is the differential gain, $\partial n/\partial T$ (=1×10^{-4} K^{-1}) is the temperature coefficient of the refractive index, ΔT is the change of temperature, and ΔN is the change of carrier concentration above threshold. F_1^+ and F_1^- are the propagation fields at the interface between the first and second matrix sections.

The boundary conditions for the v-polarized light at the right and left surfaces are given by

$$F_n^- = r_{r,v}F_n^+ \quad \text{and} \quad F_{out} = t_{r,v}F_n^+, \tag{9A.13}$$

$$F_1^+ = t_{f,v}e^{\gamma_{v,1}\Delta L_1}F_{in} + r_{f,v}e^{2\gamma_{v,1}\Delta L_1}F_1^-, \tag{9A.14}$$

where $r_{f(r),v}$ and $t_{f(r),v}$ are the reflection and transmission coefficients at the front (rear) surfaces, respectively. Substituting the boundary conditions into (9A.10) gives

$$\begin{bmatrix} mv_{11} & mv_{12} \\ mv_{21} & mv_{22} \end{bmatrix}^{-1} \begin{bmatrix} F_n^+ \\ r_{r,v}F_n^+ \end{bmatrix} = \begin{bmatrix} r_{vL}e^{2\gamma_{v,1}\Delta L_1} & 0 \\ 1 & 0 \end{bmatrix} \begin{bmatrix} F_1^- \\ 0 \end{bmatrix} + \begin{bmatrix} t_{t,v}e^{\gamma_{v,1}\Delta L_1}F_{in} \\ 0 \end{bmatrix}. \tag{9A.15}$$

Using the fact that

$$\begin{bmatrix} -1 & r_{f,v}e^{2\gamma_{v,1}\Delta L_1} \\ 0 & 0 \end{bmatrix} \begin{bmatrix} r_{f,v}e^{2\gamma_{v,1}\Delta L_1} & 0 \\ 1 & 0 \end{bmatrix} \begin{bmatrix} F_1^- \\ 0 \end{bmatrix} = \begin{bmatrix} 0 \\ 0 \end{bmatrix}, \tag{9A.16}$$

gives

$$\begin{bmatrix} -1 & r_{f,v}e^{2\gamma_{v,1}\Delta L_1} \\ 0 & 0 \end{bmatrix} \begin{bmatrix} mv_{11} & mv_{12} \\ mv_{21} & mv_{22} \end{bmatrix}^{-1} \begin{bmatrix} F_n^+ \\ r_{r,v}F_n^+ \end{bmatrix} = \begin{bmatrix} -t_{f,v}e^{\gamma_{v,1}\Delta L_1}F_{in} \\ 0 \end{bmatrix}. \tag{9A.17}$$

Further simplification and using the boundary condition $F_{\text{out}} = t_{\text{r},v} F_n^+$ gives

$$\frac{F_{\text{out}}}{F_{\text{in}}} = \frac{t_{\text{r},v} t_{\text{f},v} e^{\gamma_{v,1}\Delta L_1}(mv_{11}mv_{22} - mv_{12}mv_{21})}{mv_{22} + mv_{21} r_{\text{f},v} e^{2\gamma_{v,1}\Delta L_1} - r_{\text{r},v}(mv_{12} + mv_{11} r_{\text{f},v} e^{2\gamma_{v,1}\Delta L_1})}. \tag{9A.18}$$

F_1^- can also be expressed in terms of F_{in}. Using a transformation similar to that in (9A.16), we can write

$$\begin{bmatrix} -r_{\text{r},v} & 1 \\ 0 & 0 \end{bmatrix} \begin{bmatrix} 1 & 0 \\ r_{\text{r},v} & 0 \end{bmatrix} \begin{bmatrix} F_n^+ \\ 0 \end{bmatrix} = \begin{bmatrix} 0 \\ 0 \end{bmatrix}. \tag{9A.19}$$

Substituting (9A.19) into (9A.10), the characteristic equation of the resonant v mode is written as

$$\begin{bmatrix} 0 \\ 0 \end{bmatrix} = \begin{bmatrix} -r_{\text{r},v} & 1 \\ 0 & 0 \end{bmatrix} \begin{bmatrix} mv_{11} & mv_{12} \\ mv_{21} & mv_{22} \end{bmatrix} \begin{bmatrix} r_{\text{f},v} e^{2\gamma_{v,1}\Delta L_1} F_1^- + t_{\text{f},v} e^{\gamma_{v,1}\Delta L_1} F_{\text{in}} \\ F_1^- \end{bmatrix}. \tag{9A.20}$$

After further simplification, the expression of F_1^-/F_{in} is given by

$$\frac{F_1^-}{F_{\text{in}}} = \frac{(mv_{11} r_{\text{r},v} - mv_{21}) t_{\text{f},v} e^{\gamma_{v,1}\Delta L_1}}{mv_{22} - r_{\text{r},v} mv_{21} + (mv_{21} - r_{\text{r},v} mv_{11}) r_{\text{f},v} e^{2\gamma_{v,1}\Delta L_1}}. \tag{9A.21}$$

The photon density inside the kth layer can be expressed as

$$\begin{bmatrix} F_k^+ \\ F_k^- \end{bmatrix} = \prod_{i=1}^{k-1} \begin{bmatrix} e^{\gamma_{v,i+1}\Delta L_{i+1}} & 0 \\ 0 & e^{-\gamma_{v,i+1}\Delta L_{i+1}} \end{bmatrix} \begin{bmatrix} \dfrac{nv_{i+1} + nv_j}{2nv_{j+1}} & \dfrac{nv_{i+1} - nv_i}{2nv_{i+1}} \\ \dfrac{nv_{i+1} - nv_j}{2nv_{i+1}} & \dfrac{nv_{i+1} + nv_i}{2nv_{i+1}} \end{bmatrix}$$

$$\begin{bmatrix} F_1^+ \\ F_1^- \end{bmatrix} = \begin{bmatrix} p_{11} & p_{12} \\ p_{21} & p_{22} \end{bmatrix} \begin{bmatrix} F_1^+ \\ F_1^- \end{bmatrix}. \tag{9A.22}$$

Using the boundary conditions, F_k^+ and F_k^- can be written as

$$F_k^+ = (r_{vL} e^{2\gamma_{v,1}\Delta L_1} p_{11} + p_{12}) F_1^- + p_{11} t_{\text{f},v} e^{\gamma_{v,1}\Delta L_1} F_{\text{in}}, \tag{9A.23}$$

$$F_k^- = (r_{vL} e^{2\gamma_{v,1}\Delta L_1} p_{21} + p_{22}) F_1^- + p_{21} t_{\text{f},v} e^{\gamma_{v,1}\Delta L_1} F_{\text{in}}. \tag{9A.24}$$

REFERENCES

1. *VCSELs Meet Demands for Datacom Bandwidth*, FiberSystems International, Feb./March 2000, pp. 39–44.
2. Y. Suematsu, K. Iga, and S. Arai, "Advanced semiconductor lasers," *Proc. IEEE*, Vol. 80, no. 3, pp. 383–397, 1992.
3. D. Wiedenmann, R. King, C. Jung, R. Jager, R. Michalzik, P. Schnitzer, M. Kicherer, and K. J. Ebeling, "Design and analysis of single mode oxidized VCSEL's for high

speed optical interconnects," *IEEE J. Select. Topics Quantum Electron.*, Vol. 5, no. 3, pp. 503–511, 1999.

4. C. Wilmsen, H. Temkin, and L. A. Coldren, eds., *Vertical Cavity Surface Emitting Lasers: Design, Fabrication, Characterization and Application*, Cambridge Univ. Press, Cambridge, U.K., 1999.
5. C. A. Brackett, "Dense wavelength division multiplexing networks: Principles and applications," *IEEE J. Select. Areas Commun.*, Vol. 8, no. 6, pp. 948–964, 1990.
6. K. D. Choquette, H. Q. Hou, K. L. Lear, H. C. Chui, K. M. Gerib, A. Mar, and B. E. Hammons, "Self-pulsing oxide confined vertical cavity surface emitting lasers with ultralow operating current," *Electron. Lett.*, Vol. 32, no. 5, pp. 459–460, 1996.
7. D. G. Deppe, C. Lei, T. J. Roger, and B. G. Streetman, "Bistability in an AlAs/GaAs/InGaAs vertical cavity surface emitting laser," *Appl. Phys. Lett.*, Vol. 58, no. 23, pp. 2616–2618, 1991.
8. X. Tang, J. P. van der Ziel, B. Chang, R. Johnson, and J. A. Tatum, "Observation of bistability in GaAs quantum well vertical cavity surface emitting lasers," *IEEE J. Quantum Electron.*, Vol. 33, no. 6, pp. 927–932, 1997.
9. C. J. Chang-Hasnain, J. P. Harbison, C. E. Zah, M. W. Maeda, L. T. Florez, N. G. Stoffel, and T. P. Lee, "Multiple wavelength tunable surface emitting laser arrays," *IEEE J. Quantum Electron.*, Vol. 27, no. 6, pp. 1368–1376, 1991.
10. C. Harder, K. Y. Lau, and A. Yariv, "Bistability and pulsations in semiconductor lasers with inhomogeneous current injection," *IEEE J. Quantum Electron.*, Vol, 18, no. 9, pp. 1351–1361, 1982.
11. M. Yuri, J. S. Harris Jr., T. Takayama, O. Imafuji, H. Naito, M. Kume, K. Itoh, and T. Baba, "Two-dimensional analysis of self-sustained pulsation for narrow-stripe AlGaAs lasers," *IEEE J. Select. Topics Quantum Electron.*, Vol. 1, no. 2, pp. 473–479, 1995.
12. S. F. Lim, J. A. Hudgings, G. S. Li, W. Yuen, K. Y. Lau, and C. J. Chang-Hasnain, "Self-pulsation and bistable VCSEL with controllable intracavity quantum well saturable absorber," *Electron. Lett.*, Vol. 33, no. 20, pp. 1708–1710, 1997.
13. H. M. Chen, K. Tai, K. F. Huang, Y. H. Kao, and J. D. Wynn, "Instability in surface emitting lasers due to external optical feedback," *J. Appl. Phys.*, Vol. 73, no. 1, pp. 16–20, 1993.
14. J. Y. Law and G. P. Agrawal, "Effects of optical feedback on the static and dynamic characteristics of vertical cavity surface emitting lasers," *IEEE J. Select. Topics Quantum Electron.*, Vol. 3, no. 2, pp. 353–358, 1997.
15. S. F. Yu, "Analysis and design of vertical cavity surface emitting lasers for self-sustained pulsation operation," *IEEE J. Quantum Electron.*, Vol. 34, no. 3, pp. 497–505, March 1998.
16. W. T. Tsang, in *Semiconductor and Semimetals*, Vol. 22B, R. K. Willardson and A. C. Beer, eds., Academic Press, New York, 1985, pp. 257.
17. P. Pellandini, R. P. Stanley, R. Houdre, U. Oesterie, M. Ilegems, and C. Weisbuch, "Dual-wavelength laser emission from a coupled semiconductor microcavity," *Appl. Phys. Lett.*, Vol. 71, no. 7, pp. 864–866, 1997.
18. S. F. Yu, "Theoretical analysis of polarization bistability in vertical cavity surface emitting semiconductor lasers," *J. Lightwave Technol.*, Vol. 15, no. 6, pp. 1032–1041, 1997.
19. K. Petermann, *Laser Diode Modulation and Noise*, Kluwer, Amsterdam, 1988.

20. E. Hechit, *Optics*, 2nd ed., Addison-Wesley, Reading, MA, 1987.
21. Y. Yeh, *Optical Waves in Layered Media*, Wiley, New York, 1988, Chapter 6.
22. J. K. Butler and J. Zoroofchi, "Radiation fields of GaAs-(AlGa)As Injection lasers," *IEEE J. Quantum Electron.*, pp. 809–815, 1974.
23. S. F Yu and E. H. Li, "Influence of lateral field on the relaxation oscillation frequency of semiconductor lasers," *IEEE J. Quantum Electron.*, Vol. 32, no. 1, pp. 1–3, 1996.
24. R. R. Burton, M. S. Stern, P. C. Kendall, and P. N. Robson, "Modeling of diffraction in pillar vertical cavity surface emitting lasers with embedded Bragg layers," *Opt. Quantum Electron.*, Vol. 28, pp. 1677–1684, 1996.
25. M. Yamada, "A theoretical analysis of self-sustained pulsation phenomena in narrow stripe semiconductor laser," *IEEE J. Quantum Electron.*, Vol. 29, no. 25, pp. 1330–1336, 1993.
26. H. Kawaguchi, "Absorptive and dispersive bistability in semiconductor injection lasers," *Opt. Quantum Electron.*, Vol. 19, pp. S1–S36, 1987.
27. S. F. Yu, "Dynamic behavior of vertical cavity surface emitting lasers," *IEEE J. Quantum Electron.*, Vol. 32, no. 7, pp. 1168–1179, 1996.
28. N. K. Dutta, "Analysis of current spreading, carrier diffusion and transverse mode guiding in surface emitting lasers," *J. Appl. Phys.*, Vol. 68, no. 5, pp. 1961–1963, 1990.
29. P. C. Chui and S. F. Yu, "Second harmonic distortion in vertical cavity surface emitting lasers with lateral loss effect," *IEEE J. Select. Topics Quantum Electron.*, Vol. 5, no. 3, pp. 546–552, 1999.
30. S. F. Yu, W. N. Wong, P. Shum, and E. H. Li, "Theoretical analysis of modulation response and second harmonic distortion in vertical cavity surface emitting lasers," *IEEE J. Quantum Electron.*, Vol. 32, no. 12, pp. 2139–2147, 1996.
31. J. Singh, *Optoelectronics—an Introduction to Materials and Devices*, McGraw-Hill, New York, 1996, Chapter 10.
32. G. Keiser, *Optical Fiber Communication*, McGraw-Hill, New York, 1991, Chapter 4.
33. M. Ueno and R. Lang, "Conditions for self-sustained pulsation and bistability in semiconductor lasers," *J. Appl. Phys.*, Vol. 58, no. 4, pp. 1689–1692, 1985.
34. D. Baoxun, "A theory on stability of double section semiconductor lasers," *IEEE J. Quantum Electron.*, Vol. 25, no. 5, pp. 847–849, 1989.
35. M. Kuznetsov, "Picosecond switching dynamics of a bistable wavelength latch two segment distributed feedback laser," *IEEE Photon. Technol. Lett.*, Vol. 2, no. 9, pp. 623–625, 1990.
36. J. P. van der Ziel, J. L. Merz, and T. L. Paoli, "Study of intensity pulsations in proton-bombarded stripe geometry double heterostructure $Al_xGa_{1-x}As$ lasers," *J. Appl. Phys.* Vol. 50, pp. 4620–4637, 1979.
37. P. J. Anthony, T. L. Paoli, and R. L. Hartman, "Observations of negative resistance associated with superlinear emission characteristics of (Al,Ga) As double heterostructure lasers," *IEEE J. Quantum Electron.*, Vol. 16, no. 7, pp. 735–739, 1980.
38. D. G. H. Nugent, R. G. S. Plumb, M. A. Fisher, and D. A. O. Davies, "Self-pulsations in vertical cavity surface emitting lasers," *Electron. Lett.*, Vol. 31, no. 1, pp. 43–44, 1995.
39. J. A. Hudgings, R. J. Stone, C. H. Chang, S. F. Lim, K. Y. Lau, and C. J. Chang-Hasnain, "Dynamic behavior and application of a three contact vertical cavity surface

emitting laser," *IEEE J. Select. Topics Quantum Electron.*, Vol. 5, no. 3, pp. 512–519, 1999.

40. J. A. Hudging, R. J. Stone, S. F. Lim, G. S. Li, W. Yuen, K. Y. Lau, and C. J. Chang-Hasnain, "The physics of negative differential resistance of an intracavity voltage controlled absorber in a vertical cavity surface emitting laser," *Appl. Phys. Lett.*, Vol. 73, no. 13, pp. 1796–1798, 1998.
41. P. S. Zory, ed., *Quantum Well Lasers*, Academic Press, New York, 1993, Chapter 1.
42. M. Kuznetsov, "Theory of bistability in two-segment diode laser," *Opt. Lett.*, Vol. 10, no. 8, pp. 399–401, 1985.
43. M. J. Adams, P. E. Barnsley, and J. Chen, "Effects of spontaneous emission on hysteresis loops in bistable laser diodes," *Electron. Lett.*, Vol. 28, no. 4, pp. 395–396, 1992.
44. R. P. Stanley, R. Houdre, U. Oesterie, M. Ilegems, and C. Weisbuch, "Coupled semiconductor microcavities," *Appl. Phys. Lett.*, Vol. 65, no. 16, pp. 2093–2095, 1994.
45. P. Michler, M. Hilert, and G. Reiner, "Dynamics of dual-wavelength emission from a coupled semiconductor microcavity laser," *Appl. Phys. Lett.*, Vol. 70, no. 16, pp. 2073–2075, 1997.
46. J. F. Carlin, R. P. Stanley, P. Pellandini, U. Oesterie, and M. Ilegems, "The dual wavelength Bi-vertical cavity surface emitting laser," *Appl. Phys. Lett.*, Vol. 75, no. 7, pp. 908–910, 1999.
47. C. J. Chang-Hasnain, J. P. Harbison, G. Hasnain, A. C. Von Lehmen, L. T. Florez, and N. G. Stoffel, "Dynamic, polarization, and transverse mode characteristics of vertical cavity surface emitting lasers," *IEEE J. Quantum Electron.*, Vol. 27, no. 6, pp. 1402–1409, 1991.
48. F. Koyama, K. Morito, and K. Iga, "Intensity noise and polarization stability of GaAlAs-GaAs surface emitting lasers," *IEEE J. Quantum Electron.*, Vol. 27, no. 6, pp. 1410–1416, 1991.
49. K. D. Choquette, K. L. Lear, R. E. Leibenguth, and M. T. Asom, "Polarization modulation of cruciform vertical cavity surface emitting lasers," *Appl. Phys. Lett.*, Vol. 64, no. 21, pp. 2767–2769, 1994.
50. G. Verschaffelt, W. van der Vleuten, M. Ceusen, E. Smalbrugge, T. G. van de Roer, F. Karouta, R. C. Strijbos, J. Danckaert, I. Veretennicoff, B. Ryvkin, H. Thienpont, and G. A. Acket, "Polarization stabilization in vertical cavity surface emitting lasers through asymmetric current injection," *IEEE Photon. Technol. Lett.*, Vol. 12, no. 8, pp. 945–947, 2000.
51. K. D. Choquette, R. P. Schneider, Jr., K. L. Lear, and R. E. Leibenguth, "Gain dependent polarization properties of vertical cavity lasers," *IEEE J. Select. Topics Quantum Electron.*, Vol. 1, no. 2, pp. 661–666, 1995.
52. A. Valle, L. Pesquera, and K. A. Shore, "Polarization behavior of birefringent multi-transverse mode vertical cavity surface emitting lasers," *IEEE Photon. Lett.*, Vol. 9, no. 5, pp. 557–559, 1997.
53. J. Martin-Regalado, J. L. A. Chilla, J. J. Rocca, and P. Brusenbach, "Polarization switching in vertical cavity surface emitting lasers observed at constant active region temperature," *Appl. Phys. Lett.*, Vol. 70, no. 25, pp. 3350–3352, 1997.
54. K. Panajotovo, B. Ryvkin, J. Danckaert, M. Peeters, H. Thienpont, and I. Veretennicoff, "Polarization switching in VCSELs due to thermal lensing," *IEEE Photon. Technol. Lett.*, Vol. 10, no. 1, pp. 6–8, 1998.

55. N. Badr, I. H. White, M. R. T. Tan, Y. M. Houng, and S. Y. Wang, "Enhanced polarization self-switching in a vertical cavity surface emitting laser by gain saturation of transverse mode," *Electron. Lett.*, Vol. 30, no. 15, pp. 1227–1229, 1994.
56. G. Ropars, P. Langot, M. Brunel, M. Vallet, F. Bretenaker, L. Le Floch, and K. D. Choquette, "Experimental evidence of single round-trip oscillation in polarization self-modulation vertical cavity surface emitting lasers," *Appl. Phys. Lett.*, Vol. 70, no. 20, pp. 2661–2663, 1997.
57. S. Jiang, Z. Pan, M. Dagenais, R. A. Morgan, and K. Kojima, "High frequency polarization self-modulation in vertical cavity surface emitting lasers," *Appl. Phys. Lett.*, Vol. 63, no. 26, pp. 3545–3547, 1993.
58. F. Robert, P. Besnard, M. L. Chares, and G. M. Stephan, "Switching of the polarization state of a vertical cavity surface emitting laser using polarized feedback," *Opt. Quantum Electron.*, Vol. 27, pp. 805–811, 1995.
59. T. Numai, "Theoretical analysis of switching times in polarization switching lasers," *J. Appl. Phys.*, Vol. 87, no. 4, pp. 1610–1613, 2000.
60. Z. G. Pan, S. Jiang, M. Dagenais, R. A. Morgan, K. Kojima, M. T. Asom, and R. E. Leibenguth, "Optical injection induced polarization bistability in vertical cavity surface emitting lasers," *Appl. Phys. Lett.*, Vol. 63, no. 22, pp. 2999–3001, 1993.
61. M. Kajita, T. Kawakami, M. Nide, A. Kimura, T. Yoshikawa, K. Kuribara, Y. Sugimoto, and K. Kasahara, "Temperature characteristics of a vertical cavity surface emitting laser with a broad gain bandwidth," *IEEE J. Select. Topics Quantum Electron.*, Vol. 1, no. 2, pp. 654–670, 1995.
62. H. Kawaguchi, I. S. Hidayat, Y. Takahashi, and Y. Yamayoshi, "Pitchfork bifurcation polarization bistability in vertical cavity surface emitting lasers," *Electron. Lett.*, Vol. 31, no. 2, pp. 109–111, 1995.
63. H. Kawaguchi and I. S. Hidayat, "Gigahertz all optical flip flop operation of polarization—bistable vertical cavity surface emitting lasers," *Electron. Lett.*, Vol. 31, no. 14, pp. 1150–1151, 1995.
64. Y. C. Chen and J. M. Liu, "Polarization bistability in semiconductor laser: Rate equation analysis," *Appl. Phys. Lett.*, Vol. 50, no. 20, pp. 1406–1408, 1987.
65. C. J. Chang-Hasnain, J. P. Harbison, C. E. Zan, L. T. Florez, and N. C. Andreadakis, "Continuous wavelength tuning of two electrode vertical cavity surface emitting lasers," *Electron. Lett.*, Vol. 27, no. 11, pp. 1002–1003, 1991.
66. L. Fan, M. C. Wu, H. C. Lee, and P. Grodzinski, "10.1 nm range continuous wavelength tunable vertical cavity surface emitting lasers," *Electron. Lett.*, Vol. 30, no. 17, pp. 1409–1410, 1994.
67. T. Wipiejewski, K. Panzlaff, E. Zeeb, and K. J. Ebeling, "Tunable extremely low threshold vertical cavity laser diodes," *IEEE Photon., Lett.*, Vol. 5, no. 8, pp. 889–891, 1993.
68. L. E. Eng, K. Bacher, W. Yuen, M. Larson, G. Ding, J. S. Harris, Jr., and C. J. Chang-Hasnain, "Wavelength shift in vertical cavity laser arrays on a patterned substrate," *Electron. Lett.*, Vol. 31, no. 7, pp. 562–563, 1995.
69. N. K. Dutta, W. S. Hobson, J. Lopata, and G. Zydzik, "Tunable InGaAs/GaAs/InGaP laser," *Appl. Phys. Lett.*, Vol. 70, no. 10, pp. 1219–1221, 1997.
70. N. Yokouchi, T. Miyamoto, T. Uchida, Y. Inaba, F. Koyama, and K. Iga, "40 Å continuous tuning of a GaInAsP/InP vertical cavity surface emitting laser using an external mirror," *IEEE Photon. Lett.*, Vol. 4, no. 7, pp. 701–703, 1992.

71. M. S. Wu, E. C. Vail, G. S. Li, W. Yuen, and C. J. Chang-Hasnain, "Tunable micromachined vertical cavity surface emitting laser," *Electron. Lett.*, Vol. 31, no. 19, pp. 1671–1672, 1995.

72. M. S. Wu, E. C. Vail, G. S. Li, W. Yuen, and C. J. Chang-Hasnain, "Wide and continuously tunable micromachined resonant cavity detector with wavelength tracking," *IEEE Photon. Lett.*, Vol. 8, no. 1, pp. 98–100, 1996.

73. E. C. Vail, G. S. Li, W. Yuen, and C. J. Chang-Hasnain, "High performance and novel effects of micromechanical tunable vertical cavity lasers," *IEEE J. Select. Topics Quantum Electron.*, Vol. 3, no. 2, pp. 691–697, 1997.

74. M. Y. Li, W. Yuen, G. S. Li, and C. J. Chang-Hasnain, "High performance continuously tunable top emitting vertical cavity laser with 20 nm wavelength range," *Electron. Lett.*, Vol. 33, no. 12, pp. 1051–1052, 1997.

75. M. Y. Li, W. Yuen, G. S. Li, and C. J. Chang-Hasnain, "Top-emitting micromechanical VCSEL with a 31.6 nm tuning range," *IEEE Photon. Lett.*, Vol. 10, no. 1, pp. 18–20, 1998.

76. M. C. Larson, A. R. Massengale, and J. S. Harris, "Continuously tunable micromachined vertical cavity surface emitting laser with 18 nm wavelength range," *Electron. Lett.*, Vol. 32, no. 4, pp. 330–331, 1996.

77. G. L. Christenson, A. T. T. D. Tran, Z. H. Zhu, Y. H. Lo, M. Hong, J. P. Mannaerts, and R. Bhat, "Long wavelength resonant vertical cavity LED/photodetector with a 75 nm tuning range," *IEEE Photon. Lett.*, Vol. 9, no. 6, pp. 725–727, 1997.

78. F. Sugihwo, M. C. Larson, and J. S. Harris, Jr., "Low threshold continuously tunable vertical cavity surface emitting lasers with 19.1 nm wavelength range," *Appl. Phys. Lett.*, Vol. 70, no. 5, pp. 547–549, 1997.

79. P. Tayebati, P. Wang, D. Vakhshoori, C. C. Lu, M. Azimi, and R. N. Sacks, "Half-symmetric cavity tunable micromechanical VCSEL with single spatial mode," *IEEE Photon. Lett.*, Vol. 10, no. 12, pp. 1679–1681, 1998.

80. F. Sugihwo, M. C. Larson, and J. S. Harris, Jr., "Micromachined widely tunable vertical cavity laser diodes," *IEEE J. Microelectromech. Sys.*, Vol. 7, no. 1, pp. 48–55, 1998.

81. P. Wang, P. Tayebati, D. Vakhshoori, C. C. Lu, and M. Azimi, "Half symmetric cavity microelectromechanically tunable vertical cavity surface emitting lasers with single spatial mode operating near 950 nm," *Appl. Phys. Lett.*, Vol. 75, no. 7, pp. 897–899, 1999.

INDEX

Analysis and Design of Vertical Cavity Surface Emitting Lasers By Siu Fung Yu
ISBN 0-471-39124-7

Printed and bound by CPI Group (UK) Ltd, Croydon, CR0 4YY
16/09/2026
14963442-0006